Models, Molecules and Mechanisms in Biogerontology

Pramod C. Rath
Editor

Models, Molecules and Mechanisms in Biogerontology

Cellular Processes, Metabolism and Diseases

Springer

Editor
Pramod C. Rath
School of Life Sciences
Jawaharlal Nehru University
New Delhi, India

ISBN 978-981-32-9004-4   ISBN 978-981-32-9005-1 (eBook)
https://doi.org/10.1007/978-981-32-9005-1

© Springer Nature Singapore Pte Ltd. 2020
This work is subject to copyright. All rights are reserved by the Publisher, whether the whole or part of the material is concerned, specifically the rights of translation, reprinting, reuse of illustrations, recitation, broadcasting, reproduction on microfilms or in any other physical way, and transmission or information storage and retrieval, electronic adaptation, computer software, or by similar or dissimilar methodology now known or hereafter developed.
The use of general descriptive names, registered names, trademarks, service marks, etc. in this publication does not imply, even in the absence of a specific statement, that such names are exempt from the relevant protective laws and regulations and therefore free for general use.
The publisher, the authors, and the editors are safe to assume that the advice and information in this book are believed to be true and accurate at the date of publication. Neither the publisher nor the authors or the editors give a warranty, expressed or implied, with respect to the material contained herein or for any errors or omissions that may have been made. The publisher remains neutral with regard to jurisdictional claims in published maps and institutional affiliations.

This Springer imprint is published by the registered company Springer Nature Singapore Pte Ltd.
The registered company address is: 152 Beach Road, #21-01/04 Gateway East, Singapore 189721, Singapore

*Purna Chandra Rath
(8 October 1889–1 May 1982)
This book is dedicated to Kaviraj Purna Chandra Rath (1889–1982), an exponent of Ayurveda, the ancient Indian traditional science of life, health, and medicine, who lived up to the age of 93 without any major health issue in the holy city of Puri situated in the state of Odisha on the east coast of the Bay of Bengal in India. He was educated in Sanskrit, Ayurveda, and prepared large number of ayurvedic medicines from natural medicinal herbs and other materials according to the ancient ayurvedic scripts at his home dispensary. He had successfully practiced Ayurveda for 60 years for the treatment of thousands of patients suffering*

from various diseases across the state of Odisha. He had played key role in the establishment of ayurvedic schools and trained next generations of experts, many of them learnt the age-old science by staying with him at his home as a "gurukul" free of cost. For these exemplary contributions in Ayurveda and public life, he was awarded the prestigious title of "Vaidyaratna" by the then viceroy of British-India in 1938. Purna Chandra had written many books on Ayurveda and poetry in Odia, the language of Odisha which later became the sixth classical language of India, for the benefit of people. His biography was written by a professor of Odia literature, describing the journey of his life beginning as a village boy to a highly acclaimed authority in Ayurveda, who demonstrated the unmatched value and utility of this ancient Indian traditional medicine system for the society at a nominal cost. This is an ideal example of healthy aging of a socially relevant person who lived in India several decades ago.
Original Address: Late Vaidyaratna Kaviraj Purna Chandra Rath, South Gate, Puri-752001, Odisha

Preface

Since life emerged in cellular form, two things must have become imminent, the cell must propagate itself and there should be a limited lifespan of a cell. If cells do not die, new cells will gradually become less and less in number and the population of cells would soon stop making any more new cell. Such a population of cells will ideally contain "senescent" cells or old cells. Thus, we have precursor cells ("stem cells"), which are ready to divide; dividing cells; cells that do not divide any more – hence becoming old cells; and cells undergoing the process of cell death. Therefore, if old cells accumulate in a tissue or organ, it manifests expression of aging. If we can remove old cells from tissues or organs, it would result in "rejuvenation" – a process of reverse aging. Biological differences among stem cells, dividing cells, senescent cells, dying or dead cells and rejuvenated cells constitute the entire spectrum of aging. Cellular aging is planned at the birth of a cell, so also for the tissues, organs and organisms. Aging may finally eliminate a cell or an individual but keeps the rest of the system or population young and active. Possibly, that is why aging has been selected through evolution. However, we would like that the aging cell or individual should remain free of major faults until the end. Hence, health-span is the new name for longevity or lifespan. Health is wealth at all ages.

Biology of aging is both inherent and intrinsic to individual cells, the genetic material (DNA) of cells, the cellular genomic information (DNA and RNA), non-genomic component (proteins and membrane systems) and the network of events (cell signalling pathways). The cellular compartmentalization (nucleus, mitochondria, endoplasmic reticulum, lysosome and peroxisome) presents an organizational and architectural design, where the processes of both life and death are executed. Since biological processes have to be carried out by macromolecular assembly of units of structure and function, be it chromatin-DNA, RNA-protein, protein-protein complexes or structural designs of bio-membranes or polymers of sugar, lipid and peptide, restrictions on temperature, pH, ionic concentration and osmolarity have to be imposed to ensure native conformation, lower activation energy, efficient catalysis, sustainable positive-negative regulations and recycling of resources. Optimization of structure gives way to efficiency of catalysis and regulation. Metabolic pathways establish foundation of cellular physiology. Utilization of nutrients generates energy and releases free radicals such as reactive oxygen species (ROS) and reactive nitrogen species (RNS). These cause oxidative stress in cells due to oxidation of proteins, lipids and DNA/RNA. In order to keep it under control, cellular scavenger molecules have evolved either as enzymes (Glutathione-S-transferase, catalase, super oxide dismutase, etc.) or as metabolites (glutathione, ascorbic acid, etc.). A small amount of ROS activates cellular signalling and gene expression but higher amounts of ROS cause disease.

During aging of cells and tissues in almost all organisms studied so far, oxidative stress increases, telomeres at the chromosomal ends get shortened, the stem cell reserves of the tissues get exhausted, the gene expression and regulation mechanisms become less efficient, the metabolic pathways lose their balance, the protein synthesis and regulation machinery gets affected, DNA damage increases and DNA repair systems fall, mutations accumulate in the genomic DNA, low grade chronic inflammation becomes persistent and immune responses against infections become weaker, and senescent cells accumulate in tissues and organs. The old organism suffers from decline of almost all physiological functions and becomes more prone to age-related diseases ultimately leading to death. Thus although aging is not a disease, it invites many diseases. Diseases and human suffering are a curse but drugs and medical treatments are a big market. Social values and cultural virtues in human civilizations have always stood for humanity and nature. Scientific advancements brought new knowledge into existence and gave birth to technology for further development. Appropriate policies ensure development to be inclusive and plural in human society. The need of the hour is to give a patient listening to the problems of the elderly and extend a helping hand to those who have spent their life in handholding in a socially relevant lifetime.

Models, Molecules and Mechanisms in Biogerontology: Cellular Processes, Metabolism and Diseases contains two parts: (I) Alterations in Cellular Mechanisms During Aging and (II) Alterations in Metabolism During Aging and Diseases. There are 21 chapters written by authors from nine Indian university/institute(s) and three university/institute(s) in the USA, Germany and Italy. Section I deals with 15 chapters: protein structure and function, DNA replication, transcription and gene expression, ribosome and translation, transcription factors (p53 & Pax), telomere and telomerase, epigenome and epigenetic rejuvenation, mitochondria and free radicals (ROS), stem cells and model systems (*Dictyostelium, Drosophila* & Killifish) in relation to aging and age-related diseases. Section II deals with six chapters: muscle, metabolic diseases, nutrient sensing, protein aggregation, biological rhythms and cancer in relation to aging. The concepts, molecules and mechanisms involved in cellular processes linked to aging, longevity and age-related diseases are highlighted. There has been an attempt to link the basic biology with gerontology and geriatrics in order to provide contemporary and cutting-edge information for basic research and clinical situations. The text is designed for graduate students and researchers. It could also help policy makers as well as public and private bodies in the field of biomedical gerontology and geriatrics.

The editor of this book thanks all the authors who have contributed the chapters unconditionally. Apologies are due for not being able to make a mention of all the work done in this field. Some of us writing chapters for this book have been either

Master's or Ph. D. students of Prof. M. S. Kanungo, our teacher, who started research on biology of aging after founding the Biochemistry Section in the Department of Zoology at Banaras Hindu University, Varanasi, India, as early as 1960s. We express our sincere regards to his departed soul. The School of Life Sciences at Jawaharlal Nehru University, New Delhi, India is acknowledged for all help.

New Delhi, India								Pramod C. Rath

Contents

Part I Alterations in Cellular Mechanisms During Aging

1. **Protein Structure and Function in Aging and Age-Related Diseases**..................... 3
   Anshumali Mittal and Pramod C. Rath

2. **DNA, DNA Replication, and Aging**........................... 27
   Bhumika Sharma, Meetu Agarwal, Vijay Verma, and Suman Kumar Dhar

3. **Transcription and Aging**.................................... 43
   Malika Saint and Pramod C. Rath

4. **Ribosome, Protein Synthesis, and Aging**..................... 67
   Reetika Manhas and Pramod C. Rath

5. **p53 and Aging**... 89
   Nilambra Dogra and Tapas Mukhopadhyay

6. **Paired Box (Pax) Transcription Factors and Aging**............ 109
   Rajnikant Mishra

7. **Telomeres, Telomerase, and Aging**........................... 119
   Deepak K. Mishra, Ramraj Prasad, and Pramod Yadava

8. **The Epigenome of Aging**..................................... 135
   Rohini Muthuswami

9. **Attaining Epigenetic Rejuvenation: Challenges Ahead**......... 159
   Jogeswar S. Purohit, Neetika Singh, Shah S. Hussain, and Madan M. Chaturvedi

10. **Mitochondria as a Key Player in Aging**..................... 181
    Rupa Banerjee and Pramod C. Rath

11. **Aging, Free Radicals, and Reactive Oxygen Species: An Evolving Concept**..................................... 199
    Shyamal K. Goswami

12	Stem Cells and Aging	213

Jitendra Kumar Chaudhary and Pramod C. Rath

13	Can Autophagy Stop the Clock: Unravelling the Mystery in *Dictyostelium discoideum*	235

Priyanka Sharma, Punita Jain, Anju Shrivastava, and Shweta Saran

14	Aging: Reading, Reasoning, and Resolving Using *Drosophila* as a Model System	259

Nisha, Kritika Raj, Pragati, Shweta Tandon, Soram Idiyasan Chanu, and Surajit Sarkar

15	*Nothobranchius furzeri* as a New Model System for Ageing Studies	303

Eva Terzibasi Tozzini

Part II Alterations in Metabolism During Aging and Diseases

16	Aging in Muscle	319

Sunil Pani and Naresh C. Bal

17	Metabolic Diseases and Aging	347

Arttatrana Pal and Pramod C. Rath

18	Interplay Between Nutrient-Sensing Molecules During Aging and Longevity	393

Ibanylla Kynjai Hynniewta Hadem, Teikur Majaw, and Ramesh Sharma

19	Protein Aggregation, Related Pathologies, and Aging	419

Karunakar Kar, Bibin G. Anand, Kriti Dubey, and Dolat Singh Shekhawat

20	Biological Rhythms and Aging	443

Anita Jagota, Kowshik Kukkemane, and Neelesh Babu Thummadi

21	The Biology of Aging and Cancer: A Complex Association	465

Mohit Rajput, Lalita Dwivedi, Akash Sabarwal, and Rana P. Singh

About the Editor

Pramod C. Rath is a Professor of Molecular Biology at the School of Life Sciences, Jawaharlal Nehru University, New Delhi. He received his Ph.D. in Zoology (Biochemistry) in 1988 from the Banaras Hindu University, Varanasi, on the topic "Gene Expression During Aging" under the supervision of Prof. M. S. Kanungo, who started research on "biology of aging" in India. He completed his postdoctoral research at the Institute of Molecular Biology I, University of Zurich, Switzerland, with Prof. Charles Weissmann, a well-known molecular biologist.

He has 29 years of teaching and research experience, having mentored 18 Ph.D. and 4 M.Phil. students. He has published his research in respected international journals, such as *Ageing Research Reviews, Molecular Neurobiology, Journal of Molecular Neuroscience, International Journal of Developmental Neuroscience, RNA Biology, PLOS ONE, International Journal of Biological Macromolecules, Molecular Biology Reports, Journal of Biosciences, Biochemical and Biophysical Research Communications, Biochimica et Biophysica Acta, FEBS Letters, Journal of Clinical Immunology*, etc. He has also published two Springer books, titled *Topics in Biomedical Gerontology* (2017) and *Models, Molecules and Mechanisms in Biogerontology: Physiological Abnormalities, Diseases and Interventions* (2019).

He teaches molecular biology, molecular genetics and genetic engineering and cell signaling to master's and Ph.D. students. Research in his laboratory is focused on cytokines, transcription factors, cell signaling and diseases; genomic biology of repetitive DNA and noncoding RNA; bone marrow stem cells and molecular aging in mammals. He has received numerous awards and fellowships and has been the Vice-President of the Association of Gerontology (India) and Acting Dean at the School of Life Sciences. He is a Member of several national academic and scientific committees.

Part I

Alterations in Cellular Mechanisms During Aging

Protein Structure and Function in Aging and Age-Related Diseases

Anshumali Mittal and Pramod C. Rath

Introduction

The central dogma of molecular biology, first stated by Francis Crick, is an explanation of flow of genetic information that "DNA makes RNA and RNA makes protein" [1]. However, recent advances in biology have suggested that the central dogma of molecular biology is more complex than envisaged by Crick. For example, RNA is not just a passive intermediary link between DNA and protein but has many structural and regulatory roles, such as catalyst (ribozyme), scaffolds, decoy, and guiding and signaling molecules (rRNA, tRNA and noncoding RNAs) [2, 3]. However, proteins are the predominant product of gene expression and are the most abundant and diverse biomolecules in living cells. They perform wide variety of roles including catalysis, immunity, transportation, scaffold, signaling, and as structural component. These functions suggest that proteins are responsible in part for maintaining functional stability and homeostasis of tissues and cells and, therefore, may occupy central stage in the aging and longevity process. During aging, there are many opportunities for an appositely transcribed peptide and protein to become structurally altered and, hence, accumulation of altered proteins may be correlated with a loss of function or, in some cases, a gain of inappropriate or toxic functions [4, 5]. In general, protein molecules need to fold into a unique three-dimensional structure to perform its function. The functional conformation of newly encoded polypeptides in the cell is achieved by the regulated folding during/after protein

---

A. Mittal (✉)
Department of Molecular Biology and Microbiology, Tufts University School of Medicine, Boston, MA, USA
e-mail: mittalans@gmail.com

P. C. Rath
Molecular Biology Laboratory, School of Life Sciences, Jawaharlal Nehru University, New Delhi, India
e-mail: pcrath@mail.jnu.ac.in

© Springer Nature Singapore Pte Ltd. 2020
P. C. Rath (ed.), *Models, Molecules and Mechanisms in Biogerontology*,
https://doi.org/10.1007/978-981-32-9005-1_1

synthesis, which is often assisted by chaperones. Chaperones bind and stabilize unfolded or partially folded polypeptides along the folding pathway that leads to correctly folded conformation, which otherwise fold incorrectly or aggregate into insoluble complexes [6, 7]. However, if chaperones fail to refold a protein and extent of protein damage is too large or the cellular conditions are not adequate for refolding, then the unfolded protein is delivered to the ubiquitin–proteasome or the lysosomal systems for degradation. Chaperone malfunctions or alterations in the cellular degradative pathways, at least in part, result in intracellular accumulation of damaged proteins and organelles, which form the basis of different human pathologies. Accumulation of damaged biomolecules is one of the characteristics of tissues in all organisms as they age and results in their functional loss during aging [8].

In addition to protein folding, modifications of amino acids and their side chains contribute significantly to the structural and functional diversity of proteins, and these modifications substantially increase the complexity of the eukaryotic proteome than coded by the genome. These posttranslational modifications of proteins influence the enzyme activity, gene expression and silencing, DNA repair, protein turnover, localization, protein–protein interactions, many cell signaling cascades, and cell division process [9]. As we age, our cells encounter an alteration in these pathways listed. Thus, it is necessary to identify and understand the alterations of specific proteins during aging and their roles in age-related pathologies.

Recently, Lopez-Otin et al. have proposed nine candidate hallmarks that contribute to the aging process and thus determine the aging phenotype [10]. These hallmarks are genomic stability, telomere attrition, epigenetic alterations, loss of proteostasis, deregulated nutrient sensing, mitochondrial dysfunction, cellular senescence, stem cell exhaustion, and altered intercellular communication [10]. In these cellular and molecular hallmarks, protein molecules occupy prominent position and during aging or in age-related pathologies, alteration in the rate of protein synthesis, posttranslational modifications, folding, turnover, and cell signaling are accelerated. In this chapter, we are discussing these topics listed and starting with the metabolic signaling pathways that regulate aging and life span.

Metabolic Pathways and Aging

Aging is a complex process driven by diverse molecular pathways and biochemical events, which are contributed by both environmental and genetic factors. Caloric restriction is one of the most robust interventions for extending life span in mammals due to its ability to affect different pathways including sirtuin 1 (SIRT1) activation [11], insulin/IGF-1 pathway [12], AMP-activated protein kinase (AMPK) pathway [13], and mammalian target of rapamycin (mTOR) signaling [14] (Table 1.1). In these pathways, the longevity response is under active control by specific regulatory proteins, and the collective aim of these pathways is to promote cell survival, stress defense mechanisms, autophagy activation, protein synthesis, and ultimately life span extension [15]. The nutrient-sensing TOR pathway regulates protein synthesis as well as protein degradation by autophagy,

Table 1.1 Examples of conserved pathways that are known to regulate lifespan in different model organisms

Pathway	Characteristics	Change that extends lifespan	Conservation
Insulin/IGF-1	Endocrine signaling	Inhibition	Worms, flies, mice, humans
TOR	Nutrient/amino acid sensing	Inhibition	Yeast, worms, flies, mice
AMPK	Nutrient/energy sensing	Overexpression	Worms, mice (indirect study)
SIR2	NAD+-dependent histone deacetylase	Overexpression	Yeast, worms (in some cases), flies (in some cases)
Mitochondrial electron transport chain	Respiration	Inhibition	Worms, flies, mice
Germline stem cells	Reproduction	Inhibition	Worms, flies

TOR target of rapamycin, *AMPK* AMP-activated protein kinase, *SIR2* silent information regulator-2
Reprinted from the permission from Elsevier (License Number 4006491099104)
Histone methylation makes its mark on longevity, Shuo Han and Anne Brunet, Trends in Cell Biology, Vol. 22, No.1, January 2012

and these processes are known to be influenced during aging [16], and therefore, here, we are focusing on the role of the TOR pathway in aging, and the other pathways are described elsewhere [17].

TOR-Signaling Network

Almost a decade back, Mikhail V. Blagosklonny proposed that the main driver of aging is TOR signaling rather than reactive oxygen species [18], as inhibition of the TOR pathway extends life span in many species, from yeast to mice [14, 19, 20]. mTOR protein is a 289-kDa multidomain serine–threonine kinase that belongs to the phosphoinositide 3-kinase (PI3K)-related kinase family. mTOR interacts with other proteins to form two main types of complexes, mTOR complexes 1 and 2 (mTORC1 and mTORC2). Raptor and Rictor are unique proteins associated with the mTORC1 and mTORC2 complexes, and they have been linked with longevity, suggesting a role of TOR pathway signaling in age-specific functions [21]. mTOR is essential for viability in organism ranging from yeast to mammals, and disruption of the mouse *mTOR* gene results in embryonic lethality and inhibition of embryonic stem cell development [22].

mTORC1 is an important node in cellular regulation impacting on cell growth that is linked to aging. Signaling through mTORC1 is activated by a variety of agents including hormones, mitogens and growth factors, and amino acids, and is negatively regulated by stressful conditions, such as decreased energy (ATP) availability [16]. It is a master regulator of cell metabolism (such as autophagy and protein synthesis), growth, proliferation, and survival. In the following section, we are discussing in brief the ways in which autophagy [23] and protein synthesis [24] may contribute to modulate longevity by the TOR pathway. The overall thrust of these studies is that inhibiting mTOR signaling or protein synthesis can extend life span [16].

mTOR, Protein Synthesis, and Life Span

Protein synthesis is an essential part of cellular metabolism during which living cells build proteins. Accumulating evidence of research in many organisms, including human, have suggested that aging is accompanied by marked alterations in the total protein synthesis, indicating a link between aging process and the regulation of protein synthesis. Protein synthesis is a conserved process involving three main steps: initiation, elongation, and termination, and the major components involved in this multistep process are ribosomes, tRNAs, aminoacyl-tRNA synthetases, and translation initiation factors (eIFs). However, protein synthesis is principally regulated at the initiation stage, which requires coordinated assembly of a large number of proteins known as initiation factors, and it appears that the activity of these proteins decline with age, resulting in reduced protein synthesis [25, 26].

Two well-known substrates of mTORC1, namely, 4E-BP1 and ribosomal protein S6 kinase (S6K), are linked to the regulation of protein synthesis at translational initiation and elongation step and are coupled to the aging process in different organisms [27–29]. TOR-mediated phosphorylation of 4E-BPs disrupts an inhibitory interaction between 4E-BP and eIF4E and promotes increased translation initiation. Reduced expression of eIF4E or enhanced inhibition between 4E-BP and eIF4E leads to life span extension in both worms and flies [30, 31]. A similar observation was also reported in *C. elegans* where inhibition of ifg-1 and rsks-1 (the worm homologue of eIF4G and S6 kinase) results in reduction of global protein synthesis and extends life span [32]. Ribosomal S6 kinase, a downstream effector of mTOR, is one of the most conserved modulators of aging, and reduced expression leads to life span extension. mTOR also regulates protein synthesis by phosphorylating S6K, which, in turn, leads to the phosphorylation and activation of translation elongation factor 2 kinase (eEF2K), which could ultimately affect global translation [30, 33]. In mice, it has been observed that deletion of S6 kinase 1 (S6K1) led to increased life span and as well as resistance to age-related pathologies, such as bone, immune, and motor dysfunction [34].

In all cases, the rate of protein synthesis in "long-lived" animals is reduced, compared to wild-type control animals. Since protein synthesis consumes approximately two-thirds of the total energy produced by a cell, a reduction in protein synthesis would lead to significant conservation of energy, and the surplus energy can be directed for the DNA repair and cellular maintenance, therefore, extending longevity. During aging, lowering the rate of protein synthesis may additionally reduce the risk of accumulation of misfolded, aggregated, and mutated proteins, which occurs in many age-related neurodegenerative diseases, such as Alzheimer's and Parkinson's disease [35].

Autophagy, mTOR, and Life span

Autophagy, regulated by mTORC1 signaling, is a self-degradative process in which misfolded or aggregated proteins and damaged cytoplasmic organelles are removed. Three different types of autophagy have been described in mammalian cells: chaperone-mediated autophagy (CMA), microautophagy, and macroautophagy, with each of them promoting proteolytic degradation of cytosolic components at the lysosome [36, 37]. In CMA, all substrate proteins contain a pentapeptide motif (Q, K/R, F/VL/I and E/D and fifth amino acid can be either K/R or F/VL/I) that is recognized by a 70 kDa heat-shock protein, and the complex formed is targeted to the surface of the lysosomes for its translocation and followed by degradation [38]. Macroautophagy involves sequestrations of random cytoplasm and dysfunctional organelles by the expanding phagophore leading to the formation of a double-membrane vesicle referred to as autophagosome that subsequently fuses with lysosomes for degradation. In microautophagy, cytosolic components are directly taken up by the lysosome itself through invagination of the lysosomal membrane [37].

Defects in autophagy have been associated with age-related cellular changes, such as intracellular accumulation of damaged proteins and organelles [39, 40]. In a study to understand a link between autophagy and aging, loss-of-function mutants in the autophagy genes (*bec-1*, *unc-51* and *atg-18*) were analyzed, which showed a clear acceleration in tissue aging and a reduced life span in *C. elegans* [16, 41]. Similar observations were also obtained from the RNAi knockdown of atg-7 that reduced the life span of wild-type *C. elegans* [42]. Additionally, in *Drosophila*, mutation in autophagy gene Atg7 and reduced expression of Atg8 also decreased longevity, while upregulation of Atg8 increased longevity [43, 44]. As indicated by the studies in *C. elegans*, inactivation of autophagy genes, such as *bec-1*, *unc-51*, and *atg-18* resulted in the inhibition of TOR activity. Taken all together, this indicates that the TOR and autophagy function via the same signaling pathways to affect life span. A decrease in the autophagic activity is observed in almost all cells and tissues during aging and age-related diseases. Cancer, neurodegenerative diseases, and metabolic disorders, such as diabetes, are some of the examples that have been identified due to alterations in autophagy. Alterations in macroautophagy and CMA have been associated with early changes in Alzheimer's disease- and Parkinson's disease–affected neurons, respectively. Autophagy has also been identified in playing a role in tumor suppression by removing damaged organelles and reducing chromosome instability [8]. Sarcopenia, an aging-related disease with a decline in mass and strength of skeleton muscle due to the imbalance between protein synthesis and protein degradation, is also associated with inaccurate regulation of Autophagy. In sarcopenia, loss of skeleton muscle mass is caused by many factors including accumulation of denatured, misfolded and aggregated proteins [45]. Detailed role of autophagy in aging can be read elsewhere [8, 23].

Posttranslational Modifications of Proteins During Aging

Posttranslational modifications (PTMs) modulate protein function in most eukaryotes and contribute to the functional diversity of the proteome. These modifications may alter physical and chemical properties of proteins, and consequently determine their correct folding, targeting, activity, and stability. It includes (a) covalent modifications of amino acids, such as methylation of lysine, arginine, and methionine and phosphorylation of serine, threonine, and tyrosine residues; (b) proteolytic cleavage of the polypeptide backbone; and (c) nonenzymatic modifications, such as deamidation of asparagine.

Phosphorylation

Phosphorylation of serine, threonine, and tyrosine residues is one of the best-studied modifications of proteins. The coordinated activities of protein kinases and protein phosphatases result into phosphorylation and dephosphorylation of protein molecules, which are central to many biological processes, including transcription, translation, protein synthesis, cell division, signal transduction, cell growth, and development [25, 46, 47]. Phosphorylation of proteins involved in the regulation of cell cycle progression has been shown to play a significant role in the progression of aging. For example, senescent fibroblast cells are unable to pass the restriction point (R) in the G1 stage of the cell cycle due to the absence of phosphorylation of retinoblastoma protein resulted due to the accumulation of inactive cyclin E/Cdk2 and possibly cyclin D/Cdk4 kinases. Additionally, senescent cells contain high amounts of p21, which is a potent cyclin-dependent kinase inhibitor whose levels are also elevated in cells arrested in G1 following DNA damage, indicating a common mechanism involved in both arrests [48]. Various proteins involved in protein synthesis undergo phosphorylation and dephosphorylation and thus regulate *protein synthesis* through a variety of mechanisms [49]. For example, phosphorylation and dephosphorylation of many translation initiation factors (eIF-2), elongation factors (eEF-1, eEF-2), ribosomal proteins (S6), and amino acid-tRNA synthetases affect the rate of protein synthesis, and there are indirect evidences indicating that the phosphorylation/dephosphorylation by specific protein kinases of these regulatory proteins may affect the protein synthesis during aging [50–53].

At the molecular level, protein phosphorylation-mediated signal transduction is accompanied by a series of biochemical events comprising phosphorylation and dephosphorylation by specific protein kinases and phosphatases, respectively. Several protein kinases have been reported to play critical roles in brain functions. In particular, protein kinase C, a family of serine- and threonine-specific protein kinase, expresses in brain tissues and plays an important role in short-term processes (ion fluxes, neurotransmitter release), midterm process (receptor modulation), as well as long-term processes (cell proliferation, synaptic remodeling, and gene expression). Based on sequence homology and sensitivity to activators 10, isoforms of PKC have been described [54–59]. Each isoform of PKC may have unique role in subcellular localization, and substrate phosphorylation and the specificity in these pathways are achieved by protein–protein [60] and protein–lipid [61] interactions, and any alteration in these interactions may impair the PKC pathway functioning, which may cause physiological or pathological aging (i.e., Alzheimer's disease) [58].

Methylation

Methylation occurs on all three components of the central dogma of molecular biology that include RNA, DNA, and protein molecules. In DNA, methylation takes place at typically at CpG sites by DNA methyltransferases [62]; in RNA, including mRNAs, tRNAs, and noncoding RNAs, methylation modification is found on adenosine at the N6 position (m6A) catalyzed by RNA methyltransferases [63] and in protein molecules by the specific methyltransferases protein complexes [64, 65]. Methylation either in DNA, RNA, or histone proteins is maintained over cell divisions and help in establishing the cell type–specific gene expression patterns. Using microarray methodology, it has been shown recently that age-related DNA methylation changes occur in human genome. For example, promoter-associated CpG islands contain low DNA methylation, and intergenic non-island CpGs contain high DNA methylation, which tend to gain and lose methylation, respectively, with age. In addition to general pattern of DNA methylation alteration with age, DNA methylation at specific sites in the genome are highly associated with age and have been used to predict chronological age accurately [66–69]. A recent report suggests that the methylation of a cytosine residue (C2381/C2278) within a conserved region of 25S rRNA, located in the vicinity of the peptidyl transferase center of the large ribosomal subunit, is critical for the rRNA-mediated translational regulation. They discovered that NSUN5 and Rcm1, conserved m^5C-rRNA methyltransferase enzymes found in flies and yeast, are responsible for the methylation of C2381/C2278 of ribosomal RNA. Interestingly, it was found that the modification of 25S rRNA either by a reduced levels of RNA methyltransferase or a point mutation of cytosine C2381/C2278 modulates longevity and is able to influence the aging process in flies/yeast [70]. The role of RNA methyltransferases in tissue renewal and pathology can be read elsewhere [71].

In proteins, methylation is one of the most observed posttranslational modifications in mammalian cell. Addition of methyl groups to the side chains of arginine, lysine, and histidine (at amino group) and glutamate and aspartate (at carboxyl group) residues increases the hydrophobicity of the protein and can neutralize a negative amino acid charge when bound to carboxylic acids [72]. Methylation, occurring to the side chain of arginine and lysine residues in the N-terminal histone tails, is catalyzed by multiprotein complexes referred to as arginine/lysine methyltransferases [73]. MLL/PcG family members are lysine methyltransferases, which modify histone H3 at lysine 4 (H3K4)/H3 at lysine 27 (H3K27) by adding mono- (Me1), di- (me2), or tri-methyl (me3) groups. These H3K4 $^{me1/me2/me3}$ and H3K27$^{me1/me2/me3}$ marks are responsible for maintaining transcriptional state of HOX genes important for the development and stem cell function [74–76]. In mammals, H3K4 methyltransferase core complex is composed of RbBP5, Ash2L, WDR5, Dpy30, and MLL subunits. Recent work in *C. elegans* has reported that a deletion of ASH-2, WDR5, or SET-2 reduces genome-wide levels of H3K4 trimethylation and significantly extends life span in worms. Interestingly, overexpression of RBR-2, which specifically demethylates H3K4-me2/me3, also extends life span in *C. elegans* [77]. As H3K4 $^{me1/me2/me3}$ is generally associated with transcriptional activation, and therefore, it appears from these studies that limiting H3K4 methylation level increases longevity in *C. elegans* through reducing the transcription of longevity-related genes (Table 1.2) [78]. The role of H3K4 $^{me1/me2/me3}$ histone mark in longevity has different effects on life span in different species (such as *Drosophila* and human) suggesting that the answer is likely complex [78]. For extensive reviews focusing on key aspects on histone methylation-mediated epigenetic regulators in aging and longevity, please see [78, 79].

Table 1.2 Histone methylation regulators in aging and longevity

	C. elegans	*Drosophila*	Mammals
H3K4me3			
Change with age?	NT	Decrease (in the brain) ♀	Change in landscape
KD or mutation of methyltransferase	Increase lifespan ♀ COMPASS complex	No effect on lifespan (Trx complex) ♂	(in the brain) NT
OE of methyltransferase	NT	NT	NT
KD or mutation of demethylase	Decrease lifespan ♀	Decrease lifespan ♂	NT
OE of demethylase	Increase lifespan ♀	NT	NT
H3K4me1/me2			
Change with age?	NT	NT	NT
KD or mutation of methyltransferase	NT	NT	NT
OE of methyltransferase	NT	NT	NT
KD or mutation of demethylase	Increase lifespan ♀	NT	NT
OE of demethylase	NT	NT	NT
H3K27me3			
Change with age?	Decrease ♀	NT	NT
KD or mutation of methyltransferase	NT	Increase lifespan ♂	NT
OE of methyltransferase	NT	NT	NT
KD or mutation of demethylase	Increase lifespan ♀	NT	NT
OE of demethylase	NT	NT	NT

NT not tested, *KD* knockdown, *OE* overexpression
Reprinted from the permission from Elsevier (License Number 4006491099104)
Histone methylation makes its mark on longevity, Shuo Han and Anne Brunet, Trends in Cell Biology, Vol. 22, No.1, January 2012

Barber et al. studied age-dependent modifications in the membrane proteins of human erythrocytes and observed three- to sevenfold increments in carboxyl methylation with age for cytoskeleton proteins, such as bands 2.1, 3, and 4.1. They proposed that age-dependent methylation at carboxyl end is different from the other form of covalent modifications because carboxyl methylation is observed only in "abnormal proteins," accumulated with age, formed by spontaneous racemization of L-aspartyl or L-asparaginyl residues [80].

Deamidation, Racemization, and Isomerization

Deamidation is a posttranslational modification observed in large percentage of proteins, where asparagine (Asn or N) and glutamine (Gln or Q) residues lose ammonium molecule to form aspartic (Asp or D) and glutamic acid (Glu or E), respectively. Deamidation, a hydrolytic reaction, introduces an additional negative charge in the altered protein by losing an ammonium group from the neutral amide side chain of N/Q amino acid to produce a negatively charged amino acid D/E, respectively. Although this modification does not alter the amino acid chain length, change in primary sequence may affect the secondary and tertiary structure ultimately affecting the stability and functionality of the altered protein [81, 82]. However, deamidation of Asn is found more frequently than that of Gln, both in vitro and in vivo during development and aging of cells [83]. Whether and to what extent Asn or Glu is deamidated depends primarily on primary sequence and three-dimensional structure of the protein, as well as on the pH, temperature, buffer ions, and other solution properties [82, 84]. Protein deamidation rates are encoded in the protein structure, and therefore, they serve as molecular clock in biological events, such as protein turnover, development, and aging [81, 85].

One aspect of human aging involves the degradation of long-lived proteins, and therefore, different methodologies, such as radioisotope pulse labelling and L/D aspartic acid racemization, were used for estimating the protein turnover rate, which identified elastin, tooth enamel, tooth dentine, and eye lens crystallins to have half-lives on the order of years [86–89]. The α-, β-, and γ-crystallins, major structural proteins of human lens responsible for maintaining transparency of lens, are excellent examples for studying age-dependent alteration of proteins. The α-crystallin occurs as large aggregates of about 800 kDa, comprising two types of αA- and αB-crystallin. Both forms of α-crystallins β-, and γ-crystallins get deamidated in an age-dependent manner, namely, at Asn101 in human αA-crystallin, Asn149 in chicken αA-crystallins, Asn146 in bovine αB-crystallin, and Gln92 and Gln170 in γ-crystallin [83, 90]. Deamidation reaction can cause changes in charge and mass of crystallins that potentially decrease their stability, enhance aggregation, alter crystalline–crystalline interactions and finally alter the physical properties of lens and accelerate cataract formation [91].

In an effort to examine structural changes in another lens protein, major intrinsic protein (MIP), MIP-enriched membrane fractions were prepared from human lenses with age varied from 7 to 86 years, followed by chemical cleavage by CNBr combined with reverse-phase HPLC to fractionate the protein into fragments for mass spectrometric analysis. It was found that the human MIP is heterogeneously modified by truncation at the N- and C-terminal, phosphorylation, and deamidation, which result in decreased levels of native intact MIP with age. These modifications can be used as a marker for cell aging [92]. Deamidation also plays a role in the progression of Alzheimer's disease and Prion diseases.

Oxidation

In 1956, D. Harman conceived the "free radical theory" of aging [93], which was further modified by him in 1972 and referred to as mitochondrial theory of aging [94]. This theory proposes that reactive oxygen species produced in the mitochondria causes aging through random molecular damage of DNA and macromolecules including proteins and lipids. Random modification of protein side chains, induced by oxidative damage, is estimated to be a dominant source of protein stability loss in aging cells [95]. Proteins, among other biological molecules such as nucleic acids, are most important targets for oxidative damage because they are most abundant macromolecule in the cell and mediate cellular processes necessary for an organism to survive. Studies focusing on understanding the effects of oxidative damage on proteins in the context of aging suggested that approximately less than one amino acids per protein molecule are modified but account for large phenotypic changes in aged organisms [95, 96]. Oxidation of positively charged side chains (Lys, Arg) and neutral side chains (cysteine, methionine, proline, threonine, and histidine) is most frequent and results in significant protein instability in aging cells and organisms. Thus, these untargeted modifications can perturb catalytic activity as well as stability of proteins [95]. Other than prominent methods of oxidative modifications to amino acids side chains, shown in Table 1.3, and oxidation of guanine and methylation of cytosine in DNA also become abundant with age.

Proteins implicated in aging are often may have the following malfunctions: [1] altered chromatin packing (DNA–histones proteins interactions), (2)abnormal histone modification, [3] decreased telomere stability, [4] decreased transcriptional response to stress, and (5) decreased protein translation and degradation. Histone proteins are highly basic and are required for packaging the nuclear DNA into a compact structure known as nucleosome, which further assembles to form higher-order chromatin structures. Oxidative damage to DNA or histone proteins affects specific charge–charge interactions and could alter stability, gene expression patterns, cell metabolism, and aging. Heat shock transcription factor-1, histone H2AX, histone-binding RbBP7, chromatin remodeling factor RbBP4, telomerase reverse transcriptase TERT, and ribosomal protein S6 are some of examples of protein relevant in aging and aging-related diseases and are at high risk of oxidative destabilization [95].

Aging-related oxidative stress is a potential source of protein carbonyl modifications and is the well-used biomarker of severe oxidative protein damage. Human

Table 1.3 Oxidative modifications on amino acids

Amino acid	Products
Cysteine	Oxidation of a sulfydryl group (Cys-SH) to form sulfenic (Cys-SO2H) or sulfonic (Cys-SO3H) derivative
	Formation of a disulfide bond (Cys-S-S-Cys) between two nearby Cys residues within a protein (intramolecular cross-linking) or between two proteins (intermolecular cross-linking)
	Formation of a mixed disulfide (Cys-S-S-glutathione) between a sulfydryl group and glutathione (S-glutathionylation)
Glutamic acid, tyrosine, lysine	Hydroperoxide
Leucine, valine, proline	
Isoleucine	
Histidine	2-oxo-histidine
Lysine, arginine, proline Threonine	Formation of carbonyl derivatives by direct oxidative attack on amino acid side chains (α-aminoadipic semialdehyde from Lys, glutamic semialdehyde from Arg, 2-pyrrolidone from Pro, and 2-amino-3-ketobutyric acid from Thr)
Lysine, cysteine, histidine	Formation of carbonyl derivatives by secondary reaction with reactive carbonyl compounds derived from
	Oxidation of carbohydrates (glycoxidation products), lipids (malondialdehyde, 4-hydroxynonenal, acrolein)
	And advanced glycation and lipoxidation and products
Methionine	Methionine sulfoxide
Phenylalanine	O-Tyrosine, m-tyrosine
Trypthophan	N-Formylkynureine, kynureine, 5-hydroxytryptophan, 7-hydroxytryptophan
Tyrosine	3,4-dihydroxyphenylalanine, 3-chlorotyrosine, 3-nitrotyrosine, dityrosine (Tyr-Tyr cross-links)

Reprinted from the permission from Elsevier (License Number 4006571384145)
Protein carbonylation in human diseases, Isabella Dalle-Donne, Daniela Giustarini, Roberto Colombo, Ranieri Rossi and Aldo Milzani, Trends in Molecular Medicine Vol.9 No.4 April 2003

diseases associated with protein carbonylation include Alzheimer's disease, chronic lung disease, chronic renal failure, diabetes, and sepsis [97]. Furthermore, the carbonylation of side chain amino acids can induce protein aggregation by promoting unfolding and formation of non-covalent (hydrophobic, electrostatics, and hydrogen interactions) as well as covalent bonds among proteins. In a study, Tanase et al. used molecular biophysics tools for establishing a relationship between aging-related oxidative stress and protein aggregation phenomenon. They concluded that oxidative stress-induced protein carbonylation favors increased protein aggregation and observed formation of denser and more compact aggregates in aging organism, which were different from the aggregates present in the cellular proteome of young and adult mice [98]. Interestingly, protein aggregate size and conformation can have an influence in the activation of ubiquitin machinery, macroautophagy, microautophagy, or chaperone-mediated autophagy. For example, soluble and 'rod shape' aggregates can be more easily unfolded and disposed of by the proteasome or through chaperone-mediated autophagy, whereas compact aggregates tend to

Table 1.4 Human diseases associated with carbonylated proteins

Disease	Selected references
Acute/adult respiratory distress syndrome	
Alzheimer's disease	[109–114]
Amyotrophic lateral sclerosis	
Cataractogenesis	
Chronic lung disease and bronchopulmonary dysplasia	
Chronic renal failure (CRF) and chronic uremia	
Cystic fibrosis	
Dementia and with Lewy bodies	
Diabetes	
Ischemia-reperfusion	
Parkinson's disease	[130, 131]
Pre-eclampsia	
Psoriasis	
Rheumatoid arthritis and juvenile chronic arthritis	
Severe sepsis	
Systemic amyloidosis	
Varicocele	

Reprinted from the permission from Elsevier (License Number 4006571384145)
Protein carbonylation in human diseases, Isabella Dalle-Donne, Daniela Giustarini, Roberto Colombo, Ranieri Rossi and Aldo Milzani, Trends in Molecular Medicine Vol.9 No.4 April 2003

accumulate in the cytosol and compromise cellular functions [98]. Human diseases associated with carbonylated proteins are listed in Table 1.4.

ADP-Ribosylation

ADP-ribosylation is a posttranslational modification, in which ADP-ribose is transferred from NAD to specific amino acid residues in proteins, catalyzed by specific poly-ADP-ribosyltransferases (PARPs). There are 17 genes encoding PARP enzymes in mammals including PARP-1, PARP-2, PARP-3, PARP-4, PARP-5/tankyrases-1, and PARP 6/tankyrases-2 [99, 100]. The areas of involvement of PARPs range from the regulation of cell signaling, chromatin structure, DNA repair, telomere maintenance, gene transcription, and many others [101–103]. Among all other PARPs, PARP-1 is best studied and a DNA nick sensor that responds very rapidly to single-stranded DNA breaks. The PARP-1 knockout mice (PARP-1−/−) exhibit reduction in life span and a significant increase of population aging rate. In human fibroblast cells, indirect evidences suggest a decrease in poly-ADP-ribosylation with age due to decreased activity of PARP [104]. There are evidences suggesting a reduced ADP-ribosylation of eEF-2 (an elongation factor which catalyzes the translocation of peptidyl-tRNA from the A site to the P site of the ribosome) levels in aged and SV40-transformed human cell cultures [105]. In conclusion, PARP activity influences the organism aging process by modulating the immune system via NFκB pathway and in the maintenance of the genomic stability by several pathways [100].

Protein Misfolding, Aggregation, and Associated Diseases

Proteins are the most diverse biomolecule, both structurally and functionally, and central to the biology. In cellular milieu, protein–protein and protein–nucleic acid interaction networks are the chief component of communication within and between cells. To mediate diverse arrays of cellular functions, most protein molecules after their synthesis as a polypeptide chain need to achieve a specific three-dimensional structure through a process referred to as protein folding. Protein folding essentially brings key functional groups into close proximity, otherwise located distantly on the primary structure of the protein, and enables living systems to develop structural diversity, substrate specificity, catalysis, and many other biological processes. Cells have molecular chaperones which interact with nascent chain emerging from the ribosome and stabilize non-native conformations and facilitate correct protein folding. However, failure of proteins to fold correctly or remain correctly folded is at the origin of a wide variety of pathological conditions, such as cystic fibrosis and Alzheimer's and Parkinson's diseases [106–108]. Thus, the accumulation of aberrant proteins in these pathologies places a burden on protein homeostasis machinery and may accelerate aging (Fig. 1.1).

Neurodegenerative Disorders

Alzheimer's Disease
Alzheimer's disease (AD) is a chronic irreversible neurodegenerative disease associated with dementia that gets worse with age. The first description of AD was reported as the case of "presenile dementia" by a German psychiatrist Dr. Alois Alzheimer [109]. Since then the cause of AD is still unknown, but genetic heritability [110], oxidative stress [111], amyloid generation [112], and tau protein abnormalities [113] are some of the key factors essential in the pathogenesis. AD pathogenesis is believed to be induced either by the accumulation of the amyloid-β peptide (Aβ) or the failure of clearance mechanism. They are hydrophobic peptides, 39 to 42 residues long, and derived from the amyloid precursor proteins by sequential proteolysis by β- and γ-secretases. The major species generated are Aβ40 and Aβ42 peptides, with Aβ42 being the dominant in neuritic plaques of AD patients. These peptides rapidly aggregate to form oligomers, protofibrils, and fibrils that can deposit outside neurons in dense formations known as senile plaques [114]. In addition to amyloid plaques, hyperphosphorylation of Tau protein leads to the formation of aggregates, which form neurofibrillary tangles, a hallmark of Alzheimer's disease. The abnormal hyperphosphorylation of Tau is also associated

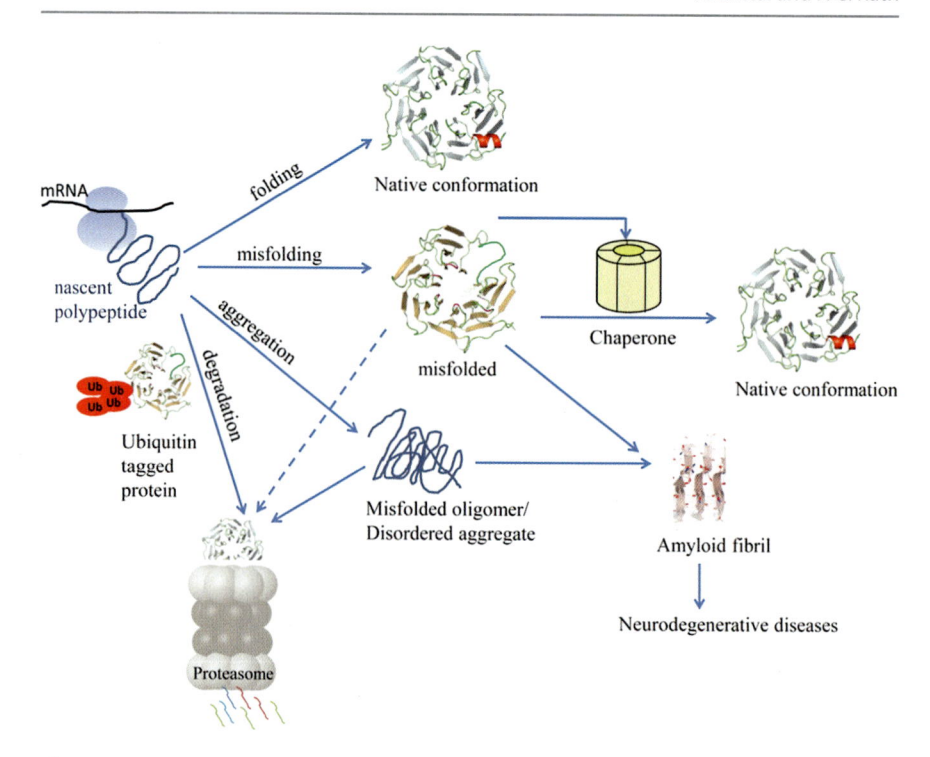

Fig. 1.1 Protein homeostasis. A newly synthesized polypeptide can assemble into various kinds of structures, such as native conformation, a misfolded structure that can be refolded by Chaperone proteins to its native conformation. Failure to fold correctly can result in misfolding and aggregation, which are associated with wide range of human neurodegenerative disorders. The ubiquitin code links misfolded/aggregated protein to lysosome or proteasome for degradation

with several other neurodegenerative disorders referred to as tauopathies [115]. These pathologies might arise due to disruption in proteostasis network due to aging and that may increase kinase or decrease phosphatase activity. Solid-state nuclear magnetic resonance (NMR) spectroscopy and electron microscopy have been used widely to study the molecular structures of these aggregates. Direct measurements on aggregates obtained from brain tissue are not possible because NMR experiments requires milligram-scale quantities of N^{15} and C^{13} labelled protein, and therefore, Aβ fibrils formed in vitro have been used in the past [116–118]. Aβ fibrils have a cross-β structure where individual β-strands are oriented perpendicular to the fibril axis [119]. Models of Aβ40 and Aβ42 fibrils have been compared based on data from NMR and scanning-tunneling electron microscopy. In brief, NMR experiments of Aβ40 fibrils have shown that residues 1–10 are unstructured and 11–40 adopt a β-turn fold, and in Aβ42 fibrils [118, 120], residues 1–17 may be unstructured and residues 18–42 adopt a β-turn-β fold [121]. In both Aβ40 and Aβ42 fibrils, the turn conformation is stabilized by hydrophobic interactions and by a salt bridge between Asp23 and Lys28 (Fig. 1.2).

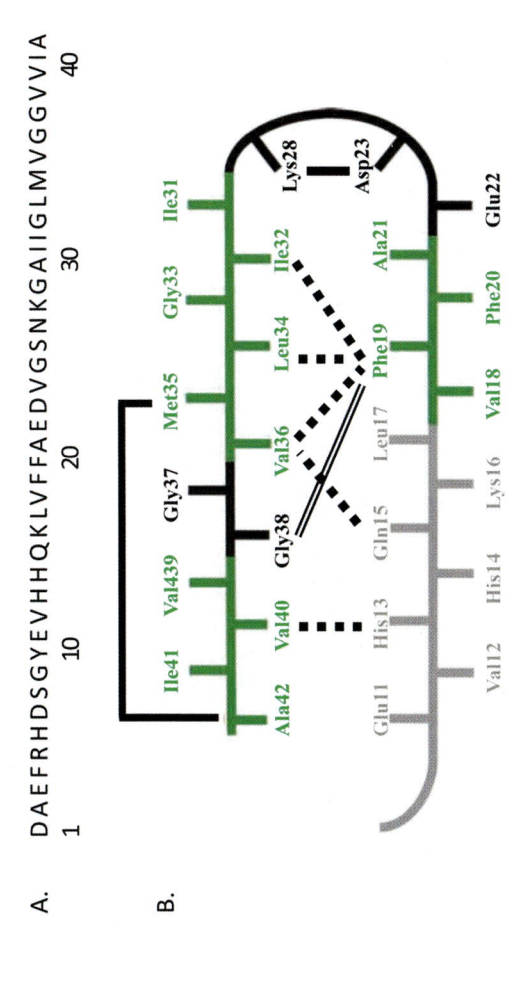

Fig. 1.2 Sequence and structure of the monomer unit in Aβ40 and Aβ42 fibrils (**a**). Sequence of Aβ42 derived from human amyloid precursor protein, (**b**). In Aβ40 model, residues 1–10 are unstructured and 11–40 adopt a β-turn-β fold stabilized by a salt bridge between Asp23 and Lys28 (solid black line). Side chain packing is observed between Phe19 and Ile32, Leu34 and Val36, and between Gln15 and Val36 as well as between His13 and Val40 (black square dot). In Aβ42 fibrils, residues 1–17 may be unstructured, with residues 18–42 forming a β-turn-β fold. In the Aβ42 model, Phe19 is in contact with Gly38 (double line), and between Met35 and Ala42 (black line). In both Aβ40 and Aβ42, the turn conformation is stabilized by hydrophobic interactions (green residues) and by a salt bridge between Asp23 and Lys28 (black line)
Reprinted from the permission from Nature Publishing Group (License Number 400748045O378)
Structural conversion of neurotoxic amyloid-β[1–42] oligomers to fibrils (Reference 127)
Nature structural & molecular biology VOLUME 17 NUMBER 5 MAY 2010

Oxidative damage is common in the aging brain but is more severe in AD. Protein oxidation, lipid peroxidation and ROS formation are some of the features that cause oxidative stress and play a major role in the pathogenesis of AD. Proteomic analysis has identified several proteins, such as creatine kinase BB, β-actin, ubiquitin C-terminal hydrolase L-1, and dihydropyrimidinase-related-protein-2 that appear to be more cabonylated in AD brain than age-matched controls, particularly in those regions which contain severe histopathological alterations [98, 122–125]. Identification of these carbonylated proteins have suggested a plausible role of these proteins in neurodegeneration in AD brain, for example, ubiquitin C-terminal hydrolase L-1 (UCH-L1), a deubiquitylase enzyme, is an extremely abundant protein in the brain (1–5% of total neuronal proteins) and is necessary for maintaining the axonal integrity. UCH-L1 is part of the ubiquitin system, and carbonylation of UCH-L1 will affect proteasomal function, which is one of the drivers for degradation of damaged, misfolded, or aggregated proteins [126]. Taken together, these carbonylated or altered proteins have suggested plausible mechanism for neurodegeneration in AD brain.

Parkinson's Disease

Parkinson's disease (PD) is the second most common age-related neurodegenerative disorder as a result of protein misfolding and aggregation. The most prominent signs of disease are akinesia and tremor. At molecular level, PD is characterized by the accumulation of neuron-associated aggregates of proteins, such as α-synuclein. These aggregates are central to the formation of Lewy bodies, which is a pathological hallmark of the disease and is associated with neuronal damage [127, 128]. Giasson and colleagues have suggested a link between oxidative damage and the neurodegeneration diseases. They showed an extensive and widespread accumulation of nitrated α-synuclein at selective tyrosine residues in the signature inclusions of Parkinson's disease [129]. Though most cases of PD are sporadic, a subset of PD cases is inheritable and attributable to mutations in specific genes including *parkin, α-synuclein, PINK1,* and *LRKK2* [130, 131]. Loss-of-function mutations in *parkin*, an E3 ubiquitin ligase that functions to promote the ubiquitin–proteasome system of protein degradation, have been implicated in heritable forms of Parkinson's disease [132]. Two mutations, namely, A30P [133] and A53T [134] in *α-synuclein* gene, have been linked to early-onset PD. Remarkably, at high local concentration, α-synuclein forms amyloids [135], and its aggregation is central to the formation of Lewy bodies. α-synuclein is a 140-amino acid intrinsically disordered protein, and residues 61–95 were referred to as NAC region (non-amyloid-β component) that was reported to be deposited with amyloid-β in the brains of Alzheimer's disease patients and is sufficient for the aggregation and toxicity of α-synuclein [136–138]. Though segments outside NAC also influence the aggregation of α-synuclein and have been associated with fibril structure, residues 68–78 are critical in both aggregation and cytotoxicity [127, 139]. A study by Conway et al. demonstrated that α-synuclein mutants (A30P and A53T) are disordered as the wild-type α-synuclein in dilute solution. However, at higher concentration, Lewy bodies like fibrils and discrete spherical assemblies were observed most rapidly for A53T. These observations suggested that mutant forms fibril more rapidly and accelerate its transition into amyloid state and hence early onset of PD [140].

Protein Turnover During Aging

The balance between protein synthesis and protein degradation in a cell is referred to as protein turnover, where old and damaged proteins are degraded and replaced by newly synthesized polypeptides. Protein turnover is believed to decrease significantly with age in all organisms including humans [141] and rats [142], which is marked by an increased concentration of damaged protein within the body. Additionally, the rate of protein synthesis also declines with age, and it must be counterbalanced by the decreasing protein degradation to maintain a functional proteome [143]. In fact, the half-life of an average protein increases about tenfold in old nematodes (*Turbatrix aceti*) in comparison to young worms [144]. Age-dependent decline in protein turnover can also contribute to the development of neurodegenerative disorders, such as Parkinson's and Alzheimer's diseases, which are associated with the accumulation of protein aggregates [143, 145]. Indeed, there are experimental evidences suggesting a decrease in the proteolytic activities of lysosome and proteasome with age [146, 147]. However, ubiquitin-mediated proteolysis, in aging fibroblasts, did not decline, and no change in the levels of ubiquitin mRNA and free ubiquitin pools were detected [148].

Conclusions

Understanding the molecular and cellular basis of aging progression is very intricate and complex and is subject to regulation by multiple factors ranging from protein synthesis, protein folding, posttranslational modifications, protein turnover, and involvement of many cell signaling pathways. Though it is widely accepted that aging derives from the malfunctioning of multiple maintenance mechanisms, the magnitude of the effects of each factor listed above regarding the onset and progression of aging and age-related diseases remains unclear. Therefore, a comprehensive understanding of protein-folding pathway and posttranslational modifications in the context of aging and role in cell signaling can help in preventing the early onset of aging and age-related diseases.

Epigenetic dysregulation has emerged as one of the hallmarks of the aging process. Recently, in an exciting research, Belmonte and his colleagues have demonstrated that short-term expression of Yamanaka factors (Oct4, Sox2, Klf4 and c-Myc) has the capability to rejuvenate cellular phenotype of aging in mouse and human cells [149]. Since the Yamanaka factors have the capability to reset the epigenetic landscape of differentiated cells, and therefore, Belmonte's lab findings indicate that cellular aging, in part, is driven by epigenetic shift and reprogramming can correct these epigenetics errors. With exciting advances in the genome editing, analysis methods, and emergence of noncoding RNA, epigenetics research is entering into an exciting era and is a promising area of investigation in the context of aging and neurodegenerative diseases. In fact, several long noncoding RNAs (lncRNA) have emerged as major regulators of transcription and chromatin-modifying complexes and in telomere maintenance and have also been implicated in aging process and aging-related diseases [2, 150–154]. There are questions remaining to be answered: Why do protein aggregation diseases primarily affect the brain and why older cells respond less effectively to clear these aggregations? Further studies are needed for a complete understanding of the protein structure function in the context of aging process, which can facilitate the development of effective interventions and therapy to increase "health span."

Acknowledgement We thank Dr. Vikash Verma (University of Massachusetts, Amherst, USA) for critical reading of the manuscript and valuable comments.

References

1. Crick F. Central dogma of molecular biology. Nature. 1970;227:561–3.
2. Rinn JL, Chang HY. Genome regulation by long noncoding RNAs. Annu Rev Biochem. 2012;81:145–66.
3. Lewin R. RNA can be a catalyst. Science. 1982;218:872–4.
4. Elsersawi A. Biochemistry of aging: wellness and longevity. Bloomington: AuthorHouse; 2010.
5. National Institute on Aging;1999.
6. Hartl FU, Bracher A, Hayer-Hartl M. Molecular chaperones in protein folding and proteostasis. Nature. 2011;475:324–32.
7. Horwich AL. Molecular chaperones in cellular protein folding: the birth of a field. Cell. 2014;157:285–8.
8. Cuervo AM. Autophagy and aging: keeping that old broom working. Trends Genet : TIG. 2008;24:604–12.
9. Karve TM, Cheema AK. Small changes huge impact: the role of protein posttranslational modifications in cellular homeostasis and disease. J Amino Acids. 2011;2011:207691.
10. Lopez-Otin C, Blasco MA, Partridge L, Serrano M, Kroemer G. The hallmarks of aging. Cell. 2013;153:1194–217.
11. Wood JG, Rogina B, Lavu S, Howitz K, Helfand SL, Tatar M, Sinclair D. Sirtuin activators mimic caloric restriction and delay ageing in metazoans. Nature. 2004;430:686–9.
12. Honjoh S, Yamamoto T, Uno M, Nishida E. Signalling through RHEB-1 mediates intermittent fasting-induced longevity in C. elegans. Nature. 2009;457:726–30.
13. Greer EL, Dowlatshahi D, Banko MR, Villen J, Hoang K, Blanchard D, Gygi SP, Brunet A. An AMPK-FOXO pathway mediates longevity induced by a novel method of dietary restriction in C. elegans. Current Biol : CB. 2007;17:1646–56.
14. Kapahi P, Zid BM, Harper T, Koslover D, Sapin V, Benzer S. Regulation of lifespan in Drosophila by modulation of genes in the TOR signaling pathway. Current Biol : CB. 2004;14:885–90.
15. Barzilai N, Huffman DM, Muzumdar RH, Bartke A. The critical role of metabolic pathways in aging. Diabetes. 2012;61:1315–22.
16. Hands SL, Proud CG, Wyttenbach A. mTOR's role in ageing: protein synthesis or autophagy? Aging. 2009;1:586–97.

17. Kenyon CJ. The genetics of ageing. Nature. 2010;464:504–12.
18. Blagosklonny MV. Aging: ROS or TOR. Cell Cycle. 2008;7:3344–54.
19. Kaeberlein M, Powers RW 3rd, Steffen KK, Westman EA, Hu D, Dang N, Kerr EO, Kirkland KT, Fields S, Kennedy BK. Regulation of yeast replicative life span by TOR and Sch9 in response to nutrients. Science. 2005;310:1193–6.
20. Vellai T, Takacs-Vellai K, Zhang Y, Kovacs AL, Orosz L, Muller F. Genetics: influence of TOR kinase on lifespan in C. elegans. Nature. 2003;426:620.
21. Jia K, Chen D, Riddle DL. The TOR pathway interacts with the insulin signaling pathway to regulate C. elegans larval development, metabolism and life span. Development. 2004;131:3897–906.
22. Gangloff YG, Mueller M, Dann SG, Svoboda P, Sticker M, Spetz JF, Um SH, Brown EJ, Cereghini S, Thomas G, et al. Disruption of the mouse mTOR gene leads to early post-implantation lethality and prohibits embryonic stem cell development. Mol Cell Biol. 2004;24:9508–16.
23. Cuervo AM, Wong E. Chaperone-mediated autophagy: roles in disease and aging. Cell Res. 2014;24:92–104.
24. Gingras AC, Raught B, Sonenberg N. eIF4 initiation factors: effectors of mRNA recruitment to ribosomes and regulators of translation. Annu Rev Biochem. 1999;68:913–63.
25. Rattan SI. Synthesis, modifications, and turnover of proteins during aging. Exp Gerontol. 1996;31:33–47.
26. Kimball SR, Vary TC, Jefferson LS. Age-dependent decrease in the amount of eukaryotic initiation factor 2 in various rat tissues. Biochem J. 1992;286(Pt 1):263–8.
27. Laplante M, Sabatini DM. mTOR signaling at a glance. J Cell Sci. 2009;122:3589–94.
28. Choo AY, Blenis J. Not all substrates are treated equally: implications for mTOR, rapamycin-resistance and cancer therapy. Cell Cycle. 2009;8:567–72.
29. McCormick MA, Tsai SY, Kennedy BK. TOR and ageing: a complex pathway for a complex process. Philos Trans R Soc Lond Ser B Biol Sci. 2011;366:17–27.
30. Hansen M, Taubert S, Crawford D, Libina N, Lee SJ, Kenyon C. Lifespan extension by conditions that inhibit translation in Caenorhabditis elegans. Aging Cell. 2007;6:95–110.
31. Zid BM, Rogers AN, Katewa SD, Vargas MA, Kolipinski MC, Lu TA, Benzer S, Kapahi P. 4E-BP extends lifespan upon dietary restriction by enhancing mitochondrial activity in Drosophila. Cell. 2009;139:149–60.
32. Pan KZ, Palter JE, Rogers AN, Olsen A, Chen D, Lithgow GJ, Kapahi P. Inhibition of mRNA translation extends lifespan in Caenorhabditis elegans. Aging Cell. 2007;6:111–9.
33. Wang X, Li W, Williams M, Terada N, Alessi DR, Proud CG. Regulation of elongation factor 2 kinase by p90(RSK1) and p70 S6 kinase. EMBO J. 2001;20:4370–9.
34. Selman C, Tullet JM, Wieser D, Irvine E, Lingard SJ, Choudhury AI, Claret M, Al-Qassab H, Carmignac D, Ramadani F, et al. Ribosomal protein S6 kinase 1 signaling regulates mammalian life span. Science. 2009;326:140–4.
35. Troulinaki K, Tavernarakis N. Protein synthesis and ageing. New York: Nova Science Publishers, Inc; 2008.
36. Cuervo AM. Autophagy: many paths to the same end. Mol Cell Biochem. 2004;263:55–72.
37. Glick D, Barth S, Macleod KF. Autophagy: cellular and molecular mechanisms. J Pathol. 2010;221:3–12.
38. Chiang HL, Terlecky SR, Plant CP, Dice JF. A role for a 70-kilodalton heat shock protein in lysosomal degradation of intracellular proteins. Science. 1989;246:382–5.
39. Komatsu M, Waguri S, Chiba T, Murata S, Iwata J, Tanida I, Ueno T, Koike M, Uchiyama Y, Kominami E, et al. Loss of autophagy in the central nervous system causes neurodegeneration in mice. Nature. 2006;441:880–4.
40. Cuervo AM, Bergamini E, Brunk UT, Droge W, Ffrench M, Terman A. Autophagy and aging: the importance of maintaining "clean" cells. Autophagy. 2005;1:131–40.
41. Toth ML, Sigmond T, Borsos E, Barna J, Erdelyi P, Takacs-Vellai K, Orosz L, Kovacs AL, Csikos G, Sass M, et al. Longevity pathways converge on autophagy genes to regulate life span in Caenorhabditis elegans. Autophagy. 2008;4:330–8.

42. Hars ES, Qi H, Ryazanov AG, Jin S, Cai L, Hu C, Liu LF. Autophagy regulates ageing in C. elegans. Autophagy. 2007;3:93–5.
43. Juhasz G, Erdi B, Sass M, Neufeld TP. Atg7-dependent autophagy promotes neuronal health, stress tolerance, and longevity but is dispensable for metamorphosis in Drosophila. Genes Dev. 2007;21:3061–6.
44. Simonsen A, Cumming RC, Brech A, Isakson P, Schubert DR, Finley KD. Promoting basal levels of autophagy in the nervous system enhances longevity and oxidant resistance in adult Drosophila. Autophagy. 2008;4:176–84.
45. Fan J, Kou X, Jia S, Yang X, Yang Y, Chen N. Autophagy as a potential target for sarcopenia. J Cell Physiol. 2016;231:1450–9.
46. Rattan SI. Synthesis, modification and turnover of proteins during aging. Adv Exp Med Biol. 2010;694:1–13.
47. Dephoure N, Zhou C, Villen J, Beausoleil SA, Bakalarski CE, Elledge SJ, Gygi SP. A quantitative atlas of mitotic phosphorylation. Proc Natl Acad Sci U S A. 2008;105:10762–7.
48. Stein GH, Dulic V. Origins of G1 arrest in senescent human fibroblasts. Bioessays. 1995;17:537–43.
49. Merrick WC. Mechanism and regulation of eukaryotic protein synthesis. Microbiol Rev. 1992;56:291–315.
50. Riis B, Rattan SI, Palmquist K, Nilsson A, Nygard O, Clark BF. Elongation factor 2-specific calcium and calmodulin dependent protein kinase III activity in rat livers varies with age and calorie restriction. Biochem Biophys Res Commun. 1993;192:1210–6.
51. Riis B, Rattan SI, Palmquist K, Clark BF, Nygard O. Dephosphorylation of the phosphorylated elongation factor-2 in the livers of calorie-restricted and freely-fed rats during ageing. Biochem Mol Biol Int. 1995;35:855–9.
52. Meinnel T, Mechulam Y, Blanquet S. Aminoacyl-tRNA Synthetases: occurrence, structure, and function. Washington, DC: ASM Press; 1995.
53. Kihara F, Ninomiya-Tsuji J, Ishibashi S, Ide T. Failure in S6 protein phosphorylation by serum stimulation of senescent human diploid fibroblasts, TIG-1. Mech Ageing Dev. 1986;37:27–40.
54. Battaini F. Protein kinase C isoforms as therapeutic targets in nervous system disease states. Pharmacol Res. 2001;44:353–61.
55. Amadio M, Battaini F, Pascale A. The different facets of protein kinases C: old and new players in neuronal signal transduction pathways. Pharmacol Res. 2006;54:317–25.
56. Van der Zee EA, Compaan JC, de Boer M, Luiten PG. Changes in PKC gamma immunoreactivity in mouse hippocampus induced by spatial discrimination learning. J Neurosci. 1992;12:4808–15.
57. Sacktor TC, Osten P, Valsamis H, Jiang X, Naik MU, Sublette E. Persistent activation of the zeta isoform of protein kinase C in the maintenance of long-term potentiation. Proc Natl Acad Sci U S A. 1993;90:8342–6.
58. Battaini F, Pascale A. Protein kinase C signal transduction regulation in physiological and pathological aging. Ann N Y Acad Sci. 2005;1057:177–92.
59. Pascale A, Amadio M, Govoni S, Battaini F. The aging brain, a key target for the future: the protein kinase C involvement. Pharmacol Res. 2007;55:560–9.
60. Mochly-Rosen D. Localization of protein kinases by anchoring proteins: a theme in signal transduction. Science. 1995;268:247–51.
61. Nishizuka Y. Protein kinase C and lipid signaling for sustained cellular responses. FASEB J. 1995;9:484–96.
62. Bloch S, Cedar H. Methylation of chromatin DNA. Nucleic Acids Res. 1976;3:1507–19.
63. Liu J, Jia G. Methylation modifications in eukaryotic messenger RNA. *Journal of genetics and genomics* =. Yi Chuan Xue Bao. 2014;41:21–33.
64. Zhang Y, Mittal A, Reid J, Reich S, Gamblin SJ, Wilson JR. Evolving catalytic properties of the MLL family SET domain. Structure. 2015;23:1921–33.
65. Nekrasov M, Wild B, Muller J. Nucleosome binding and histone methyltransferase activity of Drosophila PRC2. EMBO Rep. 2005;6:348–53.

66. Jones MJ, Goodman SJ, Kobor MS. DNA methylation and healthy human aging. Aging Cell. 2015;14:924–32.
67. Bocklandt S, Lin W, Sehl ME, Sanchez FJ, Sinsheimer JS, Horvath S, Vilain E. Epigenetic predictor of age. PLoS One. 2011;6:e14821.
68. Hannum G, Guinney J, Zhao L, Zhang L, Hughes G, Sadda S, Klotzle B, Bibikova M, Fan JB, Gao Y, et al. Genome-wide methylation profiles reveal quantitative views of human aging rates. Mol Cell. 2013;49:359–67.
69. Florath I, Butterbach K, Muller H, Bewerunge-Hudler M, Brenner H. Cross-sectional and longitudinal changes in DNA methylation with age: an epigenome-wide analysis revealing over 60 novel age-associated CpG sites. Hum Mol Genet. 2014;23:1186–201.
70. Schosserer M, Minois N, Angerer TB, Amring M, Dellago H, Harreither E, Calle-Perez A, Pircher A, Gerstl MP, Pfeifenberger S, et al. Methylation of ribosomal RNA by NSUN5 is a conserved mechanism modulating organismal lifespan. Nat Commun. 2015;6:6158.
71. Blanco S, Frye M. Role of RNA methyltransferases in tissue renewal and pathology. Curr Opin Cell Biol. 2014;31:1–7.
72. Huang M. The University of Western Ontario;2012.
73. Bannister AJ, Kouzarides T. Regulation of chromatin by histone modifications. Cell Res. 2011;21:381–95.
74. Eissenberg JC, Shilatifard A. Histone H3 lysine 4 (H3K4) methylation in development and differentiation. Dev Biol. 2010;339:240–9.
75. Bibikova M, Laurent LC, Ren B, Loring JF, Fan JB. Unraveling epigenetic regulation in embryonic stem cells. Cell Stem Cell. 2008;2:123–34.
76. Milne TA, Briggs SD, Brock HW, Martin ME, Gibbs D, Allis CD, Hess JL. MLL targets SET domain methyltransferase activity to Hox gene promoters. Mol Cell. 2002;10:1107–17.
77. Greer EL, Maures TJ, Hauswirth AG, Green EM, Leeman DS, Maro GS, Han S, Banko MR, Gozani O, Brunet A. Members of the H3K4 trimethylation complex regulate lifespan in a germline-dependent manner in C. elegans. Nature. 2010;466:383–7.
78. Han S, Brunet A. Histone methylation makes its mark on longevity. Trends Cell Biol. 2012;22:42–9.
79. Sen P, Shah PP, Nativio R, Berger SL. Epigenetic mechanisms of longevity and aging. Cell. 2016;166:822–39.
80. Barber JR, Clarke S. Membrane protein carboxyl methylation increases with human erythrocyte age. Evidence for an increase in the number of methylatable sites. J Biol Chem. 1983;258:1189–96.
81. de la Mora-de la Mora I, Torres-Larios A, Enriquez-Flores S, Mendez ST, Castillo-Villanueva A, Gomez-Manzo S, Lopez-Velazquez G, Marcial-Quino J, Torres-Arroyo A, Garcia-Torres I, et al. Structural effects of protein aging: terminal marking by deamidation in human triosephosphate isomerase. PLoS One. 2015;10:e0123379.
82. Robinson NE, Robinson AB. Deamidation of human proteins. Proc Natl Acad Sci U S A. 2001;98:12409–13.
83. Lindner H, Helliger W. Age-dependent deamidation of asparagine residues in proteins. Exp Gerontol. 2001;36:1551–63.
84. Scotchler JW, Robinson AB. Deamidation of glutaminyl residues: dependence on pH, temperature, and ionic strength. Anal Biochem. 1974;59:319–22.
85. Robinson NE, Robinson AB. Molecular clocks. Proc Natl Acad Sci U S A. 2001;98:944–9.
86. Toyama BH, Hetzer MW. Protein homeostasis: live long, won't prosper. Nat Rev Mol Cell Biol. 2013;14:55–61.
87. Masters PM, Bada JL, Zigler JS Jr. Aspartic acid racemisation in the human lens during ageing and in cataract formation. Nature. 1977;268:71–3.
88. Helfman PM, Bada JL. Aspartic acid racemisation in dentine as a measure of ageing. Nature. 1976;262:279–81.
89. Helfman PM, Bada JL. Aspartic acid racemization in tooth enamel from living humans. Proc Natl Acad Sci U S A. 1975;72:2891–4.

90. Hooi MY, Raftery MJ, Truscott RJ. Age-dependent deamidation of glutamine residues in human gammaS crystallin: deamidation and unstructured regions. Protein Sci. 2012;21:1074–9.
91. Lampiasi N, Umezawa K, Montalto G, Cervello M. Poly (ADP-ribose) polymerase inhibition synergizes with the NF-kappaB inhibitor DHMEQ to kill hepatocellular carcinoma cells. Biochim Biophys Acta. 2014;1843:2662–73.
92. Schey KL, Little M, Fowler JG, Crouch RK. Characterization of human lens major intrinsic protein structure. Invest Ophthalmol Vis Sci. 2000;41:175–82.
93. Harman D. Aging: a theory based on free radical and radiation chemistry. J Gerontol. 1956;11:298–300.
94. Harman D. Origin and evolution of the free radical theory of aging: a brief personal history, 1954-2009. Biogerontology. 2009;10:773–81.
95. de Graff AM, Hazoglou MJ, Dill KA. Highly charged proteins: the Achilles' heel of aging proteomes. Structure. 2016;24:329–36.
96. Stadtman ER. Protein oxidation and aging. Science. 1992;257:1220–4.
97. Dalle-Donne I, Giustarini D, Colombo R, Rossi R, Milzani A. Protein carbonylation in human diseases. Trends Mol Med. 2003;9:169–76.
98. Tanase M, Urbanska AM, Zolla V, Clement CC, Huang L, Morozova K, Follo C, Goldberg M, Roda B, Reschiglian P, et al. Role of carbonyl modifications on aging-associated protein aggregation. Sci Rep. 2016;6:19311.
99. Hassa PO, Hottiger MO. The diverse biological roles of mammalian PARPS, a small but powerful family of poly-ADP-ribose polymerases. Front Biosci. 2008;13:3046–82.
100. Piskunova TS, Yurova MN, Ovsyannikov AI, Semenchenko AV, Zabezhinski MA, Popovich IG, Wang ZQ, Anisimov VN. Deficiency in poly(ADP-ribose) Polymerase-1 (PARP-1) accelerates aging and spontaneous carcinogenesis in mice. Curr Gerontol Geriatr Res. 2008;2008:754190.
101. Beneke S, Burkle A. Poly(ADP-ribosyl)ation in mammalian ageing. Nucleic Acids Res. 2007;35:7456–65.
102. Saxena A, Saffery R, Wong LH, Kalitsis P, Choo KH. Centromere proteins Cenpa, Cenpb, and Bub3 interact with poly(ADP-ribose) polymerase-1 protein and are poly(ADP-ribosyl)ated. J Biol Chem. 2002;277:26921–6.
103. O'Connor MS, Safari A, Liu D, Qin J, Songyang Z. The human Rap1 protein complex and modulation of telomere length. J Biol Chem. 2004;279:28585–91.
104. Dell'Orco RT, Anderson LE. Decline of poly(ADP-ribosyl)ation during in vitro senescence in human diploid fibroblasts. J Cell Physiol. 1991;146:216–21.
105. Riis B, Rattan SI, Derventzi A, Clark BF. Reduced levels of ADP-ribosylatable elongation factor-2 in aged and SV40-transformed human cell cultures. FEBS Lett. 1990;266:45–7.
106. Dobson CM. Protein folding and misfolding. Nature. 2003;426:884–90.
107. Vendruscolo M, Knowles TP, Dobson CM. Protein solubility and protein homeostasis: a generic view of protein misfolding disorders. Cold Spring Harb Perspect Biol. 2011;3:a010454.
108. Bukau B, Weissman J, Horwich A. Molecular chaperones and protein quality control. Cell. 2006;125:443–51.
109. Berrios GE. Alzheimer's disease: A conceptual history. Int J Geriatr Psychiatry. 1990;5:355–65.
110. Wilson RS, Barral S, Lee JH, Leurgans SE, Foroud TM, Sweet RA, Graff-Radford N, Bird TD, Mayeux R, Bennett DA. Heritability of different forms of memory in the late onset Alzheimer's disease family study. J Alzheimers Dis. 2011;23:249–55.
111. Butterfield DA, Drake J, Pocernich C, Castegna A. Evidence of oxidative damage in Alzheimer's disease brain: central role for amyloid beta-peptide. Trends Mol Med. 2001;7:548–54.
112. Hardy J, A.D. Amyloid deposition as the central event in the aetiology of Alzheimer's disease. Trends Pharmacol Sci. 1991;12:383–8.

113. Mudher A, Lovestone S. Alzheimer's disease-do tauists and baptists finally shake hands? Trends Neurosci. 2002;25:22–6.
114. Saido TC. Metabolism of amyloid beta peptide and pathogenesis of Alzheimer's disease. Proc Jpn Acad Ser B Phys Biol Sci. 2013;89:321–39.
115. Wang JZ, Xia YY, Grundke-Iqbal I, Iqbal K. Abnormal hyperphosphorylation of tau: sites, regulation, and molecular mechanism of neurofibrillary degeneration. J Alzheimers Dis. 2013;33(Suppl 1):S123–39.
116. Lu JX, Qiang W, Yau WM, Schwieters CD, Meredith SC, Tycko R. Molecular structure of beta-amyloid fibrils in Alzheimer's disease brain tissue. Cell. 2013;154:1257–68.
117. Ahmed M, Davis J, Aucoin D, Sato T, Ahuja S, Aimoto S, Elliott JI, Van Nostrand WE, Smith SO. Structural conversion of neurotoxic amyloid-beta(1-42) oligomers to fibrils. Nat Struct Mol Biol. 2010;17:561–7.
118. Schmidt M, Sachse C, Richter W, Xu C, Fandrich M, Grigorieff N. Comparison of Alzheimer Abeta(1-40) and Abeta(1-42) amyloid fibrils reveals similar protofilament structures. Proc Natl Acad Sci U S A. 2009;106:19813–8.
119. Kirschner DA, Abraham C, Selkoe DJ. X-ray diffraction from intraneuronal paired helical filaments and extraneuronal amyloid fibers in Alzheimer disease indicates cross-beta conformation. Proc Natl Acad Sci U S A. 1986;83:503–7.
120. Paravastu AK, Leapman RD, Yau WM, Tycko R. Molecular structural basis for polymorphism in Alzheimer's beta-amyloid fibrils. Proc Natl Acad Sci U S A. 2008;105:18349–54.
121. Luhrs T, Ritter C, Adrian M, Riek-Loher D, Bohrmann B, Dobeli H, Schubert D, Riek R. 3D structure of Alzheimer's amyloid-beta(1-42) fibrils. Proc Natl Acad Sci U S A. 2005;102:17342–7.
122. Hensley K, Hall N, Subramaniam R, Cole P, Harris M, Aksenov M, Aksenova M, Gabbita SP, Wu JF, Carney JM, et al. Brain regional correspondence between Alzheimer's disease histopathology and biomarkers of protein oxidation. J Neurochem. 1995;65:2146–56.
123. Aksenov MY, Aksenova MV, Butterfield DA, Geddes JW, Markesbery WR. Protein oxidation in the brain in Alzheimer's disease. Neuroscience. 2001;103:373–83.
124. Aksenov M, Aksenova M, Butterfield DA, Markesbery WR. Oxidative modification of creatine kinase BB in Alzheimer's disease brain. J Neurochem. 2000;74:2520–7.
125. Castegna A, Aksenov M, Aksenova M, Thongboonkerd V, Klein JB, Pierce WM, Booze R, Markesbery WR, Butterfield DA. Proteomic identification of oxidatively modified proteins in Alzheimer's disease brain. Part I: creatine kinase BB, glutamine synthase, and ubiquitin carboxy-terminal hydrolase L-1. Free Radic Biol Med. 2002;33:562–71.
126. Bishop P, Rocca D, Henley JM. Ubiquitin C-terminal hydrolase L1 (UCH-L1): structure, distribution and roles in brain function and dysfunction. Biochem J. 2016;473:2453–62.
127. Rodriguez JA, Ivanova MI, Sawaya MR, Cascio D, Reyes FE, Shi D, Sangwan S, Guenther EL, Johnson LM, Zhang M, et al. Structure of the toxic core of alpha-synuclein from invisible crystals. Nature. 2015;525:486–90.
128. Spillantini MG, Schmidt ML, Lee VM, Trojanowski JQ, Jakes R, Goedert M. Alpha-synuclein in Lewy bodies. Nature. 1997;388:839–40.
129. Giasson BI, Duda JE, Murray IV, Chen Q, Souza JM, Hurtig HI, Ischiropoulos H, Trojanowski JQ, Lee VM. Oxidative damage linked to neurodegeneration by selective alpha-synuclein nitration in synucleinopathy lesions. Science. 2000;290:985–9.
130. Thomas B, Beal MF. Parkinson's disease. Hum Mol Genet. 2007;16(2):R183–94.
131. Tan JM, Wong ES, Lim KL. Protein misfolding and aggregation in Parkinson's disease. Antioxid Redox Signal. 2009;11:2119–34.
132. Rana A, Rera M, Walker DW. Parkin overexpression during aging reduces proteotoxicity, alters mitochondrial dynamics, and extends lifespan. Proc Natl Acad Sci U S A. 2013;110:8638–43.
133. Kruger R, Kuhn W, Muller T, Woitalla D, Graeber M, Kosel S, Przuntek H, Epplen JT, Schols L, Riess O. Ala30Pro mutation in the gene encoding alpha-synuclein in Parkinson's disease. Nat Genet. 1998;18:106–8.

134. Polymeropoulos MH, Lavedan C, Leroy E, Ide SE, Dehejia A, Dutra A, Pike B, Root H, Rubenstein J, Boyer R, et al. Mutation in the alpha-synuclein gene identified in families with Parkinson's disease. Science. 1997;276:2045–7.
135. Singleton AB, Farrer M, Johnson J, Singleton A, Hague S, Kachergus J, Hulihan M, Peuralinna T, Dutra A, Nussbaum R, et al. Alpha-Synuclein locus triplication causes Parkinson's disease. Science. 2003;302:841.
136. Ueda K, Fukushima H, Masliah E, Xia Y, Iwai A, Yoshimoto M, Otero DA, Kondo J, Ihara Y, Saitoh T. Molecular cloning of cDNA encoding an unrecognized component of amyloid in Alzheimer disease. Proc Natl Acad Sci U S A. 1993;90:11282–6.
137. Giasson BI, Murray IV, Trojanowski JQ, Lee VM. A hydrophobic stretch of 12 amino acid residues in the middle of alpha-synuclein is essential for filament assembly. J Biol Chem. 2001;276:2380–6.
138. Periquet M, Fulga T, Myllykangas L, Schlossmacher MG, Feany MB. Aggregated alpha-synuclein mediates dopaminergic neurotoxicity in vivo. J Neurosci. 2007;27:3338–46.
139. Der-Sarkissian A, Jao CC, Chen J, Langen R. Structural organization of alpha-synuclein fibrils studied by site-directed spin labeling. J Biol Chem. 2003;278:37530–5.
140. Conway KA, Harper JD, Lansbury PT. Accelerated in vitro fibril formation by a mutant alpha-synuclein linked to early-onset Parkinson disease. Nat Med. 1998;4:1318–20.
141. Young VR, Steffee WP, Pencharz PB, Winterer JC, Scrimshaw NS. Total human body protein synthesis in relation to protein requirements at various ages. Nature. 1975;253:192–4.
142. Lewis SE, Goldspink DF, Phillips JG, Merry BJ, Holehan AM. The effects of aging and chronic dietary restriction on whole body growth and protein turnover in the rat. Exp Gerontol. 1985;20:253–63.
143. Ryazanov AG, Nefsky BS. Protein turnover plays a key role in aging. Mech Ageing Dev. 2002;123:207–13.
144. Prasanna HR, Lane RS. Protein degradation in aged nematodes (Turbatrix aceti). Biochem Biophys Res Commun. 1979;86:552–9.
145. Alves-Rodrigues A, Gregori L, Figueiredo-Pereira ME. Ubiquitin, cellular inclusions and their role in neurodegeneration. Trends Neurosci. 1998;21:516–20.
146. Friguet B, Bulteau AL, Chondrogianni N, Conconi M, Petropoulos I. Protein degradation by the proteasome and its implications in aging. Ann N Y Acad Sci. 2000;908:143–54.
147. Cuervo AM, Dice JF. When lysosomes get old. Exp Gerontol. 2000;35:119–31.
148. Pan JX, Short SR, Goff SA, Dice JF. Ubiquitin pools, ubiquitin mRNA levels, and ubiquitin-mediated proteolysis in aging human fibroblasts. Exp Gerontol. 1993;28:39–49.
149. Ocampo A, Reddy P, Martinez-Redondo P, Platero-Luengo A, Hatanaka F, Hishida T, Li M, Lam D, Kurita M, Beyret E, et al. In vivo amelioration of age-associated hallmarks by partial reprogramming. Cell. 2016;167(1719–1733):e1712.
150. Kour S, Rath PC. Age-dependent differential expression profile of a novel intergenic long noncoding RNA in rat brain. Int J Dev Neurosci. 2015;47:286–97.
151. Kour S, Rath PC. Age-dependent differential expression profile of a novel intergenic long noncoding RNA in rat brain. Int J Dev Neurosci. 2015;46:55–66.
152. Kour S, Rath PC. Age-related expression of a repeat-rich intergenic long noncoding RNA in the rat brain. Mol Neurobiol. 2016;54:639–60.
153. Kour S, Rath PC. Long noncoding RNAs in aging and age-related diseases. Ageing Res Rev. 2016;26:1–21.
154. Kour S, Rath PC. All-trans retinoic acid induces expression of a novel intergenic long noncoding RNA in adult rat primary hippocampal neurons. J Mol Neurosci. 2016;58:266–76.

DNA, DNA Replication, and Aging

Bhumika Sharma, Meetu Agarwal, Vijay Verma, and Suman Kumar Dhar

DNA and Aging

Cells are the building block of an organism. Each cell in the body consists of genetic material in the form of DNA. The genes are arranged in long twisted double-stranded DNA molecule called chromosomes. Linear chromosomes have a stretch of hundreds of hexameric repeats at either ends which constitute the telomere. DNA replication particularly telomere replication is associated with aging. Apart from DNA replication and DNA damage, epigenetic modifications are also responsible for aging (Fig. 2.1a) [1, 2]. Additionally, many disorders which may lead to premature aging are directly or indirectly linked with mutations in the particular gene segment involved in the maintenance of DNA. *Saccharomyces cerevisiae*, defective strain of TEL1 gene (ataxia telangiectasia mutated ortholog), shows chromosomal aberrations and short telomeres, causing premature aging [3]. The caretaker genes which protect the genome integrity against mutations (e.g., transcription-coupled repair genes, nucleotide excision repair genes) may cause premature aging phenotypes after decreasing its activity [4]. On the other hand, increased activity of caretaker genes may increase the life span. Defects in nucleotide excision repair (NER) pathways cause premature aging phenotypes in xeroderma pigmentosum and trichothiodystrophy diseases [4].

Authors Bhumika Sharma, Meetu Agarwal, and Vijay Verma have contributed equally to this chapter.

B. Sharma · M. Agarwal · V. Verma · S. K. Dhar (✉)
Special Centre for Molecular Medicine, Jawaharlal Nehru University, New Delhi, India
e-mail: skdhar@mail.jnu.ac.in

DNA Replication Stress and Aging

The term "replication stress" covers a wide variety of events that change the spatial and temporal DNA replication programs. These set of events that can be of endogenous or exogenous origins have impact on the dynamics of DNA replication. Replication stress occurs when replication fork is stalled due to DNA damage, excessive chromatin compacting which prevents replisome access, and oncogene overexpression [5]. Replication stress may lead to genomic instability, aging, and cancer [6]. Genomic instability may increase during aging due to replication pausing and stalling. Generally, different forms of DNA damage and DNA breakage increase in pausing/stalling of replication (Fig. 2.1b) [7, 8].

Interestingly, it has been found recently that in aging yeast cells, the reduced replication pause sites (RPSs) are found at different ribosomal DNA (rDNA) and in non-ribosomal DNA (non-rDNA) locations [9]. These locations are rich in nonhistone proteins complexes which may reduce replication pausing and potential DNA damage. Less efficient nuclear protein import during aging and the presence of non-stoichiometric amounts of nonhistone protein complexes in the nucleus cause defect in proper formation of the mature complexes that can bind to DNA [10].

Secondary structure of DNA and proteins bound to DNA also cause replication stress. A number of overlapping mechanism have been proposed to explain age-related genomic instability that may include oxidative damage to DNA, mitochondrial oxidative stress damaging cellular constituents, mutations in proteins required for efficient DNA replication, and altered DNA repair and other genomic maintenance programs (Fig. 2.1b) [11, 12].

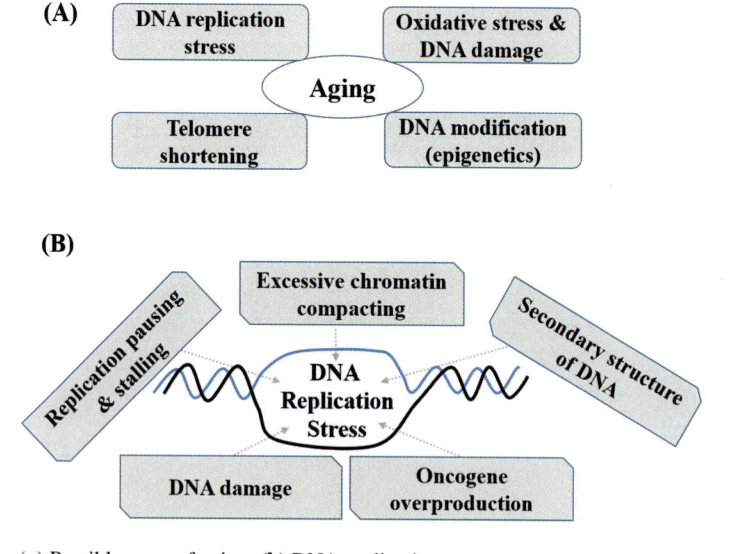

Fig. 2.1 (a) Possible ways of aging. (b) DNA replication stress covers a wide variety of events

Telomeres and Aging

During the process of replication, the leading strand is copied continuously; however, there is discontinuous synthesis of the lagging strand which is formed as Okazaki fragments. Each Okazaki fragment has its own set of RNA primer, which is later replaced with DNA nucleotides in order to make the strand continuous. RNA primers which are synthesized during replication are removed with DNA strand except 5′ end of newly synthesized strand [13]. This 5′ end RNA can be removed by RNase and FEN1, but DNA polymerase can add new DNA only if it has existing strand 5′ behind it to extend. As a result, the newly replicated DNA is shorter at its 5′ end. This creates 3′ overhang at one end of each daughter DNA strand resulting two daughter DNAs having their 3′ overhang at opposite ends (Fig. 2.2a). In telomeric ends, there is no DNA template in the 5′ direction after the final RNA primer. In this condition, DNA polymerase cannot replace the RNA with DNA. Consequently, both the daughter DNA strands have an incomplete 5′ strand with 3′ overhang which can be digested with cellular nucleases, and each daughter strand will become shorter than parental DNA. The eukaryotic chromosome has telomere to prevent this shortening. In telomeric DNA replication, G-rich strand is synthesized as leading strand, and C-rich strand is synthesized as lagging strand. Due to the noncoding nature of telomere, their loss during successive replication cycle may not adversely affect the cell. However, after adequate rounds of the replication of telomere, all the telomeric repeats will be lost creating risks of losing coding sequences with subsequent rounds of DNA replication [14, 15].

Telomeres play important role in maintaining genome integrity. The length of telomeres is maximum at birth (~8000 bp) and decreases with increasing age up to ~3000 bp in adults and as low as ~1500 bp in old age people. After each division, on an average, cell loses 30–200 bp sequences from end of the telomeres [16]. In nondividing cells such as heart muscles, the telomeres do not shorten. Cells undergoing progressive shortening of the telomere may become senescent or die, a process known as replicative senescence.

Embryonic stem cells and germ cells have a mechanism to evade replicative senescence owing to the presence of an enzyme, telomerase. It is a reverse transcriptase with its own RNA template that maintains the chromosome ends by adding complementary RNA bases to 3′ end of the DNA strand [17] (Fig. 2.2b). After 3′ end elongation by telomerase, the DNA polymerase adds the complementary DNA sequences to the ends of the chromosomes. This enzyme is absent in adult somatic cells. As a result, telomeres are continually shortened with subsequent rounds of cell division in somatic cells. Telomeres (tandem repeats motif of 5′ TTAGGG 3′) along with its associated proteins (sheltering complex) are the endcaps of chromosomes that prevent the degradation of coding DNA [18]. A telomere-associated six-subunit protein complex known as sheltering complex (TRF1, TRF2, RAP1, TIN2, TPP1, and POT1) plays very crucial role in the protection of chromosome ends [19, 20]. The sheltering protein TRF1 and TRF2 specifically bind to telomeric repeats of dsDNA and different accessory proteins of the telomere. TRF2 helps in formation of single-stranded 3′ overhang, while POT1 helps in the formation of t-loop by

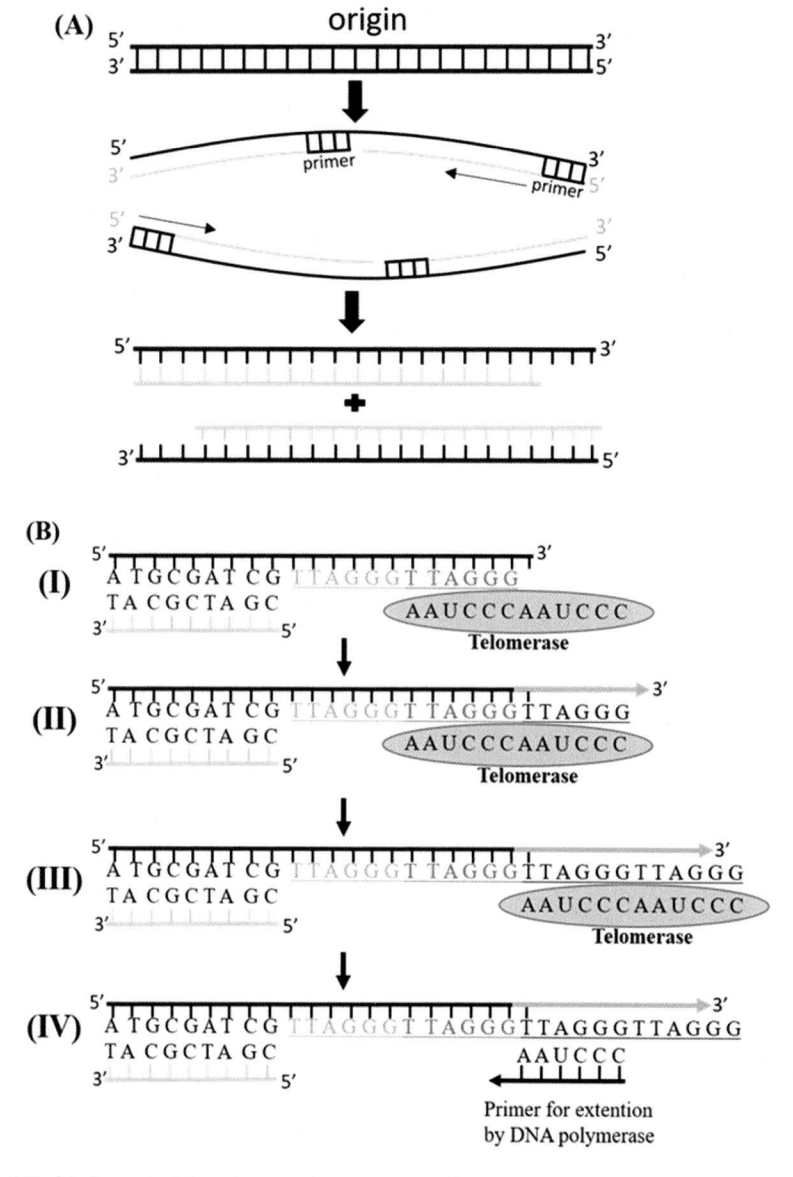

Fig. 2.2 (a) shows the 3′ overhang region at one end of each daughter DNA strand (two daughter DNAs have their 3′ overhang at opposite ends). (b) Replication of 3′ overhang region of telomere with the help of enzyme telomerase

binding to single-stranded 3′ overhang region [21]. Both POT1 and its interacting partner TPP1promote the recruitment of telomerase to the telomere [22].

Telomere length shortening during cell proliferation is a combined effect of various activities and factors that may include repeated cell replication, oxidative stress, inflammation, and generation of reactive oxygen species (ROS) [23]. ROS produced during mitochondrial respiration have major effect on chromosomal DNA by oxidizing nucleotides. ROS-induced loss of telomere is more (~50–200 bp per cell division) as compared to loss due to end replication problem (~10 bp per cell division) [16, 24, 25]. However, the rate of telomere shortening varies among species. For example, the shortening is slower in slow-aging animals than in fast-aging animals [26].

It has been found that shorter telomere length is associated with increased age [27]. They have shown that older people (60 plus) who have shorter telomeres are three times more prone to heart disease and eight times more prone to infectious disease as compared to young with larger telomeres [28]. At this stage, we do not know whether shorter telomeres are just a sign of aging or really contributing to aging. In one interesting finding, scientists have found that telomerase can reverse the age-related conditions. They have taken the telomerase deficient mice with tissue atrophy, organ failure, impaired tissue injury, and stem cell depletion conditions. In these mice, reactivation of telomerase led to reduced DNA damage, improvement of the function of the spleen, intestine and reversal of neurodegeneration suggesting that telomerase reactivation has the potential to cure the age-related disease in human [29].

Telomeres, Aging, and Cancer

Since cancer cells divide more frequently, theoretically their telomeres should become very short, and finally cells would die. On the contrary, these cells escape death by producing more telomerase enzyme which prevents telomere length shortening. Measuring the level of telomerase or length of telomere could be a way to detect cancer. If the telomerase converts cancer cells to immortal cells, will it be possible to prevent the normal cells from aging? What would be the possibility to extend the life span by restoring the telomere length using induced telomerase? If it is possible then, what would be the risk of getting cancer? At present, we are not sure about extending the life span by increasing the telomere length. Scientists are using telomerase to keep human cells dividing beyond their normal limit without converting it to cancerous cells [30, 31]. Telomerase-directed immortalized human cells can be used in transplantation, for treating arthritis, for insulin production, for treating muscular dystrophy, and healing from severe burns. It will also open the path of using laboratory grown human cells for the testing the drug and gene therapy [32, 33].

DNA Damage and Aging

All the macromolecules within the cells (proteins, lipids, and nucleic acid) are exposed to endogenous and exogenous reactive agents [34]. Damage caused to proteins and lipids may be taken care of by fast turnover and will not accrue in the long run. However, DNA (nucleic acid) is the carrier of the hereditary information of an organism, and any DNA insult will be critical to cellular function. Cells have devised DNA repair pathways to rectify DNA damage to an extent [35] but beyond that cellular senescence and cell death (apoptosis) will occur which promotes aging.

Theories of Aging

Various theories were put forth to answer how DNA damage can account for aging. These broadly belong to two classes – one theory considers aging as a programmed process in which changes in gene regulation are preset to reduce viability and the other which takes into account the stochastic changes (wear and tear) of the cellular macromolecules.

Theory of Intrinsic Mutagenesis

In 1974, Burnet introduced the *theory of intrinsic mutagenesis* which proposes that it is the organism which plays with the balance between beneficial mutations which are essential for adapting to the changes in the niche and the somatic mutations that accumulate with time and lead to senescence [36]. It is the genetic constituent and the gene products that determine the faithfulness of DNA and its replication. He advocated that short-lived mammals tend to accumulate more somatic mutations as compared to long-lived mammals. The DNA replication mechanism of short-lived organisms is more error-prone. However, there is little experimental proof for this theory [36].

Somatic Mutation Theory of Aging

A lot of reports were published which dealt with the accumulation of somatic mutations in the genetic material, collectively called as the *somatic mutation theory of aging*. These mutations could be lethal or nonlethal. Lethal mutations in the nuclear DNA impair cellular functions leading to cell death. These mutations mount up with increasing age and lessen the cellular viability, ultimately responsible for shortening the life span of the organism [37, 38].

The Free Radical Theory of Aging

The *free radical theory of aging*, lately termed as the *mitochondrial theory of aging* has gained substantial acceptance [39, 40]. Mitochondria are specialized organelle required for cellular respiration and energy production. Electron transport chain and oxidative phosphorylation are responsible for ATP production along with by-products like superoxide anion, hydrogen peroxide, and hydroxyl radicals. The mitochondrion has its own DNA (mtDNA), which encodes for 13 proteins required in respiratory chain. Since mitochondria are the main site of production of free radicals, the mtDNA is constantly exposed to damaging agent like free radicals as mentioned above (Fig. 2.3). Further, unlike the nuclear DNA, mtDNA is devoid of any

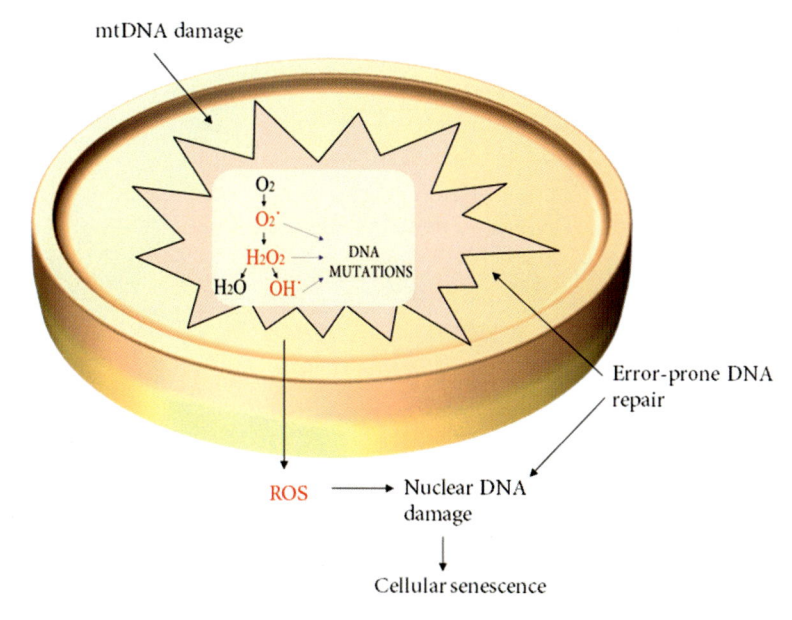

Fig. 2.3 Production of ROS and mitochondrial DNA (mtDNA) damage, followed by nuclear DNA damage and cellular senescence

protection in the form of histone coat. This makes it even more prone to insults. Damage to mtDNA is detrimental to cellular function because mutations incorporated in mitochondrial genome result in incorrect or abridged proteins that will affect the sustainability of the respiratory chain and generation of ATP. These mutations are kept in check by proteins involved in DNA repair like XRCC4, RAD23A, and mitochondrial DNA polymerase θ [41]. Many studies have linked inability of cell to repair mtDNA to faster or premature aging [42, 43].

DNA Damage Theory of Aging

Another theory which comprehends the abovementioned facts is the *DNA damage theory of aging*. DNA carries the genetic information of an organism and needs to be maintained throughout its lifetime. DNA can be either physically damaged which results in the distortion of the double helix, or it can be mutated by exogenous or endogenous damaging agents. Mutations primarily modify the nucleotide sequence of the DNA; these can be classified as deletions, insertions, rearrangements, or substitution of nucleotides. In essence, this theory identifies DNA damage and subsequent alterations in its sequence as a major cause of cellular senescence (Fig. 2.4). Accumulation of these senescent cells in the body gradually leads to decline in growth and repair and thus aggravates aging (reviewed in [44]). We will focus on the DNA damage-related aspect of aging in this chapter.

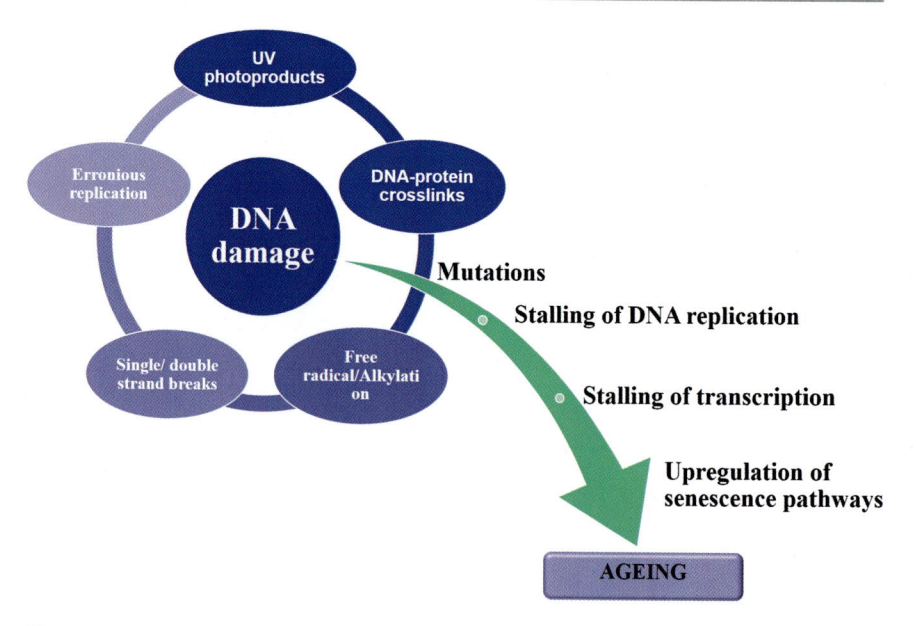

Fig. 2.4 Different ways of DNA damage following mutation and finally upregulation of senescence pathways

Sources and Types of DNA Damage

As mentioned earlier, DNA damage can result from intrinsic (within the cells) or extrinsic (foreign) sources. Most common endogenous agents are chemicals, ultraviolet rays (UV), or ionizing radiations, and in some cases even viruses [45]. Chemical mutagen like 5-bromo-deoxyuridine (5BU) is a *base analogue* which mimics thymine in keto form and cytosine in enol form, respectively. Thus, it can pair with both adenine and guanine in different tautomeric forms. Pairing of enolic 5BU to guanine will lead to base transition in the subsequent DNA replication cycles. Besides base analogues, there is another class of chemical mutagens termed *alkylators*. These chemicals directly bind to the G-rich regions and modify the guanine residue. The modified G-residues lead error-prone repair, most often leading to base degradation or double-strand breaks. Examples of alkylating mutagens are ethyl methane sulfonate (EMS), methyl methane sulfonate (MMS), diethyl sulfate (DES), and nitrosoguanidine. Using yeast cells as model system, it has been reported that MMS damages both nuclear DNA and mitochondrial DNA, making DNA heat-labile and more prone to DNA breaks [46, 47].

UV rays are classified according to their wavelength components; UV-A (320–400 nm) is the most energetic and lethal but is absorbed by the ozone layer; UV-B (280–320 nm) and UV-C (200–280 nm) have distinct mutagenic effects in the form of generation of pyrimidine dimers (covalent bond between adjacent thymine–thymine or thymine–cytosine) and oxygen-free radicals, respectively. Most common DNA lesions produced by UV rays are *cis-syn* cyclobutane pyrimidine dimers (CPDs) and the pyrimidine (6–4) pyrimidone photoproducts [(6–4) PPs] [48]. These are bulky in structure and result in stalling of DNA replication and transcription at these sites, unless repaired. UV-B radiations are also responsible for deamination of cytosine and 5-methylcytosine within the dimer site [45]. In addition to UV rays, ionizing radiations like X-rays or gamma rays also create free radicals after interacting with biological molecules like water. There can be direct effect exhibited in the form of single-stranded or double-stranded DNA breaks giving rise to deletions, rearrangement, or chromosome loss if unrepaired. Radiations are also capable of cross-linking DNA to itself or a protein. However, it has been shown that damaging effect of these radiations is dose-dependent. Recently, the role of ionizing radiations in inducing mutagenesis in the germ line has been established. The offspring of male parents exposed to radiations had higher frequency de novo copy number variants (CNVs) in which certain sections of genome are repeated as well as more insertion/deletions of nucleotide sequences are found in the genome [49].

DNA Damage and Senescence Pathway

As described above, cells undergo aging as a result of replicative senescence (telomere-dependent) or DNA damage-induced premature aging. Signal transduction pathways are induced if any DNA lesion is not taken care of by DNA repair system. Several reports have shown p53 protein as the mediator of cell arrest at G1-phase or apoptosis post-DNA damage. Upon irreparable DNA damage, phosphorylated p53 protein levels increase, and it translocates to the nucleus to bind at specific DNA sequences and activates transcription of genes like *mdm2*, *gadd45* (the growth arrest and DNA damage response gene), and CIPl (cyclin-dependent kinase (Cdk) inhibitor) [50]. Phosphorylation status of retinoblastoma protein (pRb) has been known to control cell cycle progression. Phosphorylated pRb dissociates itself from E2F, which then activates transcription of genes required for DNA replication. Cells undergoing senescence express Cdk inhibitors like CIP1 which facilitate hyperphosphorylation of pRb and keep E2F in bound state [50, 51]. This is the mechanism by which p53 and pRb arrest the cells whose DNA damage cannot be fixed.

Further, a p53-related protein named p63 has been shown to directly prevent the cells from undergoing senescence. Loss of p63 leads to widespread cellular senescence as well as accelerated aging [52]. Cells cultured from patients having premature aging syndrome showed increased senescence in vitro [53]. Moreover, many studies on human samples and animal models have established that senescent cells accumulate in the organism with aging. These reports clearly establish link between cellular senescence and organismal aging.

DNA Repair and Associated Diseases

The primary mechanisms of DNA repair are base excision repair and nucleotide excision repair for single-stranded DNA breaks and homologous recombination and nonhomologous end-joining method for double-strand breaks. It has been shown that inefficiency of these pathways leads to accumulation of mutations and thus cellular senescence. Double-strand breaks have been held responsible for high-scale genomic instability [54]. A prime example is Werner syndrome which is characterized by symptoms of normal aging like cataracts, graying of hair, and skin aging but at very early age. Patients exhibit high number of DNA strand breaks along with dysfunctional telomeres due to mutation in gene coding for helicase protein WRN [55]. WRN helps in maintenance and repair of DNA by processing replicating DNA for cell division.

Another example of aging-related disease due to defects in DNA damage repair is Cockayne syndrome. It is caused due to mutations in either excision repair 8 (ERCC8gene) or the ERCC6 gene (CSB). Proteins encoded by these genes are involved in DNA repair. Other examples are trichothiodystrophy (TTD), xeroderma pigmentosum (XPE), Rothmund–Thomson syndrome (RTS), dyskeratosis congenita, atypical Werner syndrome, restrictive dermopathy (RD), and mandibuloacral dysplasia (MAD) (reviewed in [56]).

Thus, there are evidences that DNA damage, cellular senescence, and organismal aging are interrelated. But a new school of thought is whether DNA damage occurs first which contributes to senescence or it is the loss of efficiency and fidelity of DDR system which allows DNA damage to accumulate with time.

DNA Modification and Aging

With the current understanding, aging is contemplated as a complex process during which many changes accumulate at the molecular, cellular, and organismal levels. Change in chromatin status during aging and role of chromatin modifiers in modulating life span bring the idea of studying aging and epigenetic research together. DNA modifications (epigenetics) include numerous changes such as reduced level of bulk histone proteins, altered pattern of DNA methylation and histone posttranslational modifications, altered noncoding RNA expression, and replacement of canonical histones with variants of histone and other processes that are reversible and heritable (Fig. 2.5), [57–60]. Epigenetic modifications are heritable phenotypic changes without any alteration in the sequence of DNA and protein [61]. These are influenced by lifestyle, environmental exposure, aging, and many complex diseases. Environmental stimuli like stress and dietary changes can also influence life span and health span of an individual [62].

During the past two decades, it has been shown clearly that epigenetic processes play a major role in aging [63, 64]. For instance, there is a tenfold difference between the life span of a queen bee as compared to workers, even though the genetic composition is same. This sheds light on the importance of nongenetic factors in aging [65].

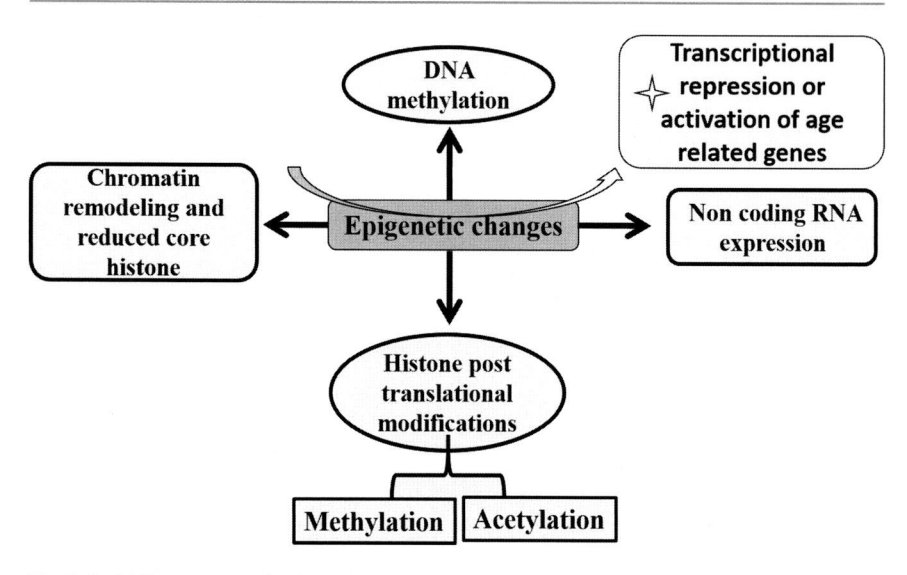

Fig. 2.5 Different ways of epigenetic changes involved in aging process. DNA methylation, histone posttranslational modification like acetylation or methylation, noncoding RNA expression and changes in histone levels, and replacement with variants causes activation or repression of genes related to aging and senescence

DNA Methylation and Aging

DNA methylation is the basis of epigenetics, and it occurs through addition of methyl group at cytosine-5 position that is usually present in CpG dinucleotides [66, 67]. There are three DNA methyltransferases (DNMTs) which are responsible for most of the DNA methylation that happens in the genome. DNA methylation patterns that are inherited are carried out by DNMT1 which causes majority of DNA methylation [68]. Methylation pattern of DNA influences various biological processes like genomic imprinting, cellular differentiation, X chromosome inactivation, and control of gene expression. However, control of gene expression is the most significant biological effect due to DNA methylation where hypermethylation in a gene regulatory region is generally associated with transcriptional repression and hypomethylation is usually associated with transcriptional activity with some notable exceptions [69]. DNA methylation decreases with age; maximum methylation is present in tissues of embryos and newborn animals [70]. Lower levels of 5-methylcytosine are present in senescent cells compared to actively cycling cells [71]. Hypomethylation of DNA is not only related to process of aging but also associated with many chronic age-associated pathologies. Loss of heterochromatin due to demethylation leads to changes in nuclear architecture, and the expression of genes in those regions causes aging and cellular senescence. Loss of transcriptional silencing has been observed in yeast to humans due to loss of heterochromatin. DNA hypomethylation is observed in many age-related diseases such as cancer, atherosclerosis, Alzheimer's disease, autoimmunity, and macular degeneration.

Gene- or region-specific hypermethylation also occurs during aging; tumor suppressor genes that play an important role in cancer and key genes from the lung, colon, breast, and kidney undergo hypermethylation during aging [72].

Histone Posttranslational Modification and Aging

Histone components of chromatin are subject of variety of posttranslational modifications (PTMs). Levels of histone methylation, acetylation, and expression change with aging. PTMs of histones such as acetylation and methylation are major epigenetic processes that can control gene expression either by disrupting chromatin organization or by creating sites for binding of other proteins to specific regions of genome. Histone acetyltransferases (HATs) and histone deacetylases (HDACs) are the major enzymes that cause histone modifications. Generally, histone acetylation caused by HATs is associated with increased transcriptional activity, while histone deacetylation performed by HDACs represses transcription by compacting DNA. Histone acetylation directly affects the physical association of histones with DNA [73]. This balance between HATs and HDACs is lost with age leading to either overactivity of gene, for example, in cancer cells or gene repression that can lead to neurodegeneration during the aging process [74]. The role of HATS and HDACs in aging is exemplified by a condition termed sarcopenia. It is characterized by involuntary loss of skeletal muscle mass and strength in older adults. Exact mechanisms involved in sarcopenia are not clearly understood, and no pharmaceutical treatment exists for sarcopenia. Involvement of HDACs in controlling muscle atrophy, dysfunction due to denervation, and muscle dystrophy further suggest the role of epigenetic modifiers in the process of aging. Therefore, inhibitors of HDACs are potential target for intervention in sarcopenia and already in use in the clinic for the treatment of sarcopenia to delay aging [75].

Histone methylation might have different effects either active or repressed genome regions depending upon the specific residue that is methylated (e.g., H3–K9 leads transcriptional repression, whereas H3–K4 is often found in active chromatin regions) [76]. The nature of methylation is dynamically regulated by histone methyltransferases and demethylases, and hence manipulation in these enzymes can modulate the life span of organism. Redistribution of active histone modification mark (H3K4me3) is observed during aging and in cellular senescence [77, 78].

Epigenetics and Age-Related Diseases

Epigenetic alterations affect many of the disease associated with age. Decrease in methylation is one global characteristic of all human cancer cells causing genome instability [79]. DNA methylation also plays an important role in Alzheimer's disease where many genes undergo epigenetic changes that may contribute to the disease [80]. Epigenetic regulation has a major role in adaptive response of the immune system. Age-related demethylation causes reduction in immune competence and decrease in autoimmunity which can contribute to diseases like rheumatoid arthritis [81]. Differential expression of genes important for T cell differentiation with aging affects immune response. Epigenetic changes have a role in premature aging diseases like Werner syndrome and progeria. The WRN gene responsible for Werner syndrome gets epigenetically silenced in different types of tumor, leading to the idea that this gene plays a role in pathogenic feature of Werner syndrome [82]. Overall, epigenetic mechanisms do not only influence fundamental aspects of aging but also play a crucial role in age-associated diseases.

Summary

Aging is an inevitable multifaceted process principally associated with the vicissitudes of the genetic material, DNA. Errors in DNA replication, telomere shortening, replication stress, DNA damage, and DNA modification all contribute to gradual aging of the cells and the organism. Different forms of DNA damage and DNA breakage increase the stalling of replication fork that causes replication stress and cellular dysfunction. Replicative senescence proves to be an intrinsic mechanism that prevents cells from dividing indefinitely and serves as an efficient means of tumor suppression. From evolutionary aspect, the shortening of telomeres during each DNA replication cycle leads to altered differentiation of senescent cells that contribute to aging, a path distinct from tumorigenesis. Other major factor of aging is oxidative stress. The reactive oxygen species which are generally generated from breath, infection, consumption of alcohol, cigarettes, and inflammation damage the DNA, proteins, and lipids. It has been shown that on exposure to substances that neutralize oxidant, the life span of worms is increased suggesting that oxidative stress is one of important cause for aging. Besides DNA replication stress and damage, another important cause of aging is glycation. Binding of glucose to DNA, protein, and lipids leads to interference of their actual function. Restriction on calorie intake can extend the life span explaining the role of glucose in aging. This chapter describes how telomere shortening, replication stress and DNA damage, epigenetic modifications, chronological age, and glycation of macromolecules work together to cause aging.

References

1. Ioannidou A, Goulielmaki E, Garinis GA. DNA damage: from chronic inflammation to age-related deterioration. Front Genet. 2016;7:187.
2. Robin JD, Magdinier F. Physiological and pathological aging affects chromatin dynamics, structure and function at the nuclear edge. Front Genet. 2016;7:153.
3. Di Domenico EG, et al. Multifunctional role of ATM/Tel1 kinase in genome stability: from the DNA damage response to telomere maintenance. Biomed Res Int. 2014;2014:787404.
4. van Heemst D, den Reijer PM, Westendorp RG. Ageing or cancer: a review on the role of caretakers and gatekeepers. Eur J Cancer. 2007;43(15):2144–52.
5. Mazouzi A, Velimezi G, Loizou JI. DNA replication stress: causes, resolution and disease. Exp Cell Res. 2014;329(1):85–93.

6. Burhans WC, Weinberger M. DNA replication stress, genome instability and aging. Nucleic Acids Res. 2007;35(22):7545–56.
7. Branzei D, Foiani M. Maintaining genome stability at the replication fork. Nat Rev Mol Cell Biol. 2010;11(3):208–19.
8. Vijg J, Suh Y. Genome instability and aging. Annu Rev Physiol. 2013;75:645–68.
9. Cabral M, et al. Absence of non-histone protein complexes at natural chromosomal pause sites results in reduced replication pausing in aging yeast cells. Cell Rep. 2016;17(7):1747–54.
10. Janssens GE, et al. Protein biogenesis machinery is a driver of replicative aging in yeast. elife. 2015;4:e08527.
11. Li Z, et al. Impaired DNA double-strand break repair contributes to the age-associated rise of genomic instability in humans. Cell Death Differ. 2016;23(11):1765–77.
12. Melov S. Mitochondrial oxidative stress. Physiologic consequences and potential for a role in aging. Ann N Y Acad Sci. 2000;908:219–25.
13. Chow TT, et al. Early and late steps in telomere overhang processing in normal human cells: the position of the final RNA primer drives telomere shortening. Genes Dev. 2012;26(11):1167–78.
14. Maestroni L, Matmati S, Coulon S. Solving the telomere replication problem. Genes (Basel). 2017;8(2):55.
15. Sampathi S, Chai W. Telomere replication: poised but puzzling. J Cell Mol Med. 2011;15(1):3–13.
16. Muraki K, et al. Mechanisms of telomere loss and their consequences for chromosome instability. Front Oncol. 2012;2:135.
17. Blackburn EH, Greider CW, Szostak JW. Telomeres and telomerase: the path from maize, Tetrahymena and yeast to human cancer and aging. Nat Med. 2006;12(10):1133–8.
18. Meyne J, Ratliff RL, Moyzis RK. Conservation of the human telomere sequence (TTAGGG)n among vertebrates. Proc Natl Acad Sci U S A. 1989;86(18):7049–53.
19. Palm W, de Lange T. How shelterin protects mammalian telomeres. Annu Rev Genet. 2008;42:301–34.
20. Xin H, Liu D, Songyang Z. The telosome/shelterin complex and its functions. Genome Biol. 2008;9(9):232.
21. Hockemeyer D, et al. POT1 protects telomeres from a transient DNA damage response and determines how human chromosomes end. EMBO J. 2005;24(14):2667–78.
22. Xin H, et al. TPP1 is a homologue of ciliate TEBP-beta and interacts with POT1 to recruit telomerase. Nature. 2007;445(7127):559–62.
23. von Zglinicki T. Oxidative stress shortens telomeres. Trends Biochem Sci. 2002;27(7):339–44.
24. Monaghan P. Telomeres and life histories: the long and the short of it. Ann N Y Acad Sci. 2010;1206:130–42.
25. Proctor CJ, Kirkwood TB. Modelling telomere shortening and the role of oxidative stress. Mech Ageing Dev. 2002;123(4):351–63.
26. Dantzer B, Fletcher QE. Telomeres shorten more slowly in slow-aging wild animals than in fast-aging ones. Exp Gerontol. 2015;71:38–47.
27. Cribbet MR, et al. Cellular aging and restorative processes: subjective sleep quality and duration moderate the association between age and telomere length in a sample of middle-aged and older adults. Sleep. 2014;37(1):65–70.
28. Cawthon RM, et al. Association between telomere length in blood and mortality in people aged 60 years or older. Lancet. 2003;361(9355):393–5.
29. Jaskelioff M, et al. Telomerase reactivation reverses tissue degeneration in aged telomerase-deficient mice. Nature. 2011;469(7328):102–6.
30. Tucey TM, Lundblad V. Regulated assembly and disassembly of the yeast telomerase quaternary complex. Genes Dev. 2014;28(19):2077–89.
31. Ramunas J, et al. Transient delivery of modified mRNA encoding TERT rapidly extends telomeres in human cells. FASEB J. 2015;29(5):1930–9.
32. Shay JW, Wright WE. Use of telomerase to create bioengineered tissues. Ann N Y Acad Sci. 2005;1057:479–91.

33. Jafri MA, et al. Roles of telomeres and telomerase in cancer, and advances in telomerase-targeted therapies. Genome Med. 2016;8(1):69.
34. Kirkwood TB. Understanding the odd science of aging. Cell. 2005;120(4):437–47.
35. Sancar A, et al. Molecular mechanisms of mammalian DNA repair and the DNA damage checkpoints. Annu Rev Biochem. 2004;73:39–85.
36. Burnet FM. Intrinsic mutagenesis: a genetic basis of ageing. Pathology. 1974;6(1):1–11.
37. Morley AA. Is ageing the result of dominant and co-dominant mutations? J Theor Biol. 1982;98(3):469–74.
38. Morley AA. The somatic mutation theory of ageing. Mutat Res. 1995;338(1–6):19–23.
39. Harman D. Aging: a theory based on free radical and radiation chemistry. J Gerontol. 1956;11(3):298–300.
40. Harman D. The biologic clock: the mitochondria? J Am Geriatr Soc. 1972;20(4):145–7.
41. Wisnovsky S, Jean SR, Kelley SO. Mitochondrial DNA repair and replication proteins revealed by targeted chemical probes. Nat Chem Biol. 2016;12(7):567–73.
42. Trifunovic A, et al. Premature ageing in mice expressing defective mitochondrial DNA polymerase. Nature. 2004;429(6990):417–23.
43. Pinto M, Moraes CT. Mechanisms linking mtDNA damage and aging. Free Radic Biol Med. 2015;85:250–8.
44. Freitas AA, de Magalhaes JP. A review and appraisal of the DNA damage theory of ageing. Mutat Res. 2011;728(1–2):12–22.
45. Pfeifer GP, You YH, Besaratinia A. Mutations induced by ultraviolet light. Mutat Res. 2005;571(1–2):19–31.
46. Lundin C, et al. Methyl methanesulfonate (MMS) produces heat-labile DNA damage but no detectable in vivo DNA double-strand breaks. Nucleic Acids Res. 2005;33(12):3799–811.
47. Stumpf JD, Copeland WC. MMS exposure promotes increased MtDNA mutagenesis in the presence of replication-defective disease-associated DNA polymerase gamma variants. PLoS Genet. 2014;10(10):e1004748.
48. Pfeifer GP. Formation and processing of UV photoproducts: effects of DNA sequence and chromatin environment. Photochem Photobiol. 1997;65(2):270–83.
49. Adewoye AB, et al. The genome-wide effects of ionizing radiation on mutation induction in the mammalian germline. Nat Commun. 2015;6:6684.
50. Di Leonardo A, et al. DNA damage triggers a prolonged p53-dependent G1 arrest and long-term induction of Cip1 in normal human fibroblasts. Genes Dev. 1994;8(21):2540–51.
51. Zhang H. Molecular signaling and genetic pathways of senescence: its role in tumorigenesis and aging. J Cell Physiol. 2007;210(3):567–74.
52. Keyes WM, et al. p63 deficiency activates a program of cellular senescence and leads to accelerated aging. Genes Dev. 2005;19(17):1986–99.
53. Davis T, et al. The role of cellular senescence in Werner syndrome: toward therapeutic intervention in human premature aging. Ann N Y Acad Sci. 2007;1100:455–69.
54. Vijg J, Dolle ME. Large genome rearrangements as a primary cause of aging. Mech Ageing Dev. 2002;123(8):907–15.
55. Kipling D, et al. What can progeroid syndromes tell us about human aging? Science. 2004;305(5689):1426–31.
56. Garinis GA, et al. DNA damage and ageing: new-age ideas for an age-old problem. Nat Cell Biol. 2008;10(11):1241–7.
57. Feser J, Tyler J. Chromatin structure as a mediator of aging. FEBS Lett. 2011;585(13):2041–8.
58. O'Sullivan RJ, Karlseder J. The great unravelling: chromatin as a modulator of the aging process. Trends Biochem Sci. 2012;37(11):466–76.
59. Brunet A, Berger SL. Epigenetics of aging and aging-related disease. J Gerontol A Biol Sci Med Sci. 2014;69(Suppl 1):S17–20.
60. Pal S, Tyler JK. Epigenetics and aging. Sci Adv. 2016;2(7):e1600584.
61. Waddington CH. The epigenotype. 1942. Int J Epidemiol. 2012;41(1):10–3.
62. Fontana L, Partridge L, Longo VD. Extending healthy life span--from yeast to humans. Science. 2010;328(5976):321–6.

63. Liu L, et al. Aging, cancer and nutrition: the DNA methylation connection. Mech Ageing Dev. 2003;124(10–12):989–98.
64. Fraga MF, Esteller M. Epigenetics and aging: the targets and the marks. Trends Genet. 2007;23(8):413–8.
65. Remolina SC, Hughes KA. Evolution and mechanisms of long life and high fertility in queen honey bees. Age (Dordr). 2008;30(2–3):177–85.
66. Holliday R. The significance of DNA methylation in cellular aging. Basic Life Sci. 1985;35:269–83.
67. Yoder JA, Walsh CP, Bestor TH. Cytosine methylation and the ecology of intragenomic parasites. Trends Genet. 1997;13(8):335–40.
68. Okano M, et al. DNA methyltransferases Dnmt3a and Dnmt3b are essential for de novo methylation and mammalian development. Cell. 1999;99(3):247–57.
69. Lai SR, et al. Epigenetic control of telomerase and modes of telomere maintenance in aging and abnormal systems. Front Biosci. 2005;10:1779–96.
70. Romanov GA, Vanyushin BF. Methylation of reiterated sequences in mammalian DNAs. Effects of the tissue type, age, malignancy and hormonal induction. Biochim Biophys Acta. 1981;653(2):204–18.
71. Wilson VL, et al. Genomic 5-methyldeoxycytidine decreases with age. J Biol Chem. 1987;262(21):9948–51.
72. Kulis M, Esteller M. DNA methylation and cancer. Adv Genet. 2010;70:27–56.
73. Moazed D. Enzymatic activities of Sir2 and chromatin silencing. Curr Opin Cell Biol. 2001;13(2):232–8.
74. Xu F, et al. Sir2 deacetylates histone H3 lysine 56 to regulate telomeric heterochromatin structure in yeast. Mol Cell. 2007;27(6):890–900.
75. Walsh ME, Van Remmen H. Emerging roles for histone deacetylases in age-related muscle atrophy. Nutr Healthy Aging. 2016;4(1):17–30.
76. Groth A, et al. Chromatin challenges during DNA replication and repair. Cell. 2007;128(4):721–33.
77. Shah PP, et al. Lamin B1 depletion in senescent cells triggers large-scale changes in gene expression and the chromatin landscape. Genes Dev. 2013;27(16):1787–99.
78. Pu M, et al. Trimethylation of Lys36 on H3 restricts gene expression change during aging and impacts life span. Genes Dev. 2015;29(7):718–31.
79. Wilson AS, Power BE, Molloy PL. DNA hypomethylation and human diseases. Biochim Biophys Acta. 2007;1775(1):138–62.
80. Qazi TJ et al. Epigenetics in Alzheimer's disease: perspective of DNA methylation. Mol Neurobiol. 2018; 55(2):1026–1044
81. Gray SG. Epigenetic-based immune intervention for rheumatic diseases. Epigenomics. 2014;6(2):253–71.
82. Heyn H, Moran S, Esteller M. Aberrant DNA methylation profiles in the premature aging disorders Hutchinson-Gilford Progeria and Werner syndrome. Epigenetics. 2013;8(1):28–33.

Transcription and Aging

Malika Saint and Pramod C. Rath

Introduction

Gene expression changes are a measure to indicate the various cellular events occurring during aging. The decline in organismal fitness due to impaired molecular processes that change with age can be queried on a quantitative genome-wide scale using transcriptomics technologies, among which microarrays and RNA sequencing have proven popular. This global transcriptomic view provides an unbiased screen for transcript changes accompanying aging. Gene expression studies also help compare the aging process across various species, one goal of which is to identify biomarkers for aging.

Given that the exact nature of aging is unknown, various molecular mechanisms have been put forth to explain aging such as DNA transposition; telomere shortening; DNA, RNA and protein damage accumulation; transcriptional deregulation; etc. These changes are expected to be accompanied by gene expression changes over time, and in this sense, expression biomarkers are often discussed in terms of identifying physiological states for an aging organism. There are limitations to transcriptional profiling serving as an outcome for identifying biomarkers and therapeutic targets of aging. First, the profiles are correlative, and causation cannot be derived from them. Second, transcript changes do not inform about possible post-transcriptional alterations downstream, even though they provide a global view of the processes affected by aging. Since aging is a multifactorial process, results from transcriptional profiling combined with studies of signalling pathways, genetics, translational mechanisms and metabolism would be able to provide a more holistic picture of the molecular events that contribute to aging.

M. Saint (✉) · P. C. Rath (✉)
Molecular Biology Laboratory, School of Life Sciences, Jawaharlal Nehru University, New Delhi, India
e-mail: malikasaint@gmail.com; pcrath@mail.jnu.ac.in

While studying changes in transcript levels in aging organisms, it is important to keep in mind that aging is not considered to be an evolutionarily conserved process; that is, it does not follow a predetermined programme [1, 2]. The force of natural selection decreases with increasing age of an individual, because with time, the expected reproductive output, on which selection can act, declines. Hence, a deterministic aging programme is unlikely to be selected for on an evolutionary level, because it is not beneficial to the individual [2]. However, while the evolutionary theory accommodates genes as the basis for functional decline, it does not postulate precise mechanisms for the decline observed with age.

Aging then, is an effect of the failure of to maintain normal function of organs and tissues in individuals who survive to old age. Since there are no specific pathways for aging, longevity maintenance is likely to be a polygenic trait, as it is non-specific to a particular process. Changes occurring during aging would be reflected in changing gene expression patterns, which have been widely studied. Several mechanistic theories for aging have been proposed which could provide a framework for the various gene expression changes observed with age in organisms [3, 4]. For example, the mutation accumulation theory postulates that the weakening force of selection with increasing age could allow mutations that may be deleterious but late acting to accumulate in the germ line, while the antagonistic pleiotropy theory proposes that beneficial alleles may be favoured by selection to accumulate for function in early life, but in late life, these allele functions may turn detrimental [5].

The disposable soma theory of aging postulates that the limited resources available to organisms are partitioned between maintenance and reproduction in a trade-off. There is evidence for longevity induced by infertility, for example, by castration. However, it is not clear why organisms supplied with unlimited food resources and are thus resource rich have shorter lifespans than those subject to dietary restriction, which live longer. Also, it is not clear why in a post-reproductive organism, age-related decline is not countered more efficiently. Damage theories of aging cover a wide variety of damages that a cell could endure during its lifespan that could contribute to aging. From the free radical theory of aging to mutation accumulation or mitochondrial dysfunction, many types of damage have been postulated to serve as the prime factor in the aging process. It is unclear though if one form of damage should be considered more important than others and also why cells are less able to surmount these challenges with increasing age [6].

A newer theory of aging postulates that the progressive decline in aging is due to the imperfect physicochemical properties of biological systems [7]. According to this theory, not all reactions in a cell are perfectly balanced, and virtually all cellular processes, whether replication, transcription, translation or metabolic, are error prone. The inherent errors in these deterministic processes, along with mutation-induced variations and stochastic elements, will add up over time, leading to physiological decline. This could result in variable lifespan even within genetically identical individuals, due to the inevitable age-related decay in non-dividing somatic cells [7]. Thus, stochastic, genetic as well as environmental factors precipitate a decline in cellular function.

Though the precise mechanisms of aging are as yet unclear, increase in longevity across various model organisms has been demonstrated via manipulation of genetic and/or environmental variables. Among environmental variations, caloric or dietary restriction has been shown to increase lifespan in various species. Calorie restriction (CR) is a dietary intervention, where either the total caloric intake for an organism is reduced or specific restrictions on dietary components, like proteins, carbohydrates, etc., are made without causing malnutrition. This manipulation has been shown to increase lifespan across a wide range of species, from yeast to mice [8, 9]. The effect on longevity due to CR is known to be mediated by the sirtuin family of deacetylases in yeast, flies, worms and mice. Sirtuins act via the insulin/IGF1 pathway to modulate lifespan. Other pathways such as AMPK, TOR, etc. also function as nutrient-sensing pathways to modulate aging [10].

Among genetic interventions, mutations in various processes have been known to increase longevity, for example, in the insulin and insulin-like growth factor signalling (IIS) pathways in *C. elegans*. Some of these mechanisms for enhancing longevity are similar in diverse model organisms, which suggests that it may be possible for various aspects of the aging phenotype to be delayed by interventions targeting the specific proteins that enhance longevity [4]. Expression profiles of calorie-restricted organisms as well as genetic mutants have been queried via transcription profiling in various species, giving a snapshot of the underlying molecular changes during interventions that enhance longevity. These will be described in the next few sections.

A Brief Introduction to Transcription Initiation

A variety of signals can cause changes in gene expression, such as external stimuli, developmental processes, stress responses, etc. Changes in gene expression are regulated events, usually mediated by one or more transcription factors, which can be either transcriptional activators or repressors. These factors bind to or associate with cognate DNA sequences or other proteins on DNA, to effect a change in transcriptional output. For example, a transcriptional activator such as Gcn4p in *S. cerevisiae* recruits multiple general transcription factors (GTFs) and chromatin-modifying activities to assemble a pre-initiation complex (PIC) in association with RNA polymerase II (RNAP II) to activate gene expression [11]. Since studies on mechanisms of RNAP I– and III–based transcription reactions during aging are currently sparse or lacking, only an RNAP II–based mechanistic outline for transcription will be described here.

Transcription factors are usually modular proteins with a DNA-binding domain and an activation/repression domain that recruit GTFs to promoter sequences of genes [12]. The GTFs comprise multi-protein complexes including TFIIA, TFIIB, TFIID, TFIIE, TFIIF, TFIIH and others such as mediator, which form the PIC at gene promoters [12]. Gene promoters contain a variety of DNA motifs that facilitate transcription. Besides activator/repressor-binding sequences, core promoter sequences present at most gene promoters add to transcription efficiency or specificity. Common core promoter elements include TATA boxes, initiator elements (INR), downstream promoter elements (DPEs), the B-recognition element, etc. [13]. PIC assembly occurs over a ~200 bp nucleosome free region (NFR) in the promoter DNA which is marked by the flanking presence of the histone variant H2A.Z at the −1 and + 1 nucleosome positions, which directs the positioning of transcription initiation factors and complexes [13]. Nucleosomes in the promoter and gene body are liable to remodelling by a variety of chromatin remodelling factors or complexes such as SAGA (*S*pt-*A*da-*G*cn5-*a*cetyltransferase complex) and SWI/SNF which facilitate PIC assembly and RNAP II recruitment. Factor recruitment is also influenced by epigenetic modification of histone tails by complexes such as the polycomb-group proteins, methyltransferases and histone acetyltransferases among others, which in turn influence gene expression outcomes [11].

Various transcription factors are associated with aging, for example, sirtuins that are NAD+−dependent lysine deacetylases which influence longevity under DR, and are involved in DNA damage repair, p53 regulation, telomere function, etc. The primary mode of action of the sirtuin SIRT1 found in mammals is via deacetylation of histones H3K9, H4K16 and H1K26. Additionally, the role of epigenetic modifications is closely linked to the availability of metabolites such as acetyl-CoA, NAD+, S-adenosyl methionine (SAM), etc., which are known cofactors for enzymatic complexes that establish chromatin modifications. Aging is also correlated with changes in heterochromatin structure, histone loss, increased transposition and decreased nucleosome occupancy at telomeres. Changing levels of metabolites with age as well as nucleosome loss could have a close impact on the epigenetics of aging, indicating once again that it is a multifactorial process [14].

Thus, various signalling pathways, epigenetic changes and metabolic and other physiological changes together impinge on the gene expression outcome. Expression microarrays as well as RNA sequencing studies have been employed to decipher changing expression patterns with age. The studies have been performed across a wide variety of organisms, from yeast, worms, flies, mice and humans, giving an insight into progression patterns in aging within as well as across species.

Gene Expression Studies in Yeast

The yeast *S. cerevisiae* has proved to be an excellent model system to study aging, given its short lifespan, genetic tractability and a well-annotated genome. Many yeast genes have been shown to have human orthologs, including those involved in age-related disease-causing genes such as for Bloom's and Werner's syndromes [15]. The lifespan of *S. cerevisiae* is measured in terms of either its replicative lifespan (RLS) or its chronological lifespan (CLS). Yeast has been an excellent model system to identify markers for aging as well as helping understand the causes associated with aging, with insights from both RLS and CLS aging mechanisms.

Replicative Aging

Yeast cells divide asymmetrically, and the number of times a single cell produces a daughter cell by budding is called its RLS. The RLS is generally studied as a model for aging in dividing cells in higher organisms. Normally, yeast cells do not divide indefinitely and stop after a median lifespan of 20–25 divisions, after which they enter a post-replicative state before they lyse. Cells become progressively larger with age and also show a decline in mitochondrial function, linearly decreasing protein and ribosomal activity, accumulation of extrachromosomal rDNA circles, oxidative stress and an increasingly fragmented nucleolus in mother cells, changes which are not passed onto daughter cells [16]. This asymmetry in partitioning, however, decreases with age, and symmetry is restored with age. A symmetric partitioning of increasingly damaged molecules would lead to the progressive decline and eventual demise of the species, and thus, even in symmetrically dividing organisms like fission yeast and *E. coli* [17, 18], there is evidence that they are functionally asymmetric. Hence, even in symmetrically dividing organisms, the rejuvenated daughter cells would continue to divide, while the cell accumulating the damaged macromolecules would die after a certain number of divisions [3].

RLS studies are typically performed via microdissection every 2–3 h with cells being maintained at 30 °C during the day and kept at 4 °C overnight to prevent cells from overgrowing [19, 20]. For large-scale studies, microfluidic platforms or elutriation is also used. Genes that are differentially expressed between young and old cells have been queried during various stages of the yeast RLS. In one such RLS study using elutriation to separate young newly budded cells from larger and older yeast cells, a trend for genes related to glucose storage and gluconeogenesis to be upregulated was observed, due to a shift away from glycolysis to gluconeogenesis and energy storage (glyoxylate cycle and glycogen synthesis). Genes involved in gluconeogenesis, environmental stress response (ESR), DNA repair, chaperone genes, the glyoxylate cycle, various cell cycle regulatory genes, mitochondrial genes and lipid metabolism were upregulated in older cells relative to younger daughter cells [21, 22]. Metabolism, stress response and genome stability genes are the most altered in older yeast cells, and these changes are perhaps interlinked, given that DNA damage may cause ESR genes to be upregulated. Given that genes for DNA damage are also upregulated in older cells without any exogenous intervention with chemical agents, it would indicate that genomic instability increases with age, as is evidenced with an increase of rDNA circles in yeast.

Older yeast cells show differences in their cell cycle, and there is evidence that this may be linked with entry into senescence. Genes involved in cell cycle regulation, and particularly the G2/M phase, are upregulated in older yeast cells. Additionally, an increase in the proportion of cells in the G2 phase of the cell cycle indicates alterations to cell cycle regulation in older cells [22]. The median replicative lifespan of *S. cerevisiae* is 20–25 generations, and even as early as the 11th generation, genes upregulated in stationary phase such as *SNZ1*, *HSP12* and *SNO1* as well as ESR genes were already beginning to be expressed. Older cells also display a shift to increasing aerobic respiration from a glycolytic mode of energy generation in younger cells. In addition, they show decreased amino acid biosynthesis and lower intracellular glucose corresponding to decreased hexose transporter expression [23].

Increased glycogen production and accumulation were found to be upregulated in older cells in what has been described as a pseudo-stationary phase. Older cells show an increase in hexose transporter expression and a diauxy-like shift, which is interesting because during RLS studies, cells are not kept in a glucose-limiting environment while getting older. This implies that there may be a glucose-sensing problem in older cells, which has been attributed to mother cells growing in volume as they get older [24]. However it could also indicate that the older cells may have a shifted metabolic cycle [25], in concert with altered cell cycles which are also observed in older cells.

Aged yeast cells show an increase in expression of stress response genes which is linked with a concerted downregulation of genes involved in ribosome synthesis, regulation of protein levels, protein folding and degradation. The causative reason has been proposed to be the decline of the protein quality control system with age [21, 22, 26]. Ribosome protein expression genes are downregulated around the fourth division of the mother cell, and their transcript levels steadily decrease thereafter [22]. However, another study found that gene expression of ribosomal and most other genes increases with increasing age and cell size. There was decreased correlation between RNA and protein expression levels for ribosomal biogenesis genes in older cells, where the protein level for ribosomal genes was higher than the corresponding transcript levels [27]. It was proposed that this uncoupling, accompanied with an imbalance in stoichiometry in many proteins and complexes may be one of the proximal causes for the degenerative effects of aging in yeast. However, this loss of stoichiometry was not observed in the ribosome biogenesis protein machinery expression, indicating that this majorly expressed protein group may not be the primary driver of aging-related degeneration. Further, ribosome protein concentrations as a function of cell size were not determined, given that the cells become larger with age [27].

The question of cell size and ribosome protein levels was addressed in a separate study by Janssens et al. [28]. According to their results, ribosome concentrations showed an absolute increase but a relative decline in concentration with age in larger cells. Dividing mother yeast cells increase in size with age. At the point of greatest cell size change during yeast RLS, individual cells entered senescence and showed slowed cell cycles. Deletion of ribosome components increases lifespan, and correspondingly, lower ribosome concentrations increased lifespan in cells past their senescence entry point (SEP). Younger cells showed the opposite trend of longer lifespans with higher ribosome concentrations [28]. An uncoupling between the levels of transcript levels for ribosomal proteins and their corresponding protein levels could occur due to a loss in regulation of protein production or degradation, or both. This could have a cascading response downstream, as ribosomes are responsible for translation of the proteome, further exacerbating the uncoupling effect. This in turn may impact transcription and lead to greater increases in the stress response, as is evidenced with aging yeast cells [27].

Another set of changes that lead to lifespan regulation in yeast are histone levels, which are lowered in aging yeast cells due to their reduced synthesis. This leads to a global increase in gene expression with age. Conversely, overexpression of histones increases lifespan, indicating that histone loss could also be a contributing factor to yeast aging [29]. Nucleosome occupancy is less precise in older cells and reduced by almost 50%; as a consequence, most genes across the genome have an elevated expression in older cells, with repressed genes and those containing TATA-promoters being highly induced [30]. Besides histone levels, certain histone modifications levels are correlated with increasing age in yeast. As an example, overexpression of the Sir2 H4K16 deacetylase leads to an increase in yeast longevity by inducing telomere and sub-telomere silencing. The levels of H4K16 increase with age, perhaps due to decreasing Sir2 levels. Other modifications such as H3K56 are also known to decrease with age [31].

Since dividing *S. cerevisiae* cells ensure retention of damaged molecules with the mother cell, various mechanisms have been proposed for this asymmetric segregation. Dividing mother yeast cells retain rDNA circles that are formed due to intrachromosomal recombination of repeat elements and are thought to contribute to RLS in yeast. The asymmetric nature of the budding process ensures that rDNA circles remain with the mother cell, allowing the propagation of daughter cells with restored replicative potential. The SAGA transcription complex promotes aging in yeast during its RLS by anchoring rDNA circles to the nuclear pore complexes (NPCs) of the mother cell. The NPCs cluster around the ERCs, causing selective retention of ERCs by mother cells [32]. In addition, the SAGA/SLIK complex is also known to act in a Sir2-dependent manner to increase lifespan in yeast, and deletion of the SAGA components *SGF73*, *SGF11* and *UBP8* is known to increase lifespan [33].

Given that aging is not evolutionarily conserved, 'public' or shared mechanisms of longevity across species such as calorie restriction imply that there may be a redundancy in how certain interventions may contribute to aging. Interestingly enough, the response to moderate CR in yeast is mediated via the NAD-dependent Sir2 deacetylase, which also prevents replicative senescence by reducing ERC formation, the latter being a 'private' mechanism of aging specific to yeast [34], indicating that the effects of CR maybe different even within the same organism. Since aging-related mechanisms and interventions in yeast show similarities with higher organisms, it remains a popular tool for aging research.

Chronological Aging

The chronological lifespan or CLS is defined as the amount of time a non-dividing cell in a post-diauxic state can maintain viability in a non-replicative environment that is depleted of extracellular glucose. These are cells that have exited the cell cycle and are metabolically active though non-dividing. The yeast CLS serves as a model for post-mitotic mammalian cell aging. Chronologically aging cells in stationary culture accumulate and utilize glycogen and trehalose, while being relatively more thermotolerant and resistant to multiple stresses. Damaged proteins accumulate during chronological aging, and interventions such as an increase in superoxide dismutase expression enhance CLS [16].

Yeast cells during chronological aging employ mitochondrial respiratory pathways for energy production and use ethanol formed from glucose fermentation as a carbon source, a metabolic state that is thought to resemble that of non-mitotic human cells [19]. The high ethanol levels contribute to acetic acid formation which acidifies the culture medium and leads to apoptotic cell death of the yeast cells. If CR is imposed on cells while they are growing exponentially, it decreases culture acidification later as it induces cells to undergo respiration at an earlier stage [17, 34].

Microarray studies on a time-course studying stationary phase aging relative to exponentially growing cells have shown that genes involved in β-oxidation of fatty acids, alcohol catabolic process and glyoxylate metabolism are upregulated during stationary phase. In early stationary phase, cells upregulated genes for trehalose biosynthesis, pyruvate metabolism, glucose transport, pentose metabolic process and cytogamy [35]. During early stationary phase, cells shift from glycolysis to oxidative metabolism and accumulate storage carbohydrates like trehalose. About 4 days later, a shift to aerobic respiration occurs with genes from the TCA cycle being upregulated, while rRNA export from the nucleus, nucleobase biosynthesis, pentose metabolic process and cytoplasmic translation genes are downregulated, indicating that growth is slowed or almost stopped. The slow growth upon entry to stationary phase is correlated with a downregulation of translation, ribosome biogenesis, tRNA synthesis, nucleotide biosynthesis and other processes due to nutrient depletion, with the largest shift occurring in cells after 6 days in stationary phase. By day 10, few genes remain differentially expressed relative to log phase. Various transcription factors like Yap1, Xbp1 and Adr1 integrate signals from the MAPK and Ras/PKA pathways and upregulate stress response signals like heat shock proteins, repression of nucleotide metabolism, autophagy and amino acid metabolism [35].

In a screen for genes that contribute to CLS in yeast, genes that reduced culture acidification due to accumulation of acetic acid in the medium containing stationary phase cells were found to be significantly enriched. There is little homology between CLS yeast genes that contribute to yeast RLS and aging in *C. elegans*, indicating that longevity-influencing aspects of these genes perhaps evolved later [36, 37]. Dietary restriction increases lifespan in both RLS and CLS of yeast. DR induces a shift in metabolism from fermentation towards respiration while also signalling downstream via the TOR/Sch9 and PKA pathways.

Aging in *Caenorhabditis elegans*

The soil nematode *C. elegans* plays an important role in aging studies owing to its short life cycle, well-defined tissues, a tractable genome and a well-studied developmental programme. *C. elegans* has proven to be a useful model in studies on the effect of dietary restriction on longevity, as well as in analysing the effect of various gene mutations influencing longevity. Screening for genes and pathways that alter expression levels with age has been pursued using transcriptomics as well as RNAi methodologies on wild-type worms as well as mutants displaying altered lifespan durations. Early studies using whole-genome microarrays have been used to identify senescence patterns in roundworms, and about 164 responsive genes were identified. These included genes for heat shock response which showed decreased expression, while various transposases showed increased expression with age in adult worms. These and other studies were performed with worms of similar chronological age spanning different ages, using pooled populations of worms [38, 39]. While this approach is useful for a systems-level averaged view of changing gene expression patterns, there is considerable heterogeneity in physiological age between worms of similar chronological lifespans, based on studies in age-related motility and related mortality patterns [40].

Using single worms for microarray analysis of aging patterns, Golden et al. [40] found that genes that show increasing expression with age include those associated with apoptosis or necrotic cell death, mitochondrial and respiratory chain genes, collagen metabolism, etc. In worms with impaired motility and coordination with age, genes for amino acid metabolism and actin-related genes were increased in expression, even though muscle function was declining. This indicated that physiologically old worms were trying to compensate for impaired muscle function. Gene expression patterns in worms were highly correlated with their physiological age, which closely corresponds to motility patterns in the organisms. Post-reproductive worms were enriched in stress response, ER-nuclear signalling, chaperones, various transcription factors and aging-related genes. Downregulated genes included those for metabolism of tRNA, DNA, carbohydrates, peptidoglycan and lipid metabolism [40]. While this study did not address tissue-specific temporal aging patterns, such differences are prominent in *C. elegans* tissues, for example, as seen in the increased autophagy in intestinal tissues of worms relative to others during aging [41].

Various genes have been shown to regulate lifespan in *C. elegans*. Mutations in the genes involved in the insulin/IGF-1 pathway when deleted extended lifespan [19, 42]. Transcriptional profiles of mutant worms having mutations in genes, such as *sir2.1*, *daf16*, *daf2* and others involved in the insulin/IGF-1 pathway, which altered longevity have been generated to identify downstream targets that influence lifespan [43–45]. In these studies, genes involved in stress response, metabolism, nutrient uptake and energy generation were found to be altered, and specific expression patterns relating to longevity in each of these mutants identified.

Dietary restriction in worms has been shown to increase longevity. The mTOR pathway is linked to longevity and DR in worms, with reduced mTOR activity in calorie-restricted worms leading to increases in lifespan [19]. CR in worms is autophagy dependent, with a number of genes involved in the process. TFs required for CR- or TOR-mediated lifespan extension like PHA-4 or HLH-30 induce autophagy-dependent gene expression, and PHA-4 overexpression can also extend lifespan and improve resistance to stress. Several other transcription factors have been found to mediate lifespan extension during DR such as KLF-1, ELT-3, WWP-1, HIF-1 and SKN-1 [46–50], usually acting in specific tissues. In addition to TFs, various epigenetic modifications have been correlated with aging in *C. elegans*. For example, the histone mark H3K36me3 is deposited by elongating RNAP II and has been shown to be important for suppressing cryptic transcription [51]. H3K36me3 modification has been found to be inversely correlated with gene expression during aging in worms, with genes having low levels of the histone mark showing the most dramatic increases in expression with age. A reduction in H3K36me3 levels due to inactivation of the methyltransferase *met-1* also resulted in a shorter lifespan, indicating an involvement in longevity regulation [52]. A more detailed analysis of epigenetic modifications during aging will be considered elsewhere in this volume.

MicroRNAs play an important role in regulating gene expression across a variety of processes, from development, metabolism and aging. Various RNAi studies have been done to uncover aging regulators in worms. For example, components of the electron transport chain when silenced produced long-lived worms, indicating a role for oxidative phosphorylation in determining lifespan [53]. Mutations in the *lin-4* miRNA and its target *lin-14* affect aging in worms, and they act in the insulin/IGF-1 signalling pathway. Other miRNAs have been found to play roles in the TOR nutrient signalling pathway, while miRNAs such as *let-7* interact with a homolog of the amyloid precursor protein and are thus implicated in playing a role in Alzheimer's disease [54, 55]. The miRNAs miR-71, miR-238 and miR-246 extend lifespan in worms, while miR-239 limits it. miR-71 appears to have multiple functions; it promotes resistance to oxidative and heat stress in *C. elegans* and interacts with the DNA damage checkpoint pathway and the insulin signalling pathway [56]. miR-71 also specifically acts in the neurons to extend lifespan after germ line removal, acting via the TF DAF16/FOXO in the worm intestine [57]. About 50 of the 200 odd identified miRNAs in *C. elegans* are differentially expressed during aging, with most being downregulated with age, in contrast with aging studies in mice, indicating differences in miRNA regulation across species [58].

Aging in *Drosophila melanogaster*

Gene expression profiling across various tissues in *Drosophila melanogaster* has revealed variable expression patterns in response to aging, indicating that different tissues have differing physiological aging mechanisms. While analysis of tissues from flies of similar chronological age does not prevent homogenisation of individual fly tissue expression patterns, it remains an improvement over whole body comparisons. A summary of transcriptomic profiling across various tissues in *Drosophila* is provided (Table 3.1).

From the above Table 3.1, it is clear that very few genes or processes are found to be common across tissues in flies, indicating that transcriptomic changes with age relate to tissue-specific functions. In calorie-restricted flies, many signatures for

Table 3.1 Examples of changes in gene expression during aging

Tissue	Upregulated genes	Downregulated genes	References
Brain	Immune response, alcohol metabolism, carbohydrate metabolism, amino acid metabolism, neurotransmitter secretion, protein catabolism	Microtubular processes, TCA cycle, oxidative stress response, immune function, neuronal function, proton transport, light reception	[57–60]
Muscle	Antimicrobial humoral response, protein catabolism, autophagy, signalling and secretory pathways (protein membrane targeting, secretion pathway, RTK signalling), cell motility	Energy generation, oxidative phosphorylation, signalling and metal ion transport genes	[57]
Malpighian tubule	Anion transport, metal ion homeostasis, tRNA metabolism, chromosome organization, amino acid metabolism	Metabolism and vesicle exocytosis	[57]
Gut	Amino acid metabolism, nutrient intake and transport	Oxidative phosphorylation, metabolism, muscle contraction, apoptosis and vesicle transport	[57]
Accessory gland	Cytoskeleton, cellular homeostasis, protein targeting, stress-activated protein kinase signalling pathway, ROS metabolism, intracellular transport	Protein targeting, glycoprotein biogenesis	[57]
Testis	Cellular morphogenesis, tRNA aminoacylation, mRNA metabolism, protein folding	Metabolism genes	[57]
Adipose tissue	Immune response, signalling and metabolism genes	JNK cascade, glucose catabolism, protein catabolism, signal transduction	[57]
Thorax	Immune response, morphogenesis, proteasome complex, actin filament processes, JNK signalling	Muscle fibre components, mitochondrial components, oxidoreductase activity, calcium ion binding	[58, 59]
Heart	Extracellular matrix, DNA repair, mitochondrial ATP synthesis, mitochondrial β-oxidation Transcription factor *Odd* and its targets, *miR-1* target genes	Carbohydrate metabolism, microRNA *miR-1*	[61]

aging are alleviated and an increase in longevity is observed with transcriptome signatures of young flies resembling that of calorie-restricted older ones. CR caused upregulation of genes involved in cellular growth, metabolism and genes responsible for light perception, which are downregulated in aging flies. Conversely, genes for cell cycle, DNA repair, RNA metabolism and processing as well as protein metabolism and modification were downregulated in CR conditions [59].

Aging in Mice

Mus musculus or the common mouse is a popular model for aging studies, despite its relatively long lifespan of up to 2–3 years in laboratory conditions. Mouse tissues display distinct physiological aging signatures. In a comparison of expression patterns in the brain, heart and kidney tissues of mice, the mouse brain shows less age-related gene regulation. Few common regulatory processes across tissues that reflect aging such as innate immunity–related functions were common to all three tissues [60]. The brains of old mice show differentially expressed genes for glutathione metabolism and chemokine signalling, while genes for focal adhesion and cancer pathways are downregulated. Glutathione counteracts oxidative stress due to ROS and hence may be upregulated as part of a protective mechanism against degenerative processes, while chemokine signalling is involved in pathways that mediate stress and immune responses. Additionally, cancer pathways are downregulated in the brains of old mice, indicating that their brains are protected against aging-related diseases, and this may impact longevity of the animals [61].

The mouse brain shows an increased immune response, complement cascade induction and an increase in stress response with a concomitant decrease in developmental regulated genes and ubiquitin-proteasome pathway genes with age. These changes are reversible with calorie restriction [62]. CR influences a wide variety of tissues and cell types. Genes as well as gene modules are affected, as seen with mRNA processing activities such as splicing factors, mRNA processing and stability factors as well as immune response genes and histone clusters. For some biological processes, there is an inverse relation between CR and aging, such as for immunity and inflammation. For most modules, however, there is a weak correlation between aging and CR on a genome-wide level, and this correlation is stronger in some tissues than others [62]. In a comparison of lung, liver and kidney tissues, growth-promoting genes show decreased expression during the development past the juvenile stage, which continues into adulthood. Cell cycle genes show the most change, and this could be due to a decline in proliferative capacity with aging [63].

The aging mouse heart shows upregulation for genes involved in processes related to cellular structural proteins (ECM, collagen deposition, cell adhesion) and immune response. Downregulated genes include those related to beta-oxidation and protein synthesis initiation factors [64]. In the murine heart, a loss of myocytes ensues with age (cardiomyopathy), with a concomitant hypertrophy of remaining myocytes, increasing fibrosis of the heart, increased apoptosis, calcium transport and stem cell loss. There is accelerated collagen deposition with age, as also reduced fatty acid metabolism and protein synthesis, effects which are reversible with induced calorie restriction [65].

Aging results in muscle mass loss called sarcopenia, a general effect of aging. Genes in skeletal muscle upregulated due to aging include those responsive to stress (oxidative, heat shock, DNA damage), and a reduction is seen in glycolysis and mitochondrial genes. Most age-related changes in gene expression in the muscle are partially or completely reversible by CR, with a shift towards increased protein and energy metabolism. A reduction in damage due to reduced expression of genes involved in oxidative stress response, heat stress and DNA repair due to reduced oxidative damage in calorie-restricted mice is observed [66]. In mouse skeletal tissue, p53-dependent genes show altered mRNA and protein expression profiles with age. Older muscle cells showed increased p53 levels, and this was correlated with an increase in p53-dependent apoptotic gene expression. Changes in p53-dependent transcript changes have been shown to be reversible via caloric restriction, indicating a role for p53 in mediating aging [67].

Older mice show decreased co-expression of genes relative to younger ones, indicating an increase in heterogeneous expression in older cells. This could result from a variety of mechanisms, such changes in chromatin dynamics with age, changes in TF-dependent activation, etc. The loss of co-expression is not a uniform phenomenon for all co-expressed genes and is evident for smaller expression modules only [68]. Besides gene expression profiling, studies on DNA methylation changes in the mouse muscle, brain and liver showed that the methylation changes in CpG sites during transition from juvenile to adult and from adult to aged period are correlated. Genes affected in the aging process may be a continuation of growth processes as no significant expression change was found for genes which showed methylation changes, except for some stress genes [69].

In a meta-analysis of 27 microarray profiling datasets across mice, rats and humans to identify common gene signatures correlated with aging, genes from processes like immune response, complement activation, lysosomal genes, apoptosis-related and cell cycle regulation, among others, were upregulated. Genes decreased in expression with age included those involved in mitochondrial and oxidative phosphorylation functions, negative regulation of transcription, and collagen-related genes, which may relate to reduced elastin and collagen deposition with age [70]. Few gene modules are commonly affected due to aging across species as disparate as worms, flies, mice and humans. Components of the electron transport chain have been shown to decrease in expression with age across all these species, indicating that these could serve as a biomarker for aging. Deletion of ETC components in worms additionally has been shown to increase lifespan, and this may apply to other species as well. Few other mechanisms have shown such widespread conservation across species, except for CR-induced anti-aging effects [9, 71]. It is not clear why mitochondrial components responsible for oxidative damage would limit their expression with age and also affect different survival times such as 2 years in a lab mouse but nearly 80 years in case of humans, even though the mechanisms of mitochondrial function are largely conserved. This indicates also that aging mechanisms are different across species with similar outcomes and that most aging-related mechanisms are private and very few are public, making it difficult to envision a general mechanistic theory of aging.

Aging in Humans

Genome-wide studies on aging in humans have been performed using tissue samples across the lifespan of individuals and also comparing to other species. The underlying molecular mechanisms that cause aging appear to be diverse, and various transcriptomic studies have been performed to dissect tissue-specific and common mechanisms for aging. These analyses hold promise for identifying normal organ/tissue processes during aging, as well as mechanisms for the development of age-associated degenerative diseases, such as Alzheimer's disease.

Differences arise within the same tissue as well; for example, the human brain is functionally differentiated into various regions. These have been shown to exhibit distinct expression signatures as well as different aging rates. For example, the cerebellum showed less-differential gene expression than the cortex in the human brain and aged slower compared to the cerebral cortex in humans, perhaps due to slower metabolism [72], while the converse was seen in the mouse [72]. Despite brain region heterogeneity, various pathways have been shown to be involved in human brain aging across regions, including the apoptotic signalling pathway, glutathione derivative biosynthesis, neurotransmitter transport, spliceosome complex assembly, ROS metabolism, etc. [73]. Genes related to oxidative phosphorylation and mitochondrial function are among the most consistently downregulated genes in the brain, indicating that a general decrease in energy metabolism accompanies old age [74].

The prefrontal cortex in human brains shows a maximal rate of expression change during foetal development with genes for cell division showing decreases during foetal development and infancy, while axonal function genes are upregulated. These changes reflect on the processes occurring during the development of the human brain. Some genes that show reversals from foetal expression patterns, such as the declining expression of ATP synthesis, energy metabolism, ion binding and transport genes during foetal development, are reversed after birth. Changes in gene expression declines postnatally and continues to decline till the 40s. In more mature brains over 40 years of age, expression change increases again [75]. The human brain cerebral cortex and cerebellum showed 56 genes in the frontal cortex and 7 genes in the cerebellum with positive correlations for age, including genes for transcription, histone acetylation, nuclear lumen, DNA metabolic process and RNA processing. Some of the age-related changes found in bulk tissue are also reflected in isolated neurons. Mitochondria- and metabolism-related genes are downregulated, indicating that metabolism decreases with age in these brain regions with age [76]. Differential gene expression in the prefrontal cortex brains region during old age is initiated in early childhood and is linked to developmental processes, suggesting a link between development and aging. This reflected in protein expression as well, with various miRNAs extending a regulatory influence on gene expression for both mRNAs and proteins [77]. Activation of the brain immune system is increased in old age. In brain regions such as the entorhinal cortex, hippocampus, superior frontal gyrus and postcentral gyrus, immune-related genes, toll-like receptor signalling and complement cascade showed a general increase in expression across the regions examined. Activation of microglia and perivascular macrophages which are part of the brain immune system is observed, along with a downregulation of factors that curtail activation of microglia and macrophages such as TOLLIP and fractalkine. However, long-term innate immune system activation can be deleterious, because of the release of factors such as pro-inflammatory cytokines [78].

An interesting aspect of gene expression related to Alzheimer's is the correlation with expression of genes which are increased during foetal development (and reversed after birth) and decreased during Alzheimer's and vice versa, indicating a link between development and aging processes [75]. Increasingly, evidence has been uncovered implying that brain aging and cognitive decline are related to synaptic decline and remodelling and not neuronal loss as was previously thought. Genes influencing brain aging overlap with those expressed during some neurodegenerative diseases, such as Alzheimer's [79]. The transcription factors SP1, SOX2, KLF15 and ESR1 regulate about half of aging-related genes, based on motif-enrichment scores on regulated gene promoters. Multiple gene modules are enriched for differential expression of genes involved in DNA repair, telomere maintenance, apoptosis, immune response, NF-kappaB signalling, growth regulation, RNA metabolism and modification, extracellular matrix, cytokine secretion, signalling pathways and some developmental processes, among others. Genes enriched between Alzheimer's and aging have a number of pathways in common between the two processes, such as those involved in apoptosis, growth signalling, neuron development, synaptic vesicle trafficking, cytoskeleton and cell adhesion [79, 80].

Human brain aging is sexually dimorphic, with greater changes occurring in the male brain for gene expression than the female brain during aging in certain brain regions [81]. Males have enrichment for GO categories of differentially expressed genes for energy production, RNA processing, ribonucleotide metabolism, protein synthesis, ribosome biogenesis, translation, protein processing and mitochondrial function. These categories for energy production were not enriched in the female brain. Instead, the female brain had greater enrichment for genes involved in immune function and inflammation relative to male brains. Downregulated genes in the female brain included those for neural morphogenesis, signal transduction, intracellular signalling, etc. Some responses are common though: Both sexes showed decreased expression of genes involved in synaptic transmission and an increase for expression of cell death, angiogenesis and cell growth responses globally [81].

Various sex-specific gene expression responses to aging have also been observed in other diverse tissues. In the lung, gene expression signatures could be grouped by age as well as by sex. The age-specific signatures had an overrepresentation of genes that were differentially expressed in old age for extracellular matrix and pro-inflammatory response [82]. Similar responses were observed in the human skin with age, with inflammatory gene expression increasing with age in the females, along with transcripts associated with T-cells, macrophages, dendritic cells, monocytes and neutrophils. In males, T- and B-cell associated transcripts alone increased with age, highlighting sex-related differences in aging. Interestingly, these sex-specific differences were not observed in mouse tail skin, indicating once again that aging mechanisms among species vary [83]. Changes in gene expression and metabolite levels in the skin occur in concert. Various metabolites such as coenzyme Q10, DHEA sulphate, proline betaine, etc. are reduced, along with expression pathways for amino acid and central carbon metabolism, tRNA biosynthesis and glycolytic flux, while gluconeogenesis shows an increase [84].

Cellular stress response pathways in male skin analysis from healthy adults showed an increased expression with old age without a corresponding increase in cellular repair, indicating that cell stress and inflammation were increased in older individuals. Most age-related genes showed a regulatory dependence on transforming growth factor beta 1 (TGFβ1), tumour necrosis factor (TNF) and mitogen-activated protein kinases (MAPKs), with apoptosis, angiogenesis and cell proliferation among important age-regulated gene processes. In male skin, midlife was marked by transient change in expression of thousands of genes involved in protein and cell metabolism, translational modification, cell cycle, etc. There was reduced enrichment for genes involved in telomere maintenance, growth and development processes and oxidoreductase activity, which reverted to earlier expression levels with age, with TNF and p53 being important regulators for these genes. Though skin composition is assumed to be more or less constant throughout life, it would appear that age-related changes in expression are not necessarily progressive throughout life [83, 85, 86].

Peripheral blood leukocytes show few age-associated transcripts, and the main pathways involved are related to mRNA processing and maturation, ribosome biogenesis and assembly, and chromatin remodelling. Transcripts with the largest effect sizes such as for immunity and inflammation, muscle development and vascularization were good predictors of young and aged individuals and can perhaps be used as biomarkers. Alterations in splicing patterns increased with age, even though gene expression levels remain relatively consistent across ages. Because these studies were done with arrays, RNA-Seq analysis would further bring out different splice patterns related to aging [87]. Genes involved in cancer, growth, maintenance and apoptosis change in gene expression as early as the late 20s/30s, with a critical shift observed in the age range of 49–56 years. Age is associated with decline in immune function due to reduced production of B- and T-cells and increase in immune senescence, leading to a greater involvement of inflammation and immune response genes with age [88].

Peripheral blood mononuclear cell (PBMC) gene expression shows sex-specific differences, implying gender-specific immune differences also exist. For example, IL6 plasma levels in males was associated with 62 differentially expressed transcripts related to processes such as inflammatory response, cell adhesion and secretion. Seven genes were found to harbour DNA methylation levels correlating with IL6 levels, indicating that some of the differences may be epigenetic in nature [89]. However, irrespective of ethnicity and gender, purified monocytes in middle-aged to old individuals showed a reduction in expression of ribosomal protein synthesis, autophagy and oxidative phosphorylation genes. A similar result was seen for T-cells, with the ribosomal protein synthesis machinery being downregulated in both cell types, though the oxidative phosphorylation machinery was not downregulated in T-cells. Additionally, the immune response pathway genes are upregulated in T-cells but not in monocytes, highlighting cell-specific differences [90].

Aging patterns across tissues are not always concordant. The human kidney has very few age-related genes in common with other tissues such as the muscle and did not show much similarity for aging processes across species. The human muscle showed about 250 genes involved in age-dependent differential expression, with differences in expression for most genes between old and young individuals being small [91]. Common processes related to aging in muscle, kidney and brain showed a common signature gene set involving extracellular matrix, cell growth, cytosolic ribosomal genes, electron transport chain, chloride transport and complement activation genes. The increased expression of ribosomal genes was peculiar, given that with age, a decline in protein synthesis occurs. The increase in expression was attributed to offset this effect [92]. Not all gene expression changes with age occur linearly across tissues. Cells in the brain, kidney cortex and medulla and muscle undergo a marked shift at two distinct age stages, around 40 and 70 years in humans. In muscle, the sixth decade shows a decrease in muscle strength though these changes start in early midlife. After 70 years of age, sarcopenia or muscle waste is known to set in [93]. In tissues such as heart, lung and blood, there is greater correlation between aging gene signatures than muscle, indicating a role for functional connection in leading to co-regulation of gene signatures across differing tissues. Mitochondrial functions are frequently downregulated, indicating that they have a central role in human aging. Many gene signatures are tissue-type correlated. A comparison of the AGEMAP and GTEx studies for heart and lung tissues showed that there are very few commonly differentially expressed homologous genes between humans and mice. Part of the discrepancy has been attributed to the study designs and experiment conditions [94].

In addition to gene expression, DNA methylation changes with age also show differences across various tissues. Promoter hyper-methylation is known to increase for specific genes in various tissues, which is also seen in diseases such as Alzheimer's and type II diabetes [95]. Differential DNA methylation between young and old subjects in leukocytes includes hyper-methylation at CpG islands (CGIs), mostly at TSSs and the first exon of genes. Hypo-methylation of DNA at non-CGIs in gene bodies and inter-genic regions increases with age, indicating that de novo methylation as well as DNA methylation maintenance changes with age. The methylation levels do not always correspond to expression levels, and most of the increasing hyper-methylation occurs at genes associated with developmental processes, transcription and DNA binding. Hypo-methylated sites do not impair very specific functions or processes, indicating that hyper-methylation may be a more regulated occurrence. Hyper-methylation-associated GO terms in old individuals include processes for transcription, growth, development, signalling processes and metabolism. Across various tissues, some functions are more affected by methylation than others, indicating that there may be some programmed features that correlate with aging. In leukocytes, genes that are affected due to expression-methylation correlation belong to immunological processes, signalling pathways, endocytosis and cytoskeletal remodelling, which may affect immune system function in old age [95].

DNA methylation pattern changes with age as observed from blood were found to be independent of the composition of the blood cell types. Hypo-methylation was seen in large blocks, containing many CGIs within them which were found to be hyper-methylated with age. These hypo-methylated blocks were also hypo-methylated in cancer, indicating that cancer-associated DNA hypo-methylation could be related to aging or may already be present in aged tissues. Gene expression at differentially methylated regions was sparsely affected, with the direction of change acting to stabilize expression levels of genes. About 58 TFs were enriched with age-dependent methylation of DNA, indicating a global change in chromatin patterning with age. TFs enriched in hypo-methylated regions include *NANOG, POU5F1, OCT4, HDAC2, BCL11A*, etc., while polycomb factors, *TAF1, REST* and *CTBP2*, were enriched among hyper-methylated DMRs [96].

Transcriptome Heterogeneity During Aging

Cellular heterogeneity is a well-studied phenomenon and is known to be responsible for cell-to-cell variation and may arise due to factors intrinsic or extrinsic to the cell. The wide variation in cellular lifespans observed across model organisms indicates that not all cells and organisms respond to the same environmental stimuli in an identical manner. Studies in *S. cerevisiae* have shown that during the RLS of single yeast cells, there is a marked variation in age of cells at death [28]. Cellular heterogeneity arising due to underlying variable gene expression patterns in single cells could contribute to a decline in tissue functioning. For instance, murine cardiomyocyte heterogeneity was found to be increased in old mice, with genes for β-actin, β-2 microglobulin and lipoprotein lipase showing variable transcript levels. The highly expressed mitochondrial COX1, COX2, COX3 genes were not noisy, perhaps because of their relative abundance [97]. Oxidative stress induced greater noise in gene expression and combined with other factors like genome arrangements has been an indicative cause for a decline in heart function due to dramatic variation in cell function [97].

Although cellular heterogeneity within tissues is a commonly observed phenomenon [98], increased cellular heterogeneity with old age could lead to mosaicism in cellular function within the same tissue, which could be a possible reason for functional decline. Besides transcriptomic heterogeneity arising due to intrinsic factors such as stochasticity in molecular processes or extrinsic factors such as the surrounding environment of the cell, accumulating mutations with age could also cause expression heterogeneity increase with age. Mutations accumulate with age in post-mitotic cells as well as dividing cells [99]. Gene expression heterogeneity was found to be increased in correlation with genome rearrangement mutations and was greater in older mice compared to young ones [100]. Mutation frequency increases with age, with replication-induced mutations being mainly point mutations and replication-independent genome rearrangements occurring in dividing as well as non-dividing cells. The increase of these genome errors with age enhances the probability of gene expression alterations leading to the functional decline of organs harbouring postmitotic cells such as brain and heart cells. This could be due to complete loss of gene function or even heterozygosity induced due to rearrangements in one allele [101]. More research is required to uncover the link between stochastic gene expression leading to cellular heterogeneity and aging phenotypes, with the above studies underscoring a possible mechanism for tissue function decline with age.

Therapeutic Approaches for Aging

Human life expectancy is on the rise, due to improvements in public health and biomedical interventions for disease. This also means that larger numbers of individuals are reaching old age today, and this proportion is expected to rise. Aging phenotypes vary across tissues, and a decline in function correlated with age ensues with increasing age. Though aging is not an adaptive process, key interventions that have prolonged lifespan have been gene mutations in select genes and calorie restriction.

While many of the mutations enhancing longevity relate to nutrient signalling pathways such as the insulin/insulin-like growth factor-like signalling (IIS) pathway in invertebrates [102, 103], not all lifespan increasing mutations have resulted in an increase in the health span of the animals under study [104]. Unlike developmental programmes which are quite precise, lifespan extension seen in genetic mutants has extremely variable effects even in genetically closely related individuals. Some studies have shown that long-lived mutants suffered from reduced resistance to various stresses, indicating that they were compromised in surmounting the challenges faced under adverse conditions [105]. The increased lifespan may also be due to a diversion of resources to maintain the somatic component of cells while compromising on reproductive fitness, whether in *D. melanogaster*, *C. elegans* or *M. musculus*. In a natural environment then, these mutants would not be favoured by selection. Hence, anti-aging interventions need to take cognisance of the conditions under which a genetic modification is judged as long-lived. Animals in the wild usually do not survive to the old age spans observed under laboratory conditions, and the majority die due to external causes like predation, disease, accidents, etc. This means that natural selection may have limited opportunity to act on the processes of senescence.

Caloric restriction has been shown to increase lifespan as well as the health span in model organisms, and there is indirect evidence that it may be beneficial in humans too as an intervention against aging [106]. Key pharmacological interventions to inhibiting nutrient response pathways like mTOR and insulin-like signalling (ILS) also appear to have a similar response to longevity as CR, and this has been visible across species. For example, rapamycin, which targets the mTOR pathway, is known to inhibit aging in mice, as does CR. This indicates that at least to some degree, there may be a genetic basis to longevity, even though it has been difficult to understand the precise molecular mechanisms behind their working [4, 103]. Mimetics of CR-dependent longevity modulation have shown promise, and most of these drugs have as their target the activation of sirtuins or AMPK targets. Resveratrol is best known to be a CR mimic, with a decrease in insulin signalling, increase in mitochondrial biogenesis and increase in NAD+/NADH ratio leading to sirtuin activation, AMPK activation and increased mitophagy and autophagy. Another compound metformin has been reported to have effects similar to CR including increased sensitivity to insulin, AMPK-independent SIRT1 increase in autophagy, increased protection to ROS and reduction in chronic inflammation [53, 107]. Other compounds such as coenzyme Q10, alpha-lipoic acid and vitamin also exert partial CR-like effects offsetting age-related transcriptional changes [64]. Global gene expression studies can reveal the effects of CR mimetics, enabling comparisons across treatments in addition to longevity outcomes, and also revealing the contributing genes for mechanistic insights.

Thus, though CR appears to be a 'one-size-fits-all' intervention for aging, the processes contributing to aging have been shown to be diverse across species and even tissues within the same organism. Identification of genes and processes involved in longevity indicate that therapeutic targets for regulatory proteins could be found to intervene in age-related diseases and help enhance healthy aging. This could also help in treatment of late-onset diseases such as Alzheimer's and Parkinson's diseases, if therapy could incorporate the effect of delayed aging and be restructured to a preventive course.

References

1. Kowald A, Kirkwood TBL. Can aging be programmed? A critical literature review. Aging Cell. 2016;15:986–98.
2. Kirkwood TBL, Melov S. On the programmed/non-programmed nature of ageing within the life history. Curr Biol. 2011;21:R701–7.
3. Zimniak P. Detoxification reactions: relevance to aging. Ageing Res Rev. 2008;7:281–300.
4. Partridge L. The new biology of ageing. Philos Trans R Soc Lond Ser B Biol Sci. 2010;365:147–54.
5. Kirkwood TBL, Austad SN. Why do we age? Nature. 2000;408:233–8.
6. Weinert BT, Timiras PS. Physiology of aging. Invited review: theories of aging. J Appl Physiol. 2003;95:1706–16.
7. Gladyshev VN. Aging: progressive decline in fitness due to the rising deleteriome adjusted by genetic, environmental, and stochastic processes. Aging Cell. 2016;15:594–602.
8. Kenyon C. The plasticity of aging: insights from long-lived mutants. Cell. 2005;120:449–60.
9. Kim SK. Common aging pathways in worms, flies, mice and humans. J Exp Biol. 2007;210:1607–12.
10. Greer EL, Brunet A. Signaling networks in aging. J Cell Sci. 2008;121:407–12.
11. Weake VM, Workman JL. Inducible gene expression: diverse regulatory mechanisms. Nat Rev Genet. 2010;11:426–37.
12. Hahn S, Young ET. Transcriptional regulation in *Saccharomyces cerevisiae*: transcription factor regulation and function, mechanisms of initiation, and roles of activators and coactivators. Genetics. 2011;189:705–36.

13. Juven-Gershon T, Kadonaga JT. Regulation of gene expression via the core promoter and the basal transcriptional machinery. Dev Biol. 2010;339:225–9.
14. O'Sullivan RJ, Karlseder J. The great unravelling: chromatin as a modulator of the aging process. Trends Biochem Sci. 2013;37:466–76.
15. Watt PM, Hickson ID, Borts RH, Louis EJL. SGS1, a homologue of the Blooms and Werner's syndrome genes, is required for maintenance of genome stability in *Saccharomyces cerevisiae*. Genetics. 1996;144:935–45.
16. Bitterman KJ, Medvedik O, Sinclair DA. Longevity regulation in *Saccharomyces cerevisiae*: linking metabolism, genome stability, and heterochromatin. Microbiol Mol Biol Rev. 2003;67:376–99.
17. Stewart EJ, Madden R, Paul G, Taddei F. Aging and death in an organism that reproduces by morphologically symmetric division. PLoS Biol. 2005;3:0295–300.
18. Coelho M, Lade SJ, Alberti S, Gross T, Tolic IM. Fusion of protein aggregates facilitates asymmetric damage segregation. PLoS Biol. 2014;12:1–11.
19. Kapahi P, Kaberlein M, Hansen M. Dietary restriction and lifespan: lessons from invertebrate models. Ageing Res Rev. 2017;39:3–14.
20. Smith ED, Kennedy BK, Kaeberlein M. Genome-wide identification of conserved longevity genes in yeast and worms. Mech Ageing Dev. 2007;128:106–11.
21. Lin SS, Manchester JK, Gordon JI. Enhanced gluconeogenesis and increased energy storage as hallmarks of aging in *Saccharomyces cerevisiae*. J Biol Chem. 2001;276:36000–7.
22. Laun P, et al. A comparison of the aging and apoptotic transcriptome of *Saccharomyces cerevisiae*. FEMS Yeast Res. 2005;5:1261–72.
23. Kamei Y, Tamada Y, Nakayama Y, Fukusaki E, Mukai Y. Changes in transcription and metabolism during the early stage of replicative cellular senescence in budding yeast. J Biol Chem. 2014;289:32081–93.
24. Yiu G, McCord A, Wise A, Jindal R, Hardee J, Kuo A, Shimogawa MY, Cahoon L, Wu M, Kloke J, Hardin J, Hoopes LLM. Pathways change in expression during replicative aging in *Saccharomyces cerevisiae*. J Gerontol – Ser A Biol Sci Med Sci. 2008;63:21–34.
25. Slavov N, Airoldi EM, Van Oudenaarden A, Botstein D. A conserved cell growth cycle can account for the environmental stress responses of divergent eukaryotes. Mol Biol Cell. 2012;23:1986–97.
26. Gonskikh Y, Polacek N. Alterations of the translation apparatus during aging and stress response. Mech Ageing Dev. 2017;168:30.
27. Janssens GE, et al. Protein biogenesis machinery is a driver of replicative aging in yeast. elife. 2015;4:e08527.
28. Janssens GE, Veenhoff LM. The natural variation in lifespans of single yeast cells is related to variation in cell size, ribosomal protein, and division time. PLoS One. 2016;11:1–18.
29. Feser J, Truong d, Das C, Carson JJ, Kieft J, Harkness T, Tyler JK. Elevated histone expression promotes lifespan extension. Mol Cell. 2010;39:724–35.
30. Hu Z, et al. Nucleosome loss leads to global transcriptional up-regulation and genomic instability during yeast aging. Genes Dev. 2014;28:396–408.
31. Dang W, et al. Histone H4 lysine-16 acetylation regulates cellular lifespan. Nature. 2009;459:802–7.
32. Denoth-Lippuner A, Krzyzanowski MK, Stober C, Barral Y. Role of SAGA in the asymmetric segregation of DNA circles during yeast ageing. elife. 2014;3:e03790.
33. McCormick MA, et al. The SAGA histone deubiquitinase module controls yeast replicative lifespan via Sir2 interaction. Cell Rep. 2014;8:477–86.
34. Longo VD, Fabrizio P. Chronological aging in Saccharomyces cerevisiae. Subcell Biochem. 2012;57:101–21.
35. Wanichthanarak K, Wongtosrad N, Petranovic D. Genome-wide expression analyses of the stationary phase model of ageing in yeast. Mech Ageing Dev. 2015;149:65–74.
36. Burtner CR, Murakami CJ, Olsen B, Kennedy BK, Kaeberlein M. A genomic analysis of chronological longevity factors in budding yeast. Cell Cycle. 2011;10:1385–96.

37. Longo VD, Shadel GS, Kaeberlein M, Kennedy B. Replicative and chronological ageing in Saccharomyces cerevisiae. Cell Metab. 2013;16:18–31.
38. McCarroll SA, et al. Comparing genomic expression patterns across species identifies shared transcriptional profile in aging. Nat Genet. 2004;36:197–204.
39. Lund J, et al. Transcriptional profile of Aging in C elegans. Curr Biol. 2002;12:1566–73.
40. Golden TR, Hubbard A, Dando C, Herren MA. Age-related behaviours have distinct transcriptional profiles in C. elegans. Aging (Albany NY). 2009;7:850–65.
41. Chapin HC, Okada M, Merz AJ, Miller DL. Tissue – specific autophagy responses to aging and stress in C. elegans. Aging (Albany NY). 2015;7:419–32.
42. Dorman JB, Albinder B, Shroyer T, Kenyon C. The age-1 and daf2 genes function in a common pathway to control the lifespan of Caenorhabditis elegans. Genetics. 1995;141:1399–406.
43. Viswanathan M, Kim SK, Berdichevsky A, Guarente L. A role for SIR-2.1 regulation of ER stress response genes in determining C. elegans life span. Dev Cell. 2005;9:605–15.
44. McElwee J, Bubb K, Thomas JH. Transcriptional outputs of the Caenorhabditis elegans forkhead protein DAF-16. Aging Cell. 2003;2:111–21.
45. Mcelwee JJ, Schuster E, Blanc E, Thomas JH, Gems D. Shared transcriptional signature in Caenorhabditis elegans Dauer larvae and long-lived daf-2 mutants implicates detoxification system in longevity assurance. J Biol Chem. 2004;279:44533–43.
46. Panowski SH, Wolff S, Aguilaniu H, Durieux J, Dillin A. PHA-4/Foxa mediates diet-restriction- induced longevity of C. elegans. Nature. 2007;447:550.
47. Jain V, Kumar N, Mukhopadhyay A. PHA-4/FOXA-regulated microRNA feed forward loops during Caenorhabditis elegans dietary restriction. Aging (Albany NY). 2014;6:835–51.
48. Bishop NA, Guarente L, Al N. Two neurons mediate diet-restriction- induced longevity in C. elegans. Nature. 2007;447:545–50.
49. Chen D, Thomas EL, Kapahi P. HIF-1 modulates dietary restriction-mediated lifespan extension via IRE-1 in Caenorhabditis elegans. PLoS Genet. 2009;5:e1000486.
50. Carrano AC, Liu Z, Dillin A, Hunter T. A conserved ubiquitination pathway determines longevity in response to diet restriction. Nature. 2009;460:396–9.
51. Venkatesh S, et al. Histone exchange on transcribed genes. Nature. 2012;489:452–5.
52. Pu M, et al. Trimethylation of Lys36 on H3 restricts gene expression change during aging and impacts life span. Genes Dev. 2015;29:718–31.
53. Chalovich JM, Eisenberg E. Systems biology of ageing in 4 species. Curr Opin Biotechnol. 2007;257:2432–7.
54. Jung HJ, Suh Y. MicroRNA in aging: from discovery to biology. Curr Genomics. 2012;13:548–57.
55. Slack FJ, Kato M. Ageing and the small, non-coding RNA world. Ageing Res Rev. 2014;12:429–35.
56. De Lencastre A, et al. NIH public access. Curr Biol. 2010;20:2159–68.
57. Boulias K, Horvitz HR. The C. elegans MicroRNA mir-71 acts in neurons to promote germline-mediated longevity through regulation of DAF-16/FOXO. Cell Metab. 2012;15:439–50.
58. Smith-vikos T, Slack FJ. MicroRNAs and their roles in aging. J Cell Sci. 2012;125:7–17.
59. Pletcher SD, et al. Genome-wide transcript profiles in aging and calorically restricted Drosophila melanogaster. Curr Biol. 2002;12:712–23.
60. Brink TC, Regenbrecht C, Demetrius L, Lehrach H, Adjaye J. Activation of the immune response is a key feature of aging in mice. Biogerontology. 2009;10:721–34.
61. Frahm C, et al. Transcriptional profiling reveals protective mechanisms in brains of long lived mice. Neurobiol Aging. 2016;52:23–31.
62. Swindell WR. Genes and gene expression modules associated with caloric restriction and aging in the laboratory mouse. BMC Genomics. 2009;10:585.
63. Lui JC, Chen W, Barnes KM, Baron J. Changes in gene expression associated with aging commonly originate during juvenile growth. Mech Ageing Dev. 2011;131:641–9.
64. Park S-K. Genomic approaches for the understanding of aging in model organisms. BMB Rep. 2011;44:291–7.

65. Park SK, Prolla TA. Lessons learned from gene expression profile studies of aging and caloric restriction. Ageing Res Rev. 2005;4:55–65.
66. Park S-K, Prolla TA. Gene expression profiling studies of aging in cardiac and skeletal muscles. Cardiovasc Res. 2005;66:205–12.
67. Edwards MG, et al. Gene expression profiling of aging reveals activation of a p53-mediated transcriptional program. BMC Genomics. 2007;8:80.
68. Southworth LK, Owen AB, Kim SK. Aging mice show a decreasing correlation of gene expression within genetic modules. PLoS Genet. 2009;5:e1000776.
69. Takasugi M. Progressive age-dependent DNA methylation changes start before adulthood in mouse tissues. Mech Ageing Dev. 2011;132:65–71.
70. de Magalhães JP, Curado J, Church GM. Meta-analysis of age-related gene expression profiles identifies common signatures of aging. Bioinformatics. 2009;25:875–81.
71. Zahn JM, Kim SK. Systems biology of aging in four species. Curr Opin Biotechnol. 2007;18:355–9.
72. Fraser HB, Khaitovich P, Plotkin JB, Pääbo S, Eisen MB. Aging and gene expression in the primate brain. PLoS Biol. 2005;3:1653–61.
73. Brinkmeyer-Langford CL, Guan J, Ji G, Cai JJ. Aging shapes the population-mean and -dispersion of gene expression in human brains. Front Aging Neurosci. 2016;8:1–14.
74. Hong MG, Myers AJ, Magnusson PKE, Prince JA. Transcriptome-wide assessment of human brain and lymphocyte senescence. PLoS One. 2008;3:e3024.
75. Colantuoni C, et al. Temporal dynamics and genetic control of transcription in the human prefrontal cortex. Nature. 2011;478:519–23.
76. Kumar A, et al. Age associated changes in gene expression in human brain and isolated neurons. Neurobiol Aging. 2013;34:1199–209.
77. Somel M, et al. MicroRNA, mRNA, and protein expression link development and aging in human and macaque brain. Genome Res. 2010;20:1207–18.
78. Cribbs DH, et al. Extensive innate immune gene activation accompanies brain aging, increasing vulnerability to cognitive decline and neurodegeneration: a microarray study. J Neuroinflammation. 2012;9:179.
79. Berchtold NC, et al. Synaptic genes are extensively downregulated across multiple brain regions in normal human aging and Alzheimer's disease. Neurobiol Aging. 2013;34:1653–61.
80. Meng G, Zhong X, Mei H. A systematic investigation into aging related genes in brain and their relationship with Alzheimer's disease. PLoS One. 2016;11:1–17.
81. Berchtold NC, et al. Gene expression changes in the course of normal brain aging are sexually dimorphic. Proc Natl Acad Sci U S A. 2008;105:15605–10.
82. Dugo M, et al. Human lung tissue Transcriptome: influence of sex and age. PLoS One. 2016;11:e0167460.
83. Swindell WR, et al. Meta-profiles of gene expression during aging: limited similarities between mouse and human and an unexpectedly decreased inflammatory signature. PLoS One. 2012;7:e33204.
84. Kuehne A, et al. An integrative metabolomics and transcriptomics study to identify metabolic alterations in aged skin of humans in vivo. BMC Genomics. 2017;18:169.
85. Haustead DJ, et al. Transcriptome analysis of human ageing in male skin shows mid-life period of variability and central role of NF-κB. Sci Rep. 2016;6:26846.
86. Glass D, et al. Gene expression changes with age in skin, adipose tissue, blood and brain. Genome Biol. 2013;14:R75.
87. Harries LW, et al. Human aging is characterized by focused changes in gene expression and deregulation of alternative splicing. Aging Cell. 2011;10:868–78.
88. Irizar H, et al. Age gene expression and coexpression progressive signatures in peripheral blood leukocytes. Exp Gerontol. 2015;72:50–6.
89. Nevalainen T, et al. Transcriptomic and epigenetic analyses reveal a gender difference in aging-associated inflammation: the vitality 90+ study. Age (Omaha). 2015;37:1–13.
90. Reynolds LM, et al. Transcriptomic profiles of aging in purified human immune cells. BMC Genomics. 2015;16:333.

91. Zahn JM, et al. Transcriptional profiling of aging in human muscle reveals a common aging signature. PLoS Genet. 2006;2:e115.
92. Rodwell GEJ, et al. A transcriptional profile of aging in the human kidney. PLoS Biol. 2004;2:e427.
93. Gheorghe M, et al. Major aging-associated RNA expressions change at two distinct age-positions. BMC Genomics. 2014;15:132.
94. Yang J, et al. Synchronized age-related gene expression changes across multiple tissues in human and the link to complex diseases. Sci Rep. 2015;5:15145.
95. Marttila S, et al. Ageing-associated changes in the human DNA methylome: genomic locations and effects on gene expression. BMC Genomics. 2015;16:1–17.
96. Yuan T, et al. An integrative multi-scale analysis of the dynamic DNA methylation landscape in aging. PLoS Genet. 2015;11:e1004996.
97. Bahar R, et al. Increased cell-to-cell variation in gene expression in ageing mouse heart. Nature. 2006;441:1011–4.
98. Yuan G-C, et al. Challenges and emerging directions in single-cell analysis. Genome Biol. 2017;18:1–8.
99. Stunkard AJ. Genome dynamics and transcriptional deregulation in ageing. Neuroscience. 2007;162:214–20.
100. Busuttil R, Bahar R, Vijg J. Genome dynamics and transcriptional deregulation in ageing. Neuroscience. 2007;145:1341–7.
101. Busuttil RA, Dolle M, Campisi J, Vijga J. Genomic instability, aging, and cellular senescence. Ann N Y Acad Sci. 2004;1019:245–55.
102. Kennedy BK. The genetics of ageing: insight from genome-wide approaches in invertebrate model organisms. J Intern Med. 2008;263:142–52.
103. Pitt JN, Kaeberlein M. Why is aging conserved and what can we do about it? PLoS Biol. 2015;13:1–11.
104. Walker DW, McColl G, Jenkins NL, Harris J, Lithgow GJ. Evolution of lifespan in *C. elegans*. Nature. 2000;405:296–7.
105. Briga M, Verhulst S. What can long-lived mutants tell us about mechanisms causing aging and lifespan variation in natural environments? Exp Gerontol. 2015;71:21–6.
106. Bishop N, Guarente L. Genetic links between diet and lifespan: shared mechanisms from yeast to humans. Nat Rev Genet. 2007;8:835–44.
107. López-Lluch G, Navas P. Calorie restriction as an intervention in ageing. J Physiol. 2015;594:2043–60.

Ribosome, Protein Synthesis, and Aging

Reetika Manhas and Pramod C. Rath

Introduction

Aging is a progressive decline involving a decrease in the organism's fitness leading to disease manifestation and eventually death. The hallmarks representing the common denominators of aging in different organisms are genomic instability, telomere attrition, epigenetic alterations, loss of proteostasis, deregulated nutrient-sensing, mitochondrial dysfunction, cellular senescence, stem cell exhaustion, and altered intercellular communication [1].

The maintenance of protein homeostasis (proteostasis) dictates that at a given time, the rate of protein synthesis should be in balance with that of protein degradation. However, owing to age-related accumulation of damaged proteins due to impaired functioning of the cellular maintenance and repair pathways, loss of proteostasis occurs as a consequence of the aging process (Fig. 4.1). Even at the pre-translational level, multiple regulatory mechanisms are employed to both spatially and temporally regulate the process of protein synthesis. Dysregulation of the surveillance mechanisms at either the pre-translational or translational level has been related to aging and disease onset (Fig. 4.1) [2]. Genomic instability thus also has a direct impact on the process of mRNA translation.

Emerging research findings indicate that there exists a causative relationship between the regulation of protein synthesis and aging such that lowering the rate of protein synthesis also lowers the rate of aging and helps in life span extension in various organisms. This involves the interplay of signaling pathways which modulate protein synthesis and hence aging (Fig. 4.1). In vitro protein synthesis monitored in several studies using incorporation of radioactively labeled amino acids in cell-free extracts, using hepatocyte suspension or purified ribosomes, has revealed

---

R. Manhas (✉) · P. C. Rath (✉)
Molecular Biology Laboratory, School of Life Sciences, Jawaharlal Nehru University, New Delhi, India
e-mail: reetikamanhas@gmail.com; pcrath@mail.jnu.ac.in

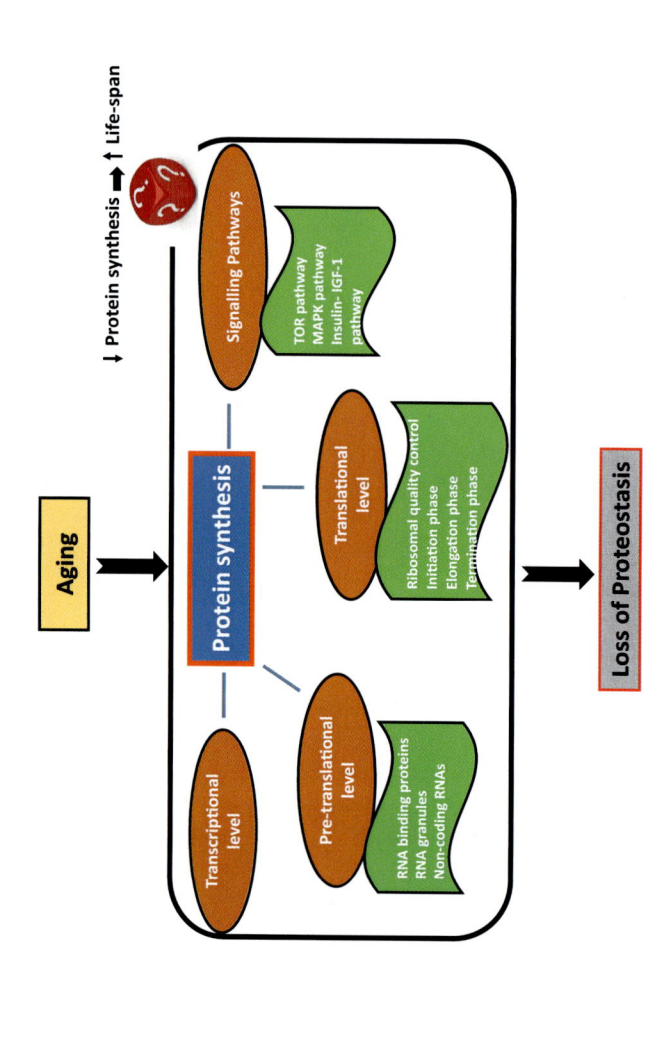

Fig. 4.1 Loss of proteostasis occurs as a consequence of the aging process. Aging and protein synthesis demonstrate a cause and effect relationship which can be overviewed at the levels of transcription, translation, and various signaling pathways which regulate protein synthesis. The downregulation of protein synthesis appears to contribute to longevity in organisms

that bulk protein synthesis slows down during aging in various cells, tissues, and organs in different organisms. Furthermore, an increase in misincorporation of amino acids into proteins with age has also been observed [3].

Among the multiple molecular mechanisms underlying the aging process, alterations in the protein synthesis machinery appear to play a crucial role. This chapter attempts to dissect the role played by various components of the translation machinery in the aging process and how aging in turn affects them. The role of the ribosome, which is at the heart of the translation process, transcriptional regulation involving RNA granules and noncoding RNAs, and translational regulation along with their alterations during the aging process, has been discussed. The signaling pathways regulating these processes and their relation to aging are also documented. The chemical changes and damage occurring to the polypeptide chains during the aging process and the contribution of these to the loss of proteostasis have been discussed. Finally, various theories explaining how downregulation of protein synthesis may contribute to longevity in different organisms have been explained.

Ribosome and the Aging Process

Ribosomes are ribozymes which constitute various subunits like different rRNAs, ribosomal proteins, and other ribosome-associated proteins. Ribosomes act as key players in translational quality control by involving transport and protein-folding components depending on the nature of polypeptide chain [2]. For example, the ribosome helps in preventing proteotoxicity arising from the misfolding of secretory proteins due to the presence of regulation of aberrant protein production (RAPP) mechanism. RAPP manifests in the form of association of the N-terminal signal sequence in the secretory polypeptide with the signal recognition particle (SRP) at the ribosomal exit terminal. This association fails in case of a mutated N-terminal sequence, and instead the polypeptide chain now interacts with another ribosome interacting protein, Ago2 (argonaute2), which directs it for degradation (Fig. 4.2) [4].

Ribosomal chaperones play an important role in regulating translational activity especially under stress conditions. For example, nascent polypeptide-associated complex (NAC) is a chaperone complex which associates reversibly with the ribosome and helps in solubilization of misfolded proteins under cellular stress conditions. However, during aging due to constitutive expression of protein aggregates, there occurs an irreversible sequestration of these chaperones leading to stalled ribosomal activity (Fig. 4.2) [5]. Acute proteotoxic stress causes ribosomal pausing on the first 50 codons of the transcript during the early elongation phase due to the limited availability of ribosome-associated chaperones like Hsc/Hsp70 which aid in translocation of the nascent polypeptide chain from the interior of ribosomes to the cytoplasm. Thus, accumulation of proteotoxic stress during aging leads to reduced protein synthesis and negatively effects ribosome functions (Fig. 4.2) [6]. It has been proposed that an increase in oxidative stress during aging may cause damage to ribosomal RNA and lead to diminished translational speed and efficiency

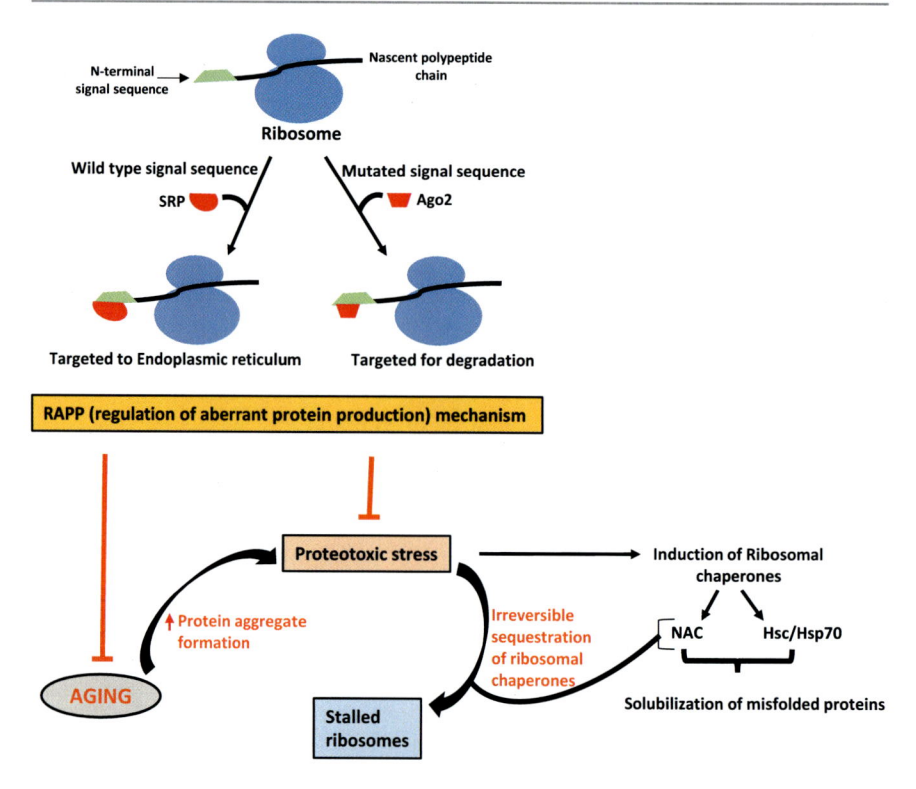

Fig. 4.2 The translational quality control mechanism at the ribosome involves RAPP (regulation of aberrant protein production). It helps in preventing proteotoxicity due to association of N-terminal signal sequence with SRP (signal recognition particle). This association fails in case of a mutated signal sequence which instead interacts with Ago2 (Argonaute 2) and is directed for degradation

Ribosomal chaperones like NAC (nascent polypeptide-associated complex) and Hsc/Hsp70 aid in solubilization of misfolded proteins under stress conditions. During aging, increased production of protein aggregates leads to acute proteotoxic stress and irreversible sequestration of ribosomal chaperones, causing stalled ribosomes. Red bar lines indicate negative (inhibitory) regulation

observed in age-associated pathologies which involve impaired neuronal function like Alzheimer's disease, etc. However, the effect of oxidation of ribosomal RNA on ribosome function and protein synthesis still awaits concrete experimental investigation [3].

Ribosomes are also regulated by small or long noncoding RNAs associated with them. These ribosome-bound noncoding RNAs can act as regulators of protein synthesis by modulating translation on a global scale under stress conditions. It has been observed that tRNA- or mRNA-derived fragments bound to the small or large ribosomal subunits, respectively, decrease the global mRNA translation rate during particular stress conditions in the archaea *H. volcanni* and yeast *S.*

cerevisiae, both in vitro and in vivo [3]. The expression of the long noncoding RNA transcribed telomeric sequence 1 (*tts-1*) helps in promoting longevity by regulation of ribosome levels in *C. elegans*. The ribosomes of the worms harboring the life-extending *daf-2* mutation have been found to be associated with *tts-1*. Depletion of *tts-1* levels in the *daf-2* mutants leads to an increase in ribosome levels and shortened worm life span [7].

Genetic evidence provides clues that stripping off the ribosome of particular ribosomal proteins affects the rate of translation and leads to life span modulation. It is observed that yeast strains harboring gene deletions for ribosomal proteins rpl3, rpl6b, rpl10, rps6, and rps18 have a significant increase in replicative life span. In another study in yeast, where the paralogous gene copies were also deleted to prevent rescue of the phenotype, 11 new ribosomal protein deletions contributing to longevity were identified. Likewise, in *C. elegans* siRNA-mediated knockdown of six ribosomal proteins of the small subunit and five of the large subunits reduced the level of protein synthesis and concurrently led to life span extension [3].

The depletion of factors involved in ribosomal biosynthesis and tRNA synthesis results in life span extension in *C. elegans* [8]. In *S. cerevisiae*, inhibiting the biogenesis of 60S ribosomal subunit leads to a significant increase in life span via induction of the nutrient-responsive transcription factor, Gcn4 [9]. The depletion of RNA methyltransferase NSUN5 which methylates 28S/25S rRNA induces increased stress resistance and life span in flies, worms, and yeast [10].

Ribosomes have been found to be dynamic structures which form distinct subpopulations with unique molecular compositions and functions under varied cellular needs. This concept of the "specialized ribosomes" explains the role played by ribosomes in response to various stimuli which require the translation of specific mRNAs. The heterogeneity in ribosome composition is contributed by post-synthetic modification of ribosomal proteins or rRNAs, association with different factors or proteins, rRNA methylations, etc. [11]. An example of the existence of "specialized ribosomes" has been documented in *E. coli*. The endonuclease MazF cleaves a 43-nucleotide fragment from the 3′ end of 16S rRNA of the 30S ribosomal subunit. This truncated fragment lacks the anti-Shine–Dalgarno sequence typically required for bacterial translation initiation. The ribosomes formed post this cleavage selectively translate leaderless mRNA that lack the canonical Shine–Dalgarno mRNA sequences [12]. The role of "specialized ribosomes" in aging is beginning to be unveiled. The lack of a single conserved C5 methylation of 25S rRNA at position C2278 (*S. cerevisiae*) reportedly extends the life span and stress resistance in yeast, worms, and flies. This occurs by locally altering the ribosome structure and "reprogramming" the ribosome toward translation of mRNAs involved in cellular stress response and aging [10]. Thus, ribosomes are dynamic components of the translation apparatus which help in ensuring protein quality control, undergo changes to accommodate various cellular stimuli, and also play a role in healthy aging of the proteome (Figs. 4.2 and 4.3).

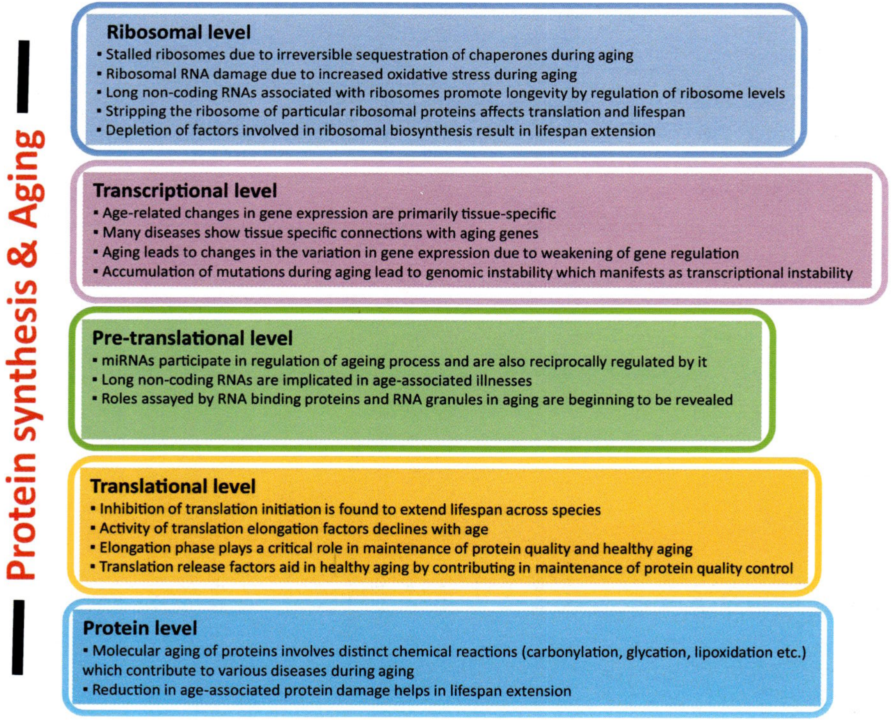

Protein Synthesis & Aging

Ribosomal level
- Stalled ribosomes due to irreversible sequestration of chaperones during aging
- Ribosomal RNA damage due to increased oxidative stress during aging
- Long non-coding RNAs associated with ribosomes promote longevity by regulation of ribosome levels
- Stripping the ribosome of particular ribosomal proteins affects translation and lifespan
- Depletion of factors involved in ribosomal biosynthesis result in lifespan extension

Transcriptional level
- Age-related changes in gene expression are primarily tissue-specific
- Many diseases show tissue specific connections with aging genes
- Aging leads to changes in the variation in gene expression due to weakening of gene regulation
- Accumulation of mutations during aging lead to genomic instability which manifests as transcriptional instability

Pre-translational level
- miRNAs participate in regulation of ageing process and are also reciprocally regulated by it
- Long non-coding RNAs are implicated in age-associated illnesses
- Roles assayed by RNA binding proteins and RNA granules in aging are beginning to be revealed

Translational level
- Inhibition of translation initiation is found to extend lifespan across species
- Activity of translation elongation factors declines with age
- Elongation phase plays a critical role in maintenance of protein quality and healthy aging
- Translation release factors aid in healthy aging by contributing in maintenance of protein quality control

Protein level
- Molecular aging of proteins involves distinct chemical reactions (carbonylation, glycation, lipoxidation etc.) which contribute to various diseases during aging
- Reduction in age-associated protein damage helps in lifespan extension

Fig. 4.3 An overview of the effects of aging process on various components of the protein synthesis machinery viz.: ribosome, transcription, translation, and the polypeptide chain, representing the cause and effect relationship between aging and protein synthesis

Age-Related Changes in Protein Synthesis

Aging has been described as degeneration resulting from progressive decline in the ability to withstand stress, damage, and disease [13]. The rate of both global and specific protein syntheses is affected with age in organisms. Aging affects protein synthesis both at the transcription and translation levels (Fig. 4.3).

Transcriptional Level

Aging is known to impact the expression level of genes in different species, and specific patterns have also been observed. The nematode *Caenorhabditis elegans* and the fruit fly *Drosophila melanogaster* though evolutionarily highly diverged share an early adulthood onset of the expression of genes involved in mitochondrial metabolism, DNA repair, catabolism, peptidolysis, and cellular transport. The alterations in mitochondrial energy production and proteasomal pathways are the conserved aging patterns observed in these organisms [14]. The electron transport chain pathway which decreases expression with age has emerged as a marker for aging across species from humans to those of the mouse and fly [15, 16].

An organism's transcriptional profile undergoes changes from the developmental to the reproductive stage during the life cycle [17–19]. A significant proportion of the age-related changes in gene expression are tissue-specific, such that different tissues are found to have different temporal patterns of transcriptomes during the aging process [13, 20]. Zhan et al. found a number of biological processes to be notably affected during aging in a tissue-specific manner in *Drosophila melanogaster* neuronal function, protein degradation, and energy production in the brain; nutrient intake and energy production in the gut; proteasome and mitochondrial function in the muscle; protein degradation, immune response, and energy metabolism in the adipose tissue; and metabolism and cell cycle-related processes in the testis [20]. Glass et al. concluded that the molecular signatures of aging predominantly reveal a tissue-specific transcriptional response in humans. Also, the skin shows most age-related gene expression changes, with many of the genes implicated in fatty acid metabolism, mitochondrial activity, cancer, and splicing [13]. Distinct and measurable patterns of gene expression changes with age have also been found in the human brain [13, 21].

Age is also a major risk factor in the development of many diseases. The interplay between the age-related gene expression changes in human tissues and manifestation of complex diseases is beginning to be revealed. A large number of diseases show tissue-specific connections with aging genes in humans [22, 23]. Age-associated changes in gene expression have been observed which lead to increased aneuploidy and the subsequent initiation of tumorigenesis [24].

Aging not only influences the level of gene expression but also leads to changes in the variation in gene expression. It is hypothesized that variation in gene expression increases with age due to weakening of gene regulation. Somel et al. observed that aging leads to heterogeneity in gene expression in rats and humans such that this heterogeneity has a minor effect on individual genes but is widespread throughout the transcriptome. This study attributed the heterogeneity to the accumulation of cellular damage and mutations in the somatic cells [25]. Bahar et al. observed a significantly elevated level of heterogeneity and increased transcriptional noise in cardiomyocytes from old mice when compared to young ones and attributed it to stochastic deregulation of gene expression as a cause of DNA damage during aging [26]. An increase in the variation of gene expression in immune function genes with advancing age in *Drosophila melanogaster* has been linked with loss of gene expression regulation [27]. The integrity of the gene expression networks has been found to decline with age in *C. elegans* [28]. The alteration of gene expression networks and pathways associated with T-cell aging leading to decline in cellular function has been identified in humans and mice [29]. At the cellular level, single-cell transcriptome analysis of human islet cells from the pancreas of old donors showed increase in transcriptional noise and inappropriate hormone expression, indicating the occurrence of age-associated transcriptional instability [30]. This age-associated transcriptional instability may be a result of genomic instability, arising from the accumulation of mutations during aging [31]. For example, the human islet cells show C–A and C–G transversions arising due to oxidative stress during aging [30]. The accumulation of epimutations (aberrant changes in epigenetic state) like changes in DNA methylation patterns also appears to stochastically increase with age thus further contributing to the prevalence of age-related genomic instability [31].

Pre-translational Level

The spatial and temporal regulation of protein synthesis occurs via multiple mechanisms at the post-transcriptional/pre-translational level. Impairment in these regulatory mechanisms causes disease onset and aging (Fig. 4.3) [2].

RNA-Binding Proteins

RNA-binding proteins (RBPs) associate with specific mRNA sequences in various pre-translation pathways to function as TTR–RBPs–mRNA turnover and translation regulatory (TTR) RBPs with roles in regulation of the expression of many age-associated proteins [32]. RBPs play an important role in maintaining the structure and function, biogenesis, stability, localization, and transport of mRNAs. The post-translational modification of RBPs like phosphorylation, arginine methylation, ubiquitination, and poly ADP-ribosylation help in regulating RNA function and localization. Also, signaling pathways like phosphoinositide-3-kinase (PI3K) and mitogen-activated protein kinases (MAPK) targeting RBPs affect mRNA translation and stability by activation of specific mRNA decay pathways [33]. Examples include the UPF proteins which regulate the nonsense-mediated decay pathway [34] and PUF proteins which promote translation repression by decapping/deadenylation and degradation across species [35]. The significance of RBP-mediated posttranscriptional regulation in controlling the progression of human aging is beginning to be highlighted [36].

The 3' UTR of 5–8% human mRNAs harbor the instability conferring AU-rich element (ARE) [37]. The RNA-binding proteins which interact with these AU-rich elements (ARE) are called ARE-binding proteins (ABPs). Examples of ABPs include HuR, AUF1, BRF, TIA-1, and TTP. HuR protein is an example of an ABP that enhances translation of target mRNAs. AUF-1 and TIA-1 bind to the 3' UTR of their target mRNAs and repress their translation. Aging differentially affects the levels of ABPs within the same or different tissues [2]. ABPs carry out physiological and pathological functions in processes like immune response and inflammation, cell cycle, and carcinogenesis where they may contribute to mRNA stability or stimulate degradation [38]. AUF-1 functions to block senescence and mediates normal aging by activating the transcription of telomerase, thereby preserving telomere length [39]. Experimental insights and research outcomes in the role assayed by ABPs in the aging process are awaited.

Regulation of mRNA Turnover by RNA Granules

Translationally inactive mRNAs and their associated RNA-binding proteins can assemble in highly dynamic non-membranous foci in the cytoplasm of eukaryotic cells called P-bodies or processing bodies. P-bodies are sites of mRNA storage and decay of translationally repressed transcripts. mRNAs found in P-bodies are subject to decapping/decay or released to the translation machinery. P-bodies are constantly present in the cell under normal conditions but increase rapidly both in size and number upon exposure to stress conditions [2, 33].

P-bodies often fuse or mature to form stress granules under conditions of stress when translation initiation is inhibited. Stress granules are cytoplasmic RNA granules which contain translationally repressed mRNAs at the initiation phase. Therefore, unlike P-bodies, they contain translation initiation factors, 40S ribosomal subunit, etc. but lack the machinery for mRNA decay. The dynamic relation between P-bodies and stress granules indicates their function in quickly responding to cellular signals. P-bodies can also interact with tissue-specific RNA granules like the neuronal transport granules and germ granules in embryonic cells. Thus, the reversible regulation of mRNA transcripts by RNA granules represents a dynamic life cycle for mRNAs in the cytoplasm. However, the impact of disruption or malfunction of the RNA granules on the aging process still needs to be deciphered [33].

Noncoding RNAs

Small Noncoding RNAs

The noncoding miRNAs are a type of *trans* acting factors that help in the regulation of steady-state transcript levels of many genes by promoting silencing and/or degradation [33]. miRNA expression is reported to change during aging in a tissue-specific pattern in mice, humans, and primates. The targets of miRNA-mediated regulation include longevity modulators like mTOR, ARE-binding proteins (ABPs), and components involved in maintenance of cellular homeostasis and stress response signaling. Also, an age-associated decline in miRNA processing factors further indicates their role in the aging process [33]. miRNAs participate in the regulation of the aging process and are also reciprocally regulated by it. For example, *Drosophila* miRNAs show distinct isoform pattern changes with age such that an increase in 2′-O-methylation of select isoforms is observed. Also, an increased loading of miRNAs into Ago2 and not Ago1 with age is observed. Mutations which affect these phenotypes result in accelerated neurodegeneration and reduced life span [40].

lin-4 and *let-7* miRNAs discovered in *C. elegans* have been found to undertake developmental regulation of cell fates and also modulate adult life span [41, 42]. *mir-71* and *mir-246* in *C. elegans* are found to be significantly upregulated during aging and act upstream in the insulin/IGF-1-like signaling as well as other known longevity pathways to aid life span extension. On the other hand, *mir-239* affects life span negatively in *C. elegans* [43]. Detailed reviews that focused on the roles of miRNAs in the aging process in different model organisms can be found elsewhere [2, 44, 45]. The role assayed by miRNAs in manifestation of the aging phenotype in humans has been reviewed separately [46].

Long Noncoding RNAs

Long noncoding RNAs (lncRNAs) are transcripts that exceed 200 nucleotides in length and are heterogeneous in their size, origin, localization, and function. lncRNAs differ from miRNAs in size, their dependence on the 2-D and 3-D structures apart from the primary sequence to undertake their function, and in being able to operate both in *cis* and *trans* [2]. They play an important role in the regulation of gene expression at the transcriptional, post-transcriptional, and post-translational levels. A detailed review of the role assayed by various lncRNAs in the processes which characterize the aging phenotype is described elsewhere [47, 48].

lncRNAs have also been implicated in age-associated illnesses. Examples include diseases arising from impairment in energy metabolism (obesity, diabetes), loss of homeostasis to proliferative and damaging stimuli (cancer, immune dysfunction), and neurodegeneration. A comprehensive review of various lncRNAs and their connection to age-related illnesses can be found documented elsewhere [49].

Translational Level

Aging affects the translation of mRNA taking place in the cytoplasm which involves the initiation phase, elongation of the polypeptide chain, and termination phase (Fig. 4.3).

Initiation Phase

Initiation is the most tightly regulated step of the translation process and is controlled by mechanisms involving the eukaryotic initiation factors (eIFs) and ribosome. The initiation phase involves the pairing of initiation codon of mRNA with the anticodon loop of initiator tRNA. The 40S ribosomal subunit and various initiation factors come together to form the 43S pre-initiation complex. This complex incorporates the initiator methionyl-tRNA bound on eIF2 and unwinds the secondary structures in mRNA as it scans it in the 5′–3′ direction. At the ATG start codon, this complex unites with the 60S ribosomal subunit to form the 80S initiation complex, and the translation initiation factors are released. Thereof, the elongation phase of protein synthesis is initiated on the 80S ribosome [50].

The regulation of initiation phase occurs at two junctures. Firstly, the activity of eIF2 is regulated by phosphorylation of its α subunit. Phosphorylation interferes with the recycling of GDP for GTP on eIF2 by eIF2B. The activity of eIF2B which is a guanine nucleotide exchange factor (GEF) is also regulated by phosphorylation. Secondly, the cap-binding protein eIF4E regulates the recruitment of 43S pre-initiation complex on the $7^{Me}G$ cap at the 5′ end of the mRNA. The activity of eIF4E is in turn controlled by direct phosphorylation and/or its association with eIF4E-binding protein (4E-BP). 4E-BP associates with eIF4E to limit its availability. Phosphorylation of 4E-BP releases eIF4E to associate with eIF4G, eIF4A, eIF4B, and poly-A binding protein (PABP) to promote the initiation of protein synthesis [50].

The regulation of translation initiation is linked to the aging process. The inhibition of translation initiation mediated through the phosphorylation of α subunit of eIF2 at Ser^{51} allows protection of cells from oxidative stress and senescence [51]. In *C. elegans*, the inhibition of eIF4G worm homologue *ifg-1*, which is a scaffold protein of the cap-binding complex, causes extension of worm life span [52, 53]. Similarly, aging in *C. elegans* is also influenced by a specific eIF4E isoform (*ife-2*) that functions in somatic tissues. Loss of *ife-2* reduces overall protein synthesis, protects against oxidative stress, and extends worm life span [54]. In another study in *C. elegans*, depletion in the levels of ribosomal proteins, translational regulator S6 kinase, and initiation factors eIF2β/*iftb-1*, eIF4E/*ife-2*, and eIF4G/*ifg-1* was found to extend life span in adult nematodes [55]. In *D. melanogaster*, overexpression of the translational repressor 4E-BP notably enhances life span upon dietary restriction [56]. These examples point to a link between rate of protein synthesis and the aging process across species wherein inhibition of protein synthesis helps in extension of life span.

Elongation and Termination Phase

The process of elongation involves two eukaryotic translation elongation factors (eEFs). eEF1 provides the ribosome with appropriate aminoacylated tRNAs and eEF2 mediates translocation of the ribosome along the mRNA. The activity of eEF2 is regulated by the calcium–calmodulin-dependent eEF2 kinase. mRNA translation is terminated when a stop codon is encountered. The eukaryotic release factors (eRFs) mediate dissociation of the ribosome from the mRNA and release of the 40S and 60S ribosomal subunits by binding and inducing hydrolytic cleavage of peptidyl-tRNA at the ribosomal P-site [33].

A decline in the activity of translation elongation factors with age has been implicated in the reduction of protein synthesis. Diminished activity of eukaryotic elongation factor 1 (eEF-1) has been observed in *Drosophila* and in the rat liver and brain [57]. Aging leads to loss of activity, increased susceptibility to oxidation, and fragmentation of eukaryotic elongation factor 2 (eEF-2) due to the effect of lipid peroxidation, as observed in the pineal gland of aged rats [58]. The efficiency of co-translational folding of the polypeptide chain being synthesized on the ribosome is influenced by translation fidelity during elongation. An increase in elongation speed due to persistent mTORC1 activation leads to reduced translation fidelity and less functional proteins. Thus, the elongation phase may be implicated to play a critical role in the maintenance of protein quality and healthy aging [59].

The translation release factors (eRFs) contribute in the maintenance of protein quality control by helping in clearance of aberrant polypeptides and rescue of stalled ribosomes. Stalling of ribosomes occurs on the polylysine segments which interact electrostatically with the ribosome tunnel and are encoded by polyadenylated non-stop mRNA. The release factor eRF3 helps in premature translation termination in ribosomes stalled at these polylysine segments in *Saccharomyces cerevisiae* [60]. Translational release factors have also been implicated as mediators in the co-translational proteasomal degradation of nascent polypeptide chains associated with 60S subunit which are released by stalled ribosomes [61]. Overall, the eRFs contribute in the maintenance of protein quality control during the termination phase and help in healthy aging.

Nonenzymatic Posttranslational Modification of Proteins and Aging

An increase in oxidative stress during aging leads to the induction of post-translational modifications in proteins like carbonylation, glycation, lipoxidation, etc. These modifications are often also observed in degenerative or metabolic disease conditions.

Glycation reaction involves the condensation of sugar with an amino acid residue on a protein to form a Schiff base. This Schiff base undergoes an Amadori rearrangement followed by subsequent rearrangements like oxidations and eliminations leading to the formation of advanced glycation end (AGE) products like carboxymethyllysine (CML). Further, the Amadori reaction can give rise to high amount of reactive carbonyl groups called α-dicarbonyls, thus causing "carbonyl stress" and generation of more AGEs [62]. Carbonylation of amino acid side chains induces formation of denser and compact protein aggregates with implications during the aging process [63].

Oxidative modifications in proteins are induced when reactive oxygen species (ROS) directly attack the protein backbone or the amino acid side-chain residues which leads to generation of protein carbonyls. Secondary by-products like oxidized sugars, aldehydes, and lipids can also cause indirect damage to proteins. Oxidized proteins are eliminated by the lysosomal or proteasomal degradation systems or other repair mechanisms. However, as these mechanisms decline during aging, oxidative modifications accumulate in proteins resulting in loss of their biochemical function. Protein oxidation has been found to occur in many age-related pathological disorders. Oxidative damage leads to the accumulation of oxidized proteins and lipids, aggregates such as lipofuscin, and advanced glycation end products (AGEs) [64].

AGEs modification alters the protein's structure, activity, and biological half-life. AGEs accumulate in tissues with age and are involved in pathophysiological processes. AGEs can act through specific receptor molecules like receptor for AGEs (RAGE), galectin-3 (AGE-R3), etc. RAGE is a multi-ligand receptor capable of inducing various signaling pathways like p38 MAP-kinase, Akt, JNK, JAK/STAT, rho-GTPases, and PI3-kinase with the major downstream target being the pro-inflammatory NF-κB pathway [62]. AGEs activate the cells for generation of ROS, which induces DNA oxidation and membrane lipid peroxidation. Extracellular proteins like collagen which are directly exposed to high glucose levels are established targets for AGE modifications, and glycation reduces cell migration and proteolytic degradation of collagen by membrane-type matrix metalloproteinases [62].

Lipofuscin constitutes of highly oxidized material from covalently cross-linked proteins (30–70%) and lipids (20–50%) which accumulates intracellularly in a time-dependent manner. As humans age, sugar residues are also found to be a part of lipofuscin. The binding of iron in mammalian cells to lipofuscin leads to generation of the highly reactive hydroxyl radical from hydrogen peroxide which leads to cytotoxicity. Lipofuscin also competitively binds the proteasomal and lysosomal proteases, thereby impairing the degradation of oxidized proteins. This results in further accumulation of cross-linked proteins in the cell [64].

Carbamylation is another form of nonenzymatic posttranslational modification caused by the binding of isocyanate to the free amino groups of proteins. Isocyanate is generated from the dissociation of urea or myeloperoxidase-mediated catabolism of thiocyanate. This leads to the accumulation of carbamylated proteins in tissues. Carbamylation brings about an adverse effect on protein structure and function and also contributes to aging in skin [65].

Other physiological chemical reactions like isomerization, deamidation, or racemization of Asn/Asp residues in proteins have been described which may induce changes in protein structure and function during aging. Molecular aging of proteins is an established complex process that involves many distinct chemical reactions which contribute to metabolic diseases during aging. The development of methods to detect the end products of these nonenzymatic chemical reactions may potentially present interesting biomarkers characteristic of the aging process [66].

Signaling Pathways That Regulate Protein Synthesis and Aging

The TOR Pathway

Target of rapamycin (TOR) is a kinase of the serine/threonine family. The TOR pathway responds to the supply of energy and nutrients, growth factors, and stress. When nutrient and energy resources are plentiful in the cell, activation of TOR triggers anabolic processes like protein and lipid biosynthesis at the expense of catabolic processes like autophagic induction [20]. TOR plays an important role in the regulation of protein translation both at the initiation and elongation phase by undertaking the phosphorylation of eIF4E-binding protein (4E-BP), which is a negative regulator of eIF4E cap-binding protein allowing the formation of a functional complex to allow cap-dependent translation initiation and eukaryotic elongation factor 2 kinase (EEF2K) which allows the activation of elongation factor 2 (eEF2), respectively. Other targets of TOR are the S6 kinases S6K1 and S6K2 which upon activation subsequently phosphorylate several components of the translation machinery including 40S ribosomal protein S6, eIF4B, and eEF2 kinase (Fig. 4.4).

Genetic inhibition of TOR is observed to promote longevity in several model organisms. Likewise, inhibition of the factors lying downstream of TOR such as S6 kinase led to life span extension in *C. elegans*, *Drosophila*, and *mice*. Also, the specific inhibitor of TORC1, rapamycin, is reported to have beneficial effect on life span. Thus, the TOR signaling pathway is pivotal and evolutionarily conserved modulator of life span across species [2]. Interestingly, mTOR inactivation mimics

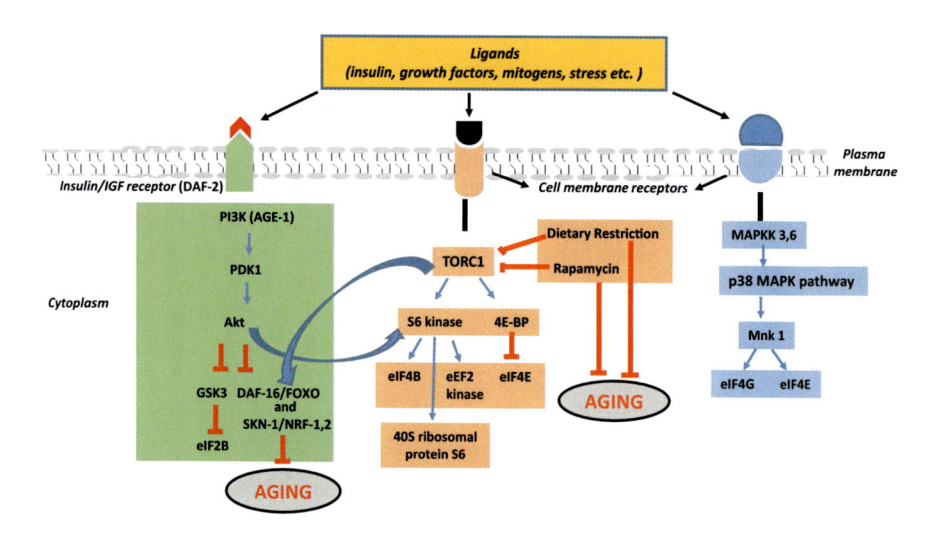

Fig. 4.4 Regulation of mRNA translation by signaling pathways: insulin-IGF-1, TOR, and MAPK pathway. Blue arrows indicate positive (stimulatory) regulation. Red bar lines indicate negative (inhibitory) regulation

Insulin-IGF-1 pathway activation leads to inhibition of forkhead (FOXO) transcription factor DAF-16 and SKN-1 (mammalian NRF-1/2 homologue) via PI3K-Akt signaling. DAF-16/FOXO and SKN-1/NRF-1,2 are essential for lifespan extension mediated by the inhibition of insulin/IGF-1 signaling pathway

TOR signaling allows cap-dependent translation initiation to progress via phosphorylation of 4E-BP, causing its dissociation from eIF4E. TOR stimulates S6 kinase which phosphorylates the 40S ribosomal protein S6 and eIF4B enhancing translation initiation. TOR also promotes translation elongation by phosphorylation and deactivation of eEF2 kinase. Inhibition of TOR signaling by dietary restriction or rapamycin leads to inhibition of aging phenotype

MAPK pathway activation causes p38 MAPK to activate Mnk1 which then phosphorylates eukaryotic initiation factor 4E (eIF4E), thereby facilitating translation initiation the effects of dietary restriction (DR) (Fig. 4.4). DR is reported to extend life span and restrict late-onset diseases in various species and also provides an overall health benefit in primates [67].

The MAPK Pathway

The mitogen-activated protein kinase (p38 MAPK) pathway plays an important role in the regulation of protein synthesis by targeting several components of the translation machinery. p38 kinases are activated by MAP kinase kinases (MKKs) such as MKK3 and MKK6. Under conditions of stress, MAPK activates the mitogen-activated protein kinase-interacting kinase Mnk1. Mnk1 is a member of the eIF4F complex and directly binds eIF4G. After activation by p38 kinase, Mnk1 phosphorylates eukaryotic initiation factor 4E (eIF4E) thereby facilitating translation initiation by stabilizing its interaction with eIF4G and 5′ cap of mRNA. The activity of Mnk1 is regulated by p38 MAPK and extracellular signal-regulated kinases 1 and 2 (ERK1 and ERK2). ERK1 and ERK2 also stimulate the S6 kinase (S6K). The involvement of the p38 MAP kinase pathway in aging can be inferred from various studies that have reported phenotypes such as cell cycle arrest and premature senescence upon activation of MKK6 and MKK3, in a p38 kinase-dependent manner (Fig. 4.4) [3, 50].

The Insulin-IGF-1 Pathway

The insulin-IGF-1 pathway is known to modulate aging across species. Many studies have reported the link between genetic inhibition of the insulin-IGF-1 signaling pathway and significant life span extension. In *C. elegans*, downregulation of the insulin-IGF-1 signaling pathway by mutations in the genes encoding the insulin-IGF-1-like receptor abnormal DAuer Formation 2 (DAF-2) or the phosphatidylinositol 3-kinase (PI3K) orthologue aging alteration 1 (AGE-1) leads to extended life span. This induction of longevity is dependent on a forkhead (FOXO) transcription factor DAF-16 which targets genes involved in metabolism, resistance to oxidative stress, detoxification, and immunity. Apart from DAF-16/FOXO, SKN-1 (the mammalian NRF-1/2 homologue) and HSF-1 are also essential for the life span extension mediated by the inhibition of insulin/IGF-1 signaling (Fig. 4.4) [2, 50]. Recent studies propose that the extended life span conferred due to inhibition of insulin/IGF-1 signaling is accompanied by reduced rate of protein synthesis. In *C. elegans*, DAF-2 mutant exhibits reduced levels of ribosomal subunits, translation initiation, and elongation factors in high-throughput proteomic studies. Polysome profiling further confirms compromised de novo protein synthesis dependent on DAF-16/FOXO in these mutants. DAF-16/FOXO and SKN-1 transcription factors have a role in TOR inhibition-mediated longevity [68].

In *Drosophila*, mutations in both the insulin-like receptor (InR) and the insulin-receptor substrate (chico) lead to extended life span which is dependent on the activity of dFOXO. Overexpression of dFOXO in the muscles of *Drosophila* increases the relative 4E-BP mRNA levels indicating a direct inhibitory relationship between insulin/IGF-1 signaling and protein synthesis [2, 50].

In case of mammals, separate receptors for insulin and IGF-1 coordinate cellular metabolism and growth and differentiation, respectively. Mutations in either the receptor gene or in upstream genes that regulate insulin or IGF-1 levels are known to extend life span. The Ames and Snell dwarf mice having low levels of insulin, IGF-1, and growth hormone have been reported to live significantly longer than the normal mice. The rate of protein synthesis is significantly downregulated in Snell dwarf mice as a result of reduced insulin-IGF-1 signaling through Akt/PKB and p38 MAPK, which in turn control translation regulators like mTOR kinase, ribosomal S6K, Mnk1, and translation initiation factors eIF4E and 4E-BP1 (Fig. 4.4). A similar mechanism mediating reduced protein synthesis operates in long-lived Ames dwarf mice in which the PI3K-Akt-mTOR signaling axis is mitigated [50].

Aging and Protein Damage

Proteostasis (protein homeostasis) is the state in which the proteome of an organism is in functional equilibrium maintained by the machineries of protein synthesis, folding, and degradation. These machineries regulate protein synthesis, autophagy, proteasomal degradation, and chaperone-mediated protein folding to prevent the accumulation of misfolded/aggregated proteins.

Aging is associated with compromised protein quality control leading to widespread protein aggregation resulting from the accumulation of aberrant protein species. This causes an imbalance in the proteome by burdening the proteostasis machinery causing interference in protein folding, clearance, etc. [69]. For example, a reduction in the activity of cytosolic proteasomal system during aging has been reported to cause an increase in the pool of oxidized proteins in human fibroblasts. The accumulation of oxidized proteins during aging causes inhibition of the proteasome and thus further contributes to the buildup of damaged proteins [70]. In *Drosophila melanogaster*, the assembly of 26S proteasome gets perturbed with age [71]. A role of the proteasomal pathway in determination of *C. elegans* life span has been deciphered under conditions of reduced insulin signaling [72].

Notably, studies on different model systems have established a connection between increased proteasomal activity and longevity as well as protection against age-related diseases [73]. In a similar way, the activation of autophagy by genetic/pharmacological means helps in life span extension and protection against neurodegeneration [74]. These observations indicate that reducing the protein damage that occurs with age can lead to extension of life span in the organism.

Hanging in the Balance: Life Span Extension by Regulation of Protein Synthesis

Several theories shed light on the benefits of reduced protein synthesis on the organism's survival and life span extension. The "antagonistic pleiotropy theory of aging" suggests that some genes that increase the odds of successful reproduction early in life may have deleterious effects later in life. Since their harmful effects do not manifest until after the reproductive phase, these genes are not eliminated by natural selection. The use of RNAi in *C. elegans* during late adulthood to reduce the expression of genes which are essential during the early reproductive life stages revealed enhanced longevity and increased stress resistance [8]. This observation points to a possible trade-off between reproduction and somatic maintenance, involving mRNA translation and its implications for longevity.

The "disposable soma" theory of aging suggests that in the somatic tissues, more energy resources are allocated to biosynthesis activities than repair, slowly leading to their senescent decline. On the other hand, in the germ line tissues, energy is mostly invested in maintenance and repair, thereby promoting immortality. Protein synthesis is one of the energy-consuming cellular processes. Taking a cue from here, it may be inferred that a decrease in the level of protein synthesis may have beneficial effects contributing to longevity by activation of stress response pathways and shifting of the cellular energy resources from anabolism to cellular maintenance and repair. This is supported by the observation that depletion of ribosomal proteins and translational factors in several model organisms increases stress resistance and extends life span [50]. This also explains the observed phenomenon of life span extension by methionine restriction in yeast, rodents, and human fibroblasts [75]. "Hormesis" manifests as mild stress-induced stimulation of maintenance and repair pathways in the cells by exposure to low doses of harmful agents like heat stress, reactive oxygen species, calorie restriction, etc. The hormetic effects extend longevity by an increased defense capacity and reduced load of damaged macromolecules leading to stress tolerance, adaptation, and survival [76]. The regulation of heat shock proteins levels and overexpression of components of the ubiquitin-proteasome system are a few molecular bases for the beneficial effects of hormesis [77]. However, whether the phenomenon of hormesis involves the regulation of mRNA translation for life span extension still needs to be unraveled.

High levels of protein biosynthesis may lead to the accumulation of damaged/misfolded proteins due to the occurrence of transcriptional/translational errors thus leading to disruption of cellular protein homeostasis. These altered proteins accumulate to form toxic aggregates in the cell resulting in cellular damage and aging. Altered proteins are associated with age-related pathologies like Alzheimer's disease, Parkinson's disease, and atherosclerosis [78]. Thus, the upregulation of chaperones in different model organisms is observed to extend life span by neutralizing the effects of protein misfolding/aggregation [79]. The damaged/misfolded proteins are recognized by the protein quality control system and either targeted for degradation or assisted folding. However, the functioning of the proteasomal system involved in the maintenance of protein quality control declines with age leading to the cellular accumulation of damaged proteins [1]. Hence, reduced protein synthesis leads to decreased levels of damaged/misfolded proteins in the aging organism with an already declining proteasomal apparatus, ultimately contributing to longevity.

However, it is important to note that perturbing protein synthesis may confer longevity benefit only till a certain threshold, beyond which it will be detrimental for the cell. Therefore, a delicate balance exists wherein impairing protein synthesis may either promote longevity or lead to senescence by impacting cellular functions.

References

1. López-Otín C, Blasco MA, Partridge L, Serrano M, Kroemer G. The hallmarks of aging. Cell. 2013;153(6):1194–217.
2. Charmpilas N, Daskalaki I, Papandreou ME, Tavernarakis N. Protein synthesis as an integral quality control mechanism during ageing. Ageing Res Rev. 2015;23(Pt A):75–89.
3. Gonskikh Y, Polacek N. Alterations of the translation apparatus during aging and stress response. Mech Ageing Dev. 2017;168:30–6.
4. Karamyshev AL, Patrick AE, Karamysheva ZN, Griesemer DS, Hudson H, Tjon-Kon-Sang S, et al. Inefficient SRP interaction with a nascent chain triggers a mRNA quality control pathway. Cell. 2014;156(1–2):146–57.
5. Kirstein-Miles J, Scior A, Deuerling E, Morimoto RI. The nascent polypeptide-associated complex is a key regulator of proteostasis. EMBO J. 2013;32(10):1451–68.
6. Liu B, Han Y, Qian S-B. Cotranslational response to proteotoxic stress by elongation pausing of ribosomes. Mol Cell. 2013;49(3):453–63.
7. Essers PB, Nonnekens J, Goos YJ, Betist MC, Viester MD, Mossink B, et al. A long noncoding RNA on the ribosome is required for lifespan extension. Cell Rep. 2015;10:339–45.
8. Chen D, Pan KZ, Palter JE, Kapahi P. Longevity determined by developmental arrest genes in Caenorhabditis elegans. Aging Cell. 2007;6(4):525–33.
9. Steffen KK, MacKay VL, Kerr EO, Tsuchiya M, Hu D, Fox LA, et al. Yeast life span extension by depletion of 60s ribosomal subunits is mediated by Gcn4. Cell. 2008;133(2):292–302.
10. Schosserer M, Minois N, Angerer TB, Amring M, Dellago H, Harreither E, et al. Methylation of ribosomal RNA by NSUN5 is a conserved mechanism modulating organismal lifespan. Nat Commun. 2015;6:6158.
11. Xue S, Barna M. Specialized ribosomes: a new frontier in gene regulation and organismal biology. Nat Rev Mol Cell Biol. 2012;13(6):355–69.
12. Vesper O, Amitai S, Belitsky M, Byrgazov K, Kaberdina AC, Engelberg-Kulka H, et al. Selective translation of leaderless mRNAs by specialized ribosomes generated by MazF in Escherichia coli. Cell. 2011;147(1):147–57.
13. Glass D, Viñuela A, Davies MN, Ramasamy A, Parts L, Knowles D, et al. Gene expression changes with age in skin, adipose tissue, blood and brain. Genome Biol. 2013;14(7):R75.
14. McCarroll SA, Murphy CT, Zou S, Pletcher SD, Chin C-S, Jan YN, et al. Comparing genomic expression patterns across species identifies shared transcriptional profile in aging. Nat Genet. 2004;36(2):197–204.
15. Zahn JM, Sonu R, Vogel H, Crane E, Mazan-Mamczarz K, Rabkin R, et al. Transcriptional profiling of aging in human muscle reveals a common aging signature. PLoS Genet. 2006;2(7):e115.
16. Zahn JM, Poosala S, Owen AB, Ingram DK, Lustig A, Carter A, et al. AGEMAP: a gene expression database for aging in mice. PLoS Genet. 2007;3(11):e201.
17. Lund J, Tedesco P, Duke K, Wang J, Kim SK, Johnson TE. Transcriptional profile of aging in C. elegans. Curr Biol. 2002;12(18):1566–73.

18. Van Driessche N, Shaw C, Katoh M, Morio T, Sucgang R, Ibarra M, et al. A transcriptional profile of multicellular development in Dictyostelium discoideum. Development. 2002;129(7):1543–52.
19. Graveley BR, Brooks AN, Carlson JW, Duff MO, Landolin JM, Yang L, et al. The developmental transcriptome of *Drosophila melanogaster*. Nature. 2011;471(7339):473–9.
20. Zhan M, Yamaza H, Sun Y, Sinclair J, Li H, Zou S. Temporal and spatial transcriptional profiles of aging in *Drosophila melanogaster*. Genome Res. 2007;17(8):1236–43.
21. Kumar A, Gibbs JR, Beilina A, Dillman A, Kumaran R, Trabzuni D, et al. Age-associated changes in gene expression in human brain and isolated neurons. Neurobiol Aging. 2013;34(4):1199–209.
22. Yang J, Huang T, Petralia F, Long Q, Zhang B, Argmann C, et al. Synchronized age-related gene expression changes across multiple tissues in human and the link to complex diseases. Sci Rep. 2015;5:15145.
23. Zierer J, Pallister T, Tsai P-C, Krumsiek J, Bell JT, Lauc G, et al. Exploring the molecular basis of age-related disease comorbidities using a multi-omics graphical model. Sci Rep. 2016. 5;6:37646.
24. Geigl JB, Langer S, Barwisch S, Pfleghaar K, Lederer G, Speicher MR. Analysis of gene expression patterns and chromosomal changes associated with aging. Cancer Res. 2004;64(23):8550–7.
25. Somel M, Khaitovich P, Bahn S, Pääbo S, Lachmann M. Gene expression becomes heterogeneous with age. Curr Biol. 2006;16(10):R359–60.
26. Bahar R, Hartmann CH, Rodriguez KA, Denny AD, Busuttil RA, Dollé MET, et al. Increased cell-to-cell variation in gene expression in ageing mouse heart. Nature. 2006;441(7096):1011–4.
27. Carlson KA, Gardner K, Pashaj A, Carlson DJ, Yu F, Eudy JD, et al. Genome-wide gene expression in relation to age in large laboratory cohorts of drosophila melanogaster. Genet Res Int. 2015;2015:1–19.
28. Viñuela A, Snoek LB, Riksen JAG, Kammenga JE. Genome-wide gene expression regulation as a function of genotype and age in C. elegans. Genome Res. 2010 Jul;20(7):929–37.
29. Chen G, Lustig A, Weng N. T cell aging: a review of the transcriptional changes determined from genome-wide analysis. Front Immunol. [Internet]. 2013 [cited 2018 July 6];4. Available from: http://journal.frontiersin.org/article/10.3389/fimmu.2013.00121/abstract
30. Enge M, Arda HE, Mignardi M, Beausang J, Bottino R, Kim SK et al. Single cell transcriptome analysis of human pancreas reveals transcriptional signatures of aging and somatic mutation patterns. 2017 [cited 2018 July 6]. Available from: http://biorxiv.org/lookup/doi/10.1101/108043
31. Benayoun BA, Pollina EA, Brunet A. Epigenetic regulation of ageing: linking environmental inputs to genomic stability. Nat Rev Mol Cell Biol. 2015;16(10):593–610.
32. Masuda K, Kuwano Y, Nishida K, Rokutan K. General RBP expression in human tissues as a function of age. Ageing Res Rev. 2012;11(4):423–31.
33. Borbolis F, Syntichaki P. Cytoplasmic mRNA turnover and ageing. Mech Ageing Dev. 2015;152:32–42.
34. Pal M, Ishigaki Y, Nagy E, Maquat LE. Evidence that phosphorylation of human Upf1 protein varies with intracellular location and is mediated by a wortmannin-sensitive and rapamycin-sensitive PI 3-kinase-related kinase signaling pathway. RNA. 2001;7(1):5–15.
35. Miller MA, Olivas WM. Roles of Puf proteins in mRNA degradation and translation. Wiley Interdiscip Rev RNA. 2011;2(4):471–92.
36. Wei Y-N, Hu H-Y, Xie G-C, Fu N, Ning Z-B, Zeng R et al. Transcript and protein expression decoupling reveals RNA binding proteins and miRNAs as potential modulators of human aging. Genome Biol [Internet]. 2015 [cited 2018 July 6];16(1). Available from: http://genomebiology.com/2015/16/1/41
37. Bakheet T, Williams BRG, Khabar KSA. ARED 3.0: the large and diverse AU-rich transcriptome. Nucleic Acids Res. 2006;34(Database issue):D111–4.
38. von Roretz C, Di Marco S, Mazroui R, Gallouzi I-E. Turnover of AU-rich-containing mRNAs during stress: a matter of survival. Wiley Interdiscip Rev RNA. 2011;2(3):336–47.

39. Pont AR, Sadri N, Hsiao SJ, Smith S, Schneider RJ. mRNA decay factor AUF1 maintains normal aging, telomere maintenance, and suppression of senescence by activation of telomerase transcription. Mol Cell. 2012;47(1):5–15.
40. Abe M, Naqvi A, Hendriks G-J, Feltzin V, Zhu Y, Grigoriev A, et al. Impact of age-associated increase in 2′-O-methylation of miRNAs on aging and neurodegeneration in Drosophila. Genes Dev. 2014;28(1):44–57.
41. Boehm M. A developmental timing microRNA and its target regulate life span in C. elegans. Science. 2005;310(5756):1954–7.
42. Shen Y, Wollam J, Magner D, Karalay O, Antebi A. A steroid receptor-microRNA switch regulates life span in response to signals from the gonad. Science. 2012;338(6113):1472–6.
43. Pincus Z, Smith-Vikos T, Slack FJ. MicroRNA predictors of longevity in *Caenorhabditis elegans*. Kim SK, editor. PLoS Genet. 2011;7(9):e1002306.
44. Garg D, Cohen SM. miRNAs and aging: a genetic perspective. Ageing Res Rev. 2014;17:3–8.
45. Jung HJ, Suh Y. MicroRNA in aging: from discovery to biology. Curr Genomics. 2012;13(7):548–57.
46. Harries LW. MicroRNAs as mediators of the ageing process. Genes (Basel). 2014;5(3):656–70.
47. Grammatikakis I, Panda AC, Abdelmohsen K, Gorospe M. Long noncoding RNAs(lncRNAs) and the molecular hallmarks of aging. Aging (Albany NY). 2014;6(12):992–1009.
48. Kour S, Rath PC. Long noncoding RNAs in aging and age-related diseases. Ageing Res Rev. 2016;26:1–21.
49. Kim J, Kim KM, Noh JH, Yoon J-H, Abdelmohsen K, Gorospe M. Long noncoding RNAs in diseases of aging. Biochimica et Biophysica Acta (BBA) Gene Regul Mech. 2016;1859(1):209–21.
50. Tavernarakis N. Ageing and the regulation of protein synthesis: a balancing act? Trends Cell Biol. 2008;18(5):228–35.
51. Rajesh K, Papadakis AI, Kazimierczak U, Peidis P, Wang S, Ferbeyre G, et al. eIF2α phosphorylation bypasses premature senescence caused by oxidative stress and pro-oxidant antitumor therapies. Aging. 2013;5(12):884–901.
52. Pan KZ, Palter JE, Rogers AN, Olsen A, Chen D, Lithgow GJ, et al. Inhibition of mRNA translation extends lifespan in Caenorhabditis elegans. Aging Cell. 2007;6(1):111–9.
53. Rogers AN, Chen D, McColl G, Czerwieniec G, Felkey K, Gibson BW, et al. Life span extension via eIF4G inhibition is mediated by posttranscriptional remodeling of stress response gene expression in C. elegans. Cell Metab. 2011;14(1):55–66.
54. Syntichaki P, Troulinaki K, Tavernarakis N. eIF4E function in somatic cells modulates ageing in Caenorhabditis elegans. Nature. 2007;445(7130):922–6.
55. Hansen M, Taubert S, Crawford D, Libina N, Lee S-J, Kenyon C. Lifespan extension by conditions that inhibit translation in *Caenorhabditis elegans*. Aging Cell. 2007;6(1):95–110.
56. Zid BM, Rogers AN, Katewa SD, Vargas MA, Kolipinski MC, Lu TA, et al. 4E-BP extends lifespan upon dietary restriction by enhancing mitochondrial activity in drosophila. Cell. 2009;139(1):149–60.
57. Tavernarakis N. Protein synthesis and aging: eIF4E and the soma vs. germline distinction. Cell Cycle. 2007;6(10):1168–71.
58. Muñoz MF, Argüelles S, Cano M, Marotta F, Ayala A. Aging and oxidative stress decrease pineal elongation factor 2: in vivo protective effect of melatonin in young rats treated with Cumene Hydroperoxide: P INEAL eEF-2 P ROTECTION BY M ELATONIN. J Cell Biochem. 2017;118(1):182–90.
59. Conn CS, Qian S-B. Nutrient signaling in protein homeostasis: an increase in quantity at the expense of quality. Sci Signal. 2013;6(271):ra24.
60. Chiabudini M, Tais A, Zhang Y, Hayashi S, Wölfle T, Fitzke E, et al. Release factor eRF3 mediates premature translation termination on polylysine-stalled ribosomes in *Saccharomyces cerevisiae*. Mol Cell Biol. 2014;34(21):4062–76.
61. Shcherbik N, Chernova TA, Chernoff YO, Pestov DG. Distinct types of translation termination generate substrates for ribosome-associated quality control. Nucleic Acids Res. 2016;44(14):6840–52.

62. Simm A. Protein glycation during aging and in cardiovascular disease. J Proteome. 2013;92:248–59.
63. Tanase M, Urbanska AM, Zolla V, Clement CC, Huang L, Morozova K, et al. Role of carbonyl modifications on aging-associated protein aggregation. Sci Rep. 2016;6:19311.
64. Höhn A, König J, Grune T. Protein oxidation in aging and the removal of oxidized proteins. J Proteome. 2013;92:132–59.
65. Gorisse L, Pietrement C, Vuiblet V, Schmelzer CEH, Köhler M, Duca L, et al. Protein carbamylation is a hallmark of aging. Proc Natl Acad Sci U S A. 2016;113(5):1191–6.
66. Jaisson S, Gillery P. Evaluation of nonenzymatic posttranslational modification-derived products as biomarkers of molecular aging of proteins. Clin Chem. 2010;56(9):1401–12.
67. Mirzaei H, Suarez JA, Longo VD. Protein and amino acid restriction, aging and disease: from yeast to humans. Trends Endocrinol Metab. 2014;25(11):558–66.
68. Stout GJ, Stigter ECA, Essers PB, Mulder KW, Kolkman A, Snijders DS, et al. Insulin/IGF-1-mediated longevity is marked by reduced protein metabolism. Mol Syst Biol. 2014;9(1):679.
69. Walther DM, Kasturi P, Zheng M, Pinkert S, Vecchi G, Ciryam P, et al. Widespread proteome remodeling and aggregation in aging C. elegans. Cell. 2015;161(4):919–32.
70. Höhn A, Weber D, Jung T, Ott C, Hugo M, Kochlik B, et al. Happily (n)ever after: aging in the context of oxidative stress, proteostasis loss and cellular senescence. Redox Biol. 2017;11:482–501.
71. Vernace VA, Arnaud L, Schmidt-Glenewinkel T, Figueiredo-Pereira ME. Aging perturbs 26S proteasome assembly in Drosophila melanogaster. FASEB J. 2007 Sep;21(11):2672–82.
72. Ghazi A, Henis-Korenblit S, Kenyon C. Regulation of Caenorhabditis elegans lifespan by a proteasomal E3 ligase complex. Proc Natl Acad Sci U S A. 2007;104(14):5947–52.
73. Chondrogianni N, Voutetakis K, Kapetanou M, Delitsikou V, Papaevgeniou N, Sakellari M, et al. Proteasome activation: an innovative promising approach for delaying aging and retarding age-related diseases. Ageing Res Rev. 2015;23:37–55.
74. Harris H, Rubinsztein DC. Control of autophagy as a therapy for neurodegenerative disease. Nat Rev Neurol. 2011;8(2):108–17.
75. Johnson JE, Johnson FB. Methionine restriction activates the retrograde response and confers both stress tolerance and lifespan extension to yeast, mouse and human cells. PLoS One. 2014;9(5):e97729.
76. Rattan SIS. Hormesis in aging. Ageing Res Rev. 2008 Jan;7(1):63–78.
77. F a C W, de S a H P, Boers-Trilles VE, AMA S. Hormesis and cellular quality control: a possible explanation for the molecular mechanisms that underlie the benefits of mild stress. Dose Response. 2012;11(3):413–30.
78. Hipkiss AR. Accumulation of altered proteins and ageing: causes and effects. Exp Gerontol. 2006;41(5):464–73.
79. Hipkiss AR. On why decreasing protein synthesis can increase lifespan. Mech Ageing Dev. 2007;128(5–6):412–4.

p53 and Aging

Nilambra Dogra and Tapas Mukhopadhyay

Introduction

Senescence can be defined as a stress response in which cells withdraw from the normal cell cycle and lose the capability to proliferate [1, 2]. Senescent cells can be distinguished on the basis of typical morphological characteristics like a flat and enlarged shape and increased expression of established senescence biomarkers. These include positive staining for β-galactosidase at pH 6.0 (senescence-associated-β-gal or SA-β-gal), decreased replication, and elevated expression of p53, p21, p16, and other cyclin-dependent kinase inhibitors, such as p27 and p15 [2–4]. Initially considered a cell culture artifact, senescence was afterwards observed in vivo in cancer lesions and during physiological aging [3, 5–8]. As senescence suppresses cell proliferation, it is likely to protect against cancer onset. Hence, in recent times, there has been considerable interest in senescence pathways as a promising cancer therapy approach.

As expected, senescence is closely related to aging as both limit lifespan. Senescent cells accumulate during the course of constant regeneration of somatic tissues. This accumulation of senescent cells gradually limits tissue renewal and affects tissue homeostasis, ultimately resulting in aging. Senescence has been characteristically associated with a state of permanent growth arrest, with the cells being unable to reenter the cell cycle. Although this concept is still widely accepted, recent studies have provided evidence that under certain conditions this cellular status is reversible. Stable suppression or changes in p53 expression in senescent fibroblasts

---

N. Dogra
Department of Experimental Medicine and Biotechnology, Post Graduate Institute of Medical Education and Research, Chandigarh, India

T. Mukhopadhyay (✉)
National Centre for Human Genome Studies and Research, Panjab University, Chandigarh, India
e-mail: tmukhopa@hotmail.com

© Springer Nature Singapore Pte Ltd. 2020
P. C. Rath (ed.), *Models, Molecules and Mechanisms in Biogerontology*,
https://doi.org/10.1007/978-981-32-9005-1_5

lead to cell-cycle reentry and immortalization, indicating that both initiation and maintenance of senescence are p53 dependent [5, 9].

The p53 gene, arguably the most widely studied tumor suppressor gene, is aptly termed "guardian of the genome" alluding to the critical role it plays in tumor suppression. p53 initiates stress responses by regulating a number of its target genes in response to diverse stress signals, including DNA damage, hyperproliferative signals, hypoxia, oxidative stress, ribonucleotide depletion, and nutrient starvation. In response to these varied stresses, p53 triggers cell-cycle arrest, apoptosis, and/or senescence. p53 is the most frequently mutated gene in human tumors with mutations occurring in more than 50% of all tumors and disruption of normal p53 function is often associated with the development and progression of tumors. p53 is highly conserved and its orthologs have been identified in worms (*Caenorhabditis elegans*), flies (*Drosophila*), zebrafish, frogs, and other species.

The p53 protein contains two amino-terminal transcriptional activation domains (TADs), a proline-rich domain (PRD), a DNA-binding domain (DBD), a tetramerization domain (TET), and a carboxy-terminal region rich in basic residues. Inactivation of p53 in human tumors typically occurs through missense mutations in the DBD of the p53 protein.

In response to stress signals, p53 becomes functionally active and initiates different tumor suppressor responses depending upon the trigger. These include a transient cell-cycle arrest, cell death (apoptosis), or cellular senescence. Although both apoptosis and cellular senescence are advantageous in that they irreversibly prevent damaged cells from undergoing neoplastic transformation, they can also diminish the pool of renewable proliferation-competent progenitor or stem cells. Such depletion can result in degenerative features of aging.

p53 is clearly involved in cancer, but the existence of p53 in short-lived organisms that do not develop cancers, such as flies and worms, suggests that tumor suppression is not its only, and probably, original function. Indeed, recent studies have shown that p53 influences development, reproduction, metabolism, and longevity. This chapter will focus on the role of p53 in cellular senescence and aging.

Molecular Mechanisms of Senescence and Aging

Aging is an extremely familiar yet poorly understood aspect of human biology. It is characterized by a progressive loss of physiological integrity, leading to impaired function and increased susceptibility to death. The process of cellular senescence was first described by Hayflick and Moorhead (1961) when they observed a state of irreversible growth arrest in normal human fibroblasts after serial passaging in in vitro culture conditions, whereas cancer cells continued to proliferate indefinitely [10]. This observation led them to speculate the existence of cellular factors which could be responsible for the limited proliferation of normal cells. Further, they contemplated the association of these factors with the process of organismal aging. Now cellular senescence is increasingly well understood at the molecular level as a response triggered by a number of mechanisms such as telomere shortening. The mechanisms contributing to cellular senescence are also involved in protection against cancer and may be responsible for aging.

The process of aging is characterized by distinct features which include genomic instability, telomere shortening, epigenetic modifications, deregulated protein homeostasis, hampered nutrient sensing, mitochondrial dysfunction, cellular senescence, stem-cell depletion, and affected intracellular communication. Several molecular pathways have been implicated in the aging process as discussed below.

Insulin/IGF-1 Signaling

The first pathway shown to be involved in aging was the insulin/insulin-like growth factor 1 (IGF-1) pathway [11]. Inhibiting insulin/IGF-1 signaling alters lifespan through changes in gene expression caused by transcription factors such as forkhead box O (FOXO), heat shock factor 1 (HSF-1), protein skinhead 1 (SKN-1), and nuclear respiratory factor (Nrf)-like xenobiotic-response factor. These transcription factors are involved in the regulation of diverse genes that collectively act to produce significant effects on lifespan. Downstream genes shown to be functionally relevant in aging include stress-response genes such as catalases, glutathione S-transferases, and metallothioneins, genes encoding antimicrobial peptides, chaperones, apolipoproteins, lipases [12], and channels [13]. In *C. elegans,* mutations that decrease the activity of *daf-2,* which encodes a hormone receptor similar to the insulin and IGF-1 receptor, extend the lifespan of the animal significantly. The same effect is observed in case of mutations affecting the downstream phosphatidylinositol 3-kinase (PI(3)K)/RAC-alpha serine/threonine kinase (AKT)/phosphoinositide-dependent kinase 1 (PDK) cascade. Surprisingly, the same genes that are normally expressed in the germ line show altered expression in the somatic tissues of *daf-2* mutants, where they contribute to longevity [14]. This indicates the involvement of germline genes in extending organismal lifespan. However, even though the germline lineage is immortal, unfertilized germ cells that express these germline genes do age, along with the rest of the animal [15]. Thus, germline genes may be only partially responsible for the longevity mechanism. Autophagy, the process of recycling cellular organelles, is involved in the inhibition of *daf-2* leading to extended lifespan [16]. Hence, autophagy is likely to be a potent antiaging mechanism.

The effect of insulin/IGF-1 pathway on the lifespan has been evolutionarily conserved [11, 17]. In *Drosophila*, inhibition of insulin/IGF-1 signaling or increased activity of FOXO in adipose tissue results in increased lifespan. Similarly, in mice, an inverse correlation between exists between IGF-1 levels and lifespan [18]. Also, mutations that inhibit the insulin receptor [17], the IGF-1 receptor [17, 19], upstream regulators [17, 19], or downstream effectors [17, 20], have all been reported to result in an extended lifespan. Interestingly, small dogs, with a mutation resulting in decreased IGF-1 levels, live longer than large dogs. Therefore, it seems likely that a mutation causing reduction in the insulin/IGF-1 signaling leads to activated DAF-16/FOXO-like proteins which consequently result in a physiological shift toward cell maintenance and longevity. Consistent with this theory, mutations in IGF-1 receptor have been reported in a cohort of Ashkenazi Jewish centenarians [21] and linked to longevity in a Japanese cohort [22]. Variants of AKT and FOXO3A have also been linked to longevity [23]. The FOXO3A cohorts include Hawaiians of Japanese descent [24], Italians [25], Ashkenazi Jews [23], Californians [23], New Englanders [23], Germans [26], and Chinese [27]. In the German cohort, the FOXO3A variants were found to be more frequent in centenarians than in 90 year olds, further strengthening the assumption that these variants extend lifespan. FOXO1 gene variants have also been linked to longevity in American and Chinese cohorts [27, 28]. Hence, FOXO variants seem to be consistently associated with longevity by affecting many cellular pathways [11].

Because the insulin/IGF-1 pathway senses nutrients, it could be involved in mediating the longevity response to dietary restriction (DR) [29]. In worms, alternate-day feeding has been shown to extend lifespan by inhibiting insulin/IGF-1 signaling [30]. Similarly, in flies, insulin/IGF-1 pathway mutants show enhanced response to DR [31, 32], suggesting that the two pathways may overlap [13].

Hence, evolution of longevity can be explained by these findings that lifespan can be increased by pathways that shift physiology toward cell protection and maintenance. This explanation does not indicate selection for longevity, as these pathways could have evolved simply for survival under harsh, unfavorable environments, however, once established, they would naturally extend lifespan by counteracting internal metabolic wear and tear that accelerates aging.

TOR Signaling

The target of rapamycin (TOR) kinase is a major amino-acid and nutrient sensor that stimulates growth and inhibits recovery pathways such as autophagy when food is abundant. Inhibiting the TOR pathway has been reported to increase lifespan in many species [33–37]. TOR inhibition increases resistance to environmental stress [38], simultaneously causing a physiological shift toward tissue maintenance. However, TOR inhibition appears to extend lifespan independently of DAF-16/FOXO3 [36, 37]. TOR modulates translation, according to the nutrient levels. Inhibition of translation by other methods also has been shown to extend lifespan [34, 38–42]. TOR inhibition also encourages autophagy, which, as in the case of mutations in insulin/IGF-1 pathway, leads to an extended lifespan [43–45]. TOR pathway is known to be linked to dietary restriction and mimics its physiological effects [33, 34, 38, 44]. In addition to translational control, the TOR pathway acts via the transcription factor SKN-1 [20]. Different diets are known to trigger different longevity responses, for example, reduced methionine or protein intake extends lifespan in mammals and flies, respectively. A low-glycemic-index diet or amino-acid limitation could trigger the response to low insulin/IGF-1 or TOR signaling [13]. Therefore, the possibility of determining the kind of diet that could lead to youthfulness and longevity is not far flung.

AMP Kinase

AMP kinase is a nutrient and energy sensor that modulates catabolic and anabolic pathways depending upon the cell's AMP/ATP (adenosine triphosphate) ratio. Overexpression of AMP kinase extends lifespan in *C. elegans* [46]. The same effect is observed in mice after administration of the antidiabetic drug metformin, which is known to activate AMP kinase [47]. AMP kinase has also been reported as a mediator for insulin/IGF-1 mutations to exert their effect on the worm lifespan [46]. AMP kinase is also involved in lifespan extension in response to dietary restriction. It appears to act directly on DAF-16/FOXO to phosphorylate and activate it [13, 48]. However, the AMP kinase and dietary restriction work independently of each other to extend lifespan under continuous low availability of food [46, 49, 50].

Altogether, the insulin/IGF-1, TOR, and AMP kinase pathways are all interlinked and may work in conjunction with or independent of each other to regulate the process of aging.

Sirtuins

Sirtuins are nicotinamide adenine dinucleotide (NAD)+-dependent protein deacetylases which have been linked to longevity in yeast, worms, and flies, however, their mode of action is not yet clear [11, 51]. One of the proposed possibilities of lifespan extension by Sir2 is the inhibition of the formation of toxic extrachromosomal ribosomal DNA circles. Another report suggests that Sir2 extends lifespan by maintaining gene silencing at telomeres during aging [52]. It has also been proposed to act through DAF-16/FOXO as mammalian SIRT1 is known to deacetylate FOXO proteins in response to oxidative stress, consequently shifting their target specificity toward genes involved in stress resistance [11, 13, 53]. However, sirtuin overexpression has not yet been reported to extend lifespan in mammals.

Inhibition of Respiration

A modest inhibition of respiration extends lifespan in a number of species including yeast, worms, flies, and mice [11, 54–57]. Perhaps, larger mammals, in general, live longer partly because of their lower metabolic rates. Inhibiting respiration activates a conserved gene expression response called the "retrograde response." The retrograde response activates alternative energy-generating as well as cell-protection pathways. Mutations that impair the retrograde response or affect individual retrograde-response genes can hamper this lifespan extension [58, 59]. Thus, inhibiting respiration may be one of the mechanisms triggering a regulated longevity response. In worms, respiration is inhibited during development in order to increase lifespan [11, 60], creating a molecular memory of the event. A similar situation might be occurring in mice [57]. Also, in flies, inhibiting respiration in neurons alone extends lifespan [54], again suggesting a regulated response [13].

ROS

Intracellular reactive oxygen species (ROS) levels are a major determining factor of cellular senescence [2, 61]. ROS are generally small, short-lived, and highly reactive molecules (e.g., oxygen anions, superoxide and hydroxyl radicals, and peroxides) formed by partial reduction of oxygen, which, if not detoxified promptly by antioxidant agents, can oxidize macromolecules and damage organelles (Fig. 5.2). Enhanced ROS production and a reduced antioxidant response are contributors to the aging process by oxidative modification of different macromolecules such as lipids, proteins, and genomic DNA [62–65]. Mitochondria, being the major intracellular source of ROS, have been implicated in oxygen radicals mediated oxidative damage eventually leading to aging. This theory has been widely accepted as the "mitochondrial free radical theory of aging." [66, 67] The mitochondrial ROS theory of aging proposes a feedback loop wherein elevated ROS levels cause senescence and aging, also resulting in damage to mitochondria, and these damaged mitochondria, in turn, produce more ROS [68–70]. Several studies have shown a negative correlation between mitochondrial ROS production and lifespan in different organisms. [71, 72]

Telomeres

Telomeres are repetitive TTAGGG sequences that protect chromosome ends by preventing them from being recognized as DNA damage [73]. Usually, human cells lack adequate levels of the telomerase enzyme which is required to maintain telomeres, and this results in gradual telomere shortening with each round of replication [74–76]. The importance of telomere length in aging was initially inferred from the work of Hayflick and Moorhead carried out in primary human fibroblasts [77]. These cells undergo a finite number of divisions, undergoing telomere shortening with each consequent passage and eventually enter a state of senescence. Telomerase reactivation results in elongation of telomeres, which, consequently, enables fibroblasts to grow indefinitely without going into senescence [78]. Since shortening of telomeres is associated with progressive aging, they have been considered as candidates for aging determination. Mice engineered to have longer telomeres are known to live longer than their normal counterparts [79]. However, for longevity, these mice require additional genetic modifications to evade cancer. Therefore, this mode of lifespan extension differs from other pathways like dietary restriction, alterations in nutrient sensors, and reduced respiration which themselves inhibit tumor formation [11, 80]. These facts suggest that unlike other mechanisms, telomere elongation does not increase lifespan by causing a protective physiological state. Instead, it may do so for other reasons like preventing stem-cell loss [13, 81, 82]. It would be of interest to know whether lifespans could be greatly extended if telomere lengthening were combined with dietary restriction or mutations in nutrient sensors.

p53 in Senescence and Aging

The tumor suppressor p53 functions as a transcription factor involved in the control of cell-cycle, DNA repair, apoptosis, and other cellular stress responses (Fig. 5.1). The role of p53 in cancer has been extensively studied, but the presence of p53 in short-lived organisms that do not develop cancer, such as flies and worms, suggests that its function is not limited to tumor suppression. Recent studies have shown that p53 influences development, reproduction, metabolism, and longevity [83–88]. p53 is also known to modulate cellular senescence and organismal aging. Senescence is a state of irreversible cell-cycle arrest that has a crucial role both in aging as well as an anticancer response, which protects cells from oncogenic insults. Therefore, depending upon the cellular context, senescence is one of the mechanisms by which p53 counteracts tumor growth.

The first evidence linking p53 to aging came from the analysis of a mutant mouse model: Tyson and colleagues obtained an aberrant serendipitous truncation of the N-terminal portion of p53 in an attempt to develop a knock-in (KI) model of the gene. Surprisingly, the truncated mutant protein resulted in an elevated constitutive p53 activity and the mutant mice manifested numerous aging-related features and drastically reduced lifespan. Consequently, a transgenic mouse model

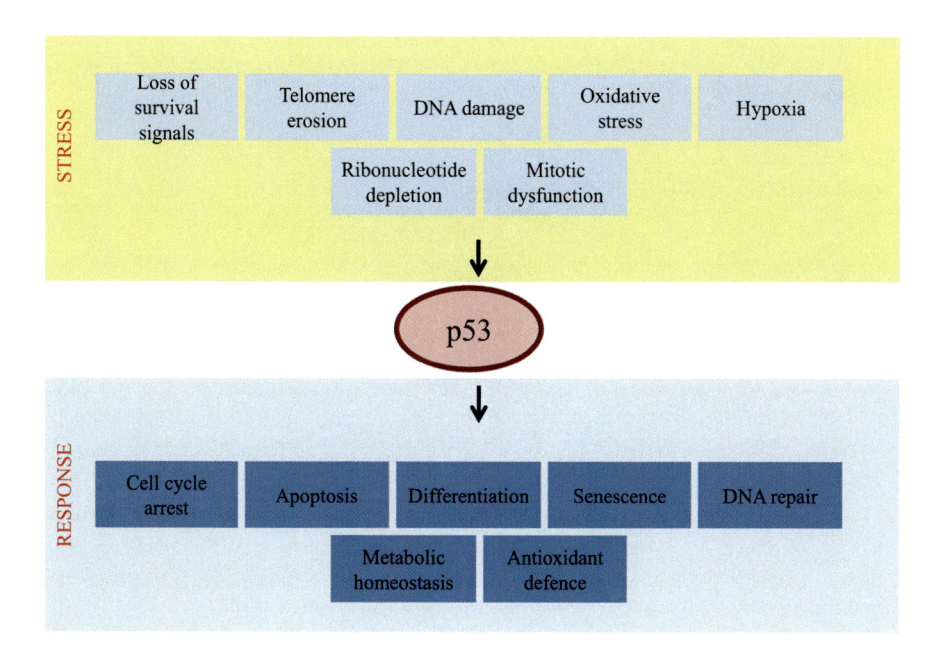

Fig. 5.1 The p53 stress response

overexpressing the truncated dominant negative (DNp53) or p44 isoform of p53 showed defective growth, reduced lifespan, and accelerated aging [75, 89]. Interestingly, p44 overexpression resulted in hyperactive p53 and increased IGF signaling, the known master regulator of aging. Further, a KI mouse model of p53 mimicking constitutively active p53 showed prominent aging features, which appeared to result from extensive apoptosis affecting the stem-cell components of several organs, hence compromising tissue self-renewal [90]. This effect was further seen to be mediated by the p53 target PUMA [90]. Thus, widespread apoptosis of stem cells may mediate p53-mediated aging, in agreement with reports highlighting stem-cell involvement in the aging process [91]. Although the mechanisms underlining this accelerated aging are still unclear, these results suggest that excessive p53 activity might compromise healthy aging. Conversely, whether absent or reduced p53 activity affects lifespan has been difficult to address as loss of p53 is known to result in severe tumor phenotype [92]. Nonetheless, some in vivo models have shed light on the issue. Serine 15 (Ser-15) (Ser-18 in mouse) is known to be phosphorylated by ATM in response to DNA damage. KI mice in which Ser-18 of p53 was replaced with nonphosphorylable alanine developed signs of accelerated aging, indicating the protective role of p53 against aging-related damage [93, 94]. In addition, the super-(Arf)/p53 mouse model bearing long genomic sequence of p53 and p19(Arf), allowing their increased expression but maintaining endogenous regulation, showed an increase in lifespan and an overall improvement in the aging-related health decline [95, 96]. Overall, these findings suggest that loss of p53 leads to accelerated aging. To summarize, it can be said that normal physiological activity of p53 protects from cancer and aging, whereas excessive p53 activation acts as a tumor suppressor, but is detrimental to the normal aging process.

In the following section, we discuss p53-mediated regulation of several physiological pathways that could explain its role in cellular senescence and aging.

p53 and the IGF-1/mTOR Pathway

As discussed earlier, the mechanistic target of rapamycin (mTOR) pathway is the major environmental nutrient-sensing pathway and regulates several processes in response to nutrient levels including cell growth and protein translation. In the absence of glucose, AMP levels rise in the cell, which in turn activates AMP kinase and results in a downstream cascade finally causing the inactivation of S6 kinase and activation of eukaryotic translation initiation factor 4E (eIF4E)-binding protein (4EBP1), both of which act to slow down protein translation of mRNAs encoding the proteins involved in ribosomal and mitochondrial biogenesis, oxidative phosphorylation, and cell growth. Simultaneously, the absence of mTOR activity also results in the activation of autophagy, a catabolic process which degrades endogenous macromolecules to provide nutrients in times of scarcity [97–99]. p53 activation under stress negatively regulates the IGF-1/mTOR promotion of cell growth and division and positively regulates apoptosis and autophagy. Four p53-regulated gene products, phosphatase and tensin homolog (PTEN), insulin-like growth factor-1-binding protein 3 (IGF-1-BP3), tuberous sclerosis complex 2 (TSC-2), and the beta subunit of AMP kinase, negatively regulate the IGF-1 and mTOR pathways creating an interpathway crosstalk that permits the cell under stress to shut down the cell growth and division, nutritional sensing, and metabolic regulation for entry into the cell cycle [100, 101]. Induction of p21 by p53 causes an irreversible senescent arrest. However, further accumulation of transcriptionally active p53 triggers inhibition of mTOR and results in a reversible cell-cycle arrest [102, 103]. The p53-target(s) responsible for this phenotype is not yet unidentified, but the ability of p53 to induce cell-cycle arrest and inhibit mTOR simultaneously could help explain the effect of moderate increase of p53 activity in protection from cancer and simultaneously prolonged lifespan. Hence, p53-mediated regulation of senescence and aging is a complex cellular process involving modulation of several additional targets and pathways. To conclude, p53 activation in a time of stress negatively regulates the IGF-1/mTOR promotion of cell growth.

p53 and E2F7

E2F7 has been described as a p53-target involved in cell-cycle arrest and senescence [104, 105]. This gene is a member of the E2F-family of transcription factors, however, unlike canonical E2Fs, it does not heterodimerize with DP1 proteins, but binds DNA as a monomer and promotes repression of several E2F target genes, including E2F1 [106, 107]. Moreover, it acts as a repressor of many genes essential for mitosis, such as cyclin A, cyclin B, and cdc2/cdk1. Hence, E2F7 arrests cell-cycle progression at the mitotic phase and is a mediator of cellular senescence.

p53 and Autophagy

Autophagy is an evolutionarily conserved catabolic mechanism by which cytoplasmic portions and organelles are delivered to the lysosome via a double-membraned vesicle called autophagosome, for degradation and recycling. Degraded cellular components are then recycled for energy production or other metabolic processes, which explains the execution of autophagy under conditions of nutrient deprivation [108]. Overall, autophagy is a cytoprotective process and can modulate aging and cancer survival [109–111]. Besides macromolecule and organelle turnover, crosstalk between the longevity pathways and autophagic process is involved in the regulation of diverse cellular functions including growth and differentiation, cell death, oxidative stress as well as response to nutrient deprivation. Mutations in genes that promote autophagy result in reduced lifespan in *C. elegans*, *D. melanogaster*, and yeast [109, 112–115]. Moreover, as discussed earlier, induction of autophagy via repression of the mTOR signaling is essential for the antiaging outcome of dietary restriction [116, 117]. p53 has a dual role in the regulation of autophagy. While nuclear p53 can induce autophagy through transcriptional upregulation of targets such as 5'-AMP-activated kinase (AMPK), PTEN, sestrins, or damage-regulated autophagy modulator (DRAM), which codes for a lysosomal protein, cytoplasmic p53 represses autophagy through a largely unknown mechanism. Kroemer's group reported enhanced autophagy upon loss of p53 activity and importantly, their data showed that cytoplasmic and not nuclear p53 is responsible for autophagy inhibition. Inducers of autophagy, such as starvation or rapamycin, induce degradation of p53 [118–121]. Functional activity of p53 is known to decline during aging in mice and a diminished response of p53 to various stress signals and reduced apoptosis has also been linked to aging progression [122]. Reduced transcriptional activity of p53 could result in impaired mitochondrial respiration and, hence, enhanced glycolysis. These observations can also provide an explanation of increased tumor incidence in older populations [122]. The inefficiency of p53 function could also lead to a decrease in autophagic degradation, a cleansing mechanism of cells. The autophagic capacity of a cell is known to decrease during aging leading to the accumulation of damaged organelles in cytoplasm [123]. An impairment in p53 transcriptional function could also hamper the expression of DRAM, a p53-dependent inducer of autophagy [124] and lead to reduced autophagy during aging.

p53 and ROS

ROS-mediated damage has been extensively implicated in the induction of cellular senescence and in the onset of aging disorders. p53, by its ability to regulate ROS levels, shows a dual role in regulating senescence [72, 102, 125]. Emerging evidence suggests that p53 plays an important role in repressing senescence by reducing intracellular ROS levels through transcriptional regulation of its target antioxidant genes, for example, mitochondrial superoxide dismutase 2 (SOD2), glutathione peroxidase 1, and mammalian sestrin homologs 1and 2 [126–130]. On the contrary, p53 induced in response to DNA damage enhances cellular ROS content, resulting in cell death or senescence [72, 125, 131, 132]. Therefore, p53 has a dual role in regulating the senescence process. Based on the studies on mouse models, it may be inferred that sustained activation of p53, caused either by interference with Mdm2 regulation or by constitutive DNA instability, is detrimental and results in premature aging possibly by exhausting the renewal capacity of tissues [96, 133–135]. Conversely, Arf/p53 transgenic mouse model, in addition to conferring cancer protection, also shows antiaging activity, probably by reducing the buildup of age-associated damage [95, 136]. Thus, overall, genetic evidence in mice indicates that deregulated constitutive activation of p53 results in accelerated aging, whereas modest increase of regulated Arf/p53 activity is antiaging.

p53 and Mitochondria

Elevated level of oxidative damage is a major cause of aging. In this context, p53, which plays an important role in cellular response to oxidative stress, is speculated to be a crucial regulator of the aging process. p53 plays a dual role depending upon the level of oxidative stress. In response to low levels of oxidative stress, p53 acts as an antioxidant to eliminate oxidative stress and ensure cell survival, whereas in response to high stress levels, it behaves as a prooxidant causing further enhancement in the stress levels, eventually leading to cell death. These context-dependent responses are brought about by differential regulation of the genes involved in cellular responses to oxidative stress and also by modulation of other cellular pathways involved in the process. p53-target genes, like sestrin, glutathione peroxidase (GPX), and aldehyde dehydrogenase (ALDH), are involved in reducing oxidative stress [127, 129, 137]. Alternatively, p53 can also reduce intracellular ROS levels indirectly by regulating cellular metabolism. For example, TIGAR (TP53-induced glycolysis and apoptosis regulator) is a p53-inducible gene that hampers glycolysis and promotes the production of nicotinamide adenine dinucleotide phosphate (NAPDH) to decrease ROS levels [138]. In addition, p53 also reduces ROS production by suppressing the expression of phosphoglycerate mutase (PGM), consequently, causing reduction of pyruvate required for oxidative respiration in mitochondria [139]. Conversely, in response to high levels of oxidative stress, p53 displays prooxidant function by upregulating prooxidative genes such as p53-induced gene-3 (PIG3) and proline oxidase [126, 140]. Overexpression of these genes enhances the levels of oxidative stress. At the same time, p53 induces the expression of Bcl-2-associated X protein (BAX) and p53 upregulated modulator of apoptosis (PUMA), which are proapoptotic genes [132, 141]. p53 can also cause elevated oxidative stress levels by inhibiting the expression of antioxidant genes like SOD2 and Nrf2 [128, 142–144]. The precise mechanism that controls the p53 pro- or antioxidant functions remains unclear, but it could explain the association of increased p53 activity with accelerated aging and prolonged lifespan in mice. Overall, a balanced pro- and antioxidant response of p53 acting to mitigate the accumulation of oxidative stress and DNA damage might be a key factor for longevity [145].

p53 and NF-kB

A number of reports have linked p53 and NF-κB crosstalk to aging. Since p53 is known to suppress NF-κB signaling through the inhibition of glycolysis [146, 147], this points toward the possibility of NF-κB signaling involvement in the regulation of the aging process by p53. A number of reports have highlighted the role of NF-κB in inducing proinflammatory aging associated changes in tissues [148]. This could, in turn, elevate glycolysis levels, since cytokines are well-established inducers of the glycolytic pathway [149–152]. In particular, macrophage migratory inhibitory factor (MIF) activates glycolysis by increasing the expression of phosphofructokinase (PFK-2). MIF can also interact with and suppress the function of p53 as well as prevent its nuclear translocation [153]. This is an interesting observation in the context of aging since cytoplasmic p53 can repress the autophagic degradation. There are various reports suggesting that NF-κB signaling increases with aging [154–157]. There appear to be several mechanisms regulating the NF-κB signaling

[123, 158, 159]. The Sirtuins SIRT1 and SIRT6 are strong inhibitors of NF-κB signaling. Sir2 proteins are known to extend the lifespan in numerous species [160]. On the other hand, free radicals and oxidative stress, which are increased during aging, are effective NF-κB signaling activators [161]. NF-κB signaling is involved in several aging-associated pathways, such as apoptosis, autophagy, muscle atrophy, immunosenescence, and inflammatory responses [123].

p53 and Sirtuins

Sirt1 is a class III histone deacetylase, having the ability to deacetylate target histone and nonhistone proteins. It can therefore regulate the chromatin structure and DNA accessibility, and also transcriptional control through deacetylation of transcription factors and cofactors. SIRT1 is crucial for the occurrence of senescence and is strongly downregulated in senescent cells [162, 163]. The regulation of p53 by SIRT1 has been implicated in senescence [164–170] and SIRT1 is strongly downregulated in senescent cells. A major substrate for SIRT1 is p53, and the deacetylation of p53 regulates cell cycle, cellular senescence, and stress resistance in various cell types. Deacetylation inhibits p53's ability to transcriptionally activate some, but not all, target genes—including those involved in apoptosis, proliferation, ROS production, and presumably also senescence [164, 171, 172]. Deacetylation of p53 by SIRT1 has been shown to impede the induction of senescence [173].

p63 and p73

Two p53 homologs, p63 and p73, have been characterized over the past two decades [174]. Although there have been several studies to unravel specific functions of these proteins, the exact roles for these different variants are still unknown. Nonetheless, like p53, both p63 and p73 genes have a role in senescence and aging. p63 null mice die shortly after birth [175, 176], whereas premature aging could be detected in heterozygous mice [177]. In inducible p63-KO (knockout) mice, depletion of p63 in the epithelium could accelerate aging, with increased senescence markers in vivo and in vitro [178]. Promyelocytic leukemia protein (PML) was found to be the major mediator of senescence caused through p63 depletion [178–180]. Flores et al. also showed that p63 isoform specific KO mice demonstrated premature aging and reduced lifespan. This could be correlated with genomic instability induced cellular senescence, which involves loss of the epithelial stem-cell population [181, 182]. However, there are additional studies suggesting that p63 itself mediates the induction of senescence independently of p53 [183]. Hence, it appears that both p63 and p53 have a dual role: In response to oncogenic stress their activation is crucial to stop transformation via senescence induction, and their absence depletes the stem-cell reservoir causing accelerated aging.

p73 has also been linked to senescence and the expression of ΔNp73 has been reported to overcome Ras-induced senescence, thereby allowing cellular transformation to occur [184]. Ras, in turn, promotes a switch from TAp73 to ΔNp73 expression in order to sustain transformation. Isoform-specific TAp73 KO models show enhanced aging-related features [185, 186]. Accelerated aging phenotype has been ascribed to mitochondrial and metabolic dysfunction ensuing as a result of loss of TAp73 in KO animals [186].

Therefore, it appears that p53 along with its family members p63 and p73 plays an important role in the regulation of senescence and aging. Altogether, this complex crosstalk interrelates stress, longevity, control over metabolic networks and pathways involved in tumor suppression and diabetes [100, 101, 187].

Conclusion

The molecular pathways involved in senescence are becoming increasingly relevant owing to its role in tumor suppression, and the possibility of its potential exploitation in cancer therapy. Many of the molecular mechanisms resulting in senescence and aging are still poorly understood, however, it has been established that the tumor suppressor p53 plays a key role in their regulation (Fig. 5.2). p53 is now known to modulate senescence at different levels and can act dually, either promoting or inhibiting the senescence program. The cause underlying this dual effect is still relatively unclear, but as is the case with its other functions, a possible explanation can be the context dependence, that is, the level and form of stress or the cellular milieu. Relatively low stress levels result in p53-mediated reparative and antioxidant mechanisms, while elevated stress leads to apoptosis and senescence, via ROS production. Since p53-mediated signaling can be manipulated to stimulate senescence, it is emerging as an alternative new therapeutic approach to eliminate cancerous cells.

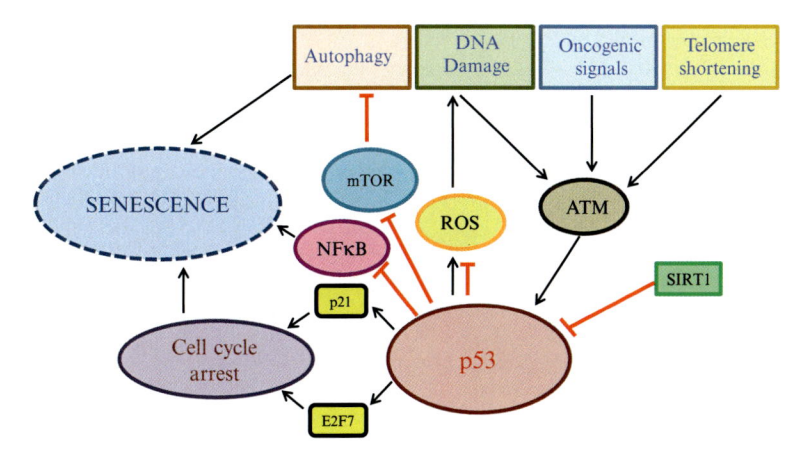

Fig. 5.2 Senescence regulation by p53

References

1. Sherwood SW, Rush D, Ellsworth JL, Schimke RT. Proc Natl Acad Sci U S A. 1988;85:9086–90.
2. Kuilman T, Michaloglou C, Mooi WJ, Peeper DS. Genes Dev. 2010;24:2463–79.
3. Romagosa C, Simonetti S, Lopez-Vicente L, Mazo A, Lleonart ME, Castellvi J, Ramon y Cajal S. Oncogene. 2011;30:2087–97.
4. Alessio N, Squillaro T, Cipollaro M, Bagella L, Giordano A, Galderisi U. Oncogene. 2010;29:5452–63.
5. Collado M, Gil J, Efeyan A, Guerra C, Schuhmacher AJ, Barradas M, Benguria A, Zaballos A, Flores JM, Barbacid M, Beach D, Serrano M. Nature. 2005;436:642.
6. Krishnamurthy J, Torrice C, Ramsey MR, Kovalev GI, Al-Regaiey K, Su L, Sharpless NE. J Clin Invest. 2004;114:1299–307.
7. Sharpless NE. Exp Gerontol. 2004;39:1751–9.
8. Caldwell ME, DeNicola GM, Martins CP, Jacobetz MA, Maitra A, Hruban RH, Tuveson DA. Oncogene. 2012;31:1599–608.
9. Beausejour CM, Krtolica A, Galimi F, Narita M, Lowe SW, Yaswen P, Campisi J. EMBO J. 2003;22:4212–22.
10. Hayflick L, Moorhead PS. Exp Cell Res. 1961;25:585–621.
11. Kenyon C. Cell. 2005;120:449–60.
12. Wang MC, O'Rourke EJ, Ruvkun G. Science. 2008;322:957–60.
13. Kenyon CJ. Nature. 2010;464:504–12.
14. Curran SP, Wu X, Riedel CG, Ruvkun G. Nature. 2009;459:1079–84.
15. Garigan D, Hsu AL, Fraser AG, Kamath RS, Ahringer J, Kenyon C. Genetics. 2002;161:1101–12.
16. Melendez A, Talloczy Z, Seaman M, Eskelinen EL, Hall DH, Levine B. Science. 2003;301:1387–91.
17. Bartke A. Cell Cycle. 2008;7:3338–43.
18. Yuan R, Tsaih SW, Petkova SB, Marin de Evsikova C, Xing S, Marion MA, Bogue MA, Mills KD, Peters LL, Bult CJ, Rosen CJ, Sundberg JP, Harrison DE, Churchill GA, Paigen B. Aging Cell. 2009;8:277–87.
19. Kappeler L, De Magalhaes Filho C, Dupont J, Leneuve P, Cervera P, Perin L, Loudes C, Blaise A, Klein R, Epelbaum J, Le Bouc Y, Holzenberger M. PLoS Biol. 2008;6:e254.
20. Selman C, Tullet JM, Wieser D, Irvine E, Lingard SJ, Choudhury AI, Claret M, Al-Qassab H, Carmignac D, Ramadani F, Woods A, Robinson IC, Schuster E, Batterham RL, Kozma SC, Thomas G, Carling D, Okkenhaug K, Thornton JM, Partridge L, Gems D, Withers DJ. Science. 2009;326:140–4.
21. Suh Y, Atzmon G, Cho MO, Hwang D, Liu B, Leahy DJ, Barzilai N, Cohen P. Proc Natl Acad Sci U S A. 2008;105:3438–42.
22. Kojima T, Kamei H, Aizu T, Arai Y, Takayama M, Nakazawa S, Ebihara Y, Inagaki H, Masui Y, Gondo Y, Sakaki Y, Hirose N. Exp Gerontol. 2004;39:1595–8.
23. Pawlikowska L, Hu D, Huntsman S, Sung A, Chu C, Chen J, Joyner AH, Schork NJ, Hsueh WC, Reiner AP, Psaty BM, Atzmon G, Barzilai N, Cummings SR, Browner WS, Kwok PY, Ziv E. Aging Cell. 2009;8:460–72.
24. Willcox BJ, Donlon TA, He Q, Chen R, Grove JS, Yano K, Masaki KH, Willcox DC, Rodriguez B, Curb JD. Proc Natl Acad Sci U S A. 2008;105:13987–92.
25. Anselmi CV, Malovini A, Roncarati R, Novelli V, Villa F, Condorelli G, Bellazzi R, Puca AA. Rejuvenation Res. 2009;12:95–104.
26. Flachsbart F, Caliebe A, Kleindorp R, Blanche H, von Eller-Eberstein H, Nikolaus S, Schreiber S, Nebel A. Proc Natl Acad Sci U S A. 2009;106:2700–5.
27. Li Y, Wang WJ, Cao H, Lu J, Wu C, Hu FY, Guo J, Zhao L, Yang F, Zhang YX, Li W, Zheng GY, Cui H, Chen X, Zhu Z, He H, Dong B, Mo X, Zeng Y, Tian XL. Hum Mol Genet. 2009;18:4897–904.

28. Lunetta KL, D'Agostino RB Sr, Karasik D, Benjamin EJ, Guo CY, Govindaraju R, Kiel DP, Kelly-Hayes M, Massaro JM, Pencina MJ, Seshadri S, Murabito JM. BMC Med Genet. 2007;8(Suppl 1):S13.
29. Kenyon C, Chang J, Gensch E, Rudner A, Tabtiang R. Nature. 1993;366:461–4.
30. Honjoh S, Yamamoto T, Uno M, Nishida E. Nature. 2009;457:726–30.
31. Clancy DJ, Gems D, Hafen E, Leevers SJ, Partridge L. Science. 2002;296:319.
32. Grandison RC, Piper MD, Partridge L. Nature. 2009;462:1061–4.
33. Kaeberlein M, Powers RW 3rd, Steffen KK, Westman EA, Hu D, Dang N, Kerr EO, Kirkland KT, Fields S, Kennedy BK. Science. 2005;310:1193–6.
34. Kapahi P, Zid BM, Harper T, Koslover D, Sapin V, Benzer S. Curr Biol. 2004;14:885–90.
35. Harrison DE, Strong R, Sharp ZD, Nelson JF, Astle CM, Flurkey K, Nadon NL, Wilkinson JE, Frenkel K, Carter CS, Pahor M, Javors MA, Fernandez E, Miller RA. Nature. 2009;460:392–5.
36. Jia K, Chen D, Riddle DL. Development. 2004;131:3897–906.
37. Vellai T, Takacs-Vellai K, Zhang Y, Kovacs AL, Orosz L, Muller F. Nature. 2003;426:620.
38. Hansen M, Taubert S, Crawford D, Libina N, Lee SJ, Kenyon C. Aging Cell. 2007;6:95–110.
39. Hamilton B, Dong Y, Shindo M, Liu W, Odell I, Ruvkun G, Lee SS. Genes Dev. 2005;19:1544–55.
40. Pan KZ, Palter JE, Rogers AN, Olsen A, Chen D, Lithgow GJ, Kapahi P. Aging Cell. 2007;6:111–9.
41. Syntichaki P, Troulinaki K, Tavernarakis N. Nature. 2007;445:922–6.
42. Steffen KK, MacKay VL, Kerr EO, Tsuchiya M, Hu D, Fox LA, Dang N, Johnston ED, Oakes JA, Tchao BN, Pak DN, Fields S, Kennedy BK, Kaeberlein M. Cell. 2008;133:292–302.
43. Hansen M, Chandra A, Mitic LL, Onken B, Driscoll M, Kenyon C. PLoS Genet. 2008;4:e24.
44. Bjedov I, Toivonen JM, Kerr F, Slack C, Jacobson J, Foley A, Partridge L. Cell Metab. 2010;11:35–46.
45. Toth ML, Sigmond T, Borsos E, Barna J, Erdelyi P, Takacs-Vellai K, Orosz L, Kovacs AL, Csikos G, Sass M, Vellai T. Autophagy. 2008;4:330–8.
46. Apfeld J, O'Connor G, McDonagh T, DiStefano PS, Curtis R. Genes Dev. 2004;18:3004–9.
47. Anisimov VN, Berstein LM, Egormin PA, Piskunova TS, Popovich IG, Zabezhinski MA, Tyndyk ML, Yurova MV, Kovalenko IG, Poroshina TE, Semenchenko AV. Cell Cycle. 2008;7:2769–73.
48. Greer EL, Dowlatshahi D, Banko MR, Villen J, Hoang K, Blanchard D, Gygi SP, Brunet A. Curr Biol. 2007;17:1646–56.
49. Lakowski B, Hekimi S. Proc Natl Acad Sci U S A. 1998;95:13091–6.
50. Greer EL, Brunet A. Aging Cell. 2009;8:113–27.
51. Kaeberlein M, Powers RW 3rd. Ageing Res Rev. 2007;6:128–40.
52. Dang W, Steffen KK, Perry R, Dorsey JA, Johnson FB, Shilatifard A, Kaeberlein M, Kennedy BK, Berger SL. Nature. 2009;459:802–7.
53. Berdichevsky A, Viswanathan M, Horvitz HR, Guarente L. Cell. 2006;125:1165–77.
54. Copeland JM, Cho J, Lo T Jr, Hur JH, Bahadorani S, Arabyan T, Rabie J, Soh J, Walker DW. Curr Biol. 2009;19:1591–8.
55. Dell'agnello C, Leo S, Agostino A, Szabadkai G, Tiveron C, Zulian A, Prelle A, Roubertoux P, Rizzuto R, Zeviani M. Hum Mol Genet. 2007;16:431–44.
56. Kayser EB, Sedensky MM, Morgan PG, Hoppel CL. J Biol Chem. 2004;279:54479–86.
57. Lapointe J, Hekimi S. J Biol Chem. 2008;283:26217–27.
58. Cristina D, Cary M, Lunceford A, Clarke C, Kenyon C. PLoS Genet. 2009;5:e1000450.
59. Kirchman PA, Kim S, Lai CY, Jazwinski SM. Genetics. 1999;152:179–90.
60. Rea SL, Ventura N, Johnson TE. PLoS Biol. 2007;5:e259.
61. Atzmon G, Cho M, Cawthon RM, Budagov T, Katz M, Yang X, Siegel G, Bergman A, Huffman DM, Schechter CB, Wright WE, Shay JW, Barzilai N, Govindaraju DR, Suh Y. Proc Natl Acad Sci U S A. 2010;107(**Suppl 1**):1710–7.
62. Chen JH, Hales CN, Ozanne SE. Nucleic Acids Res. 2007;35:7417–28.
63. Lenaz G. Biochim Biophys Acta. 1998;1366:53–67.
64. Linnane AW, Marzuki S, Ozawa T, Tanaka M. Lancet. 1989;1:642–5.

65. Toescu EC, Myronova N, Verkhratsky A. Cell Calcium. 2000;28:329–38.
66. Harman D. J Gerontol. 1956;11:298–300.
67. Harman D. J Am Geriatr Soc. 1972;20:145–7.
68. Balaban RS, Nemoto S, Finkel T. Cell. 2005;120:483–95.
69. Gough DR, Cotter TG. Cell Death Dis. 2011;2:e213.
70. Lemarie A, Huc L, Pazarentzos E, Mahul-Mellier AL, Grimm S. Cell Death Differ. 2011;18:338–49.
71. Lambert AJ, Boysen HM, Buckingham JA, Yang T, Podlutsky A, Austad SN, Kunz TH, Buffenstein R, Brand MD. Aging Cell. 2007;6:607–18.
72. Lu T, Finkel T. Exp Cell Res. 2008;314:1918–22.
73. Millis AJ, Hoyle M, McCue HM, Martini H. Exp Cell Res. 1992;201:373–9.
74. Ventura A, Kirsch DG, McLaughlin ME, Tuveson DA, Grimm J, Lintault L, Newman J, Reczek EE, Weissleder R, Jacks T. Nature. 2007;445:661–5.
75. Maier B, Gluba W, Bernier B, Turner T, Mohammad K, Guise T, Sutherland A, Thorner M, Scrable H. Genes Dev. 2004;18:306–19.
76. Kang MK, Kameta A, Shin KH, Baluda MA, Kim HR, Park NH. Exp Cell Res. 2003;287:272–81.
77. Martins CP, Brown-Swigart L, Evan GI. Cell. 2006;127:1323–34.
78. Bodnar AG, Ouellette M, Frolkis M, Holt SE, Chiu CP, Morin GB, Harley CB, Shay JW, Lichtsteiner S, Wright WE. Science. 1998;279:349–52.
79. Aubert G, Lansdorp PM. Physiol Rev. 2008;88:557–79.
80. Pinkston JM, Garigan D, Hansen M, Kenyon C. Science. 2006;313:971–5.
81. Rufini A, Tucci P, Celardo I, Melino G. Oncogene. 2013;32:5129–43.
82. Tomas-Loba A, Flores I, Fernandez-Marcos PJ, Cayuela ML, Maraver A, Tejera A, Borras C, Matheu A, Klatt P, Flores JM, Vina J, Serrano M, Blasco MA. Cell. 2008;135:609–22.
83. Kon N, Zhong J, Kobayashi Y, Li M, Szabolcs M, Ludwig T, Canoll PD, Gu W. Cell Death Differ. 2011;18:1366–75.
84. Sah VP, Attardi LD, Mulligan GJ, Williams BO, Bronson RT, Jacks T. Nat Genet. 1995;10:175–80.
85. Levine AJ, Tomasini R, McKeon FD, Mak TW, Melino G. Nat Rev Mol Cell Biol. 2011;12:259–65.
86. Gottlieb E, Vousden KH. Cold Spring Harb Perspect Biol. 2010;2:a001040.
87. Begus-Nahrmann Y, Lechel A, Obenauf AC, Nalapareddy K, Peit E, Hoffmann E, Schlaudraff F, Liss B, Schirmacher P, Kestler H, Danenberg E, Barker N, Clevers H, Speicher MR, Rudolph KL. Nat Genet. 2009;41:1138–43.
88. Feng Z, Lin M, Wu R. Genes Cancer. 2011;2:443–52.
89. Marcel V, Dichtel-Danjoy ML, Sagne C, Hafsi H, Ma D, Ortiz-Cuaran S, Olivier M, Hall J, Mollereau B, Hainaut P, Bourdon JC. Cell Death Differ. 2011;18:1815–24.
90. Liu D, Ou L, Clemenson JGD, Chao C, Lutske ME, Zambetti GP, Gage FH, Xu Y. Puma is required for p53-induced depletion of adult stem cells. Nat Cell Biol. 2010;12:993–8.
91. Pollina EA, Brunet A. Oncogene. 2011;30:3105–26.
92. Donehower LA, Harvey M, Slagle BL, McArthur MJ, Montgomery CA Jr, Butel JS, Bradley A. Nature. 1992;356:215–21.
93. Armata HL, Garlick DS, Sluss HK. Cancer Res. 2007;67:11696–703.
94. Spinnler C, Hedstrom E, Li H, de Lange J, Nikulenkov F, Teunisse AF, Verlaan-de Vries M, Grinkevich V, Jochemsen AG, Selivanova G. Cell Death Differ. 2011;18:1736–45.
95. Matheu A, Maraver A, Klatt P, Flores I, Garcia-Cao I, Borras C, Flores JM, Vina J, Blasco MA, Serrano M. Nature. 2007;448:375–9.
96. Matheu A, Maraver A, Serrano M. Cancer Res. 2008;68:6031–4.
97. Thomas G. Biol Res. 2002;35:305–13.
98. Inoki K, Zhu T, Guan KL. Cell. 2003;115:577–90.
99. Lum JJ, Bauer DE, Kong M, Harris MH, Li C, Lindsten T, Thompson CB. Cell. 2005;120:237–48.
100. Feng Z, Zhang H, Levine AJ, Jin S. Proc Natl Acad Sci U S A. 2005;102:8204–9.

101. Feng Z, Hu W, de Stanchina E, Teresky AK, Jin S, Lowe S, Levine AJ. Cancer Res. 2007;67:3043–53.
102. Demidenko ZN, Korotchkina LG, Gudkov AV, Blagosklonny MV. Proc Natl Acad Sci U S A. 2010;107:9660–4.
103. Lee JJ, Kim BC, Park MJ, Lee YS, Kim YN, Lee BL, Lee JS. Cell Death Differ. 2011;18:666–77.
104. Aksoy O, Chicas A, Zeng T, Zhao Z, McCurrach M, Wang X, Lowe SW. Genes Dev. 2012;26:1546–57.
105. Carvajal LA, Hamard PJ, Tonnessen C, Manfredi JJ. Genes Dev. 2012;26:1533–45.
106. Di Stefano L, Jensen MR, Helin K. EMBO J. 2003;22:6289–98.
107. Logan N, Delavaine L, Graham A, Reilly C, Wilson J, Brummelkamp TR, Hijmans EM, Bernards R, La Thangue NB. Oncogene. 2004;23:5138–50.
108. Klionsky DJ. Nat Rev Mol Cell Biol. 2007;8:931–7.
109. Rubinsztein DC, Marino G, Kroemer G. Cell. 2011;146:682–95.
110. Hofius D, Munch D, Bressendorff S, Mundy J, Petersen M. Cell Death Differ. 2011;18:1257–62.
111. Wu WK, Coffelt SB, Cho CH, Wang XJ, Lee CW, Chan FK, Yu J, Sung JJ. Oncogene. 2012;31:939–53.
112. Lee JH, Budanov AV, Park EJ, Birse R, Kim TE, Perkins GA, Ocorr K, Ellisman MH, Bodmer R, Bier E, Karin M. Science. 2010;327:1223–8.
113. Matecic M, Smith DL, Pan X, Maqani N, Bekiranov S, Boeke JD, Smith JS. PLoS Genet. 2010;6:e1000921.
114. Hars ES, Qi H, Ryazanov AG, Jin S, Cai L, Hu C, Liu LF. Autophagy. 2007;3:93–5.
115. Minois N, Carmona-Gutierrez D, Bauer MA, Rockenfeller P, Eisenberg T, Brandhorst S, Sigrist SJ, Kroemer G, Madeo F. Cell Death Dis. 2012;3:e401.
116. Levine B, Kroemer G. Cell Death Differ. 2009;16:1–2.
117. Kapahi P, Chen D, Rogers AN, Katewa SD, Li PW, Thomas EL, Kockel L. Cell Metab. 2010;11:453–65.
118. Gao W, Shen Z, Shang L, Wang X. Cell Death Differ. 2011;18:1598–607.
119. Maiuri MC, Galluzzi L, Morselli E, Kepp O, Malik SA, Kroemer G. Curr Opin Cell Biol. 2010;22:181–5.
120. Tasdemir E, Maiuri MC, Galluzzi L, Vitale I, Djavaheri-Mergny M, D'Amelio M, Criollo A, Morselli E, Zhu C, Harper F, Nannmark U, Samara C, Pinton P, Vicencio JM, Carnuccio R, Moll UM, Madeo F, Paterlini-Brechot P, Rizzuto R, Szabadkai G, Pierron G, Blomgren K, Tavernarakis N, Codogno P, Cecconi F, Kroemer G. Nat Cell Biol. 2008;10:676–87.
121. Liang C. Cell Death Differ. 2010;17:1807–15.
122. Feng Z, Hu W, Teresky AK, Hernando E, Cordon-Cardo C, Levine AJ. Proc Natl Acad Sci U S A. 2007;104:16633–8.
123. Salminen A, Kaarniranta K. Trends Mol Med. 2009;15:217–24.
124. Crighton D, Wilkinson S, O'Prey J, Syed N, Smith P, Harrison PR, Gasco M, Garrone O, Crook T, Ryan KM. Cell. 2006;126:121–34.
125. Kang MY, Kim HB, Piao C, Lee KH, Hyun JW, Chang IY, You HJ. Cell Death Differ. 2013;20:117–29.
126. Polyak K, Xia Y, Zweier JL, Kinzler KW, Vogelstein B. Nature. 1997;389:300–5.
127. Budanov AV, Sablina AA, Feinstein E, Koonin EV, Chumakov PM. Science. 2004;304:596–600.
128. Hussain SP, Amstad P, He P, Robles A, Lupold S, Kaneko I, Ichimiya M, Sengupta S, Mechanic L, Okamura S, Hofseth LJ, Moake M, Nagashima M, Forrester KS, Harris CC. Cancer Res. 2004;64:2350–6.
129. Tan M, Li S, Swaroop M, Guan K, Oberley LW, Sun Y. J Biol Chem. 1999;274:12061–6.
130. Olovnikov IA, Kravchenko JE, Chumakov PM. Semin Cancer Biol. 2009;19:32–41.
131. Johnson TM, Yu ZX, Ferrans VJ, Lowenstein RA, Finkel T. Proc Natl Acad Sci U S A. 1996;93:11848–52.
132. Macip S, Igarashi M, Berggren P, Yu J, Lee SW, Aaronson SA. Mol Cell Biol. 2003;23:8576–85.

133. Dumble M, Moore L, Chambers SM, Geiger H, Van Zant G, Goodell MA, Donehower LA. Blood. 2007;109:1736–42.
134. Cao L, Li W, Kim S, Brodie SG, Deng CX. Genes Dev. 2003;17:201–13.
135. Varela I, Cadinanos J, Pendas AM, Gutierrez-Fernandez A, Folgueras AR, Sanchez LM, Zhou Z, Rodriguez FJ, Stewart CL, Vega JA, Tryggvason K, Freije JM, Lopez-Otin C. Nature. 2005;437:564–8.
136. Sherr CJ. Nat Rev Cancer. 2006;6:663–73.
137. Yoon KA, Nakamura Y, Arakawa H. J Hum Genet. 2004;49:134–40.
138. Bensaad K, Tsuruta A, Selak MA, Vidal MN, Nakano K, Bartrons R, Gottlieb E, Vousden KH. Cell. 2006;126:107–20.
139. Bensaad K, Vousden KH. Trends Cell Biol. 2007;17:286–91.
140. Donald SP, Sun XY, Hu CA, Yu J, Mei JM, Valle D, Phang JM. Cancer Res. 2001;61:1810–5.
141. Liu Z, Lu H, Shi H, Du Y, Yu J, Gu S, Chen X, Liu KJ, Hu CA. Cancer Res. 2005;65:1647–54.
142. Drane P, Bravard A, Bouvard V, May E. Oncogene. 2001;20:430–9.
143. Faraonio R, Vergara P, Di Marzo D, Pierantoni MG, Napolitano M, Russo T, Cimino F. J Biol Chem. 2006;281:39776–84.
144. Pani G, Bedogni B, Anzevino R, Colavitti R, Palazzotti B, Borrello S, Galeotti T. Cancer Res. 2000;60:4654–60.
145. Liu D, Xu Y. Antioxid Redox Signal. 2011;15:1669–78.
146. Finkel T, Serrano M, Blasco MA. Nature. 2007;448:767–74.
147. Papazoglu C, Mills AA. J Pathol. 2007;211:124–33.
148. de Magalhaes JP, Curado J, Church GM. Bioinformatics. 2009;25:875–81.
149. Zentella A, Manogue K, Cerami A. Cytokine. 1993;5:436–47.
150. Benigni F, Atsumi T, Calandra T, Metz C, Echtenacher B, Peng T, Bucala R. J Clin Invest. 2000;106:1291–300.
151. Bauer DE, Harris MH, Plas DR, Lum JJ, Hammerman PS, Rathmell JC, Riley JL, Thompson CB. FASEB J. 2004;18:1303–5.
152. Atsumi T, Cho YR, Leng L, McDonald C, Yu T, Danton C, Hong EG, Mitchell RA, Metz C, Niwa H, Takeuchi J, Onodera S, Umino T, Yoshioka N, Koike T, Kim JK, Bucala R. J Immunol. 2007;179:5399–406.
153. Jung H, Seong HA, Ha H. J Biol Chem. 2008;283:20383–96.
154. Helenius M, Hanninen M, Lehtinen SK, Salminen A. Biochem J. 1996;318 (. **Pt 2**:603–8.
155. Spencer NF, Poynter ME, Im SY, Daynes RA. Int Immunol. 1997;9:1581–8.
156. Adler AS, Sinha S, Kawahara TL, Zhang JY, Segal E, Chang HY. Genes Dev. 2007;21:3244–57.
157. Kawahara TL, Michishita E, Adler AS, Damian M, Berber E, Lin M, McCord RA, Ongaigui KC, Boxer LD, Chang HY, Chua KF. Cell. 2009;136:62–74.
158. Renner F, Schmitz ML. Trends Biochem Sci. 2009;34:128–35.
159. Salminen A, Ojala J, Huuskonen J, Kauppinen A, Suuronen T, Kaarniranta K. Cell Mol Life Sci. 2008;65:1049–58.
160. Longo VD, Kennedy BK. Cell. 2006;126:257–68.
161. Gloire G, Legrand-Poels S, Piette J. Biochem Pharmacol. 2006;72:1493–505.
162. Ota H, Akishita M, Eto M, Iijima K, Kaneki M, Ouchi Y. J Mol Cell Cardiol. 2007;43:571–9.
163. Pearson M, Carbone R, Sebastiani C, Cioce M, Fagioli M, Saito S, Higashimoto Y, Appella E, Minucci S, Pandolfi PP, Pelicci PG. Nature. 2000;406:207–10.
164. Dixit D, Sharma V, Ghosh S, Mehta VS, Sen E. Cell Death Dis. 2012;3:e271.
165. Deng CX. Int J Biol Sci. 2009;5:147–52.
166. Luo J, Nikolaev AY, Imai S, Chen D, Su F, Shiloh A, Guarente L, Gu W. Cell. 2001;107:137–48.
167. Vaziri H, Dessain SK, Ng Eaton E, Imai SI, Frye RA, Pandita TK, Guarente L, Weinberg RA. Cell. 2001;107:149–59.
168. Campagna M, Herranz D, Garcia MA, Marcos-Villar L, Gonzalez-Santamaria J, Gallego P, Gutierrez S, Collado M, Serrano M, Esteban M, Rivas C. Cell Death Differ. 2011;18:72–9.
169. Feldman JL, Dittenhafer-Reed KE, Denu JM. J Biol Chem. 2012;287:42419–27.
170. Houtkooper RH, Pirinen E, Auwerx J. Nat Rev Mol Cell Biol. 2012;13:225–38.

171. Brooks CL, Gu W. Nat Rev Cancer. 2009;9:123–8.
172. Li L, Wang L, Wang Z, Ho Y, McDonald T, Holyoake TL, Chen W, Bhatia R. Cancer Cell. 2012;21:266–81.
173. Krummel KA, Lee CJ, Toledo F, Wahl GM. Proc Natl Acad Sci U S A. 2005;102:10188–93.
174. Dotsch V, Bernassola F, Coutandin D, Candi E, Melino G. p63 and p73, the ancestors of p53. Cold Spring Harb Perspect Biol. 2010;2:a004887.
175. Yang A, Schweitzer R, Sun D, Kaghad M, Walker N, Bronson RT, Tabin C, Sharpe A, Caput D, Crum C, McKeon F. Nature. 1999;398:714–8.
176. Mills AA, Zheng B, Wang XJ, Vogel H, Roop DR, Bradley A. Nature. 1999;398:708–13.
177. Flores ER, Sengupta S, Miller JB, Newman JJ, Bronson R, Crowley D, Yang A, McKeon F, Jacks T. Cancer Cell. 2005;7:363–73.
178. Keyes WM, Wu Y, Vogel H, Guo X, Lowe SW, Mills AA. Genes Dev. 2005;19:1986–99.
179. Salomoni P, Dvorkina M, Michod D. Cell Death Dis. 2012;3:e247.
180. Peche LY, Scolz M, Ladelfa MF, Monte M, Schneider C. Cell Death Differ. 2012;19:926–36.
181. Su X, Paris M, Gi YJ, Tsai KY, Cho MS, Lin YL, Biernaskie JA, Sinha S, Prives C, Pevny LH, Miller FD, Flores ER. Cell Stem Cell. 2009;5:64–75.
182. Paris M, Rouleau M, Puceat M, Aberdam D. Cell Death Differ. 2012;19:186–93.
183. Guo X, Keyes WM, Papazoglu C, Zuber J, Li W, Lowe SW, Vogel H, Mills AA. Nat Cell Biol. 2009;11:1451–7.
184. Zaika A, Irwin M, Sansome C, Moll UM. J Biol Chem. 2001;276:11310–6.
185. Tomasini R, Tsuchihara K, Wilhelm M, Fujitani M, Rufini A, Cheung CC, Khan F, Itie-Youten A, Wakeham A, Tsao MS, Iovanna JL, Squire J, Jurisica I, Kaplan D, Melino G, Jurisicova A, Mak TW. Genes Dev. 2008;22:2677–91.
186. Rufini A, Niklison-Chirou MV, Inoue S, Tomasini R, Harris IS, Marino A, Federici M, Dinsdale D, Knight RA, Melino G, Mak TW. Genes Dev. 2012;26:2009–14.
187. Feng Z, Hu W, Rajagopal G, Levine AJ. Cell Cycle. 2008;7:842–7.

Paired Box (Pax) Transcription Factors and Aging

Rajnikant Mishra

Introduction

Aging has been explained as a temporal state of cell, organ, or organisms, characterized with accumulation of damages and decline in their regenerative potential over time. Aging-related studies from several models present gradual modifications in DNA bases and genetic program of organisms, changes in the chromatin structure, alterations in activities of genes, transcription factors, mRNAs, noncoding RNAs, proteins, regulators of immune response, survival pathways, and molecular hallmarks of aging [1–4]. Cell lines, yeast, worm, flies, fish, mice, rats, and humans have been preferred for aging-associated investigations but none of the models serve suitable on all criteria. Bacteria were initially thought to be immortal, but they also age and undergo asymmetrical divisions [5]. Yeast serves a simple model [6] to explain the impact of aging on metabolism, but does not explain the effects of extracellular factors like hormones because they lack intracellular inflammatory signaling. A worm, *Caenorhabditis elegans* (*C. elegans*), shows slow metabolism of lipids and proteins [7], neurodegeneration, and uncoordinated movements [8] during aging. The age-related alterations in inflammatory and immunological response have also been supported from *Drosophila* species, fish models [9–11]. Amphibians show slow aging, continuous neurogenesis, and myogenesis [12]. However, reptiles have been poorly studied but support gradual senescence [13]. Studies from different models suggest that aging-related changes in one tissue may cause deterioration of other tissues' "contagious aging" or "bystander effects," for example, impaired kidney function can increase risk of heart disease [14] and antiaging manipulations of one tissue can retard the aging process in other tissues [15]. Several theories

R. Mishra (✉)
Biochemistry and Molecular Biology Laboratory, Department of Zoology,
Institute of Science, Banaras Hindu University, Varanasi, India
e-mail: rmishraa@bhu.ac.in

Table 6.1 Pax family genes and proteins in different models

Pax family members	Expression in tissues	Model organisms	References
Pax1	Axial skeleton, pectoral girdle	Drosophila, mice, human	[22, 66]
Pax2	Ureteric bud, kidneys, and otic vesicle, astrocytes, eyes	Mice, chicken, Zebrafish, Drosophila, C. elegans	[27, 29]
Pax3	Muscles	Splotch (Sp) mice, human	[17, 67]
Pax4	β-Cell regeneration	Mice	[68]
Pax5	B-cell development, Central Nervous System (CNS), testis, spleen, lymph node, tonsils, appendix	Mice, human	[38, 39]
Pax6	Eye, brain, and pancreas	Drosophila, Zebrafish, mice, human	[44, 69, 70]
Pax7	Muscle quiescent satellite cells	Drosophila, mice, human	[71, 72]
Pax8	Thyroid gland, kidney	Mouse, human	[57, 58, 73]
Pax9	Tooth morphogenesis, esophagus, limb formation	Zebrafish, chick, mouse, and man	[61, 62, 64]

support alteration in genes and proteins as either cause or effect of aging. However, aging does not appear only due to selection of specific gene or protein but networks of several pathways and transcriptional regulators. Among different regulators Pax family transcription factors prove critical for development and aging because they regulate cell cycle, growth, and differentiation from embryonic development to aging at the levels of molecules, cells, tissues, organs, or organisms [16–22] (Table 6.1).

Impact of Pax Family Transcription Factor in Aging

There are nine known functional forms of Pax transcription factors (Fig. 6.1), and their mutations lead to haplo-insufficiency-associated pathogenesis-matching developmental and common aging-related impairments. The management of Pax1 may be an encouraging direction in the area of biogerontology because undulated (*un*) phenotype in mouse, due to loss of Pax1, shows defects in pectoral girdle, intervertebral disks, and absence of the acromion of the scapula or its replacement with a ligament [22]. The expression of Pax1 is also required for maintenance of T lymphocytes and the level of expression decreases during aging in human thymic epithelial cells [2]. The hypomethylation of Pax1 was also observed in

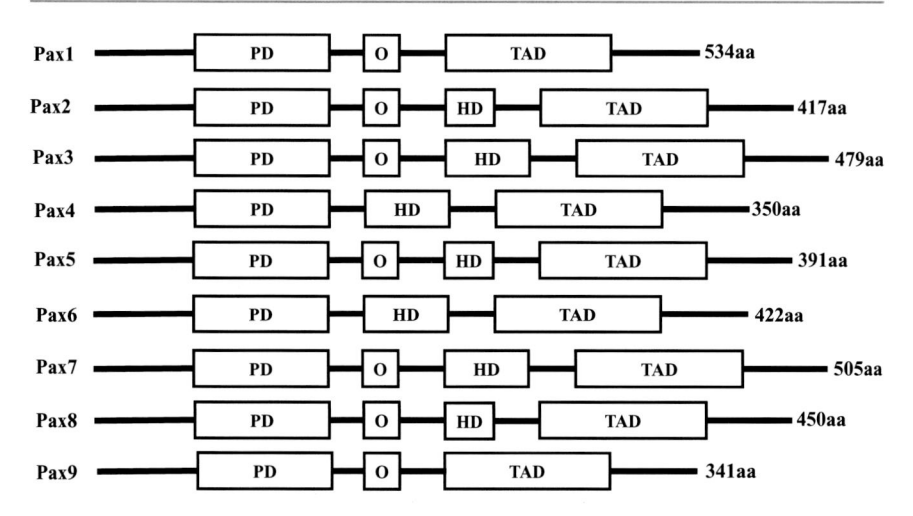

Fig. 6.1 Diagrammatic representation of structural domains of Pax family genes includes Paired domain (PD), octapeptide domain (O), homeodomain (HD), and transactivation domain (TAD)

mesenchymal stroma cells (MSC) during aging [23]. The Pax2 has been observed in ureteric bud, kidneys [24], optic vesicle [25], and astrocytes [26]. A significant decrease in the density of astrocytes and Pax2-positive cells in human retina of middle and old age [27] and an increase in expression of Pax2 in glomerular cells have been observed with aging [28]. During aging, it regulates density of astrocyte-related pathological conditions [29] but it is unknown whether loss of Pax2 expression by astrocytes during aging has any impact on their functions. The myogenic progenitors express both Pax3 and Pax7 and they are required for normal development [30] of muscles but satellite cells show reduced expression of Pax3 with aging [31]. It may have muscle regeneration potential because its expression increases in cancer patients and healthy elderly individuals in comparison to healthy middle-aged individuals [32]. However, inactivation of Pax4 by homologous recombination results in the absence of mature insulin and somatostatin-producing cells in the pancreas [33]. It promotes regeneration of β-cell by protecting from apoptosis and promoting proliferation [34]. The status of methylation in promoter region of Pax4 gene does not change but expression of Pax4 alters in pancreas with age [35]. Therefore, Pax4 may serve one of the critical markers for accessing status of insulin and glucagon-dependent glucose metabolism during aging.

Binding sites for Pax5 have been identified in the promoters of multiple genes as well as at multiple sites within the Immunoglobulin Heavy Chain(IgH) locus [36, 37]. The Pax5 is also referred to as B-cell lineage-specific activator protein (BSAP) that regulates development of B-cell but is not detectable in terminally differentiated B-cells [38, 39]. In variety of mouse strains, B-lymphopoiesis within bone marrow gets affected during senescence [40]. Age-dependent reduction in expression of early B factor (EBF) affects ability of multipotent hematopoietic stem cells to generate B-cells but potential of generating myeloid cells remains unchanged [41].

The aged mice show decreased binding activity of Pax5 to targets [42] in B-cells. The aged B-cell culture leads to impaired development of pre-B and new B-cells due to reduced expression of Pax5. It is presumed that reduction in level of Pax5 in aged B-cell precursors may be multifactorial but mainly due to increased ubiquitin/proteasome-mediated protein degradation through increased mitogen-activated protein kinase (MAPK) and Notch activity [43].

The Pax6 essentially presumed as light-sensing molecule proved a multifunctional protein because it regulates several critical downstream regulators from cell cycle to cell death [44]. Pax6 expression is restricted to pancreatic islets at birth but inactivation of Pax6 causes reduction in the level of hormone production and alteration in morphology of islets [45]. Aging-associated defects match phenotypes of mutations in Pax6 like variety of ocular defects, neuronal and endocrine malformations [46, 47], decrease in volume of regions of brain [47], and behavioral abnormalities [48]. Decrease in the level of Pax6 in old intermediate neural progenitors (INPs) allows Notch signaling to promote the dedifferentiation of INP progeny into ectopic INPs, thereby creating a proliferative mass of ectopic progenitors in the brain [49]. A direct association of Pax6 with aging-related neuronal dystrophy [50], p53-mediated cell death and neural plasticity [51], binding of Pax6 to the promoter sequence elements of genes involved in immunological surveillance and energy homeostasis [52, 53] has been observed. However, Pax6 is constantly expressed in retina throughout the lifespan of mice [54].

The postnatal muscle quiescent satellite cells express Pax7 while their proliferating progeny co-express Pax7 and MyoD. The decline in Pax7 expression, along with the onset of expression of the muscle transcription factor myogenin, marks the entry of satellite cell progeny into the differentiation phase where Pax7 proves essential for regulating the expansion and differentiation of satellite cells during both neonatal and adult myogenesis [55]. Pax7 deficiency leads to spinal muscular atrophy (SMA), characterized by loss of motor neurons in the spinal cord that results in muscle denervation and profound weakness [56].

Functions of thyroid and aging could be interdependent because Pax8 is expressed in thyroid gland, inner ear, epithelial cells of female and male genital tracts, and lineages of kidney formation [57, 58]. Mutations of Pax8 cause infertility in male and female mice because of the malformation and obstruction of reproductive organs [57, 59]. Age-dependent infertility and cardiomyopathy may be associated with levels of Pax8 because Pax8 knockout leads to the heart dysplasia as a result of excessive senescence and apoptosis in cardiac myocytes [60]. Literature is unavailable to show direct association of Pax9 with aging but it seems to be associated with age-dependent defects in immunological surveillance, esophagus, lung, and teeth because Pax9 is expressed in the adult thymus [61], and is also required for permanent tooth development [62]. It has been implicated in oral squamous cell carcinoma [63], increasing malignancy in esophageal [64], and lung cancer [65].

Conclusion

Various models from bacteria to humans have been used to study the process of aging and associated diseases but none of them proved ideal. The Pax family transcription factors are expressed in almost all vital organs and show age-dependent alterations. They have been critical during development and maintaining functional anatomy of bone, muscle, heart, eyes, brain, and endocrine systems (Fig. 6.2). Phenotypes due to mutations in Pax transcription factors match aging-associated symptoms. Studies on regulatory sequences of interactions of Pax proteins would provide specific molecular target for therapeutics in healthy aging.

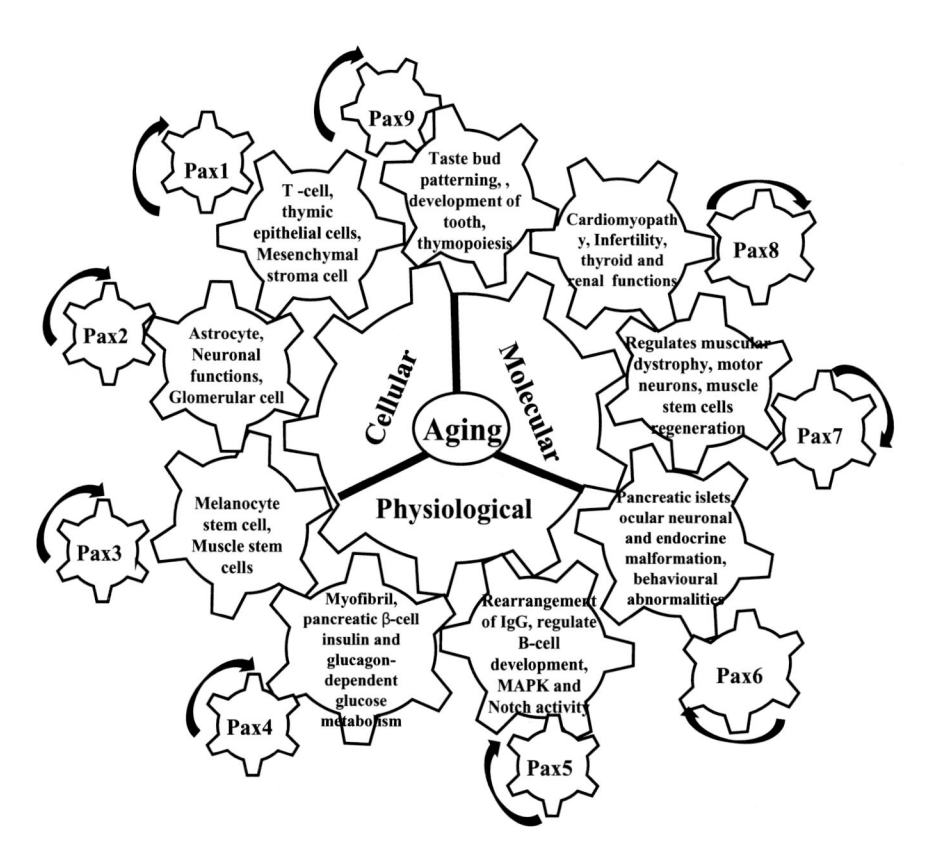

Fig. 6.2 During aging, the alteration in expression of Pax genes influences the vital functions of every respective organ in which they express all that collectively drives the process of aging. The molecular mechanism behind the modulation in the functional properties of an organ may be due to Pax genes alone respectively or it may be due to the cumulative effect of Pax genes and their interacting partners which need to be further investigated

Acknowledgments Help in literature survey and typing of this manuscript by Shashank Kumar Maurya, Ayan Banerjee, Khushboo Srivastava, and Suman Mishra is gratefully acknowledged.

References

1. Harries LW, Hernandez D, Henley W, et al. Human aging is characterized by focused changes in gene expression and deregulation of alternative splicing. Aging Cell. 2011;10:868–78.
2. Ugalde AP, Espanol Y, Lopez-Otin C. Micromanaging aging with miRNAs: new messages from the nuclear envelope. Nucleus. 2011;2:549–55.
3. Liu N, Landreh M, Cao K, et al. The microRNA miR-34 modulates ageing and neurodegeneration in Drosophila. Nature. 2012;482:519–23.
4. Smith-Vikos T, Slack FJ. MicroRNAs and their roles in aging. J Cell Sci. 2012;125:7–17.
5. Ackermann M. Bacteria as a new model system for aging studies: investigations using light microscopy. BioTechniques. 2008;44:564–7.
6. Kaeberlein M, Burtner CR, Kennedy BK. Recent developments in yeast aging. PLoS Genet. 2007;3:e84.
7. Onken B, Driscoll M. Metformin induces a dietary restriction-like state and the oxidative stress response to extend C elegans Healthspan via AMPK, LKB1, and SKN-1. PLoS One. 2010;5:e8758.
8. Zhang R, Chen HZ, Liu DP. The four layers of aging. Cell. 2015;1:180–6.
9. Gilbert MJH, Zerulla TC, Tierney KB. Zebrafish (Danio rerio) as a model for the study of aging and exercise: physical ability and trainability decrease with age. Exp Gerontol. 2014;50:106–13.
10. Terzibasi E, Valenzano DR, Benedetti M, et al. Large differences in aging phenotype between strains of the short-lived annual fish *Nothobranchius furzeri*. PLoS One. 2008;e3866:3.
11. Herrera M, Jagadeeswaran P. Annual fish as a genetic model for aging. J Gerontol A Biol Sci Med Sci. 2004;59:101–7.
12. Kara TC. Ageing in amphibians. Gerontology. 1994;40:161–73.
13. Patnaik BK. Agieng in reptiles. Gerontology. 1994;40:200–20.
14. Sarnak MJ, Levey AS, Schoolwerth AC, et al. Kidney disease as a risk factor for development of cardiovascular disease: a statement from the American Heart Association councils on kidney in cardiovascular disease, high blood pressure research, clinical cardiology, and epidemiology and prevention. Circulation. 2003;108:2154–69.
15. Lavasani M, Robinson AR, Lu A, et al. Muscle-derived stem/progenitor cell dysfunction limits healthspan and lifespan in a murine progeria model. Nat Commun. 2012;3 https://doi.org/10.1038/ncomms1611.
16. Balling R, Deutsch U, Gruss P. Undulated, a mutation affecting the development of the mouse skeleton, has a point mutation in the paired box of Pax-1. Cell. 1988;55:531–5.
17. Epstein DJ, Vekemans M, Gros P. Splotch (Sp2H), a mutation affecting development of the mouse neural tube, shows a deletion within the paired homeodomain of Pax-3. Cell. 1991;67:767–74.
18. Hill RE, Favor J, Hogan BLM, et al. Mouse small eye results from mutation in a paired-like homeobox-containing gene. Nature. 1991;354:522–5.
19. Favor J, Sandulache R, Neuhäuser-Klaus A, et al. The mouse Pax21Neu mutation is identical to a human PAX2 mutation in a family with renal-coloboma syndrome and results in developmental defects of the brain, ear, eye and kidney. Proc Natl Acad Sci U S A. 1996;93:13870–5.
20. Tassabehji M, Read AP, Newton VE, et al. Waardenburg's syndrome patients have mutations in the human homologue of the Pax- 3 paired box gene. Nature. 1992;355:635–6.
21. Sanyanusin P, Schimmenti LA, McNoe LA, et al. Mutation of the PAX2 gene in a family with optic nerve colobomas, renal anomalies and vesicoureteral reflux. Nat Genet. 1995;9:358–64.
22. Gruneberg H. Genetical studies on the skeleton of the mouse. XII. The development of undulated. J Genet. 1950;52:441–55.

23. Bork S, Pfister S, Witt H, et al. DNA methylation pattern changes upon long-term culture and aging of human mesenchymal stromal cells. Aging Cell. 2010;9:54–63.
24. Dressler GR, Deutsch U, Chowdhury K, et al. Pax2, a new murine paired-box-containing gene and its expression in the developing excretory system. Development. 1990;109:787–95.
25. Nornes S, Mikkola I, Krauss S, et al. Zebrafish Pax9 encodes two proteins with distinct C-terminal transactivating domains of different potency negatively regulated by adjacent N-terminal sequences. J Biol Chem. 1996;271:26914–23.
26. Chu Y, Hughes S, Chan-Ling T. Differentiation and migration of astrocyte precursor cells and astrocytes in human fetal retina: relevance to optic nerve coloboma. FASEB J. 2001;15:2013–5.
27. Ramirez JM, Ramirez AI, Salazar JJ, et al. A change of astrocytes in retinal ageing and age-related macular degeneration. Exp Eye Res. 2001;73:601–15.
28. Zhang J, Hansen KM, Pippin JW, et al. De novo expression of podocyte proteins in parietal epithelial cells in experimental aging nephropathy. Am J Physiol Renal Physiol. 2012;302:F571–80.
29. Takuma K, Baba A, Matsuda T. Astrocyte apoptosis: implications for neuroprotection. Prog Neurobiol. 2004;72:111–27.
30. Buckingham M, Relaix F. The role of Pax genes in the development of tissues and organs: Pax3 and Pax7 regulate muscle progenitor cell functions. Annu Rev Cell Dev Biol. 2007;23:645–73.
31. Kirkpatrick LJ, Yablonka-Reuveni Z, Rosser BWC. Retention of Pax3 expression in satellite cells of muscle spindles. J Histochem Cytochem. 2010;58:317–27.
32. Brzeszczyńska J, Johns N, Schilb A, et al. Loss of oxidative defense and potential blockade of satellite cell maturation in the skeletal muscle of patients with cancer but not in the healthy elderly. Aging (Albany NY). 2016;8:1690–702.
33. Sosa-Pineda, Chowdhury K, Torres M, et al. The Pax4 gene is essential for differentiation of insulin-producing beta cells in the mammalian pancreas. Nature. 1997;386:399–402.
34. Brun T, Gauthier BR. A focus on the role of Pax4 in mature pancreatic islet beta cell expansion and survival in health and disease. J Mol Endocrinol. 2008;40:37–45.
35. Ashapkin VV, Linkova NS, Khavinson VK, Vanyushin BF. Epigenetic mechanisms of peptidergic regulation of gene expression during aging of human cells. Biochemistry (Mosc). 2015;80:310–22.
36. Neurath MF, Strober W, Wakatsuki Y. The murine Ig 3′ alpha enhancer is a target site with repressor function for the B cell lineage-specific transcription factor BSAP (NF-HB, S alpha-BP). J Immunol. 1994;153:730–42.
37. Michaelson JS, Singh M, Birshtein BK. B cell lineage-specific activator protein (BSAP). A player at multiple stages of B cell development. J Immunol. 1996;156:2349–51.
38. Nutt SL, Morrison AM, Dörfler P, et al. Identification of BSAP (Pax-5) target genes in early B-cell development by loss- and gain-of-function experiments. EMBO J. 1998;17:2319–33.
39. Nutt SL, Heavey B, Rolink AG, Busslinger M. Commitment to the B-lymphoid lineage depends on the transcription factor Pax5. Nature. 1999;401:556–62.
40. Riley RL. Impaired B lymphopoiesis in old age: a role for inflammatory B cells? Immunol Res. 2013;57:361–9.
41. Lescale C, Dias S, Maës J, et al. Reduced EBF expression underlies loss of B-cell potential of hematopoietic progenitors with age. Aging Cell. 2010;9:410–9.
42. Anspach J, Poulsen G, Kaattari I, et al. Reduction in DNA binding activity of the transcription factor Pax-5a in B lymphocytes of aged mice. J Immunol. 2001;166:2617–26.
43. Frasca D, Nguyen D, Riley RL, Blomberg BB. Decreased E12 and/or E47 transcription factor activity in the bone marrow as well as in the spleen of aged mice. J Immunol. 2003;170:719–26.
44. Gehring WJ, Ikeo K. Pax6: mastering eye morphogenesis and eye evolution. Trends Genet. 1999;15(9):371–7.
45. Ashery-Padan R, Zhou X, Marquardt T, et al. Conditional inactivation of Pax6 in the pancreas causes early onset of diabetes. Dev Biol. 2004;269:479–88.
46. Vincent MC, Gallai R, Olivier D, et al. Variable phenotype related to a novel PAX 6 mutation (IVS4+5G>C) in a family presenting congenital nystagmus and foveal hypoplasia. Am J Ophthalmol. 2004;138:1016–21.

47. Hiraoka K, Sumiyoshi A, Nonaka H, et al. Regional volume decreases in the brain of Pax6 heterozygous mutant rats: MRI deformation-based Morphometry. PLoS One. 2016;e0158153:11.
48. Yoshizaki K, Furuse T, Kimura R, et al. Paternal aging affects behavior in Pax6 mutant mice: a gene/environment interaction in understanding neurodevelopmental disorders. PLoS One. 2016;e0166665:11.
49. Farnsworth DR, Bayraktar OA, Doe CQ. Aging neural progenitors lose competence to respond to Mitogenic notch signaling. Curr Biol. 2015;25:3058–68.
50. Tripathi R, Mishra R. Aging-associated modulation in the expression of Pax6 in mouse brain. Cell Mol Neurobiol. 2012;32:209–18.
51. Tripathi R, Mishra R. Interaction of Pax6 with SPARC and p53 in brain of mice indicates Smad3 dependent auto-regulation. J Mol Neurosci. 2010;41:397–403.
52. Maurya SK, Mishra R. Pax6 binds to promoter sequence elements associated with immunological surveillance and energy homeostasis in brain of aging mice. Ann Neurosci. 2017a;24:20–5.
53. Maurya SK, Mishra R. Pax6 interacts with Iba1 and shows age-associated alterations in brain of aging mice. J Chem Neuroanat. 2017b;82:60–4.
54. Stanescu D, Iseli HP, Schwerdtfeger K, et al. Continuous expression of the homeobox gene Pax6 in the ageing human retina. Eye. 2007;21:90–3.
55. Cheung TH, Rando TA. Molecular regulation of stem cell quiescence. Nat Rev Mol Cell Biol. 2013;14:329–40.
56. Montarras D, L'Honore A, Buckingham M. Lying low but ready for action: the quiescent muscle satellite cell. FEBS J. 2013;280:4036–50.
57. Mittag J, Winterhager E, Bauer K, Grummer R. Congenital hypothyroid female pax8-deficient mice are infertile despite thyroid hormone replacement therapy. Endocrinology. 2007;148:719–25.
58. Bouchard M, Souabni A, Mandler M, et al. Nephric lineage specification by Pax2 and Pax8. Genes Dev. 2002;16:2958–70.
59. Wistuba J, Mittag J, Luetjens CM, et al. Male congenital hypothyroid Pax8−/− mice are infertile despite adequate treatment with thyroid hormone. J Endocrinol. 2007;192:99–109.
60. Yihao W, Zhou X, Huang X, et al. Pax8 plays a pivotal role in regulation of cardiomyocyte growth and senescence. J Cell Mol Med. 2016;20:644–54.
61. Peters H, Doll U, Niessing J. Differential expression of the chicken Pax-1 and Pax-9 gene: in situ hybridization and immunohistochemical analysis. Dev Dyn. 1995;203:1–16.
62. Suda N, Ogawa T, Kojima T, et al. Non-syndromic oligodontia with a novel mutation of PAX9. J Dent Res. 2011;90:382–6.
63. Lee JC, Sharma M, Lee YH, et al. Pax9 mediated cell survival in oral squamous carcinoma cell enhanced by c-myb. Cell Biochem Funct. 2008;26:892–9.
64. Gerber JK, Richter T, Kremmer E, et al. Progressive loss of PAX9 expression correlates with increasing malignancy of dysplastic and cancerous epithelium of the human oesophagus. J Pathol. 2002;197:293–7.
65. Kendall J, Liu Q, Bakleh A, et al. Oncogenic cooperation and coamplification of developmental transcription factor genes in lung cancer. Proc Natl Acad Sci U S A. 2007;104:16663–1666810.
66. Polyakova VO, Kvetnoy IM, Konovalov SS, et al. Age-related differences in expression of signal differentiation factors for human thymic epithelial cells. Bull Exp Biol Med. 2007;144:235–7.
67. Relaix F, Rocancourt D, Mansouri A, Buckingham M. A Pax3/Pax7-dependent population of skeletal muscle progenitor cells. Nature. 2005;435:948–53.
68. Lorenzo PI, Fuente-Martín E, Brun T, et al. PAX4 defines an expandable β-cell subpopulation in the adult pancreatic islet. Sci Rep. 2015;5:15672.
69. Ashery-Padan R, Zhou X, Marquardt T, et al. Conditional inactivation of Pax6 in the pancreas causes early onset of diabetes. Dev Biol. 2004;269:479–88.
70. Tuoc TC, Radyushkin K, Tonchev AB, et al. Selective cortical layering abnormalities and behavioral deficits in cortex-specific Pax6 knock-out mice. J Neurosci. 2009;29:8335–49.

71. von Maltzahn J, Jones AE, Parks RJ, Rudnicki MA. Pax7 is critical for the normal function of satellite cells in adult skeletal muscle. Proc Natl Acad Sci U S A. 2013;110:16474–9.
72. Martínez-Hernández R, Bernal S, Alias L, Tizzano EF. Abnormalities in early markers of muscle involvement support a delay in myogenesis in spinal muscular atrophy. J Neuropathol Exp Neurol. 2014;73:559–67.
73. Plachov D, Chowdhury K, Walther C, et al. Pax8, a murine paired box gene expressed in the developing excretory system and thyroid gland. Development. 1990;110:643–51.

7. Telomeres, Telomerase, and Aging

Deepak K. Mishra, Ramraj Prasad, and Pramod Yadava

Living systems have acquired varied degrees of proficiency through evolutionary process in terms of the capability to carry on vital functions that include:

(a) Cell growth and division that would involve regulation of DNA replication, transcription, and translation
(b) Drawing of raw materials from the environment
(c) Sensing and responding to the environment in terms of availability of food, restrictive factor, and pathogens
(d) Following a developmental program as inscribed in the genome that largely depends on internal signaling
(e) Digesting and assimilating the materials to be recast into their own mass
(f) Ensuring distribution of essential requirements to all constituent cells in case of multicellular organisms
(g) Defending one's own being against invaders
(h) Sorting and disposing off waste materials resulting from all the above activities
(i) Healing injuries
(j) Maintaining redox balances
(k) Repairing damaged biomolecules, particularly nucleic acids
(l) Replacing lost cells with new ones (for multicellular organisms)
(m) Producing gametes and propagules to ensure continuity of life through generations
(n) Tuning of time in diurnal and seasonal frames

D. K. Mishra · R. Prasad
School of Life Sciences, Jawaharlal Nehru University, New Delhi, India

P. Yadava (✉)
Department of Biological Sciences, IISER, Berhampur, Odisha, India
e-mail: pkyadava1953@gmail.com

Intactness of the above functions defines good healthy living, and any irreversible deteriorations mark dwindling health, senescence, and aging. Since these functions are essential, each one of them offers a nodal point characteristically performing certain functions and often influencing functioning of other nodal points. This enables the system to modulate each group of functional biomolecules in concert with others.

Growth is an obligatory process for an organism and aging marks the limit over it. Both phenomena are kept in balance throughout life. The first half of life remained elusive to aging parameters and growth prevails. The other side of life manifests the aging process surpassing the growth rate. Growth is an outcome of both gain in mass and number of cells. Aging is a phenotypic manifestation of evolutionary adopted deregulation of our biological system. Growth and/or aging is the net result of dynamic equilibria of biological processes in an organism.

The Conflict Between Stability and Flexibility

Living systems consist of molecules with "intermediate" thermodynamic stability in order to be part of the dynamic phenomena in the cell and therefore would likely deteriorate with age. Intermediate stability here would mean a state that is characteristic and stays long enough to function but can be reorganized in response to surrounding stimuli. However, at variance from deterioration in nonliving materials, living systems are equipped to reverse to a large extent such changes that would lead to organizational and functional decline in them. This may seem to be in contrast with the earlier held view that individual cells are immortal and it is the tissue, organs, and organisms that really undergo age-dependent changes. It is true nevertheless that signs of senescence and aging are more easily discernible at the level of multicellular organisms than for individual cells. The success of Alexis Carrel in culturing heart-derived fibroblast for over three decades led to the false conclusion that cells were immortal by default [11]. Experiments of Hayflick, thoroughly reinvestigating serial culture of cells, showed that cells put to culture meet a crisis at certain point corresponding to a finite number of doublings [24]. Thus, telomere length may seem to be an intrinsic determinant of cellular lifespan. It remains debatable if there are other telomere-independent cell-intrinsic determinants of replicative lifespan [32]. It has been found that telomere length and unique telomerase alleles may be responsible for long healthy lifespan among centenarians of Ashkenazi Jewish community [3]. Telomerase reactivation increases the replicative lifespan of cells, but it is debatable if it necessarily gets translated into organism's lifespan [13]. It nevertheless does promise an aide in regenerative cell therapy, e.g. telomerase-immortalized corneal cells give rise to morphologically distinct hexagonal and functional cells [41]. Certain cells lacking telomerase activity can still maintain telomere length via alternative mechanisms involving recombination [10].

The End Replication Problem and Replicative Senescence

Eukaryotic genomes usually consist of a characteristic number of chromosomes consisting of linear double-stranded DNA molecules. The termini of such molecules, if not specially organized, would look like double-strand breaks and would recruit damage monitoring and repair molecules. In order to distinguish them from double-strand breaks, the telomeres are organized by self-looping stretch consisting of a tandem array of small hexanucleotide repeats (typically 5'TTAGGG3' for mammals spread over up to 15 kb) and are studded with several telomere-specific proteins. DNA polymerases involved in replication require a primer perfectly annealed with the template strand and extend the copy strand in 5' to 3' direction. Thus, the copy strand can grow continuously in one direction (continuous or leading strand) and by repeated primer synthesis and extension in the other direction (discontinuous or lagging strand) if one begins with a replication bubble. As the discontinuous strand approaches close to the 3' end of the template, it finds impossible to use the same stretch for synthesis of RNA primer as well as copy strand. Thus, a fair part of DNA at such ends is left unreplicated and the copy synthesized by DNA polymerase becomes shorter than the template. If this process continues, the attrition of telomere will reach crucial coding region of the chromosome (Fig. 7.1).

Extrinsic Determinants of Lifespan

Cellular senescence is typically marked by arresting the cells in a certain phase by overexpression of CDKIs which follows damages to cell membrane, proteins, or DNA. This may precede telomere shortening to a critical extent and raises the question if senescence can directly be induced by extrinsic factors bypassing the telomerase-dependent mechanisms. Figure 7.2 summarizes various pathways that can result in expression of endogenous senescence markers like b-galactosidase. Thus oncogenic activation, DNA damage, ceramide-induced stress, and overexpression of CDKIs can all confer a phenotype typical of cellular senescence. Differentiated cells have a lower proliferative potential compared to diploid fibroblasts and reach senescence via a telomerase-independent pathway. Loss of CDK inhibitor such as p16 is as important as expression of telomerase for sustained proliferation of differentiated cells in culture.

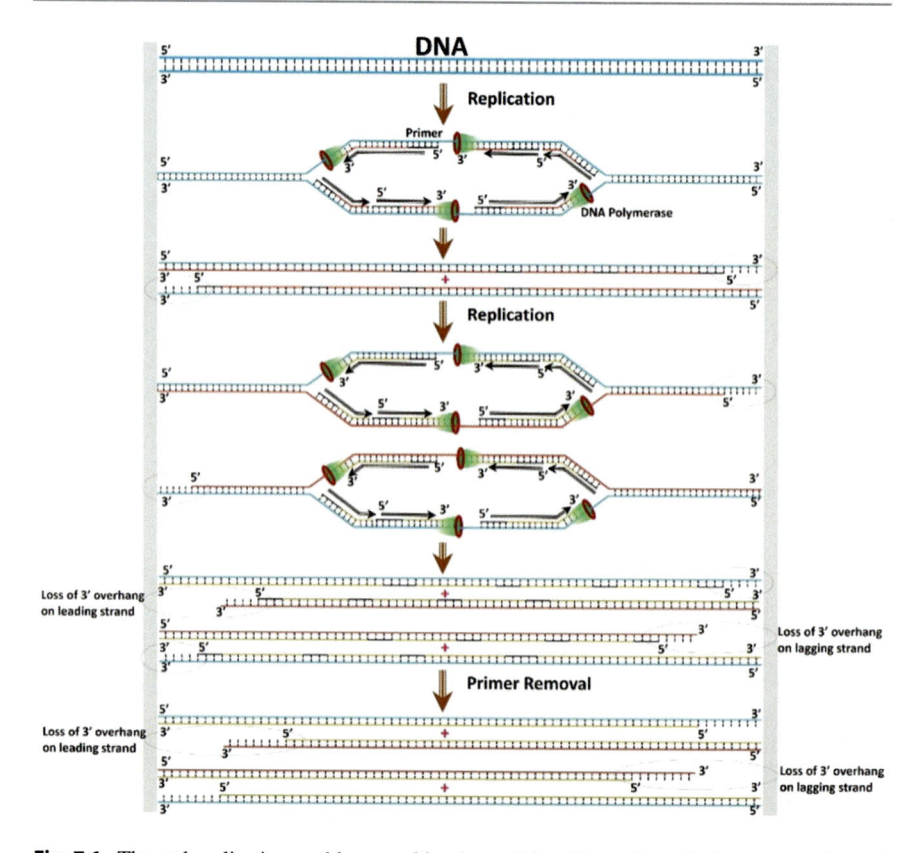

Fig. 7.1 The end replication problem resulting in partial attrition of terminal sequences through successive replication cycles

Theories of Aging

Many theories have been put forward to account for aging and senescence, but none provides an unequivocal mechanism. These theories envisage (i) accumulation of genetic errors, (ii) destabilization of the genome, (iii) accumulation of free radicals and oxidative stress, (iv) dysfunction of mitochondria, (v) accumulation of errors above bearing capacity of the system, (vii) accumulation of misfolded proteins, (viii) dysregulation of protein homeostasis, (ix) gross epigenetic alterations, and (x) hyperfunctioning of the system with advancing age [21]. Perhaps no single theory will ever be able to incorporate all the diverse parameters that show association with aging. However, it should be possible on one hand to restore some of them and prolong healthy lifespan and on the other assess in their terms the phase in aging.

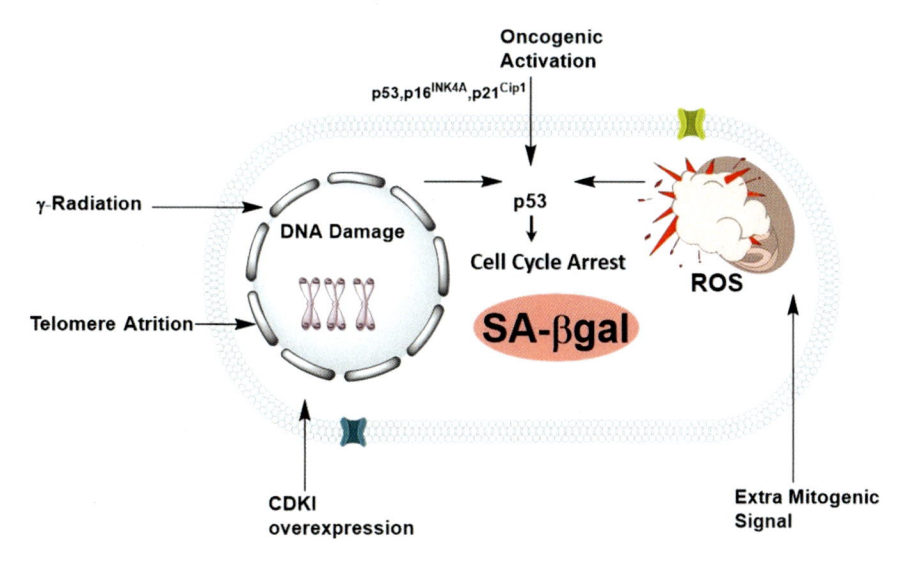

Fig. 7.2 A schematic presentation of mechanisms of inducing cellular senescence may or may not involve telomere length (Adapted from: [32])

Telomeres, Senescence, and Cancer

Senescence at the point of attrition of telomeres below a critical length probably reflects the cell's bid to avoid death. It is debatable if malignant transformation is a loss of the innate program to enter senescence or a positive avoidance of senescence by reactivating telomerase. Ectopic expression of telomerase in cells also helps bypass senescence, while disruption of telomerase in such cells leads to apoptosis. Working with Terc−/− and Terc+/− mice, Samper et al. [40] found that restoration of telomerase activity could rescue animals from chromosomal instability and premature aging[1]. Taking a comparative account of telomere length and lifespan, however, it seems unlikely that telomeres and telomerase could be primary determinants of aging process per se or of lifespan [25]. A very positive note came from studies on mice receiving AAV-mediated telomerase expression asserting that there was a significant (up to 24%) increase in lifespan, while no increase in frequency of cancer could be noticed [7].

Cells of the hematopoietic lineage do regain telomere length upon antigenic stimulation and serial transplantation [1]. Telomere elongation overtakes telomere attrition in these situations. Transgenic mice constitutively expressing telomerase were found to have increased frequency of mammary carcinoma in aging mice [2][2]. Transcriptomic analysis with specially prepared "old" yeast cells showed similarity with that of telomerase-deficient cells in terms of energy storage metabolism and induction of DNA repair genes [30]. This suggested that even unicellular organisms undergo aging-like process and the interactions of telomerase in cells affect functions beyond telomere length maintenance.

Telomerase interacts with several cellular proteins to effect telomere elongation and other extracurricular functions. Mice lacking telomerase and some of the DNA repair proteins like Ku86 and catalytic domain of DNAPK show accelerated loss of telomere length which mice lacking poly-ADP-ribose polymerase (PARP) and telomerase do not show such acceleration [18].

The Telomerase Complex

Telomeres typically consist of characteristic hexamer (may vary in some species) repeats of varied length. This can account for up to 50 kb of terminal DNA in mouse. The double-stranded region is richly studded with proteins like TRF1 and TRF2, while the single-stranded region is complexed with POT1. Many other proteins (e.g., TIN2) play a bridging role, regulating activity of the partner proteins, keep the multi-subunit telomeric complex together, and hide open end of chromosomal DNA (Fig. 7.3). PINX1, an inhibitor of telomerase, abridges TRF1 with telomerase and inhibits telomere elongation [44].

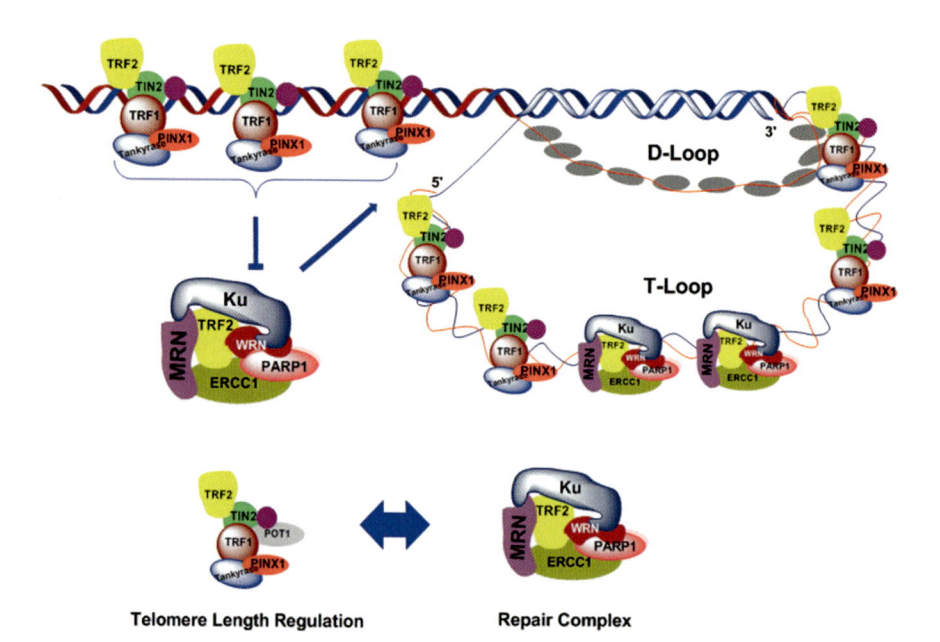

Fig. 7.3 Organization of the telomeric complex that consists of telomeric repeats (TTAGGG for mammals) over a largely double-stranded region ending in a G-rich 3′ overhang. The double-stranded region forms a T-loop, thus facilitating invasion by the single-stranded overhang in the double-stranded region forming a displacement or D-loop. (Adapted from [8])

Telomerase in Stimulated T Cells

Cells of the immune system are uniquely under the need to proliferate in a programmed manner as happens following stimulation. The activated cells upregulate telomerase expression during the first and second activation, but in the third and subsequent cycles of activation in vitro, there seems to be no major response in terms of telomerase expression in CD8+ T cells. However, these cells continue to divide in response to later stimuli and lose telomere length culminating into DNA damage and cell cycle arrest. Thus senescent CD∗ + cells accumulate in vivo. In contrast with CD∗ T cells, the CD4 cells seem to retain their responsiveness to antigen stimulation till seventh encounter. It is suggested that expressing telomerase might rejuvenate the CD8 T cells [15]. Importance of studying human subjects for association of telomerase with immune functions has been emphasized since many reprogramming factors and patterns such as psychological stress are unique to human immune system [16]. Table 7.1 shows a comparison among carriers of heterozygous mutations in either TERT or TERC showed that these individuals lost telomeric length to a larger extent than age-matched homozygous nonmutants [4].

Diseases Resulting from Telomerase Dysfunction

Heterozygous state of mutations in hTERT and hTERC results in diseases like dyskeratosis congenital, bone marrow failure, and idiopathic pulmonary fibrosis [19]. Disease manifestations can be further aggravated by environmental and genetic factors that result in telomere shortening over the basal effects of haploinsufficiency. Dyskeratosis congenita is characterized by reticulate skin pigmentation, white patches in the oral mucosa, and deformity in nails. The disease can have diverse genetic origin with some patients showing signs of X-linked disease. It has been claimed that natural product-derived drugs can activate telomerase and push the median telomere length significantly upward though the maximum telomere length is not altered to the same extent in subject showing telomere shortening following persistent viral infection [23, 46].

Telomerase Downregulation May Be a Result of Differentiation Rather than Aging of Myoblasts

Unlike differentiated myocytes, muscle stem cells (satellite cells) show a good expression of telomerase even in old mice. Myoblasts adopted for primary culture rapidly downregulate telomerase expression suggesting that the satellite cellos rather than cultured myoblasts should be used for replacement cell therapy [36].

Table 7.1 Telomere loss is more severe in individuals carrying heterozygous mutations in telomerase as compared to their age-matched relatives ([4] permission pending)

Δtel (kb)[a]	Lymphocytes	Granulocytes	CD20+	CD45RA + CD20−	CD45RA−	CD57+
Tel +/− individuals (60)	−2.7	−2.9	−2.8	−3.2	−2.3	−2.4
TERT +/− individuals (37)	−2.6	−2.9	−2.7	−3.1	−2.3	−2.5
TERC +/− individuals (23)	−2.8	−2.9	−2.9	−3.3	−2.3	−2.1
Tel +/+ relatives (36)	−0.6	−0.9	−0.6	−0.7	−0.6	−0.6
Tel +/+ parent (7)	−0.9	−0.8	−0.8	−1.1	−0.8	−0.6
Tel +/+ sibling (8)	−0.7	−0.7	−0.8	−0.9	−0.8	−0.6

Summary of telomere loss adjusted for age (Δtel) in telomerase heterozygous individuals considered as a group (*Tel*), considered as *TERT* or *TERC* heterozygous individuals separately, and unaffected relatives of telomerase heterozygous individuals considered as a group (*Tel*), or separating parents or siblings
[a]Difference between the median telomere length and the regression estimate of the telomere length distribution in healthy individuals of the same age (Δtel).
Tel: *hTERC* or *hTERT* genes. Doi: 10.1371/journal.pgen.1002696.t002

Some Food Supplements and Plant Products Enhance Telomerase Activity

While studying the effect of a food supplement (2007-0721-GX), Lin et al. [31] found a clear increase in telomerase activity and telomere length and a distinct rise in CD34+ cells. All the parameters used for assaying the antiaging effect of food supplements in these studies were related to aging. These parameters were CD34 expression, telomerase activity, and insulin-like growth factor (IGF-1). Many plant products have been demonstrated to inhibit or activate telomerase.

Multifunctional Complex with Multifunctional Partners

While TERT and TERC are associated with crucial determinants of cell function spanning over ribosome biogenesis, cell cycle control, plasminogen activator, and Kruppel-like factor to mention just a few, even the proteins that seem to be functionally involved in regulating telomerase turn out to be multifunctional. TCTP and TERT influence each other's expression in a reciprocal manner. TCTP turns out to be an evolutionarily conserved protein seeming to be an anti-oxidative molecule on one hand and being associated with TGF-β signaling [20] and inhibition of apoptosis on the other [22].

Anti-apoptotic Proteins and Inhibitor of Apoptosis Protein (IAP) Family

Telomerase-associated extracurricular activities show its role in survival and proliferation. However, enhanced expression and activity of telomerase in cells were found to be modulated by other factors like inhibitory apoptotic proteins (IAPs), anti-apoptotic proteins. Association of such factors with elevation of telomerase activity suggests its potential role as a determinant of cellular lifespan and cancer. Survivin (IAP) is one of them that maintains elevated activity of TERT in colorectal adenomas. The correlation between anti-apoptotic proteins like Bcl-2, MCL-1, and hTERT reinstates the same in head and neck cancer. Although these results are shown in cancer, the extrapolation of these observations suggests their possible role in cellular aging.

Telomere Length and Telomerase Activity Measurement

Longer telomere length during early age shows longevity in birds. However, the refuting reports show that it is inflammation that is responsible for aging rather than telomere attrition. The reports signify the assessment of telomere length. The measurement of telomere length can be assayed by both PCR-based technique, (a) terminal restriction fragment and (b) PCR with telomere-specific primers, or by microcopy. The microscope-based techniques like (c) Q-FISH with mitotically active cells, (d) Q-FISH with interphase cells, and (e) Flow-FISH have been used to quantify the extent of telomere retained in the cells. Apart from telomere length, the availability of telomerase enzyme in cells suggests to find its activity in cells. Telomerase activity is mainly assayed by PCR-based technology. It is known as telomere repeat amplification protocol. Here we check the abundance of telomere fragments by gel electrophoresis after amplification of telomere repeats using cellular lysate containing telomerase.

Telomerase, Telomere, and Redox Homoeostasis

Mitochondrial health and aging have been the talk of town since 1980. The mitochondrial theory of aging given by Miquel suggested that the DNA damage accumulation induced by ROS generation in mitochondria costs into aging. Although it is a telomere-independent aging factor, however, the mitochondrial stress-induced increased expression of telomerase has been seen as an independent function of telomerase to telomere lengthening activity. Telomerase translocation in mitochondria under H_2O_2 treatment and rescue of cells after telomerase overexpression display the antiaging effect of telomerase independent of telomere elongation. Telomerase functional mechanism in mitochondria is still obscure, but current surveys suggests its availability in cell displays the healthy organelle.

Stress and Socioeconomic Coordinates of Aging

Socioeconomic status may influence life expectancy among humans though its effect on aging remains to be established [12]. Similarly, individuals subjected to chronic stress, e.g., mothers of chronically unwell children, show symptoms of accelerated cellular aging like shortening of telomeres [17]. It is a conjecture that some of these situations may lead to rapid attrition of telomeres which may be equivalent of several decades of aging process. Under oxidative stress nuclear TERT is exported to the cytosol following tyrosine phosphorylation of TERT mediated by Src kinases. This is followed by telomere shortening and aging of cells. However, there are other factors that tend to retain TERT within the nucleus. Nuclear fibroblasts deficient in Src, Fyn, and Yes do not show such export of TERT. The tyrosine phosphatase Shp-2 is projected to counteract export of TERT from the nucleus and as probable antiaging target [27]. There seems to be a strong correlation between food habit, obesity, stress, and telomerase activity [14]).

Certain Proteins Do Modulate Pluripotency

Nuclear transplant into oocytes results in reprogramming of gene expression in a manner to produce embryo-like pattern of development accompanying overexpression of retinoic acid receptor, oct4, and nanog and a depletion of tpt1 (TCTP).

Knocking down the retinoic acid receptor results in overexpression of oct1 and nanog, while depletion of tpt1 results in lowering of oct4 and nanog transcription. Oct4 and nanog are established markers/determinants of pluripotency positively influenced by tpt1 [28]. In addition to these, many other proteins like GCNF [26], SF1 [6], LRH1 [29], and Sal1 [47] also affect the transcription of oct4. Telomerase expression is indispensable for pluripotency. The induced pluripotent stem cells need telomerase at early stage of and may be further assisted by ALT pathway.

GHRH Antagonists Reflect on Association with Telomerase in Younger Mice

Administration of GHRH antagonist in 2–4-month-old mice resulted in increased telomerase expression, improved cognitive functions, average longevity (but not maximal lifespan), and decreased incidence of cancer. Older animals (>7 month) did not show sensitivity to a similar regimen of treatment [5]. There seems to be a fine-tuning of telomerase expression with the growth and developmental program of mouse. Inhibition of cell cycle progression with aphidicolin or CGK1026 resulted in lowering of interferon-inducible protein IFI16 and overexpression of telomerase in human diploid fibroblast. Knocking down IFI16 induces c-myc expression which in turn increases the expression of hTERT, while overexpression of IFI16 resulted in suppression of c-myc expression and lower signals with hTERT-Luc reporters [43].

Telomerase RNA Component Variants Influence Telomere Length

A compilation of analyses involving 2953 white subjects showed significant association of a common variant (rs2293607, G/A) of telomerase RNA and short telomere length reflecting on role of telomerase RNA as a determinant of telomerase function and telomere length. One would expect evolutionarily selected form of TERC to be optimal for interaction with TERT and telomerase function. A change in RNA can affect general association, affinity, and half-life of the associated state and even processivity of the telomerase enzyme complex [35].

Epigenetic modification and chromatin microenvironment of the telomerase locus are an important determinant of its expression [48].

Exposure to Environmental Pollutants Can Retard Telomerase Activity and Reduce Telomere Length

Exposure to Chicago air mixed with volatile polychlorinated biphenyls resulted in lowering of telomerase activity and erosion of telomeres decreasing cell proliferation of 18 days [42]. Assaults on cells damaging crucial molecules (particularly

DNA) are responded to by cytokine release and inflammation which in turn is controlled by certain RNA-binding proteins that degrade cytokine mRNA. One such protein AUF-1 is known to activate telomerase activity and suppress senescence, thus interconnecting inhibition of both inflammatory response and cellular senescence while promoting telomere length maintenance [38].

The Dream Therapeutic Strategy

Targeted inactivation of telomerase proteins employing immunotherapeutic approach or to knock down telomerase-related RNA would be anyone's goal even though this is not as simple to achieve [37]. It is possible to transfect viral vectors delivering expression cassette of the therapeutic peptide or RNA from a TERT promoter. One may also expect to develop nucleic acid or peptide aptamers to target telomerase or to target molecules that will result in sensitization of cells to low doses of a conventional therapeutic molecule. Glyceraldehyde-phosphate dehydrogenase (GAPDH) undergoes nuclear localization under oxidative stress and induces apoptosis. It has been noticed to interact with telomerase RNA, inhibiting telomerase activity and causing telomere shortening [34]. Since the substrate GAPD and an NO donor S-nitrosoglutathione (GSNO) negatively regulate GAPDH, it is suggested that these molecules may be used to target telomerase via GAPDH.

Telomerase Has Pleiotropic Regulatory Interaction with Other Cellular Molecules

hTR has a basal expression level in most of the cells, while expression of hTERT goes in parallel with the cells undergoing division [33]. However, in knockout mice lacking telomerase RNA, restoration of RNA was found to be associated with expression of several proteins [9]. Working with cultured mammalian cell lines, Ramkrishnan et al. [39] reported that expression of several proteins involved in ribosome biogenesis, chromatin modulation, cell cycle control, and p63-dependent pathways were perturbed in cells expressing hTR-targeting hammerhead ribozyme. Likewise, genome-wide profiling of differentially expressed genes following siRNA- and shRNA-mediated knockdown of hTERT in HeLa cells showed overexpression of KLF4, FGF2, IRF-9, and PLAU by real-time PCR [45]

References

1. Allsop R, Cheshier S, Weissman I. Telomerase reactivation and rejuvenation of telomere length in stimulated T cells derived from serially transplanted hematopoietic stem cells. J Exp Med. 2002;196:1427–33.
2. Artandi SE, Alson S, Tietze MK, Sharpless NE, Ye S, Greenberg RA, Castrillon DH, Horner JW, Weiler SR, Carrasco RD, DePinho RA. Constitutive telomerase expression promotes mammary carcinomas in aging mice. Proc Natl Acad Sci. 2002;99:8191–6.

3. Atzmon G, Cho M, Cawthon RM, Budagov T, Katz M, Yang X, Siegel G, et al. Genetic variation in human telomerase is associated with telomere length in Ashkenazi centenarians. Proc Natl Acad Sci U S A. 2009;107:1–8.. Retrieved from http://www.ncbi.nlm.nih.gov/pubmed/19915151
4. Aubert G, Baerlocher GM, Vulto I, Poon SS, Lansdorp PM. Collapse of telomere homeostasis in hematopoietic cells caused by heterozygous mutations in telomerase genes. PLoS Genet. 2012;8(5):e1002696.
5. Banks W a, Morley JE, Farr S a, Price TO, Ercal N, Vidaurre I, Schally AV. Effects of a growth hormone-releasing hormone antagonist on telomerase activity, oxidative stress, longevity, and aging in mice. Proc Natl Acad Sci U S A. 2010;107:22272–7.
6. Barnea E, Bergman Y. Synergy of SF1 and RAR in activation of Oct-3/4 promoter. J Biol Chem. 2000;275:6608–19.
7. Bernardes de Jesus B, Vera E, Schneeberger K, Tejera AM, Ayuso E, Bosch F, Blasco MA. Telomerase gene therapy in adult and old mice delays aging and increases longevity without increasing cancer. EMBO Mol Med. 2012;4(8):691–704.
8. Blasco. Mice with bad ends: mouse models for studying telomeres and telomerase in cancer and aging. EMBO J. 2005;24:1095–103.
9. Blasco MA, Lee H-W, Hande MP, Samper E, Lansdorp PM, DePinho RA, Greider CW. Telomere shortening and tumor formation by mouse cells lacking telomerase RNA. Cell. 1997;91:25–34.
10. Bryan TM, Englezou A, Dalla-Pozza L, Dunham MA, Redell RR. Evidence for an alternative mechanism for maintaining telomere length in human tumors and tumor-derived cell lines. Nat Med. 1997;3:1271–4.
11. Carrel A. "On the Permanent Life of Tissues Outside of the Organism" (PDF). J Exp Med. 1912;15(5):516–28.
12. Cherkas LF, Aviv A, Valdes AM, Hunkin JL, Gardner JP, Surdulescu GL, Kimura M, Spector TD. The effects of social status on biological aging as measured by white-blood-cell telomere length. Aging Cell. 2006;5(2006):361–5.
13. Cowell JK. Telomeres and telomerase in ageing and cancer. Age (Omaha). 1999;22:59–64.
14. Daubenmier J, Lin J, Blackburn E, Hecht FM, Kristeller J, Maninger N, Kuwata M, et al. Changes in stress, eating, and metabolic factors are related to changes in telomerase activity in a randomized mindfulness intervention pilot study. Psychoneuroendocrinology. 2012;37(7):917–28.
15. Effros R. Telomerase induction in T cell: a cure for aging and disease. Exp Gerentol. 2007;42:416–20.
16. Effros RB. Telomere/telomerase dynamics within the human immune system: effect of chronic infection and stress. Exp Gerontol. 2011;46:135–40.
17. Epel ES, Blackburn EH, Lin J, Dhabhar FS, Adler NE, Morrow JD, Cawthon RM. Accelerated telomere shortening in response to life stress. Proc Natl Acad Sci U S A. 2004;101(2004):17312–5.
18. Espejel S, Klatt P, Murcia JM-D, Martín-Caballero J, Flores JM, Taccioli G, De Murcia G, Blasco MA. Impact of telomerase ablation on organismal viability aging and tumorigenesis in mice lacking the DNA repair proteins PARP1, Ku86 or DNAPKc. J Cell Biol. 2004;167:627–38.
19. Garcia CK, Wright WE, Shay JW. Human diseases of telomerase dysfunction: insights into tissue aging. Nucleic Acids Res. 2007;35:7406–16.
20. Gnanasekar M, Ramaswamy K. Translationally controlled tumor protein of Brugia malayi functions as an antioxidant protein. Parasitol Res. 2007;101:1533–40.
21. Goto S. Biological mechanisms of aging, a historical and critical overview. In: Mori N, Mook-Jung I, editors. Aging Mechanisms: Springer; 2015. https://doi.org/10.1007/978-4-431-55763-0_1.

22. Graidist P, Yazawa M, Tonganunt M, Nakatomi A, Lin CC, Chang J, Phongdara A, Fujise K. Fortilin binds Ca^{2+} and blocks Ca^{2+} dependent apoptosis in vivo. Biochem J. 2007;408:181–91.
23. Harley CB, Liu W, Blasco M, Vera E, Andrews WH, Briggs LA, Raffaele JM. A natural product telomerase activator as part of a health maintenance program. Rejuvenation Res. 2011;14(1):45–56.
24. Hayflick L, Moorhead P. The serial cultivation of human diploid cell strains. Exp Cell Res. 1961;25:585–621.
25. Hornsby P. Telomerase and aging process. Exp Gerontol. 2007;42:575–81.
26. Hummelke GC, Cooney AJ. Germ cell nuclear factor is a transcriptioinal repressor for embryonic development. Front Biosci. 2001;6:1186–91.
27. Jacob S, Shroeder P, Buechner N, Kunze K, Altschmeid J, Haendeler J. Nuclear Shp-2 keeps telomerase reverse transcriptase in the nucleus- new potential anti-aging target. Cell Commun Signal. 2009;7:A60. https://doi.org/10.1186/1478-811X-7-S1-A60.
28. Koziol MJ, Garrett N, Gurdon JB. Tpt1 activates transcription of oct4 and nanog in transplanted nuclei. Curr Biol. 2007;17:801–7.
29. LeMenuet PGD, Chung AC, Mancini M, Wheeler DA, Cooney AJ. Orphan nuclear receptor LRH-1 is required to maintain Oct4 expression at the epiblast stage of embryonic development. Mol Cell Biol. 2005;25:3492–505.
30. Lesur I, Campbell JL. The transcriptome of prematurely aging yeast cell is similar to that of telomerase-deficient cells. Mol Cell Biol. 2003;15:1297–312.
31. Lin PC, Chiou TW, Liu PY, Chen SP, Wang HI, Huang PC, Lin SZ, et al. Food supplement 20070721-GX may increase CD34+ stem cells and telomerase activity. J Biomed Biotechnol. 2012;2012:498051.
32. Lloyd AC. Limits to lifespan. Comment Nat Cell Biol. 2002;4:E25–7.
33. Meyerson M, Counter CM, Eaton ENG, Ellisen LW, Steiner P, Dickinson Caddie S, Ziaugra L, Beijersbergen RL, Davidoff MJ, Qingyun L, Bacchetti S, Haber DA, Weinberg RA. hEST2, the putative human telomerase catalytic subunit gene, is up-regulated in tumor cells and during immortalization. Cell. 1997;90:785–95.
34. Nicholls C, Pinto AR, Li H, Li L, Wang L, Simpson R, Liu J-P. Glyceraldehyde-3-phosphate dehydrogenase (GAPDH) induces cancer cell senescence by interacting with telomerase RNA component. Proc Natl Acad Sci. 2012;109(33):13308–13.
35. Njajou OT, Blackburn EH, Pawlikowska L, Mangino M, Damcott CM, Kwok PY, Spector TD, et al. A common variant in the telomerase RNA component is associated with short telomere length. PLoS One. 2010;5(9):e13048.
36. O'Connor MS, Carlson ME, Conboy IM. Differentiation rather than aging of muscle stem cells abolishes their telomerase activity. Biotechnol Prog. 2009;25:1130–7.
37. Ouellette MM, Wright WE, Shay JW. Targeting telomerase-expressing cancer cells. J Cell Mol Med. 2011;15:1433–42.
38. Pont AR, Sadri N, Hsiao SJ, Smith S, Schneider RJ. MRNA decay factor AUF1 maintains Normal aging, telomere maintenance, and suppression of senescence by activation of telomerase transcription. Mol Cell. 2012;47(1):5–15.
39. Ramakrishnan SK, Varshney A, Sharma A, Das BC, Yadava PK. Expression of targeted ribozyme against telomerase RNA causes altered expression of several other genes in tumor cells. Tumor Biol. 2014;35:5539–50.
40. Samper E, Flores J, Blasco M. Restoration of telomerase activity rescues chromosomal instability and premature aging in mice. EMBO Rep. 2001;2:800–7.
41. Schmedt T, Chen Y, Nguyen TT, Li S, Bonanno JA, Jurkunas UV. Telomerase immortalization of human corneal endothelial cells yields functional hexagonal monolayers. PLoS One. 2012;7(12):e51427.
42. Senthilkumar PK, Klingelhutz AJ, Jacobus JA, Lehmler H, Robertson LW, Ludewig G. Airborne polychlorinated biphenyls (PCBs) reduce telomerase activity and shorten telomere length in immortal human skin keratinocytes (HaCat). Toxicol Lett. 2011;204(1):64–70.

43. Song LL, Ponomareva L, Shen H, Duan X, Alimirah F, Choubey D. Interferon-inducible IFI16, a negative regulator of cell growth, down-regulates expression of human telomerase reverse transcriptase (hTERT) gene. PLoS One. 2010;5(1):e8569.
44. Soohoo CY, Shi R, Lee TH, Huang P, Lu KP, Zhou XZ. Telomerase inhibitor PinX1 provides a link between TRF1 and telomerase to prevent telomere elongation. J Biol Chem. 2011;286(5):3894–906.
45. Varshney AK, Ramkrishnan SK, Sharma AK, Santosh B, Bala J, Yadava PK. Global expression profile of telomerase-associated genes in HeLa cells. Gene. 2014;547:211–7.
46. Westin ER, Aykin-Burns N, Buckingham EM, Spitz DR, Goldman FD, Klingelhutz AJ. The p53/p21(WAF/CIP) pathway mediates oxidative stress and senescence in dyskeratosis congenita cells with telomerase insufficiency. Antioxid Redox Signal. 2011;14(6):985–97.
47. Zhang J, Tam WL, Tong GQ, Wu Q, Chan HY, Soh BS, Lou Y, Yang J, Ma Y, Chai L, et al. Sal4 modulates embryonic stem cell lpluripotency and early embryonic development by the transcriptional regulation of Pou5f1. Nat Cell Biol. 2006;8:1114–23.
48. Zhu J, Zhao Y, Wang S. Chromatin and epigenetic regulation of the telomerase reverse transcriptase gene. Protein Cell. 2010;1:22–32.

The Epigenome of Aging

Rohini Muthuswami

Introduction

The gradual decline of normal physiological functions in a time-dependent manner is a defining hallmark of the aging process. A consequence of aging, of course, is that the organism becomes increasingly prone to diseases including cancer, cardiovascular disorders, diabetes, and neurodegenerative disorders. The aging process is similar yet different from the cellular or replicative senescence wherein the normal diploid cells stop dividing once they reach a maximum limit defined as Hayflick limit [1]. In contrast, nine hallmarks have been described for the aging process: genomic instability caused due to DNA damage, telomere shortening, epigenetic alterations, loss of proteostasis, deregulated nutrient sensing, mitochondrial dysfunction, cellular senescence, stem cell exhaustion, and altered intracellular communication [2]. These nine hallmarks have been further classified into primary hallmarks, antagonistic hallmarks, and integrative hallmarks. The primary hallmarks – DNA damage, telomere shortening, epigenetic alterations, mitochondrial dysfunction, and loss of proteostasis – are considered as negative hallmarks. Accumulation of DNA damage, epigenetic alterations, and mitochondrial dysfunction are strongly associated with aging. The effects of the antagonistic hallmarks, on the other hand, depend on their intensity. Thus, at low intensity these hallmarks might have beneficial effects that become deleterious at higher intensity. Finally, the integrative hallmarks like altered intracellular communications affect tissue homeostasis and function [2]. In this review, I will be focusing on epigenetic alterations, a primary hallmark of aging. The epigenetic markers regulate both gene expression as well as protein function and thus, modulate all the hallmarks of aging.

R. Muthuswami (✉)
Chromatin Remodeling Laboratory, School of Life Sciences, Jawaharlal Nehru University, New Delhi, India
e-mail: rohini_m@mail.jnu.ac.in

© Springer Nature Singapore Pte Ltd. 2020
P. C. Rath (ed.), *Models, Molecules and Mechanisms in Biogerontology*,
https://doi.org/10.1007/978-981-32-9005-1_8

The DNA Damage Theory of Aging

Aging is long known to be associated with the accumulation of DNA damage [3, 4]. Damage to DNA can be induced both by cellular metabolites like reactive oxygen species and exogenous species like radiation. The damage to DNA leads to loss of information and thus the formation of mutated proteins leading to loss of function. Therefore, the damage caused to the DNA must be repaired and the cell responds to this signal through an intricate network of DNA damage repair proteins and cell cycle checkpoints [5, 6]. The failure to repair the damage DNA, therefore, leads to genomic instability and is now known to be the underlying cause of both cancer and aging-related disorders [6]. Indeed, the DNA damage theory of aging postulated by Leo Szilard states that the accumulation of DNA damage and ensuing loss of function as well as cellular homeostasis are the leading cause of aging [7]. The hypothesis is borne out by the progeroid diseases as WS, XP, CS, and AT are caused by mutations in the DNA repair genes [8, 9].

Depending on the type of DNA damage, the cell uses one of three main repair pathways – the nucleotide excision repair (NER), the base excision repair (BER), and the double-strand DNA damage repair (DDR) pathway. The NER repair pathway consists of two subsystems: global genome repair (GGR) pathway that repairs damage occurring in the transcriptionally inactive regions of the genome; and the transcription-coupled repair (TCR) pathway that repairs the damage occurring in the transcribing DNA [10]. The BER pathway is initiated by DNA glycosylase that removes the damaged DNA base, thus creating an abasic site. The abasic site is subsequently repaired by enzymes that break of phosphodiester backbone, fill the gap, and then ligate the ends [11, 12]. The third pathway, DDR, is activated whenever double-strand breaks occur. The DDR pathway too consists of two subsystems: the non-homologous end joining (NHEJ) pathway involving DNA-dependent protein kinase (DNA-PK) and Ku; and the homologous recombination (HR) pathway involving the MRN complex as well as Ataxia-telangiectasia mutated (ATM) and ATM- and Rad3-related (ATR) kinases [13, 14].

The efficiency of NER has been shown to decrease with age in cells as well as tissues [15]. The transcriptome analysis of old mice and progeroid mouse model induced by mutations in genes involved in NER pathway shows highly significant correlation [16]. The free radical theory proposes that accumulation of the reactive oxygen species that are produced in the mitochondria through the lifetime of an organism contributes to aging as they can damage the nucleic acids as well as proteins. Thus, reducing the production of reactive oxygen species results in lengthening the lifespan of an organism [17]. In yeast, a defective BER pathway results in reduced lifespan [18]. The senescent cells as well as tissues of old mice also show an accumulation of double-strand breaks [19, 20]. Thus, increasing DNA damage appears to strongly correlate with aging.

Further, it has been well established that the three repair pathways are inhibited by the presence of nucleosomes, and therefore, chromatin remodeling is essential for the repair to be effected in the in vivo milieu [21–23].

The Basis of Epigenetics

The packaging of chromosomes into a tiny nucleus using histones and other non-histone proteins is essential as the length of the genomic DNA far exceeds that of the size of the nucleus necessitating at least 10,000-fold compaction. The first level of compaction involves wrapping of the DNA molecule around histone octamers composed of H2A, H2B, H3, and H4. The crystal structure of the nucleosome shows that the histones possess a structured core with disordered N-terminal tails protruding outwards from the protein–DNA complex [24]. The protein–DNA interaction is mainly between the positively charged amino acids of the histone molecules and the phosphate backbone of the DNA. ~147 bp of DNA is wrapped around the histone octamer with two nucleosomes spaced approximately 200 bp apart. The second level of compaction involves in the fifth histone, H1, that binds to the 200 bp linker DNA and nucleosome–nucleosome interaction that results in the formation of solenoid structure [25]. The higher-order compactions require non-histone proteins including cohesion, S/MAR proteins, and topoisomerases [26].

The packaging of DNA into a compact structure ensures that the long piece of DNA can fit into the cell. However, it also restricts access to proteins that mediate transcription, repair, replication, and recombination. Therefore, the chromosomal structure must be unwrapped and re-wrapped as and when required – a process known as chromatin remodeling.

Histone Modifications

The idea of chromatin remodeling was advanced as early as 1960s, when Allfrey showed that acetylation was associated with an increase in transcription [27]. Studies in yeast showed that *GCN5* was required for transcription [28–30]. The isolation of the histone acetyltransferase from Tetrahymena and its similarity with Gcn5 led to the consolidation of the idea that histone acetylation is associated with transcription activation [31]. Since these studies, histones have been shown to be phosphorylated, methylated, ADP-ribosylated, SUMOylated, Ubiquitinated, and glycosylated [32, 33]. These modifications regulate the basic DNA metabolic processes either by activating or repressing them in a context-dependent manner leading to the hypothesis of histone code [34].

Histone Variants

In addition, the core histone molecules can be replaced by non-canonical histones that are known as histone variants [35]. The histone variants, unlike histones, are expressed throughout the cell cycle and can be deposited both in replication-independent and replication-dependent manner [36]. Histone variants are known for almost all histones except for H4 and can be modified by histone modifying enzymes [36].

ATP-Dependent Chromatin Remodeling Proteins

The modifications on histones per se do not alter the nucleosome spacing. The nucleosomes are repositioned by ATP-dependent chromatin remodeling proteins that use the energy released by ATP hydrolysis to alter the spaces between nucleosomes either by repositioning them or by evicting them [37, 38]. In addition, these proteins can also catalyze the exchange of core histones with histone variants [39].

The ATP-dependent chromatin remodeling proteins belong to the helicase superfamily due to the seven helicase motifs that they possess [40, 41]. The prototype, Snf2, was identified in a genetic screen from *S. cerevisiae* and remains the best-characterized ATP-dependent chromatin remodeling protein [42, 43]. The Snf2 protein, and its orthologs BRG1 and BRM in higher eukaryotes, regulate transcription by remodeling the nucleosome spacing [44]. These proteins form the SWI/SNF complex interacting with a wide range of proteins that help in modulating gene expression [44].

DNA Methylation

The architecture of the chromatin can also be regulated by DNA methylation and non-coding RNA. Methylation of the cytosine residues present as CpG dinucleotide is catalyzed by DNA methyltransferases [45]. The CpG dinucleotide can be present both within the gene body as well as upstream in the promoter regions. Generally, in mammalian cells, the CpG dinucleotide within the gene body is methylated while the dinucleotide stretches (also known as CpG islands) present in the promoter region are unmethylated [46, 47]. The DNA methylation is catalyzed by two classes of DNA methyltransferases: (1) Maintenance methyltransferase or DNMT1 which catalyzes the methylation of cytosine residues during S-phase; (2) De novo methyltransferase or DNMT3A and 3B that catalyzes the methylation of cytosine residues during embryogenesis. In general, DNA methylation is strongly correlated with transcription silencing [48, 49].

DNA methylation is erased by both passive as well as enzyme-catalyzed reactions [50, 51]. Passive demethylation occurs when DNMT1 fails to methylate DNA during replication process either because the enzyme has been inactivated or because its expression has been downregulated [52]. A specific DNA demethylase has not been identified but the 5-methylcytosine has been shown recently to be converted into 5-hydroxymethylcytosine in an enzyme-catalyzed reaction [53]. It has been well established that DNMT1 does not recognize 5-hydroxylmethylcytosine and thus, the conversion of 5-methylcytosine to 5-hydroxymethylcytosine results in the loss of DNA methylation mark, and consequently alterations in the gene expression pattern [54, 55].

Non-Coding RNA

The recent advances in the sequencing technologies have revealed that almost 90% of the human genome is transcribed [56]. While the protein-coding genes still only comprise 1%, the vast majority of the genome appears to be coding for RNA molecules that are not translated into protein. These RNA molecules, termed as noncoding RNA (ncRNA), include not only tRNA and rRNA but also a large variety of small and large RNA molecules [57, 58]. The small RNA molecules, typically 20–25 nucleotides long, include microRNA(miRNA) [59, 60], small interfering RNA (siRNA), and piRNA [61]. The long RNA molecules are 200 nucleotides or more in length and include the Xist RNA that silences the X-chromosomes [62]. Some of the RNA molecules like siRNA in *S. pombe* [63] and Xist RNA [62] directly regulate the formation of the chromatin structure while others, like the miRNA, regulate the expression of histone modifying enzymes [64, 65] and thereby, indirectly regulate the architecture of the chromatin.

Disease Models for Aging

The rare genetic disorders – Werner's syndrome, Hutchinson–Gilford progeria syndrome, Cockayne syndrome, Ataxia-telangiectasia, and Xeroderma pigmentosum – serve as excellent model systems to gain insights into the process of human aging [66].

Werner's Syndrome (WS)

This disease is caused by mutations in the *WRN* gene and is inherited in an autosomal recessive manner. WRN is a bifunctional protein possessing both helicase [67] and exonuclease activity [68] that physically interacts with RPA [69], PCNA [70], PARP-1 [71, 72], Ku [73], DNA-PK [74], p53 [75], and SMARCAL1 [76]. The protein has been proposed to play a role in DNA replication by resolving alternate DNA structures that could impede the movement of the replication fork [77]. In addition, as it also interacts with many DNA damage response proteins, it plays a role in DNA repair [77]. The patients with WS display normal growth until adolescence when the symptoms of the disease that includes disorders seen in aging individuals start manifesting [78].

Hutchinson–Gilford Progeria Syndrome (HGPS)

The disease, a rare autosomal dominant disorder, also known as progeria (Greek: *gerias* meaning old age), has an incidence rate of 1 in 8 million live births [79]. Studies have shown that the primary defect is in the *Lmna* gene that encodes for Lamin-type A protein that is required for the formation of the nuclear envelope [79–81]. The cells derived from HGPS patients exhibit many chromatin defects as well as increased DNA damage, two characteristics that have also been observed in healthy aged patients, thus making it an attractive model to study the epigenetic changes associated with aging [82, 83].

Cockayne Syndrome (CS)

This is an autosomal recessive disorder, caused by mutations either in *CSA*, *CSB*, or *XPD* genes [84, 85]. *CSB* gene encodes for ERCC6, a member of the ATP-dependent chromatin remodeling protein family that forms a complex with the CSA protein encoded by *CSA* gene. These two proteins further interact with TFIIH (one subunit is encoded by *XPD*), a general transcription factor, and the complex thus formed helps in repairing the transcriptionally active genes [86, 87]. The patients show neurodegeneration, growth retardation, hearing loss, and retinal degeneration [88, 89]. However, despite mutations in CSA or CSB, the patients are not predisposed to cancer.

Ataxia-Talengiectsia (A-T)

This disease is caused by mutations in the *ATM* gene [90, 91]. The ATM protein is a serine/threonine kinase that is present as an inactive dimer in the nucleus of eukaryotic cells [92, 93]. On DNA damage, the protein is converted into the active monomer by phosphorylation and in turn, phosphorylates a plethora of substrates resulting in cell cycle arrest and effecting DNA damage response [93]. Patients with mutations in ATM show early-onset cancer, neurodegeneration, accelerated telomere loss, sensitivity to radiation, and growth retardation [94].

Xeroderma Pigmentosum (XP)

Mutations in any one of the seven genes (*XPA*, *XPB*, *XPC*, *XPD*, *XPE*, *XPF*, and *XPG*) results in the autosomal recessive disorder called XP [84]. The XP genes are involved in UV-induced DNA damage repair [95, 96]. Clinically, the defect outcomes include photosensitivity, actinic damage to skin, neurodegeneration, and predisposition to cancer [97].

Trichothiodystrophy (TTD)

This is another autosomal recessive disorder that is caused by mutations in the *XPD* gene [84]. Thus, the *XPD* gene can give rise to XP, TTD, or CS, depending upon the localization of the mutation. XPD is a subunit of TFIIH, a general transcription factor that plays a role both in transcription and DNA repair [98]. In transcription, TFIIH helps in the transition of the closed promoter complex to an open promoter complex, facilitating the binding of RNA polymerase II at the promoter, leading to promoter escape and initiation of transcription. In DNA repair, TFIIH functions as a helicase that unwinds the DNA either side of the damage. XPD is a helicase but its helicase activity is required only for DNA repair and not for transcription [98, 99].

Patients suffering from TTD have sulfur-deficient brittle hair along with ichthyotic skin, and mental retardation [100].

The Role of Histone Modifications in Aging

Of all the histone modifications known, the role of histone acetylation in transcription has been most characterized. Histone acetylation catalyzed by histone acetyltransferases is strongly correlated with transcriptional activation while deacetylation catalyzed by histone deacetyltransferases is associated with transcription repression [101]. In contrast, the role of histone methylation is known to be context-dependent functioning both as activation as well as a repression mark [34, 101].

Loss of Histones and Heterochromatin During Aging

Studies have shown that in aging yeast cells ~50% histones are lost and about 60% of the lost histones could be restored by overexpressing these proteins [102]. The loss of histone acetyltransferases, Asf1 and Rtt109, has also been found to shorten lifespan in yeast [103]. Deletion of these genes results in downregulation of histone transcripts, and thus, a reduction in the histone protein levels [103]. The decreased level of histones has an impact on the chromatin landscape and thus, the gene expression. Indeed, Hu et al. found that the nucleosome phasing at the transcription start site as well as at the transcription termination site was altered in aging yeast cells leading to increased gene expression [102]. Further, they found that genes that are normally repressed were activated during aging [102]. In 1997, Villepontaeu proposed the heterochromatin loss model of aging [104]. He proposed that the heterochromatin domain established during early embryogenesis were lost with aging. The major assumption of his model was that the euchromatic domains were the default stable structures while the heterochromatin were metastable and subject to decay with perturbations like DNA damage [104]. The heterochromatic loss model of aging is supported by experimental data from yeast as well as humans. In aging yeast cells, the heterochromatic silencing was lost from the *HM* loci [105]. An increased activity with aging of transposable elements that are normally present as heterochromatin has also been reported [106].

Histone Acetyltransferases in DNA Damage Repair

The role of chromatin in DNA repair was first detected in experiments performed by Lieberman and Smerdon who showed that repair occurs primarily in nuclease-sensitive regions immediately after UV irradiation [107]. Subsequently, they found that some of the nuclease-sensitive regions became nuclease-resistant leading to the formulation of Access-Repair-Restore model [21, 107]. In this model, histone-modifying enzymes and ATP-dependent chromatin remodeling factors are proposed to reposition nucleosomes creating access for the repair proteins and later, after the repair is effected, reposition the nucleosomes to their canonical positions. The histone acetyltransferase, GCN5, is a subunit of TBP-free TAF_{II} complex that binds UV-damaged DNA preferentially resulting in acetylation of histone H3 [108]. Another histone acetyltransferase-containing complex, STAGA, has also shown to interact with DDB1, a UV-damaged DNA-binding protein that has a role in NER [109]. DDB, the UV-damaged DNA-binding complex, is composed of two subunits DDB1 and DDB2. This complex has also been shown to interact with another histone acetyltransferase complex, CBP/p300 [110]. Downregulation of p300 is associated with senescence, forging a probable link between acetyltransferases and aging [111].

Histone Deacetylases: The Role of Sirtuins in Ageing

Of the known histone deacetylases, the sirtuins have been shown to play an important role in aging [112]. This class of enzymes deacetylase histones in a NAD^+-dependent manner and functionally play a role in heterochromatin maintenance [113]. In yeast, *S. cerevisiae*, the heterochromatin is known to serve two functions: (1) stabilize regions such as telomeres and rDNA repeats, and (2) maintain loci such as mating-type loci in a permanently silent state. The Sir2 protein in yeast has been shown to regulate both these functions. At the mating type loci as well as at the telomeres, the Sir2 protein interacts with Sir3 and Sir4 that direct and regulate its deacetylase activity [114]. At the rDNA loci, Sir2 is a component of the Regulator of Nucleolar silencing and Telophase exit (RENT) complex along with Net1 and Cdc14 in a 1:1:1 ratio [115]. Genetic screens show that a gain-of-function mutation in *SIR4* (*SIR4-42*) resulting in a truncated Sir4 protein abrogates the protein–DNA interaction such that it no longer can bind to the telomeres or the mating-type loci [116, 117]. Further, the mutant Sir4 protein also targets the Sir2 and Sir3 protein to the nucleolus, an act that seems to increase the lifespan of the yeast by ~40% [118]. The role of the nucleolus in aging became clear when extrachromosomal rDNA circles (ERC) was discovered [119]. The rDNA loci in all eukaryotic cells consist of tandem repeats of rDNA genes all of which are not expressed all the time. The repetitive sequences comprising rDNA loci, thus, are prone to undergo homologous recombination resulting in the formation of ERC that is replicated during the S phase. Over the course of cell divisions, the nucleus of a yeast cell can become packed with >1000 ERC molecules that have the potential to titrate away essential protein from the rest of the genome leading to cell death. The Sir2 protein by binding to the rDNA loci prevents the formation of ERC and thus, increases the lifespan. Indeed, experimental evidence has substantiated the hypothesis as activators of Sir2 as well as overexpression of Sir2 has been shown to result in increased lifespan [120, 121].

In mammalian cells, there are seven sirtuins – SIRT1 to SIRT7 – classified as Class III histone deacetylases sharing ~250 amino acid catalytic domain. Of these, SIRT1, SIRT6, and SIRT7 localize to the nucleus while SIRT3, SIRT4, and SIRT5 localize to the mitochondria [122]. These are, like the yeast Sir2 protein, NAD$^+$-dependent deacetylases and the ratio of NAD$^+$ to NADH is a critical regulator of their activity [123]. The relationship between NAD+ levels and SIRT1 activity was elegantly demonstrated using a long-lived human vascular SMC cell line. In this cell line, the activity of nicotinamide phosphoribosyltransferase (Nampt) was found to be upregulated under serum starvation condition resulting in upregulation of SIRT1 activity and consequently, resulting in increased lifespan [124]. Contrarily, the levels of SIRT1 have been found to decrease during replicative senescence in the primary human fibroblast cell line. However, overexpression of SIRT1 does not increase the replicative lifespan of the normal human cell lines [122]. Human SIRT1 has been found to deacetylase both p53 as well as histones (H3K9 and H4K16) [122, 125]. As p53 is an important regulator of senescence during cell cycle, it is possible that the regulation of its activity by SIRT1 determines lifespan. However, the significance of deacetylation of p53 has not been fully resolved.

SIRT1 also interacts with XPA, APE-1, and NBS1- proteins that are involved in DNA repair and deacetylates them, thereby activating the repair process [126–128]. Downregulation of SIRT1 results in hyperacetylation of these proteins and the cells, therefore, become more sensitive to genotoxic stress.

SIRT6 deacetylates H3K56 as well H3K9 in an NAD$^+$-dependent manner and regulates the telomere length [129, 130]. SIRT6 has been shown to stabilize DNA-PK and thus, help in mediating NHEJ pathway to effect repair of DSBs [131]. Deficiency of SIRT6 at cellular as well as organismal level leads to slower growth and increased sensitivity to DNA damaging agents. The SIRT6$^{-/-}$ cells show increased genomic instability and the mutant mice develop progeroid degenerative syndrome, indicating the importance of this deacetylase in maintaining normal lifespan [132]. In addition to playing a role in DNA repair and maintenance of telomere length, SIRT6 also regulates the metabolism directly by influencing the insulin signaling pathway (IIS). SIRT6 interacts with NF-κB and SIRT6-deficient mice are small and exhibit severe metabolic defects. Further, they develop age-related defects by 2–3 weeks of age. Overexpression of SIRT6 resulted in increased life span for male mice as compared to their female littermates. Overexpression also protected mice from diet-induced obesity seen in normal mice by middle age [133].

SIRT7 localizes to the nucleoli [122] and regulates rDNA transcription. This protein localizes to the rDNA promoter along with RNA polymerase I and activates transcription of the rDNA genes [134]. Proteomic studies have shown that SIRT7, in fact, interacts with a wide range of protein complexes including SWI/SNF and regulates transcription by all the three RNA polymerases [135]. Unlike SIRT1 and SIRT6, SIRT7 has no detectable deacetylase activity on purified histones but can deacetylase histones when present as nucleosomes [136].

SIRT7 also possesses NAD$^+$-dependent desuccinylase activity and has been shown to descuccinylate H3K122 at DSB sites [137]. Succinylation of H3K122 is associated with relaxed chromatin while desuccinylation leads to heterochromatin formation. The SIRT7-mediated transient desuccinylation of H3K122 is required for efficient double-strand break repair as abrogation of either succinylation or desuccinylation results in impaired repair process and thus, cell survival [137].

Targeted gene knockout of SIRT7 in mice resulted in progressive heart hypertrophy and decreased stress resistance [138]. In another study, the SIRT7-knockout mice were found to exhibit embryonic lethality. The mice that survived to adulthood exhibited progeroid-like symptoms and the SIRT7-deficient cells showed increased replication stress and impaired DNA damage repair due to increased H3K18 acetylation levels [139].

Thus, the sirtuins are important regulators of lifespan through the regulation of DNA damage response, metabolic signaling, and transcription regulation.

Histone Methylation in Aging

As stated earlier, the role of histone methylation in transcription is context-dependent [140]. Yet, it is generally accepted that H3K4me2/3 and H3K36me3 are 'activating' marks while H3K27me3 and H3K9me2/3 are 'repressive marks' [63, 140, 141]. It is within this context that the experimental results from HGPS patients as well as from model systems have been interpreted.

In SAMP8 mice, an accelerated aging model of mouse that is prone to neurodegenerative disorders, an increase in H3K27me3 and H3K9me2/3 marks has been observed with a concomitant decrease in H3K20me1 and H3K36me3 suggesting that the genome becomes increasingly heterochromatinized with age [142]. These results contrast with the experimental observations reported in HGPS patients and in Drosophila. In HGPS patients, an upregulation of H4K20me3 and a downregulation of H3K9me3 and H3K27me3 have been observed [143, 144]. A similar result has been obtained in Drosophila wherein a reduction in H3K9me3 levels correlated with loss in HP1-associated heterochromatin and affecting lifespan [145]. Taken together, these two experimental systems suggest that aging is associated with a loss in heterochromatin.

The levels of methylated histones have also been investigated in *C. elegans* in a time-dependent manner. RNAi screens in *C. elegans* have provided direct evidence that histone methyltransferases and demethylases like ASH2, SET-9 LSD-1 are involved in regulating the lifespan of the worm [146–148]. The H3K4me3 is catalyzed by the Trithorax complex and in *C. elegans* loss of any of the component of this complex leads to decreased H3K4me3 and therefore, an increase in the lifespan [148]. Further, in the worms, increased activity of UTX-1 demethylase has been reported with age [147]. This demethylase modulates the expression of insulin/IGF-1 signaling (IIS) pathway by regulating the epigenetic status of *daf-2* [147, 149]. In 'younger' worms, the UTX-1 expression is low and therefore, *daf-2* is in a repressed state. As the worms age, a dramatic increase in UTX-1 expression is observed that correlates with decreased H3K4me3 levels leading to activation of the IIS pathway via an increase in *daf-2* expression. RNAi of the *utx-1* leads to restoration of H3K27me3 levels during aging and thereby, increasing the lifespan by 30% [149]. Thus, like in HGPS patients, aging in worms too appear to be modulated by the levels of H3K27me3.

Histone Variants in Aging

There is no good correlation between histone variant deposition and aging. The studies at best can be described as providing tenuous connection between different histone variants and the aging process [150]. One of the most interesting observations has been the accumulation of H3.3 variant in the cells that are no longer replicating like, for example, neurons [151]. While this could be because H3.1 and H3.2, the canonical H3 histones, are not expressed in post-mitotic cells, yet it has also been noticed that H3.3 in generally accumulates in time-dependent fashion. Thus, H3.3 abundance increases during development in chicken and mice [152]. H3.3 is deposited on the chromatin by a histone chaperone known as HIRA whose levels have been found to increase with age in baboons [153]. Thus, it is quite possible that aging is associated with increased deposition of H3.3 on the chromatin. It should be noted that the nucleosome with H3.3 is a much open structure as compared to the one with the canonical histone, thus, it is possible that aging is associated with a loss in more compact structure of the chromatin whether it be the loss of heterochromatin or the formation of loose nucleosome-DNA structure [154, 155]. However, a direct experimental evidence for this hypothesis is still awaited.

The only other histone variant that has been studied with respect to aging is the incorporation of H2A.Z. This variant is incorporated in a replication-independent manner into the chromatin [156]. The abundance of the variant does not change with age but loss of H2A.Z, for example by knockdown, results in premature entry of human fibroblasts into senescence [157]. However, stress in general hastens the entry of fibroblasts into senescence [158]; therefore, it is not clear whether it is the loss of the histone variant or the stress induced by the knockdown that results in senescence. Further, as stated previously, cellular senescence, though often used as a model for studying aging, does not exactly mimic the process and therefore, it would premature to conclude that H2A.Z plays a role in the aging process.

ATP-Dependent Chromatin Remodeling Proteins and Aging

DAF-16/FOXO, a transcription factor, in *C. elegans* is required for entry into dauer state, a development arrest or diapause state when the insulin-like signaling is inactivated [159]. It is also required for lifespan extension during decreased insulin-like signaling. Studies have shown that DAF-16/FOXO interacts with the SWI/SNF protein complex and this interaction is required for activation of downstream target genes [160]. Inactivation of the SWI/SNF subunits by RNAi fully suppresses the lifespan extension mediated by DAF-16/FOXO [160].

PASG or Lsh, another member of the ATP-dependent chromatin remodeling protein family, is needed for global DNA methylation [161, 162]. The Lsh$^{-/-}$ mice display prenatal mortality in mice suggesting that the gene is essential for embryo development. Targeted disruption of *Lsh* in mice resulted in global hypomethylation and therefore, altered gene expression pattern. In particular, senescence-related genes like p16^{INK4a} and bmi-1 expression increased resulting in premature aging [163].

A yeast two-hybrid system identified RBBP4 and RBBP7 as interacting partners of Lamin A, the product of *Lmna* gene that is mutated in HGPS patients [164]. RBBP4 and RBBP7 are components of the Nucleosome Remodeling and Deacetylate (NuRD) complex [165]. The NuRD complex contains at least two subunits – CHD3 (Mi-2α) and CHD4 (Mi-2β) – which are ATP-dependent chromatin remodeling proteins. In addition, the complex also contains histone deacetylase 1 and 2 (HDAC1 and 2) and thus, coupling ATP-dependent chromatin remodeling with histone deacetylation [166]. This complex modulates transcription by promoting the formation of heterochromatin [165]. A loss of expression of RBBP4 and RBBP7 was observed in cells obtained from HGPS patients and this downregulation correlated with loss of HP1 and therefore, the heterochromatin structure [164].

The fourth ATP-dependent chromatin remodeling protein identified to play a role in aging is the ISWI complex in *S. cerevisiae* and *C. elegans*. The ISWI subfamily promotes heterochromatin formation by catalyzing chromatin assembly. Deletion of *iswi2Δ* was found to extend lifespan in a genetic screen in *S. cerevisiae* by altering the expression of stress-response genes [167]. The gene was also found to regulate lifespan in *C. elegans* [6]. In vitro studies have shown the role of ISWI [168] and the yeast homolog of CSB, Rad26, are involved in the NER pathway.

DNA Methylation and Aging

The role of DNA methylation in the aging process was elegantly demonstrated as early as 1967 by Berdyshev et al. wherein they showed a loss in global DNA methylation levels with age in spawning humpbacked salmon [169]. This was later augmented by Vanyushin et al. who showed that the cytosine methylation decreases in rat brain and heart [170]. Similar decreases have been reported in other organisms leading to a general conclusion that the aging process is correlated with reduction in methylated cytosine content. Changes in DNA methylation patterns have been also found to associate with the chronological aging across the entire human lifespan. These changes encompassed both hypo- and hyper-methylation. A comparative study of the newborn and centenarian genomes using whole genome bisulfite sequencing found differences in the methylation pattern [171]. In general, the genome of the centenarian cohort was found to be more hypomethylated especially in the promoters, intronic, exonic, and intragenic regions [171]. A methylome-wide association study involving ~700 subjects ranging from 25 to 92 years in age found more than 70 age-associated differentially methylated regions (DMR) [172]. Of these, 42 were found to be hypomethylated while 28 were found to be hypermethylated with age. An interesting finding was that many of the DMR were at genes that have been implicated in age-related diseases [172].

It needs to be remembered that the methylation pattern is dependent on the influence of the environmental factors and thus, the methylation signatures vary between individuals. Comparison of DNA methylation in monozygotic twins has shown that the patterns are similar at a younger age but become dissimilar as the age progresses [68]. The differences are pronounced if the monozygotic twins are exposed to different environment, thus clearly demonstrating that these factors can affect the lifespan both positively as well as negatively [173]. Environment, in fact, plays a predominant role in determining the extent of DNA methylation. Cigarette smoke, arsenic, bismuth, selenium, and catechol-containing compounds can impact the extent of DNA methylation [54, 174–176]. Arsenic, for example, alters the DNA methylation by altering the ratio of S-adenosylmethionine (SAM) to S-adenosylhomocysteine (SAH) ratio [174].

Many of the studies have focused only on global patterns leading researchers to ask a more specific question as to what sequence/genes are targeted for demethylation as a function of age. Using restriction enzymes that can differentiate between methylated and non-methylated cytosines, early studies showed that demethylation occurs in a gene-specific and tissue-specific manner. Thus, the β-*actin* gene was found to be demethylated with age in rat spleen but not in brain or liver [177]. A study done by Ono et al. found that the degree of methylation changes with age, but the change was not uniform across tissues [178]. They examined the methylation status of 25 genes and found that only 7 of them showed age-related changes in methylation level. For example, the CpG residues of c-*myc* gene were found to become hypermethylated in the murine liver [178]. Further, the CpG islands present on the promoters of MYOD1, estrogen Receptor, IgF2, and N33 too have been found to be hypermethylated [179].

The repetitive sequences have also been found to be increasingly methylated in older rat brains as compared to the younger rat brains [180]. The rDNA loci too have been found to become hypermethylated with age in liver and germ cells of senescent rats. In contrast, the Alu repeats were found to be hypomethylated with progress in age in a study involving elderly subjects between the ages of 55 and 92 years [181]. Alu as well as HERV-K was found to be hypomethylated in another study involving subjects ranging in age from 20 to 88 years [182]. However, no correlation was found between methylation levels and aging in LINE-I elements suggesting that methylation does not affect all repetitive elements equally [182].

In addition to methylation, the cytosine residues can also undergo hydroxymethylation to generate 5-hydroxymethylcytosine (5hmC). This reaction is possibly catalyzed by the TET enzymes using 5-methylcytosine as the substrate [55]. This modification is surprisingly enriched in the nuclei of the cerebellar Purkinje cells [183]. Though not much is known about the role of the 5hmC in aging, yet few studies have shown an increase in the levels of 5hmC with aging in the neuronal tissues [184, 185].

Non-Coding RNA in Aging

The role of non-coding RNA (ncRNA) in lifespan maintenance encompasses both direct and indirect effect on epigenetics. The long ncRNA molecules can directly affect the structure of the chromatin while the small ncRNA molecules like miRNA direct their effect on the chromatin structure through modulating the expression of histone modifying enzymes as well as DNA methyltransferases [186].

miRNA in Aging

It is generally agreed that miRNA regulates three key pathways – the insulin/insulin growth factor (IGF), target of rapamycin (mTOR), and the sirtuin family – that determine the lifespan [187]. Studies have also attempted to profile the global miRNA profile in young and aged tissues, but these studies have shown contradictory results. Variation in the miRNA profile in fetal, adult, and Alzheimer hippocampal brains have been documented wherein it was found that the expression of certain miRNAs like miR-9 and miR-128 was upregulated in Alzheimer's hippocampus as compared to normal adult brain [188]. However, it is difficult to assess the impact of these changes in miRNA levels as the effect on cognate mRNA expression was not reported. In another study, four miRNAs (miR-17, miR-196, miR-20a, and miR-106a) were found to be downregulated in four replicative aging and three organismal aging models [189]. The downregulation of these four miRNAs correlated with increased transcript levels of target genes including that of cdk inhibitor, p21/CDKN1, leading the authors of the study to propose that these miRNAs could be used as markers of cellular aging [189]. However, in contrast, no difference was found in the miRNA profile in lung tissues between adult and aged lungs of mouse models [190].

Role of Long Non-Coding RNA in Aging

TERRA The telomeres encode for two long ncRNA – *TERC* and *TERRA*. TERC is an essential component of the telomerase enzyme and acts as the primer during replication of the chromosomal ends [191]. This RNA molecule, thus, maintains the telomere length and prevents premature senescence and aging. Indeed, *TERC*-deficient mice show both chromosomal instability as well as premature aging [192]. *TERRA*, on the other hand, suppresses telomere elongation. This RNA molecule is transcribed by RNA polymerase II and can vary in length from 100 to >9000 nucleotides [193]. Studies have shown that *TERRA* interacts with TERT, origin recognition complex (ORC), HP1 and TRF2, thus, mediating the formation of heterochromatin at the telomeres [194, 195]. Further, by interacting with TERT, it prevents telomere elongation [194]. In cells obtained from immunodeficiency, centromeric instability, and facial dysmorphism (ICF syndrome), the CpG methylation at sub-telomeric regions is lost due to mutation in DNMT3b, leading to elevated levels of *TERRA* and, therefore, shortening of telomeres [196]. The patients suffering from ICF, hence, exhibit both premature senescence and aging.

H19 The long ncRNA *H19* regulates the imprinting of the H19/IGF2 cluster present on chromosome 1. *H19* interacts with methyl-CpG binding protein1 (MBD1), forming a ribonucleoprotein complex, which in turn recruits the histone methyltransferases to create a heterochromatin structure on the maternal allele [197]. Loss of imprinting has been reported during aging in normal human prostate tissues as well as in cancer, thus connecting the role of *H19* to aging [198].

pRNA The rDNA loci consist of tandemly repeated rDNA genes separated by intergenic spacers. The rDNA genes exist in two conformations: (1) active euchromatin conformation, and (2) the inactive heterochromatin conformation. The active–inactive ratio is maintained by the epigenetic modifiers and the long ncRNA called pRNA synthesized from the intergenic spacer [199]. The heterochromatin state is mediated by the ATP-dependent chromatin remodeling complex NoRC in conjunction with pRNA, recruiting DNA methyltransferase to the rDNA promoter and triggering de novo DNA methylation. The loss of rDNA coding genes has been reported in aging animals [200] but not in WRN syndrome [201]. However, the methylation of the rDNA genes was significantly higher in the fibroblasts isolated from WS patients as compared to the normal thus, indicating heterochromatinization of these genes in WS [201]. A similar result has been documented in a study done on patients suffering from Alzheimer's disease (AD). Comparison of methylation levels in the parietal and prefrontal cortex showed hypermethylation of CpG islands present on rDNA promoter in AD patients as compared to normal controls [202]. However, these studies have not correlated the expression of pRNA in WS and AD patients.

Conclusion

The epigenome of an organism undoubtedly undergoes tremendous change over the lifespan with the entire panoply of epigenetic regulators being associated with age-related changes. The change is impacted by the environment and lifestyle choices as experiments have shown that calorie restriction leads to increased lifespan due to modulation of the epigenome by the sirtuins [203, 204]. There are still lacunae in our understanding and paradoxes that need to be resolved. For example, the heterochromatic loss model of aging while supported by experimental results cannot explain the observation of senescence-associated heterochromatin foci (SAHF) in aging cells [205, 206].

The epigenetic modulators are increasingly finding favor as therapeutic targets and understanding the role of these proteins in the pathophysiology of aging-related disorders could lead to the development of therapies that, if not ameliorate, at least can ease the pain and suffering associated with these diseases.

References

1. Hayflick L. The limited in vitro lifetime of human diploid cell strains. Exp Cell Res. 1965;37:614–36.
2. López-Otín C, Blasco MA, Partridge L, Serrano M, Kroemer G. The hallmarks of aging. Cell. 2013;153:1194–217.
3. Chen J-H, Hales CN, Ozanne SE. DNA damage, cellular senescence and organismal ageing: causal or correlative? Nucleic Acids Res. 2007;35:7417–28.
4. Garinis GA, van der Horst GTJ, Vijg J, Hoeijmakers HJ, J. DNA damage and ageing: new-age ideas for an age-old problem. Nat Cell Biol. 2008;10:1241–7.
5. Zhou BB, Elledge SJ. The DNA damage response: putting checkpoints in perspective. Nature. 2000;408:433–9.
6. Jackson SP, Bartek J. The DNA-damage response in human biology and disease. Nature. 2009;461:1071–8.
7. Szilard L. On the nature of the aging process. Proc Natl Acad Sci USA. 1959;45:30–45.
8. Epstein J, Williams JR, Little JB. Deficient DNA repair in human progeroid cells. Proc Natl Acad Sci USA. 1973;70:977–81.
9. Kyng KJ, Bohr VA. Gene expression and DNA repair in progeroid syndromes and human aging. Ageing Res Rev. 2005;4:579–602.
10. Marteijn JA, Lans H, Vermeulen W, Hoeijmakers JHJ. Understanding nucleotide excision repair and its roles in cancer and ageing. Nat Rev Mol Cell Biol. 2014;15:465–81.
11. David SS, O'Shea VL, Kundu S. Base-excision repair of oxidative DNA damage. Nature. 2007;447:941–50.
12. Krokan HE, Bjoras M. Base excision repair. Cold Spr Harb Perspect Biol. 2013;5:a012583.
13. Jackson SP. Sensing and repairing DNA double-strand breaks. Carcinogenesis. 2002;23:687–96.
14. Kanaar R, Hoeijmakers JH, van Gent DC. Molecular mechanisms of DNA double-strand break repair. Trends Cell Biol. 1998;8:483–9.
15. Guo Z, Heydari A, Richardson A. Nucleotide excision repair of actively transcribed versus nontranscribed DNA in rat hepatocytes: effect of age and dietary restriction. Exp Cell Res. 1998;245:228–38.
16. Niedernhofer LJ, Garinis GA, Raams A, Lalai AS, Robinson AR, Appeldoorn E, Odijk H, Oostendorp R, Ahmad A, van Leeuwen W, et al. A new progeroid syndrome reveals that genotoxic stress suppresses the somatotroph axis. Nature. 2006;444:1038–43.
17. Harman D. The aging process. Proc Natl Acad Sci USA. 1981;78:7124–8.
18. Maclean MJ, Aamodt R, Harris N, Alseth I, Seeberg E, Bjoras M, Piper PW. Base excision repair activities required for yeast to attain a full chronological life span. Aging Cell. 2003;2:93–104.
19. Chevanne M, Caldini R, Tombaccini D, Mocali A, Gori G, Paoletti F. Comparative levels of DNA breaks and sensitivity to oxidative stress in aged and senescent human fibroblasts: a distinctive pattern for centenarians. Biogerontology. 2003;4:97–104.
20. Singh NP, Ogburn CE, Wolf NS, van Belle G, Martin GM. DNA double-strand breaks in mouse kidney cells with age. Biogerontology. 2001;2:261–70.
21. Smerdon MJ. DNA repair and the role of chromatin structure. Curr Opin Cell Biol. 1991;3:422–8.
22. Bao Y, Shen X. Chromatin remodeling in DNA double-strand break repair. Curr Opin Genet Dev. 2007;17:126–31.
23. Huertas D, Sendra R, Muñoz P. Chromatin dynamics coupled to DNA repair. Epigenetics. 2009;4:31–42.
24. Richmond TJ, Davey CA. The structure of DNA in the nucleosome core. Nature. 2003;423:145–50.
25. Woodcock CL, Skoultchi AI, Fan Y. Role of linker histone in chromatin structure and function: H1 stoichiometry and nucleosome repeat length. Chromosome Res. 2006;14:17–25.

26. Woodcock CL, Ghosh RP. Chromatin Higher-order Structure and Dynamics. Cold Spring Harb Perspect Biol. 2010;2:a000596.
27. Allfrey VG, Faulkner R, Mirsky AE. Acetylation and methylation of histones and their possible role in the regulation of RNA synthesis. Proc Natl Acad Sci USA. 1964;51:786–94.
28. Georgakopoulos T, Thireos G. Two distinct yeast transcriptional activators require the function of the GCN5 protein to promote normal levels of transcription. EMBO J. 1992;11:4145–52.
29. Kuo MH, Zhou J, Jambeck P, Churchill ME, Allis CD. Histone acetyltransferase activity of yeast Gcn 5p is required for the activation of target genes in vivo. Genes Dev. 1998;12:627–39.
30. Grant PA, Duggan L, Côté J, Roberts SM, Brownell JE, Candau R, Ohba R, Owen-Hughes T, Allis CD, Winston F, et al. Yeast Gcn 5 functions in two multisubunit complexes to acetylate nucleosomal histones: characterization of an Ada complex and the SAGA (Spt/Ada) complex. Genes Dev. 1997;11:1640–50.
31. Brownell JE, Zhou J, Ranalli T, Kobayashi R, Edmondson DG, Roth SY, Allis CD. Tetrahymena histone acetyltransferase A: a homolog to yeast Gcn5p linking histone acetylation to gene activation. Cell. 1996;84:843–51.
32. Peterson CL, Laniel M-A. Histones and histone modifications. Curr Biol. 2004;14:R546–51.
33. Bannister AJ, Kouzarides T. Regulation of chromatin by histone modifications. Cell Res. 2011; https://doi.org/10.1038/cr.2011.22.
34. Cosgrove MS, Wolberger C. How does the histone code work? Biochem Cell Biol Biochim Biol Cell. 2005;83:468–76.
35. Kamakaka RT, Biggins S. Histone variants: deviants? Genes Dev. 2005;19:295–310.
36. Sarma K, Reinberg D. Histone variants meet their match. Nat Rev Mol Cell Biol. 2005;6:139–49.
37. Hargreaves DC, Crabtree GR. ATP-dependent chromatin remodeling: genetics, genomics and mechanisms. Cell Res. 2011;21:396–420.
38. Flaus A, Owen-Hughes T. Mechanisms for ATP-dependent chromatin remodelling: the means to the end. FEBS J. 2011;278:3579–95.
39. Watanabe S, Radman-Livaja M, Rando OJ, Peterson CL. A Histone Acetylation Switch Regulates H2A.Z Deposition by the SWR-C Remodeling Enzyme. Science. 2013;340:195–9.
40. Gorbalenya AE, Koonin EV. Helicases: amino acid sequence comparisons and structure-function relationships. Curr Opin Struct Biol. 1993;3:419–29.
41. Flaus A, Martin DMA, Barton GJ, Owen-Hughes T. Identification of multiple distinct Snf2 subfamilies with conserved structural motifs. Nucleic Acids Res. 2006;34:2887–905.
42. Neigeborn L, Carlson M. Genes affecting the regulation of SUC2 gene expression by glucose repression in Saccharomyces cerevisiae. Genetics. 1984;108:845–58.
43. Laurent BC, Treich I, Carlson M. Role of yeast SNF and SWI proteins in transcriptional activation. Cold Spring Harb Symp Quant Biol. 1993;58:257–63.
44. Trotter KW, Archer TK. The BRG1 transcriptional coregulator. Nucl Recept Signal. 2008;6:e004.
45. Bestor TH. The DNA methyltransferases of mammals. Hum Mol Genet. 2000;9:2395–402.
46. Jones PA, Takai D. The role of DNA methylation in mammalian epigenetics. Science. 2001;293:1068–70.
47. Smith ZD, Meissner A. DNA methylation: roles in mammalian development. Nat Rev Genet. 2013;14:204–20.
48. Boyes J, Bird A. DNA methylation inhibits transcription indirectly via a methyl-CpG binding protein. Cell. 1991;64:1123–34.
49. Kass SU, Pruss D, Wolffe AP. How does DNA methylation repress transcription? Trends Genet. 1997;13:444–9.
50. Ooi SKT, Bestor TH. The colorful history of active DNA Demethylation. Cell. 2008;133:1145–8.
51. Wu SC, Zhang Y. Active DNA demethylation: many roads lead to Rome. Nat Rev Mol Cell Biol. 2010;11:607–20.
52. Bhutani N, Burns DM, Blau HM. DNA demethylation dynamics. Cell. 2011;146:866–72.

53. Guo JU, Su Y, Zhong C, Ming G, Song H. Hydroxylation of 5-Methylcytosine by TET1 Promotes Active DNA Demethylation in the Adult Brain. Cell. 2011;145:423–34.
54. Zampieri M, Ciccarone F, Calabrese R, Franceschi C, Bürkle A, Caiafa P. Reconfiguration of DNA methylation in aging. Mech Ageing Dev. 2015;151:60–70.
55. Pastor WA, Aravind L, Rao A. TETonic shift: biological roles of TET proteins in DNA demethylation and transcription. Nat Rev Mol Cell Biol. 2013;14:341–56.
56. ENCODE Project Consortium. The ENCODE (ENCyclopedia of DNA Elements) project. Science. 2004;306:636–40.
57. Eddy SR. Non-coding RNA genes and the modern RNA world. Nat Rev Genet. 2001;2:919–29.
58. Mattick JS, Makunin IV. Non-coding RNA. Hum Mol Genet. 2006;15 Spec No 1:R17–29.
59. Winter J, Jung S, Keller S, Gregory RI, Diederichs S. Many roads to maturity: microRNA biogenesis pathways and their regulation. Nat Cell Biol. 2009;11:228–34.
60. Ha M, Kim VN. Regulation of microRNA biogenesis. Nat Rev Mol Cell Biol. 2014;15:509–24.
61. Aravin AA, Hannon GJ, Brennecke J. The Piwi-piRNA pathway provides an adaptive defense in the transposon arms race. Science. 2007;318:761–4.
62. Plath K, Mlynarczyk-Evans S, Nusinow DA, Panning B. Xist RNA and the mechanism of X chromosome inactivation. Annu Rev Genet. 2002;36:233–78.
63. Volpe TA, Kidner C, Hall IM, Teng G, Grewal SIS, Martienssen RA. Regulation of heterochromatic silencing and histone H3 lysine-9 methylation by RNAi. Science. 2002;297:1833–7.
64. Wong CF, Tellam RL. MicroRNA-26a targets the histone methyltransferase enhancer of Zeste homolog 2 during myogenesis. J Biol Chem. 2008;283:9836–43.
65. Yuan J, Yang F, Chen B, Lu Z, Huo X, Zhou W, Wang F, Sun S. The histone deacetylase 4/SP1/microrna-200a regulatory network contributes to aberrant histone acetylation in hepatocellular carcinoma. Hepatology. 2011;54:2025–35.
66. Navarro CL. Molecular bases of progeroid syndromes. Hum Mol Genet. 2006;15:R151–61.
67. Gray MD, Shen JC, Kamath-Loeb AS, Blank A, Sopher BL, Martin GM, Oshima J, Loeb LA. The Werner syndrome protein is a DNA helicase. Nat Genet. 1997;17:100–3.
68. Shen J-C. Werner syndrome protein. I. DNA helicase and DNA exonuclease reside on the same polypeptide. J Biol Chem. 1998;273:34139–44.
69. Brosh RM, Orren DK, Nehlin JO, Ravn PH, Kenny MK, Machwe A, Bohr VA. Functional and physical interaction between WRN helicase and human replication protein A. J Biol Chem. 1999;274:18341–50.
70. Lebel M, Spillare EA, Harris CC, Leder P. The Werner syndrome gene product co-purifies with the DNA replication complex and interacts with PCNA and topoisomerase I. J Biol Chem. 1999;274:37795–9.
71. Li B. Identification and biochemical characterization of a Werner's syndrome protein complex with Ku70/80 and poly(ADP-ribose) polymerase-1. J Biol Chem. 2004;279:13659–67.
72. von Kobbe C. Poly(ADP-ribose) polymerase 1 regulates both the exonuclease and helicase activities of the Werner syndrome protein. Nucleic Acids Res. 2004;32:4003–14.
73. Li B. Functional interaction between Ku and the Werner syndrome protein in DNA end processing. J Biol Chem. 2000;275:28349–52.
74. Yannone SM, Roy S, Chan DW, Murphy MB, Huang S, Campisi J, Chen DJ. Werner syndrome protein is regulated and phosphorylated by DNA-dependent protein kinase. J Biol Chem. 2001;276:38242–8.
75. Blander G, Kipnis J, Leal JFM, Yu C-E, Schellenberg GD, Oren M. Physical and functional interaction between p53 and the Werner's syndrome protein. J Biol Chem. 1999;274:29463–9.
76. Ciccia A, Bredemeyer AL, Sowa ME, Terret M-E, Jallepalli PV, Harper JW, Elledge SJ. The SIOD disorder protein SMARCAL1 is an RPA-interacting protein involved in replication fork restart. Genes Dev. 2009;23:2415–25.
77. Croteau DL, Popuri V, Opresko PL, Bohr VA. Human Rec Q helicases in DNA repair, recombination, and replication. Annu Rev Biochem. 2014;83:519–52.
78. Goto M. Werner's syndrome: from clinics to genetics. Clin Exp Rheumatol. 2000;18:760–6.
79. Pollex RL, Hegele RA. Hutchinson-Gilford progeria syndrome. Clin Genet. 2004;66:375–81.

80. Eriksson M, Brown WT, Gordon LB, Glynn MW, Singer J, Scott L, Erdos MR, Robbins CM, Moses TY, Berglund P, et al. Recurrent de novo point mutations in lamin A cause Hutchinson–Gilford progeria syndrome. Nature. 2003;423:293–8.
81. Capell BC, Collins FS. Human laminopathies: nuclei gone genetically awry. Nat Rev Genet. 2006;7:940–52.
82. Liu G-H, Barkho BZ, Ruiz S, Diep D, Qu J, Yang S-L, Panopoulos AD, Suzuki K, Kurian L, Walsh C, et al. Recapitulation of premature ageing with iPSCs from Hutchinson–Gilford progeria syndrome. Nature. 2011;472:221–5.
83. Miller JD, Ganat YM, Kishinevsky S, Bowman RL, Liu B, Tu EY, Mandal PK, Vera E, Shim J, Kriks S, et al. Human iPSC-based modeling of late-onset disease via progerin-induced aging. Cell Stem Cell. 2013;13:691–705.
84. Lehmann AR. DNA repair-deficient diseases, xeroderma pigmentosum, Cockayne syndrome and trichothiodystrophy. Biochimie. 2003;85:1101–11.
85. van Gool AJ, van der Horst GT, Citterio E, Hoeijmakers JH. Cockayne syndrome: defective repair of transcription? EMBO J. 1997;16:4155–62.
86. Henning KA, Li L, Iyer N, McDaniel LD, Reagan MS, Legerski R, Schultz RA, Stefanini M, Lehmann AR, Mayne LV, et al. The Cockayne syndrome group A gene encodes a WD repeat protein that interacts with CSB protein and a subunit of RNA polymerase II TFIIH. Cell. 1995;82:555–64.
87. Venema J, Mullenders LH, Natarajan AT, van Zeeland AA, Mayne LV. The genetic defect in Cockayne syndrome is associated with a defect in repair of UV-induced DNA damage in transcriptionally active DNA. Proc Natl Acad Sci USA. 1990;87:4707–11.
88. Rapin I, Lindenbaum Y, Dickson DW, Kraemer KH, Robbins JH. Cockayne syndrome and xeroderma pigmentosum. Neurology. 2000;55:1442–9.
89. Lehmann AR, Thompson AF, Harcourt SA, Stefanini M, Norris PG. Cockayne's syndrome: correlation of clinical features with cellular sensitivity of RNA synthesis to UV irradiation. J Med Genet. 1993;30:679–82.
90. Lavin MF. Ataxia-telangiectasia: from a rare disorder to a paradigm for cell signalling and cancer. Nat Rev Mol Cell Biol. 2008;9:759–69.
91. Savitsky K, Sfez S, Tagle DA, Ziv Y, Sartiel A, Collins FS, Shiloh Y, Rotman G. The complete sequence of the coding region of the ATM gene reveals similarity to cell cycle regulators in different species. Hum Mol Genet. 1995;4:2025–32.
92. Lee J-H. Direct activation of the ATM protein kinase by the Mre11/Rad50/Nbs1 complex. Science. 2004;304:93–6.
93. Bakkenist CJ, Kastan MB. DNA damage activates ATM through intermolecular autophosphorylation and dimer dissociation. Nature. 2003;421:499–506.
94. Lavin MF, Shiloh Y. the genetic defect in Ataxia-telangiectasia. Annu Rev Immunol. 1997;15:177–202.
95. DiGiovanna JJ, Kraemer KH. Shining a light on xeroderma pigmentosum. J Invest Dermatol. 2012;132:785–96.
96. Lehmann AR, Kirk-Bell S, Arlett CF, Paterson MC, Lohman PH, de Weerd-Kastelein EA, Bootsma D. Xeroderma pigmentosum cells with normal levels of excision repair have a defect in DNA synthesis after UV-irradiation. Proc Natl. Acad. Sci USA. 1975;72:219–23.
97. Kraemer KH, Lee MM, Scotto J. Xeroderma pigmentosum. Cutaneous, ocular, and neurologic abnormalities in 830 published cases. Arch Dermatol. 1987;123:241–50.
98. Lehmann AR. The xeroderma pigmentosum group D (XPD) gene: one gene, two functions, three diseases. Genes Dev. 2001;15:15–23.
99. Taylor EM, Broughton BC, Botta E, Stefanini M, Sarasin A, Jaspers NG, Fawcett H, Harcourt SA, Arlett CF, Lehmann AR. Xeroderma pigmentosum and trichothiodystrophy are associated with different mutations in the XPD (ERCC2) repair/transcription gene. Proc Natl Acad Sci USA. 1997;94:8658–63.
100. Faghri S, Tamura D, Kraemer KH, Digiovanna JJ. Trichothiodystrophy: a systematic review of 112 published cases characterises a wide spectrum of clinical manifestations. J Med Genet. 2008;45:609–21.

101. Berger SL. The complex language of chromatin regulation during transcription. Nature. 2007;447:407–12.
102. Hu Z, Chen K, Xia Z, Chavez M, Pal S, Seol J-H, Chen C-C, Li W, Tyler JK. Nucleosome loss leads to global transcriptional up-regulation and genomic instability during yeast aging. Genes Dev. 2014;28:396–408.
103. Feser J, Truong D, Das C, Carson JJ, Kieft J, Harkness T, Tyler JK. Elevated histone expression promotes life span extension. Mol Cell. 2010;39:724–35.
104. Villeponteau B. The heterochromatin loss model of aging. Exp Gerontol. 1997;32:383–94.
105. Smeal T, Claus J, Kennedy B, Cole F, Guarente L. Loss of transcriptional silencing causes sterility in old mother cells of S. cerevisiae. Cell. 1996;84:633–42.
106. Sturm Á, Ivics Z, Vellai T. The mechanism of ageing: primary role of transposable elements in genome disintegration. Cell Mol Life Sci CMLS. 2015;72:1839–47.
107. Smerdon MJ, Lieberman MW. Nucleosome rearrangement in human chromatin during UV-induced DNA- repair synthesis. Proc Natl Acad Sci USA. 1978;75:4238–41.
108. Brand M. UV-damaged DNA-binding protein in the TFTC complex links DNA damage recognition to nucleosome acetylation. EMBO J. 2001;20:3187–96.
109. Martinez E, Palhan VB, Tjernberg A, Lymar ES, Gamper AM, Kundu TK, Chait BT, Roeder RG. Human STAGA complex is a chromatin-acetylating transcription coactivator that interacts with pre-mRNA splicing and DNA damage-binding factors in vivo. Mol Cell Biol. 2001;21:6782–95.
110. Datta A, Bagchi S, Nag A, Shiyanov P, Adami GR, Yoon T, Raychaudhuri P. The p48 subunit of the damaged-DNA binding protein DDB associates with the CBP/p300 family of histone acetyltransferase. Mutat Res. 2001;486:89–97.
111. Bandyopadhyay D, Okan NA, Bales E, Nascimento L, Cole PA, Medrano EE. Downregulation of p300/CBP histone acetyltransferase activates a senescence checkpoint in human melanocytes. Cancer Res. 2002;62:6231–9.
112. Haigis MC, Guarente LP. Mammalian sirtuins--emerging roles in physiology, aging, and calorie restriction. Genes Dev. 2006;20:2913–21.
113. Finkel T, Deng C-X, Mostoslavsky R. Recent progress in the biology and physiology of sirtuins. Nature. 2009;460:587–91.
114. Hoppe GJ, Tanny JC, Rudner AD, Gerber SA, Danaie S, Gygi SP, Moazed D. Steps in assembly of silent chromatin in yeast: Sir3-independent binding of a Sir2/Sir4 complex to silencers and role for Sir2-dependent deacetylation. Mol Cell Biol. 2002;22:4167–80.
115. Straight AF, Shou W, Dowd GJ, Turck CW, Deshaies RJ, Johnson AD, Moazed D. Net1, a Sir2-associated nucleolar protein required for rDNA silencing and nucleolar integrity. Cell. 1999;97:245–56.
116. Kennedy BK, Austriaco NR, Zhang J, Guarente L. Mutation in the silencing gene SIR4 can delay aging in S. cerevisiae. Cell. 1995;80:485–96.
117. Kaeberlein M, McVey M, Guarente L. The SIR2/3/4 complex and SIR2 alone promote longevity in Saccharomyces cerevisiae by two different mechanisms. Genes Dev. 1999;13:2570–80.
118. Kennedy BK, Gotta M, Sinclair DA, Mills K, McNabb DS, Murthy M, Pak SM, Laroche T, Gasser SM, Guarente L. Redistribution of silencing proteins from telomeres to the nucleolus is associated with extension of life span in S. cerevisiae. Cell. 1997;89:381–91.
119. Sinclair DA, Guarente L. Extrachromosomal rDNA circles--a cause of aging in yeast. Cell. 1997;91:1033–42.
120. Howitz KT, Bitterman KJ, Cohen HY, Lamming DW, Lavu S, Wood JG, Zipkin RE, Chung P, Kisielewski A, Zhang L-L, et al. Small molecule activators of sirtuins extend Saccharomyces cerevisiae lifespan. Nature. 2003;425:191–6.
121. Kaeberlein M. Lessons on longevity from budding yeast. Nature. 2010;464:513–9.
122. Michishita E, Park JY, Burneskis JM, Barrett JC, Horikawa I. Evolutionarily conserved and nonconserved cellular localizations and functions of human SIRT proteins. Mol Biol Cell. 2005;16:4623–35.
123. Saunders LR, Verdin E. Sirtuins: critical regulators at the crossroads between cancer and aging. Oncogene. 2007;26:5489–504.

124. van der Veer E, Ho C, O'Neil C, Barbosa N, Scott R, Cregan SP, Pickering JG. Extension of human cell lifespan by nicotinamide phosphoribosyltransferase. J Biol Chem. 2007;282:10841–5.
125. Vaquero A, Scher M, Lee D, Erdjument-Bromage H, Tempst P, Reinberg D. Human SirT1 interacts with histone H1 and promotes formation of facultative heterochromatin. Mol Cell. 2004;16:93–105.
126. Yuan Z, Seto E. A functional link between SIRT1 deacetylase and NBS1 in DNA damage response. Cell Cycle Georget Tex. 2007;6:2869–71.
127. Fan W, Luo J. SIRT1 regulates UV-induced DNA repair through deacetylating XPA. Mol Cell. 2010;39:247–58.
128. Yamamori T, DeRicco J, Naqvi A, Hoffman TA, Mattagajasingh I, Kasuno K, Jung S-B, Kim C-S, Irani K. SIRT1 deacetylates APE1 and regulates cellular base excision repair. Nucleic Acids Res. 2010;38:832–45.
129. Michishita E, McCord RA, Boxer LD, Barber MF, Hong T, Gozani O, Chua KF. Cell cycle-dependent deacetylation of telomeric histone H3 lysine K56 by human SIRT6. Cell Cycle Georget Tex. 2009;8:2664–6.
130. Michishita E, McCord RA, Berber E, Kioi M, Padilla-Nash H, Damian M, Cheung P, Kusumoto R, Kawahara TLA, Barrett JC, et al. SIRT6 is a histone H3 lysine 9 deacetylase that modulates telomeric chromatin. Nature. 2008;452:492–6.
131. McCord RA, Michishita E, Hong T, Berber E, Boxer LD, Kusumoto R, Guan S, Shi X, Gozani O, Burlingame AL, et al. SIRT6 stabilizes DNA-dependent protein kinase at chromatin for DNA double-strand break repair. Aging. 2009;1:109–21.
132. Mostoslavsky R, Chua KF, Lombard DB, Pang WW, Fischer MR, Gellon L, Liu P, Mostoslavsky G, Franco S, Murphy MM, et al. Genomic Instability and Aging-like Phenotype in the Absence of Mammalian SIRT6. Cell. 2006;124:315–29.
133. Kanfi Y, Naiman S, Amir G, Peshti V, Zinman G, Nahum L, Bar-Joseph Z, Cohen HY. The sirtuin SIRT6 regulates lifespan in male mice. Nature. 2012;483:218–21.
134. Ford E, Voit R, Liszt G, Magin C, Grummt I, Guarente L. Mammalian Sir2 homolog SIRT7 is an activator of RNA polymerase I transcription. Genes Dev. 2006;20:1075–80.
135. Tsai Y-C, Greco TM, Boonmee A, Miteva Y, Cristea IM. Functional Proteomics Establishes the Interaction of SIRT7 with Chromatin Remodeling Complexes and Expands Its Role in Regulation of RNA Polymerase I Transcription. Mol Cell Proteomics. 2012;11:60–76.
136. Tong Z, Wang Y, Zhang X, Kim DD, Sadhukhan S, Hao Q, Lin H. SIRT7 Is Activated by DNA and Deacetylates Histone H3 in the Chromatin Context. ACS Chem Biol. 2016;11:742–7.
137. Li L, Shi L, Yang S, Yan R, Zhang D, Yang J, He L, Li W, Yi X, Sun L, et al. SIRT7 is a histone desuccinylase that functionally links to chromatin compaction and genome stability. Nat. Commun. 2016;7:12235.
138. Vakhrusheva O, Braeuer D, Liu Z, Braun T, Bober E. Sirt 7-dependent inhibition of cell growth and proliferation might be instrumental to mediate tissue integrity during aging. J Physiol Pharmacol. 2008;59(Suppl 9):201–12.
139. Vazquez BN, Thackray JK, Simonet NG, Kane-Goldsmith N, Martinez-Redondo P, Nguyen T, Bunting S, Vaquero A, Tischfield JA, Serrano L. SIRT7 promotes genome integrity and modulates non-homologous end joining DNA repair. EMBO J. 2016;35:1488–503.
140. Kouzarides T. Histone methylation in transcriptional control. Curr Opin Genet Dev. 2002;12:198–209.
141. Wang Z, Zang C, Rosenfeld JA, Schones DE, Barski A, Cuddapah S, Cui K, Roh T-Y, Peng W, Zhang MQ, et al. Combinatorial patterns of histone acetylations and methylations in the human genome. Nat Genet. 2008;40:897–903.
142. Wang CM, Tsai SN, Yew TW, Kwan YW, Ngai SM. Identification of histone methylation multiplicities patterns in the brain of senescence-accelerated prone mouse 8. Biogerontology. 2010;11:87–102.
143. Shumaker DK, Dechat T, Kohlmaier A, Adam SA, Bozovsky MR, Erdos MR, Eriksson M, Goldman AE, Khuon S, Collins FS, et al. Mutant nuclear lamin A leads to progressive alterations of epigenetic control in premature aging. Proc Natl Acad Sci. 2006;103:8703–8.

144. McCord RP, Nazario-Toole A, Zhang H, Chines PS, Zhan Y, Erdos MR, Collins FS, Dekker J, Cao K. Correlated alterations in genome organization, histone methylation, and DNA-lamin A/C interactions in Hutchinson-Gilford progeria syndrome. Genome Res. 2013;23:260–9.
145. Larson K, Yan S-J, Tsurumi A, Liu J, Zhou J, Gaur K, Guo D, Eickbush TH, Li WX. Heterochromatin formation promotes longevity and represses ribosomal RNA synthesis. PLoS Genet. 2012;8:e1002473.
146. Han S, Brunet A. Histone methylation makes its mark on longevity. Trends Cell Biol. 2012;22:42–9.
147. Maures TJ, Greer EL, Hauswirth AG, Brunet A. The H3K27 demethylase UTX-1 regulates C. elegans lifespan in a germline-independent, insulin-dependent manner: The H3K27me3 demethylase UTX-1 regulates worm lifespan. Aging Cell. 2011;10:980–90.
148. Greer EL, Maures TJ, Hauswirth AG, Green EM, Leeman DS, Maro GS, Han S, Banko MR, Gozani O, Brunet A. Members of the H3K4 trimethylation complex regulate lifespan in a germline-dependent manner in C. elegans. Nature. 2010;466:383–7.
149. Jin C, Li J, Green CD, Yu X, Tang X, Han D, Xian B, Wang D, Huang X, Cao X, et al. Histone demethylase UTX-1 regulates C. elegans life span by targeting the insulin/IGF-1 signaling pathway. Cell Metab. 2011;14:161–72.
150. Das C, Tyler JK. Histone exchange and histone modifications during transcription and aging. Biochim Biophys Acta. 2013;1819:332–42.
151. Piña B, Suau P. Changes in histones H2A and H3 variant composition in differentiating and mature rat brain cortical neurons. Dev Biol. 1987;123:51–8.
152. Urban MK, Zweidler A. Changes in nucleosomal core histone variants during chicken development and maturation. Dev Biol. 1983;95:421–8.
153. Jeyapalan JC, Ferreira M, Sedivy JM, Herbig U. Accumulation of senescent cells in mitotic tissue of aging primates. Mech Ageing Dev. 2007;128:36–44.
154. Jin C, Felsenfeld G. Nucleosome stability mediated by histone variants H3.3 and H2A.Z. Genes Dev. 2007;21:1519–29.
155. Thakar A, Gupta P, Ishibashi T, Finn R, Silva-Moreno B, Uchiyama S, Fukui K, Tomschik M, Ausio J, Zlatanova J. H2A.Z and H3.3 histone variants affect nucleosome structure: biochemical and biophysical studies. Biochemistry (Mosc.). 2009;48:10852–7.
156. Mizuguchi G, Shen X, Landry J, Wu W-H, Sen S, Wu C. ATP-driven exchange of histone H2AZ variant catalyzed by SWR1 chromatin remodeling complex. Science. 2004;303:343–8.
157. Lee K, Lau ZZ, Meredith C, Park JH. Decrease of p400 ATPase complex and loss of H2A.Z within the p21 promoter occur in senescent IMR-90 human fibroblasts. Mech. Ageing Dev. 2012;133:686–94.
158. Toussaint O, Medrano E, von Zglinicki T. Cellular and molecular mechanisms of stress-induced premature senescence (SIPS) of human diploid fibroblasts and melanocytes. Exp Gerontol. 2000;35:927–45.
159. Lin K, Hsin H, Libina N, Kenyon C. Regulation of the Caenorhabditis elegans longevity protein DAF-16 by insulin/IGF-1 and germline signaling. Nat Genet. 2001;28:139–45.
160. Riedel CG, Dowen RH, Lourenco GF, Kirienko NV, Heimbucher T, West JA, Bowman SK, Kingston RE, Dillin A, Asara JM, et al. DAF-16 employs the chromatin remodeller SWI/SNF to promote stress resistance and longevity. Nat Cell Biol. 2013;15:491–501.
161. Myant K, Termanis A, Sundaram AYM, Boe T, Li C, Merusi C, Burrage J, Heras JIL, Stancheva I. LSH and G9a/GLP complex are required for developmentally programmed DNA methylation. Genome Res. 2011;21:83–94.
162. Zhu H, Geiman TM, Xi S, Jiang Q, Schmidtmann A, Chen T, Li E, Muegge K. Lsh is involved in de novo methylation of DNA. EMBO J. 2006;25:335–45.
163. Sun L-Q. Growth retardation and premature aging phenotypes in mice with disruption of the SNF2-like gene, PASG. Genes Dev. 2004;18:1035–46.
164. Pegoraro G, Kubben N, Wickert U, Göhler H, Hoffmann K, Misteli T. Ageing-related chromatin defects through loss of the NURD complex. Nat Cell Biol. 2009;11:1261–7.
165. Denslow SA, Wade PA. The human Mi-2/NuRD complex and gene regulation. Oncogene. 2007;26:5433–8.

166. Lai AY, Wade PA. Cancer biology and NuRD: a multifaceted chromatin remodelling complex. Nat Rev Cancer. 2011;11:588–96.
167. Dang W, Sutphin GL, Dorsey JA, Otte GL, Cao K, Perry RM, Wanat JJ, Saviolaki D, Murakami CJ, Tsuchiyama S, et al. Inactivation of yeast Isw2 chromatin remodeling enzyme mimics longevity effect of calorie restriction via induction of genotoxic stress response. Cell Metab. 2014;19:952–66.
168. Ura K. ATP-dependent chromatin remodeling facilitates nucleotide excision repair of UV-induced DNA lesions in synthetic dinucleosomes. EMBO J. 2001;20:2004–14.
169. Berdyshev GD, Korotaev GK, Boiarskikh GV, Vaniushin BF. Nucleotide composition of DNA and RNA from somatic tissues of humpback and its changes during spawning. Biokhimiia Mosc Russ. 1967;32:988–93.
170. Vanyushin BF, Nemirovsky LE, Klimenko VV, Vasiliev VK, Belozersky AN. The 5-methylcytosine in DNA of rats. . Tissue and age specificity and the changes induced by hydrocortisone and other agents. Gerontologia. 1973;19:138–52.
171. Heyn H, Li N, Ferreira HJ, Moran S, Pisano DG, Gomez A, Diez J, Sanchez-Mut JV, Setien F, Carmona FJ, et al. Distinct DNA methylomes of newborns and centenarians. Proc Natl Acad Sci. 2012;109:10522–7.
172. McClay JL, Aberg KA, Clark SL, Nerella S, Kumar G, Xie LY, Hudson AD, Harada A, Hultman CM, Magnusson PKE, et al. A methylome-wide study of aging using massively parallel sequencing of the methyl-CpG-enriched genomic fraction from blood in over 700 subjects. Hum Mol Genet. 2014;23:1175–85.
173. Fraga MF, Ballestar E, Paz MF, Ropero S, Setien F, Ballestar ML, Heine-Suner D, Cigudosa JC, Urioste M, Benitez J, et al. From the cover: epigenetic differences arise during the lifetime of monozygotic twins. Proc Natl Acad Sci. 2005;102:10604–9.
174. Reichard JF, Schnekenburger M, Puga A. Long term low-dose arsenic exposure induces loss of DNA methylation. Biochem Biophys Res Commun. 2007;352:188–92.
175. Lee KWK, Pausova Z. Cigarette smoking and DNA methylation. Front Genet. 2013;4:132.
176. Hirner AV, Rettenmeier AW. Methylated metal (loid) species in humans. Met Ions Life Sci. 2010;7:465–521.
177. Slagboom PE, De Leeuw WJF, Vijg J. Messenger RNA levels and methlation patterns of GAPDH and β-actin genes in rat liver, spleen and brain in relation to aging. Mech Ageing Dev. 1990;53:243–57.
178. Ono T, Uehara Y, Kurishita A, Tawa R, Sakurai H. Biological significance of DNA methylation in the ageing process. Age Ageing. 1993;22:S34–43.
179. Ahuja N, Li Q, Mohan AL, Baylin SB, Issa JP. Aging and DNA methylation in colorectal mucosa and cancer. Cancer Res. 1998;58:5489–94.
180. Rath PC, Kanungo MS. Methylation of repetitive DNA sequences in the brain during aging of the rat. FEBS Lett. 1989;244:193–8.
181. Bollati V, Schwartz J, Wright R, Litonjua A, Tarantini L, Suh H, Sparrow D, Vokonas P, Baccarelli A. Decline in genomic DNA methylation through aging in a cohort of elderly subjects. Mech Ageing Dev. 2009;130:234–9.
182. Jintaridth P, Mutirangura A. Distinctive patterns of age-dependent hypomethylation in interspersed repetitive sequences. Physiol Genomics. 2010;41:194–200.
183. Kriaucionis S, Heintz N. The Nuclear DNA Base 5-Hydroxymethylcytosine Is Present in Purkinje Neurons and the Brain. Science. 2009;324:929–30.
184. Hu C, Svetlana D, Hari M. Effect of aging on 5-hydroxymethylcytosine in the mouse hippocampus. Restor Neurol Neurosci. 2012; https://doi.org/10.3233/RNN-2012-110223.
185. Szulwach KE, Li X, Li Y, Song C-X, Wu H, Dai Q, Irier H, Upadhyay AK, Gearing M, Levey AI, et al. 5-hmC–mediated epigenetic dynamics during postnatal neurodevelopment and aging. Nat Neurosci. 2011;14:1607–16.
186. Grammatikakis I, Panda AC, Abdelmohsen K, Gorospe M. Long noncoding RNAs (lncRNAs) and the molecular hallmarks of aging. Aging. 2014;6:992–1009.
187. Chen L-H, Chiou G-Y, Chen Y-W, Li H-Y, Chiou S-H. microRNA and aging: a novel modulator in regulating the aging network. Ageing Res Rev. 2010;9:S59–66.

188. Lukiw WJ. Micro-RNA speciation in fetal, adult and Alzheimer's disease hippocampus. Neuroreport. 2007;18:297–300.
189. Hackl M, Brunner S, Fortschegger K, Schreiner C, Micutkova L, Mück C, Laschober GT, Lepperdinger G, Sampson N, Berger P, et al. miR-17, miR-19b, miR-20a, and miR-106a are down-regulated in human aging. Aging Cell. 2010;9:291–6.
190. Williams AE, Perry MM, Moschos SA, Lindsay MA. microRNA expression in the aging mouse lung. BMC Genomics. 2007;8:172.
191. Artandi SE, DePinho RA. Telomeres and telomerase in cancer. Carcinogenesis. 2010;31:9–18.
192. Samper E, Flores JM, Blasco MA. Restoration of telomerase activity rescues chromosomal instability and premature aging in *Terc* $^{-/-}$ mice with short telomeres. EMBO Rep. 2001;2:800–7.
193. Luke B, Lingner J. TERRA: telomeric repeat-containing RNA. EMBO J. 2009;28:2503–10.
194. Redon S, Reichenbach P, Lingner J. The non-coding RNA TERRA is a natural ligand and direct inhibitor of human telomerase. Nucleic Acids Res. 2010;38:5797–806.
195. Wang C, Zhao L, Lu S. Role of TERRA in the regulation of telomere length. Int J Biol Sci. 2015;11:316–23.
196. Deng Z, Campbell AE, Lieberman PM. TERRA, CpG methylation and telomere heterochromatin: lessons from ICF syndrome cells. Cell Cycle Georget Tex. 2010;9:69–74.
197. Monnier P, Martinet C, Pontis J, Stancheva I, Ait-Si-Ali S, Dandolo L. H19 lncRNA controls gene expression of the Imprinted Gene Network by recruiting MBD1. Proc Natl Acad Sci. 2013;110:20693–8.
198. Fu VX, Dobosy JR, Desotelle JA, Almassi N, Ewald JA, Srinivasan R, Berres M, Svaren J, Weindruch R, Jarrard DF. Aging and cancer-related loss of insulin-like growth factor 2 imprinting in the mouse and human prostate. Cancer Res. 2008;68:6797–802.
199. Bierhoff H, Schmitz K, Maass F, Ye J, Grummt I. Noncoding transcripts in sense and antisense orientation regulate the epigenetic state of ribosomal RNA genes. Cold Spring Harb Symp Quant Biol. 2010;75:357–64.
200. Johnson R, Strehler BL. Loss of genes coding for ribosomal RNA in ageing brain cells. Nature. 1972;240:412–4.
201. Machwe A. Accelerated methylation of ribosomal RNA genes during the cellular senescence of Werner syndrome fibroblasts. FASEB J. 2000;14:1715–24.
202. Pietrzak M, Rempala G, Nelson PT, Zheng J-J, Hetman M. Epigenetic silencing of nucleolar rRNA genes in Alzheimer's disease. PLoS ONE. 2011;6:e22585.
203. Bordone L, Guarente L. Calorie restriction, SIRT1 and metabolism: understanding longevity. Nat Rev Mol Cell Biol. 2005;6:298–305.
204. Masoro EJ. Overview of caloric restriction and ageing. Mech Ageing Dev. 2005;126:913–22.
205. Zhang R, Chen W, Adams PD. Molecular dissection of formation of senescence-associated heterochromatin foci. Mol Cell Biol. 2007;27:2343–58.
206. Narita M, Narita M, Krizhanovsky V, Nuñez S, Chicas A, Hearn SA, Myers MP, Lowe SW. A novel role for high-mobility group a proteins in cellular senescence and heterochromatin formation. Cell. 2006;126:503–14.

9. Attaining Epigenetic Rejuvenation: Challenges Ahead

Jogeswar S. Purohit, Neetika Singh, Shah S. Hussain, and Madan M. Chaturvedi

Epigenetics and Aging

> The major problem, I think, is chromatin... you can inherit something beyond the DNA sequence. That's where the real excitement of genetics is now [1]

The association of DNA with histone proteins to form chromatin is indeed an artistic invention of nature, facilitating the compaction of a massive amount of genomic information into a small package. Though chromatin aids in increasing the genomic stability, its structural conformation also hinders the accessibility of DNA to various enzyme complexes, which are required for nuclear events such as replication, transcription, repair and recombination. Therefore, the structure of chromatin needs to be altered by various mechanisms to orchestrate the sequential recruitment of these enzymes. Such mechanisms exert an upstream regulation of genomic information, thereby laying the foundation of epigenetics, which means 'above genetics'. Conrad Waddington in 1942 coined the term epigenetics. Epigenetics can be broadly defined as the 'changes in phenotype without changes in the genotype'.

The various epigenetic modifications, such as DNA methylation, histone modifications, nucleosome remodelling, histone variants and non-coding RNA, work in coherence to introduce variations into the chromatin. Further, these mechanisms dynamically regulate the chromatin and dictate its 'transcriptionally on' or 'transcriptionally off' state, thereby modulating the gene expression patterns [2–4]. Various seminal discoveries, which have explored the contribution of these mechanisms, have transformed the understanding of epigenetics from an inquisitive

---

J. S. Purohit
Cluster Innovation Centre, University Stadium, G.C. Narang Marg, North Campus, Delhi University, Delhi, India

N. Singh · S. S. Hussain · M. M. Chaturvedi (✉)
Laboratory for Chromatin Biology, Department of Zoology, University of Delhi, Delhi, India
e-mail: mchaturvedi@zoology.du.ac.in

biological phenomenon to a functionally well-dissected research area. Therefore, from its inception to this time, the epistemology of epigenetics has undergone a major transition.

The dispersal of the epigenetic marks along the genome, induced by various epigenetic mechanisms, in specific environment and particular time has a functional relevance in the maintenance of the stability of a cell. However, the gain or loss of these epigenetic marks may increase the functional heterogeneity of the cell within a tissue. Such functional cellular heterogeneity in any tissue may act as driving force for altered phenotypes, a typical characteristic of aging [4].

The process of aging has attracted curiosity for long. Research on aging can be traced back to 3000–1500 BC in the Indian medical system, Ayurveda [5]. The very first attempt to link chromatin and aging can be traced back to 1960s when it was reported that DNA composition does not change with age [6]. Further the age-related increased thermal stability as a consequence of increased nucleoprotein–DNA interaction was observed in beef thymus chromatin, making it less readable by polymerase [6]. Later it was reported, in 1978, that there are tissue-specific and age-dependent changes of H1 histones variants in chromatin from mouse tissue [7]. Also, in 1980, role of disulphide bonds was reported in young and old mice chromatin structure undergoing condensation, with increase in compaction in old age [8]. In 1985, the changes in the structural conformation of chromatin during aging were reported in the aged rats, where compaction of chromatin takes place with a concomitant decline in transcription [9]. Further, recent studies have also supported the role of epigenetics in the aetiology of aging [10].

Hence, in the light of all these developments, a convincing argument can be made that epigenetics play a regulatory role in aging, which indeed has given a new dimension to the field of epigenetics.

Broadly, the life span of any organism can be divided into (i) period of growth culminating in sexual maturity; (ii) period of maximal fitness and fertility and (iii) period of aging, a multifactorial biological process, characterized by various biochemical and molecular hallmarks [11].

Aging is demarcated by a sequential loss of physiological integrity, resulting in accumulation of cellular damage and increased vulnerability to death. The associated hallmarks of aging are genomic instability, telomere exhaustion, loss of homeostasis, epigenetic alterations, dysregulated sensing nutrient, dysfunctional mitochondria, cellular senescence and altered intercellular communication [12].

In the subsequent sections, we address the age-dependent changes in various epigenetic marks to provide a unified picture of aging and epigenetic research, done so far.

DNA Methylation and Aging

DNA methylation is a heritable epigenetic mark where a DNA methyltransferase (DNMTs) methylates C-5 position of cytosine ring of DNA. DNA methylation plays a crucial role in several key processes, such as genomic imprinting,

X-chromosome inactivation, transcription and transposition. Dysregulation in DNA methylation results in cancer [13]. In mammals, methylation is confined to cytosine residues as CpG islands [14].

Methylation of DNA is one of the extensively studied epigenetic marks which play an important role during aging. There is CpG hypomethylation along with loss of heterochromatin associated with aging [15]. Also, loss of CpG methylation at repetitive sequences is associated with retrotransposition events, thereby increasing the genomic instability. Aging-associated global decrease in DNA methylation is attributed to the progressive decline in level of DNMTs [16]. Conversely, along with the general and localized DNA hypomethylation, hypermethylation is also observed at specific CpG sites, to repress expression of specific genes [15]. In *Drosophila*, overexpression of DNMT2 enhances longevity, whereas its deletion generates short lived flies [11].

Post-Translational Modifications of Histones and Aging

Nucleosomes are the fundamental repeating units of the chromatin. Structurally, a typical nucleosome spans ~200 bp of DNA, consisting of 146–147 bp of core DNA that binds to the histone protein octamer (comprising two copies each of H2A, H2B, H3 and H4) and 53–54 bp of linker DNA. The entry and exit points of the nucleosomes are sealed by linker histone H1 [17]. The N-terminal tails of all core histones and the C-terminal tail of histone H2A protrude out from the nucleosome core [18]. Mainly, the histone tails and few core residues are the sites for various post-translational modifications, thereby providing a dockyard for various protein factors. These modifications and/or bound factors alter the chromatin state by modulating the DNA accessibility to various other chromatin interacting enzymatic machineries [19].

The most prominent histone modifications that occur during aging are acetylation and methylation, along with few other modifications.

Histone Acetylation and Aging

In budding yeast, acetylation marks, such as H3K56ac (acetylation present on lysine 56 of histone H3) and H4K16ac (acetylation present on lysine 16 of histone H4), are associated with replicative aging through different mechanisms [20, 21]. H3K56ac decreases during aging in yeast, while there is an increase in H4K16ac [20]. Further, deletion of genes encoding the histone deacetylases (HDACs), which remove H3K56ac, such as Hst4 and Hst3, leads to a shortened life span [20, 22] and genomic instability in yeast [20]. Further, deletion of Rtt109, a histone acetyltransferase (HAT), which acetylates H3K56ac or of histone chaperone Asf1, results in a shortened life span in yeast [21].

The increased level of H4K16ac is correlated with progressive decline of the related HDAC sirtuin silent information regulator 2 (Sir2) during aging [20].

Overexpression of Sir2 has been associated with the extension in life span [20]. Further, SIRT6, an orthologue of Sir2, deacetylates H3K9ac and H3K56ac in mice. Deficiency of SIRT6 in mice generates progeria phenotypes, and its overexpression has been linked with aging [23]. Also, failure to up-regulate acetylation at histone H4K12 (acetylation present on lysine 12 of histone H3) has been associated with memory impairment in aged mice. Restoration of this epigenetic mark has been reported to recover both gene expression and learning behaviour in aged mice [24].

Histone Methylation and Aging

Role of histone methylation has also been studied during aging. Significant changes have been observed in trimethylation marks on lysine (K) residues of 4th, 9th, 27th and 36th positions on histone H3 (H3K4me3, H3K9me3, H3K27me3 and H3K36me3), which indicates loss of heterochromatic structure with aging [15]. Further, during aging in *Drosophila*, decrease in heterochromatin protein (HP1) has been reported with a correlated decrease in H3K9me2 and H3K4me3 marks [25]. However, an increased H3K9me3 mark has been reported in the brain of aging fly heads [26]. Further, an increased H3K9me3 mark and a correlated increase in the expression of SUV39H1 (H3K9me3 methyltransferase) are observed in progeria mouse model, which is associated with impaired DNA repair. Once there is DNA damage, the chromatin remodellers take over to provide access to DNA repair factors. Increased H3K9me3 mark plays a role in the prevention of the remodelling of heterochromatic region, leading to sustained DNA damage, early senescence and eventually early death. Depletion of SUV39H1 has been linked to restoration of the defects in DNA repair and premature aging, thereby leading to an extended life span [27].

In senescent mouse brain tissues, there is depicted decrease in H4K20me1 and H3K36me3 marks and an increase in H3K27me3 mark [28]. In *C. elegans,* overexpression of H3K27me3 demethylase results in the increase in life span. On the contrary, in *Drosophila,* down-regulation of H3K27me3 methyl transferase has been associated with increased life span [29].

H3K36me3 mark has been linked with promotion of longevity. During replicative aging in yeast, reduction of H3K36me3 results in formation of a more open chromatin conformation, thereby exposing cryptic promoters resulting in limited life span. Deletion of the H3K36me3 demethylase thus leads to chromatin compaction, resulting in life span extension in yeast [30].

Other Histone Modifications and Aging

Deletion of the genes related to histone deubiquitinase module (DUBM) components such as SGF3, SGF11 and UBP8 encoding SAGA complexes increased the life span in yeast [31]. Further, in *C. elegans*, gene promoters related with aging and stress are enriched with O-N-acetyl-glucosamine (O-GlcNac), which can be deposited on histones H2A, H2B and H4 [31]. Furthermore, in old rats, there is an age-dependent decrease of histone carbonylation marks [32].

Generation of Heterochromatin Foci

Formation of senescence-associated heterochromatin foci (SAHF) is yet another distinct feature of senescent cells and is crucial for senescence. SAHF bodies are densely stained by DAPI, show resistance to nucleases digestion, implicating a compact structure, and possess enriched H3K9me3 marks. Each focus represents condensed chromatin domains of one chromosome. Further, additional proteins, such as histone chaperone HIRA Asf1, HP1, high-mobility group A (HMGA) proteins, histone variants and macro H2A, contribute in formation and/or maintenance of the SAHF [33].

Intertwining of DNA Methylation and Histone Modification

DNA methylation is also intertwined with various post-translational modifications of histones to exert the changes observed during aging. Age-induced DNA hypermethylation of loci-containing Polycomb target protein (PcG) brings repression of gene expression by enrichment of H3K27me3 marks. Hypermethylation of DNA during aging is enriched at the poised promoter, carrying bivalent histone marks (H3K27me3 and H3K4me3) both and DNA hypermethylation co-occurring with permissive histone modification marks such as H3K9ac, H3K27ac, H3K4me1 and H3K4me3 in the enhancer regions of the genes [15].

Chromatin Remodelling and Aging

Chromatin remodelling is brought about by a set of chromatin remodellers which are multi-subunit protein complexes containing a characteristic ATPase domain. These remodelling complexes utilize the energy of ATP to disrupt nucleosome–DNA contact, move, remove or exchange nucleosomes, thereby enabling the accession of DNA or histones to various proteins. On the basis of their subunit composition, these remodelling complexes are divided into five families: switch/sucrose nonfermentable (SWI/SNF), imitation switch (ISWI), (Mi-2 protein/nucleosome remodelling factor and deacetylase(Mi-2/NURD), inositol requiring protein (INO80) and swi/snf-related protein (SWR1) families [34–36].

In *C. elegans*, LET 418/Mi2 homologue of the catalytic subunit of the NURD family has been reported to be associated with longevity, as deletion of the gene results in the prolonged life span and increased environmental stress resistance. It is evolutionarily conserved and plays similar role in plant and fruit flies also [37]. In Hutchinson–Gilford progeria syndrome (HGPS) cells, the components of the NURD complex, including RBBP4 and RBBP7, were found to be substantially reduced. Experimental depletion of RBBP4 and RBBP7 in cell culture systems leads to endogenous DNA damage similar to HGPS cells and aged cells. The endogenous DNA damage is also preceded by defects in the chromatin structure, indicating the fact that loss of the NURD function might be an early event during aging, which makes the genome more susceptible to subsequent DNA damage [37]. Further, the absence of Isw2, chromatin remodelling enzyme, leads to extended life span in yeast [38]. RNAi-mediated down-regulation of NURD results in the loss of the heterochromatin domains [39]. In *C. elegans*, a functionally active SWI/SNF is required for promoting longevity. Also, in human adrenal cortex carcinoma-derived cell lines, senescence can be induced by *BRG1* [11].

Histone Variants, Histone Exchange and Aging

Replacement of canonical S-phase synthesized histones with histone variants in a replication-independent manner has been found to alter the chromatin. Such differentiation of chromatin is quite distinct at centromeres where H3 variant, centromere protein A (CENP-A), assembles into the specialized nucleosomes laying the foundation for kinetochore assembly. Further, other than the centromeres, H3 is also replaced by another variant histone, H3.3 in a replication-independent manner. Differences in the complement of covalent modifications between H3 and H3.3 might modulate the property of chromatin at actively transcribed loci of gene. Also, H2A variants such as H2A.X and H2A.Z also regulate chromatin structure and function. Vertebrate-specific variants, macro H2A and H2A.B also have an impact on transcription with former impeding and the latter facilitating the transcription [40].

In rat brain cortical neurons, there is replication-independent enrichment of histone H3.3 [41]. Further, in chicken and mice model, there is an increased incorporation of H3.3 during aging [42]. In slowly replicating cells, there is an accumulation of H3.3 and has an aberrant effect on heterochromatization [43]. The histone variant macroH2A1 isoform is increased in the liver of old rodents and human and in hepatocellular carcinoma cells (HCC). MacroH2A1 is used as a marker for SAHF and synergizes with DNA demethylating chemotherapeutic agent, 5-aza-2′-deoxycytidine (5-aza-dC), and is responsible for silencing of tumour suppressor genes [44]. Further, there is an increased incorporation of H3.3 in human fibroblasts cells entering aging. An increase in the specific H3.3 chaperons has also been documented in aged baboons [45].

Non-coding RNA and Aging

Non-coding RNAs are the RNA transcripts that do not code for proteins. Depending on the nucleotide length, they are classified as small non-coding RNAs (20–30 nucleotides) and long non-coding RNAs (lncRNAs more than 200 nucleotides). Non-coding RNAs are involved in processing and regulation of other RNAs [46].

In budding yeast, ncRNA transcription from the rDNA locus plays an important role in the life span determination. Mutations, preventing the expression of those ncRNA from the rDNA loci, play role in the life span extension [47]. In mouse and worms, during aging, there is a declined expression of Dicer, suggesting formation of defective small ncRNA (sncRNA) during aging [48]. Further, majority of microRNAs (miRNA), a class of sncRNA that negatively control the target gene expression, have been shown to be down-regulated during aging [48–50]. In mice, owing to the absence of Dicer in mice, there is an appearance of early signs of aging. Further, in adipocytes, collected from elderly humans, a decreased Dicer level has also been reported [48]. In *C. elegans,* miRNAs are involved in the modulation of life span and in controlling tissue aging [51, 52]. Further miRNAs also contribute in neurodegeneration, a characteristic of aging [53]. Further, *mir-34* family plays an important role in aging in flies and in worms [54]. *Mir-144* is also enriched in aged brains and contributes to age-associated neurodegeneration via down-regulation of various crucial protective factors [53]. *Mir-34* targets pro-survival factor BCL2 and the anti-aging deacetylase SIRT1, implicating a role in aging [55, 56]. lncRNA, *Gas5,* has been associated with impaired learning in mice, implicating an indirect role in the modulation of aging [57]. Few other ncRNAs, mostly products of RNAi pathway, play an important role in heterochromatin assembly in repetitive DNA elements [58].

Apart from these reversible histone modifications during aging listed above, one more irreversible post-translational phenomenon of histone proteolysis [59], that is, clipping of histones by protease at N-terminal and C-terminal regions, has also been reported during aging. The site-specific proteolysis of histones has been documented in histones (H2A,H3,H1) under different physiological and developmental conditions [60]. In *Tetrahymena,* transcriptionally inactive and senescent macronuclei undergo prototype clipping of core histone at N-terminal region [61]. Proteolytic clipping of histone has been depicted in many organisms during aging [62]. There is also observation of progesterone-induced cleavage of histone H3 in Japanese quail [63, 64]. In old/aged chicken, liver histone H3 undergoes proteolytic processing [62]. Further, a H2A-specific protease activity [65] and a H3-specific protease activity are also reported in aged chicken [66].

Linking Metabolic Reprogramming to Aging Epigenetics

Metabolic reprogramming is a mechanism to accomplish the proliferative needs and utilize the metabolic products predominantly for anabolic growth [67]. Metabolic pathways are recently being delineated in the context of aging and diseases such as cancer and diabetes [68]. It has been observed that cells encounter a progressive decrease in energy production during aging. This is most likely due to declined function of mitochondria, commonly called as mitochondrial dysfunction [69]. However, the exact mechanism of the alteration in energy metabolism that leads to aging is unknown. It has been observed that aging and aging-associated diseases are linked to availability of nutrients, intermediates of TCA cycle and the expression of enzymes of the TCA pathway. To survive under nutrient-deficient conditions, cancerous cells undergo changes such as mutations in metabolic enzymes like isocitrate dehydrogenase (IDH) and display of neomorphic activity for the production of 2-hydroxyglutarate (2HG) from αKG (alpha keto-glutarate), suggesting that cancer cells select altered metabolism during tumorigenesis [20, 70]. In the tumour core region, as compared to the periphery, dramatic decrease in the glutamine levels was observed [71]. This was found to be associated with the low synthesis of αKG and hypermethylation of histone H3. αKG upon entering the nucleus acts as a substrate for many deoxygenase enzymes that modify epigenetic marks. 2HG is structural analogue of αKG, and its specificity is greater for JHDM (Jumonji C domain histone demethylases) [72]. It has been observed that differential abundance of different metabolites in microenvironment plays an important role in cell proliferation. These intermediate metabolites of different anaplerotic reactions seem also to play an important role in the histone modifications. It has been found that aging cells have alteration in glycolytic pathway and electron transport chain similar to that in cancer cells [73, 74]. Decline in mitochondrial function is one of the hallmark of aging and its associated diseases [69, 75–78]. Aging cells have restricted supply of glucose [79]. By understanding different modifications of epigenetic marks during metabolic reprogramming, the mechanistic understanding for targeted therapeutics can also be understood for anti-aging. Aging can be reversed if the cellular efficiency could be restored. There are events that lead to communication between the nucleus and mitochondria. It has been observed that interruption of this nucleo-mitochondrial communication accelerates aging. Dysregulation of mitochondria implicated in many diseases like cancer, excess reactive oxygen species and alteration in mitochondrial metabolism indicates pivotal role played by mitochondria in cellular homeostasis regulation [80–83]. Intricate cascade of molecule, nicotinamide adenine dinucleotide (NAD$^+$), plays a key role in the shuttle between nucleus and mitochondria. It has been observed that nuclear NAD$^+$, a cofactor for the sirtuins, an NAD$^+$-dependent deacetylase, plays an important role in regulation of metabolism in aging [84]. Decline in the NAD$^+$ during aging leads to the accumulation of HIFα, a pseudohypoxic condition. This has been found to be associated with decrease in the production of αKG and correlated hypermethylation of histone H3 linking metabolic reprogramming to histone modifications.

Supply of the substrate and cofactors and metabolic activity within the cells can be altered by environmental stimuli (including nutrients). Cells can easily detect the altered energy levels such as level of NAD$^+$, which decreases with aging. This decrease in the NAD$^+$ level can be utilized as a potential hallmark for the aging process in different biological models [85]. NAD$^+$ is a cofactor for metabolic enzymes and acts as the co-substrate for the various sirtuins deacetylases [84, 86]. NAD$^+$-dependent sirtuins are deacetylases group of enzymes that are conserved from bacteria to humans [87]. The role of SER genes in *Saccharomyces cerevisiae* has been reported to play role in increasing the life span. One of the consequences of metabolic reprogramming is the promotion of glycolysis and inhibition of oxidative phosphorylation (OXPHOS). A similar kind of hypoxic (pseudohypoxic) condition induced by gerometabolites leads to dysfunctional mitochondria and Warburg-like metabolic reprogramming (Fig. 9.1).

Caloric Restriction and Metabolic Reprogramming

Caloric restriction (CR), defined as reduced intake of calories not causing malnutrition, is the sole intervention that can prolong longevity in all species that have been investigated so far, from yeast to non-human primates [88]. Altered mitochondrial metabolism, circulating level of adiponectin and sensitivity to insulin signalling are common signature of CR, and it suggests that metabolic reprogramming plays important role in the caloric restriction and longevity of organism (Fig. 9.1).

Reduced calorie uptake without malnutrition is a diet regimen that decelerates aging process and extends lifespan in diverse species by unknown mechanism [89–91]. There is an inverse relationship between CR and aging which suggests that

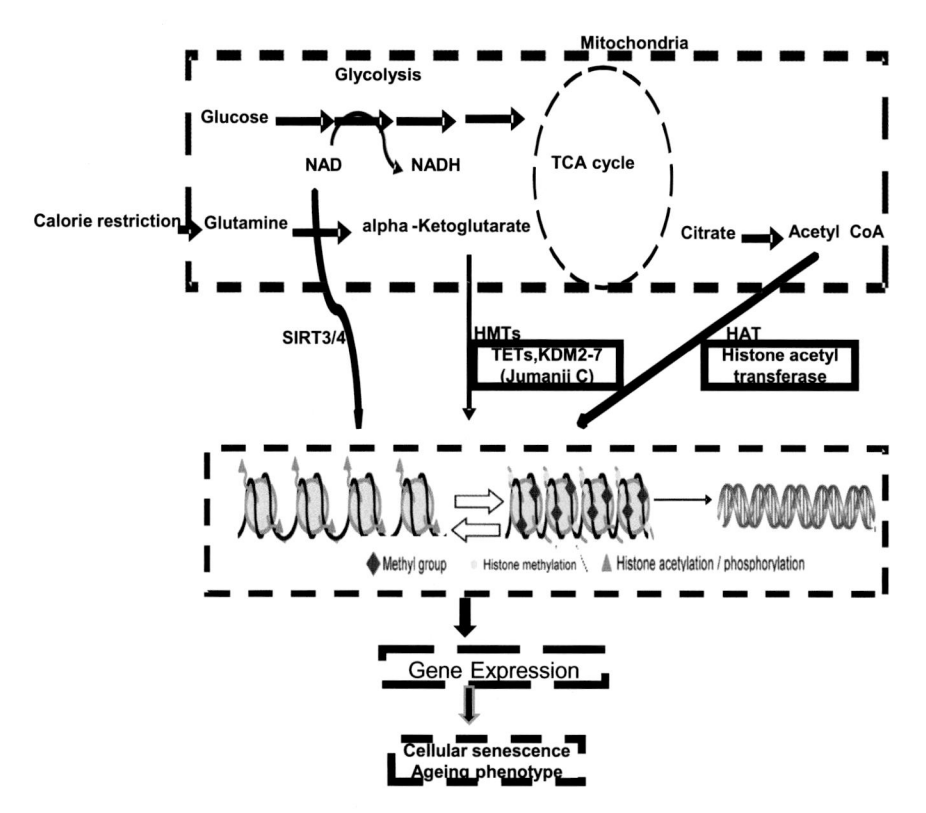

Fig. 9.1 Linking metabolic reprogramming to aging epigenetics. Glucose enters glycolytic pathway, which is converted to the end products through TCA cycle. Different intermediate metabolites can act as the substrate or co-factor for the enzymes involved in the histone modifications and thereby modulate epigenetic marks. Deregulation of the intermediates in the glycolytic pathway, thus, also modulates epigenetic marks in a different way

metabolic reprogramming plays an important role in decelerating the aging process. CR is involved in the reduced gene expression of inflammation and opposes many age-associated changes [92]. CR is associated with metabolic reprogramming that leads to increased expression of involved genes of glycolytic pathway, amino acid metabolism and mitochondrial metabolism [93]. From these studies, mitochondrial alterations can be correlated to CR [94]. Weindruch et al. (1986) proposed relation between the caloric intake and lifespan in female mice and discovered that caloric intake is inversely proportional to lifespan, which indicates that energy metabolism is key regulator of CR [95]. Anderson and Weindruch (2007) proposed that CR induces metabolic reprogramming because of availability of nutrient and CR induces an active response to altered nutrient availability [96] (Fig. 9.1).

It can be concluded that the nutritional factors and composition of food lead to interaction between caloric intake and genotype cascade of events that induces mutations and expression of genes which influences longevity. CR and optimum macronutrients (carbohydrates, fats and amino acids) lead to extension of life span in mice; thus, it is important to investigate metabolic pathways especially TCA cycle and those affecting glucose, fats and amino acids metabolites [97–99], because these metabolic intermediates and pathways have significant and complex effects on aging. Nutrient components which are intermediates of metabolic pathway play an important role in the aging by regulation of metabolism, which indicates importance of metabolic reprogramming in case of CR.

Delaying Aging

With passage of time, post-mitotic cells such as neurons undergo chronological aging due to gradual loss of normal structure and function. On the contrary, continuously dividing cells such as stem cells and yeast cells undergo replicative aging, in addition to chronological aging [99]. At least in few model organisms, in experimental conditions, the life span of these individuals can be extended. For example, mutation in the mTOR gene, the mammalian target of rapamycin gene of the insulin signalling pathway or mutation of the genes related to AMPK signalling pathways has resulted in the extension of life span in few model organisms [100–102]. Further, it has been shown that dietary restrictions have led to increase in life span in case of *C. elegans* [90]. However, it is evident that extension of life span has no correlation with the phenomenon of delaying aging.

There are few examples existing in nature, such as *C. elegans* larva, where aging clock can be arrested at least temporarily, during food scarcity conations, and the *C. elegans* larva attains a 'dauer' stage. This is relatively metabolically inactive state in the dauer survives for months in contrast to the normal life span of 2 weeks of the organism [103]. Seeds as old as 2000 years remain viable and can be regenerated [104]. There are reports of preservation of bacterial spores up to 25–30 million years in amber [105]. These are few excellent examples of uncoupling biological and chronological aging of an organism. However, most of these examples are related to lower organism and the delay in aging happens only in specific conditions. Hence, there is a need to explore the aging reversal process in higher organisms, which could be universal in nature.

Reversal of Aging: Aging Clock Resetting

With fertilization, the aging clock is reset. In many organisms, such as mammals, the chronological ages of the sperm and oocyte are erased while the zygote is formed. This process can be termed as resetting the aging clock [106]. This specific age reprogramming is mediated by factors of the oocyte cytoplasm [106].

The fertilization-mediated reprogramming process was exploited in the somatic cell nuclear transfer (SCNT) experiment. Here, an enucleated oocyte was transferred with a nucleus from a mature somatic cell [107, 108]. This resulted in reprogramming of the somatic cell nucleus by the oocyte cytoplasm and resetting the age of the somatic cell nucleus. This classical experiment further demonstrated the fact that somatic cells further can be rejuvenated and the pluripotency can be restored. This established the notion of reversing aging [106]. SCNT was the process which led to the generation of the first cloned mammal Dolly [109]. However, the level of reprogramming and the molecular basis of rejuvenation of the somatic cell nucleus are yet to be explored.

The molecular basis of reprogramming in SCNT can further be explained by the discovery of another process of creation of induced pluripotent stem cells (iPSCs) [110]. In this process, by introduction of genes of few differentiation-specific transcription factors such as Oct4, Sox2 and Klf4 (termed as Yamanaka factors), terminally differentiated adult cells can be reverted back to dedifferentiated state called iPSCs [111]. These iPSCs are similar to embryonic stem cells (ESCs), as they also like ESCs, have the potential of forming an entire organism [112] and have similar gene expression profiles and chromatin state as ESCs [113]. In contrast to SCNT, in case of iPSCs, however, the reprogramming is achieved by only few factors, and it does not require the oocyte mediates systems. The above iPSC-mediated age reprogramming is also called developmental reprogramming.

In both the above processes, the aging clock resetting mediated rejuvenation is coupled to differentiation programme reversal. The cells in these processes become young, but at the same time, they also lose their identity and become pluripotent stem cells. These methods may pose problems in therapeutic interventions such as direct reversal of aged tissues to respective young tissues. The ideal interventions would be directly converting these aged tissues into young without the intermittent pluripotent SCNT or iPSC state, as these induced stem cells have the possibility of forming tumours having features of many germ layers (teratomas) [106]. If achieved, this would experimentally uncouple the aging clock resetting from developmental reprogramming. The above hypothesized mechanism will pave the way in achieving epigenetic rejuvenation (Fig. 9.2).

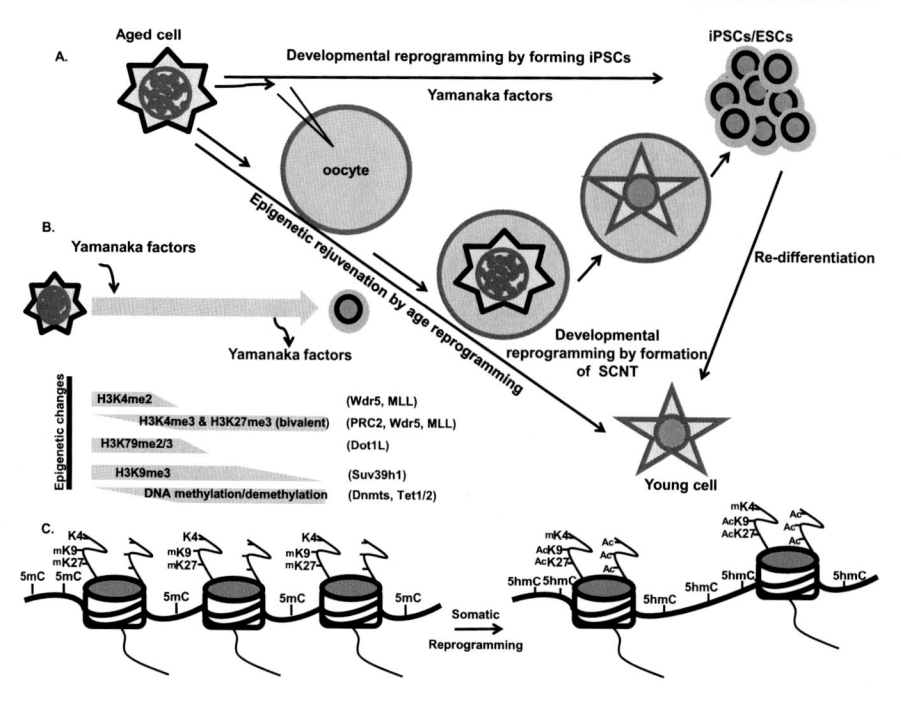

Fig. 9.2 Epigenetic rejuvenation. **A.** Old cells have distinct SAHF bodies. By using Yamanaka factors, such as Klf4, c-Myc, Sox2 and Oct4, these cells can be dedifferentiated to the induced pluripotent cells (iPSCS). These iPSCs can further be redifferentiated into young cells or rejuvenated cells. The rejuvenation can also be achieved by placing the cell inside the oocyte, which converts it to an ES state which again differentiates to form young cells. Rejuvenation can also be achieved by direct conditioning of the old cells into young cells, and this process is called as epigenetic rejuvenation. **B.** Epigenetic changes taking place between an aged and a young cell. 'Ac' refers to acetylation mark; 'me' refers to methylation marks. The symbolic representation such as H3K9me3 refers to histone H3 Lys 9th residue is trimethylated. **C.** Nucleosomal arrays representing the epigenetic marks on H3 of an aged and a rejuvenated cell. For easy understanding, methylation marks are shown on one tail of H3 and acetylation marks only are shown on the other histone H3 tail

Rejuvenation Without Differentiation

Historically by joining systemic circulations of two animals, young-to-old or old-to-young transition was achieved [114]. Recently, it has been observed that tissue-specific stem cells of old mice were rejuvenated by exposure to young environment [115]. The specific rejuvenation was also confirmed by their molecular signatures. Conversely, young stem cells can be converted to aged tissues by exposing them to an aged environment [116]. It has been further seen that the aged phenotype is maintained by increased activity of NFκB [117]. Further, when the NFκB expression in aged mice was down-regulated in the skin tissues, they returned to a more youthful state [117], with concomitant down-regulation of all senescent-associated genes. Furthermore, rapamycin (an mTOR inhibitor) delayed the aging in mice

[118]. Rapamycin also rejuvenated the aged haematopoietic stem cells in mice by inhibiting mTOR [119]. Together, these observations hint towards the plausible uncoupling of the aging clock from dedifferentiation programmes, suggesting the epigenetic-mediated rejuvenation.

Epigenetic Reprogramming Is the Basis of Epigenetic Rejuvenation

Once a cell is terminally differentiated, the epigenetic state of the cell tends to be maintained during subsequent divisions. However, the SCNT and iPSC cell experiments suggest that the epigenetic state of the terminally differentiated cell also can be reversed and reprogrammed to dedifferentiated state [106]. During development also, there is a hint of epigenetic reprogramming in the zygote [120]. Following fertilization, the DNA methylation is erased initially and then re-established in the zygote [121]. Histone modifications are also reorganized in the zygote. Many of these epigenetic resettings are also observed in iPSC reprogramming [122]. Both the zygote and the ES cell stage are observed to possess bivalent chromatin domains (containing marks for transcriptionally permissive chromatin such as H3K4me3 and transcriptionally repressive chromatin such as H3K27me3) [123]. Later, during ES cell differentiation, these H3K27me3 marks are replaced by H3K27ac mark in the gene promoter and enhancer regions [124]. A similar epigenetic signature among the zygote, SCNT, iPSC and ESCs strongly hints a common epigenetic reprogramming pathway during differentiation and dedifferentiation. Further, there are reports that P16 and P53 proteins, which are overexpressed in aged and damaged cells, pose strong barrier for iPSC formation [125]. Moreover, the halting of the aging clock in *C. elegans* larval stage further suggests that aging is mostly mediated by epigenetic forces [126]. It also argues that aging can be reversed, indicating that rejuvenation could be achieved by epigenetic mechanisms.

Epigenetic Signatures of the Aged and Rejuvenated Cells

Cells undergoing senescence accumulate facultative heterochromatin domains stained as condensed regions in DAPI staining and are popularly called senescence-associated heterochromatin foci (SAHFs) [127]. SAHFs have enrichment of HP1 protein (HP1β) and H3K27me3 and H3K9me3 marks [128]. HP1β binds to H3K9me3 and both HP1β and macro H2A co-localize in SHAF. Further, there is accumulation of heterochromatin-associated histone H2A variant called macro H2A. Telomerase shortening is another phenomenon in these senescent cells. These SAHF bodies take up the cell division–specific genes such as E2F. While in iPSCs, it has been observed that there is expression of telomerase with concomitant increase in telomere length, degeneration of the SAHF bodies and reduced expression of mH2A along with increased signal for H3K4me3 and reduction of H3K27me3. All these epigenetic signatures are characteristic of ESCs and young cells, hence strongly hinting the developmental reprogramming-mediated epigenetic basis of aging reversal.

During nuclear reprogramming, there is decompaction of the chromatin with concomitant increase in nuclear volume. This is hypothesized to be achieved by replacement of H1 with HMG [129]. This replacement is mediated by histone chaperone nucleoplasmins. The chromatin occupancy studies by the Yamanaka factors have identified their action targets, while formation of iPSCs. Genes already existing in open state (positive for DNase I hypersensitivity and with H3K4me2/3 state) are readily bound by these Yamanaka factors. Few distal regions of the gene with H3K4me1 marks, (characteristic of permissive chromatin domains), require chromatin remodelling and histone depletion. Broad heterochromatic region with H3K9me2/3 marks is extensively remodelled. There are few bivalent chromatin domains with both H3K4me2 and H3K27me3 marks in the ES cells. Genes of these domains are transcriptionally silent in the iPSCs but pose active targets for remodelling [130]. Broadly, during iPSC induction, there is global decrease in H3K4me2 marks; gradual increase in H3K3me2/3 marks; gradual decrease in H3K79me2/3 marks and restructuring of DNA methylation and demethylation. Accordingly, loss of the H3K27me3 methyl transferase (PRC2) inhibits iPSC formation, while loss of the H3K79me2 methyl transferase Dot1L favours iPSC formation [130]. Further, the iPSC terminal stage transition is favoured by activation of H3K36 demethylase and down-regulation of H3K27 demethylase. On the contrary, knockdown of H3K9 methyl transferase such as Suv39h1 or G9 favours iPSC formation. Knockdown of the CHD chromatin remodelling factors also inhibits iPSC formation, while knockdown of NURD complex favours iPSC formation [130]. All these observations strongly suggest that the reprogramming is truly mediated by underlying epigenetic phenomenon.

Are Aging Signs Reversed by Reprogramming?

There is an increase in the length of the telomeres following epigenetic reprogramming in old mice and human cells, with concomitant increase in telomerase activity. Following epigenetic reprogramming, the mitochondrial functions are also restored to some extent, similar to young ESCs [131]. There is also down-regulation of many senescence associated and apoptotic genes following reprogramming [132]. It has been observed that the fibroblast rejuvenated from a senescent fibroblast possessed a rejuvenated transcriptome, characteristic of a young cell [131]. However, the gene expression profiles of the iPSC and the hESC were similar but not identical. Further, though the reprogrammed cells such as redifferentiated fibroblasts seemed younger than the aged fibroblasts, they were distinguishable from the young fibroblasts [133]. It has further been observed that during dedifferentiation into an iPSC, the epigenome is reorganized; however, the epigenetic memory of the cell is not lost, and upon redifferentiation, the epigenome again manifests itself [134]. A comparison of the epigenetic signatures of the cells during programming is tabulated in Table 9.1.

Table 9.1 Aging characters those can be reversed or not reversed by reprogramming (Lapasset et al. 2011)

Character	Young cell	Old cell	iPSC	Redifferentiated cells
Telomere	Long	Short	Extended	Extension maintained
Mitochondria	Functional	Dysfunctional	Rejuvenated	Improved functional
Transcriptome	Youthful	Aged	Rejuvenated	Youthful like
DNA	Intact	Damaged	Persisted mutation	Persisted mutation
Epigenome	Youthful	Aged	Rejuvenated with age memory	Youthful like with age memory
Protein aggregates	No	Yes	Yes?	Yes?

Attaining Epigenetic Rejuvenation: Challenges Ahead

Aging is time-dependent function which leads to progressive loss of physiological functions and increased vulnerability to death. Aging is a complex phenomenon which is caused by integrated events of genetics, epigenetics to metabolomics factors. Alteration in any one of factor leads to cellular senescence and death of cells. All these events are correlated to the alterations in energy metabolism. The exact mechanism of the alteration in energy metabolism that leads to aging is unknown. It has been observed that CR and optimum diet intake can help in delaying aging. It might be due to the metabolic rewiring which brings the alternative pathways to repair the faulty metabolic pathways and thus plays important role in aging. Many of the intermediates of these metabolic pathways act as the substrate and cofactors for the enzymes of epigenetics marks of histones. For example, αKG acts as substrate for the JHDM domain. So by using the controlled CR, we can reprogramme the epigenetic marks and ultimately attain epigenetic rejuvenation.

Recently, the iPSC techniques have posed tremendous potential for regenerative therapies as these cells can be reprogrammed even from aged cells. However, epigenetic rejuvenation if achieved will be a better option for aging reversal programmes. This is due to the fact that by epigenetic rejuvenation, the old cells are directly reprogrammed into young and this process does not involve passage through a dedifferentiated pluripotent state. This has an advantage as this process is proposed to be achieved only by epigenetic mechanisms which avoid the genetic abnormalities that may arise during dedifferentiation-based aging reversal processes. Further, the time required for epigenetic rejuvenation of the old cells only requires few days compared to the few week durations required for the iPSC-based therapies. Still, the epigenetic rejuvenation process is at its infancy stage due to the lack of complete understanding of the mechanism and also due to lack of an appropriate model system for the study. When human fibroblast cells and Hutchinson–Gilford progeria syndrome (HGPS) cells were treated with rapamycin, there was decrease in protein aggregates and they were converted into young fibroblasts. In a recent preposition, it is also being hypothesized that curcumin also might pose as a potential anti-aging molecule. Few recent unpublished observations (Chaturvedi Group) by our group have revealed that curcumin works by modulating the epigenetic signatures of the cell. However, attaining epigenetic rejuvenation and practice of this mechanism as a therapeutic model requires further investigations.

References

1. Watson JD. Celebrating the genetic jubilee: a conversation with James D. Watson. Interviewed by John Rennie. Sci Am. 2003;288(4):66–9.
2. Jaenisch R, Bird A. Epigenetic regulation of gene expression: how the genome integrates intrinsic and environmental signals. Nat Genet. 2003;33(Suppl):245–54.
3. Weber CM, Henikoff S. Histone variants: dynamic punctuation in transcription. Genes Dev. 2014;28(7):672–82.
4. Sierra MI, Fernandez AF, Fraga MF. Epigenetics of aging. Curr Genomics. 2015;16(6):435–40.
5. Ashok BT, Ali R. Aging research in India. Exp Gerontol. 2003;38(6):597–603.
6. Pyhtila MJ, Sherman FG. Age-associated studies on thermal stability and template effectiveness of DNA and nucleoproteins from beef thymus. Biochem Biophys Res Commun. 1968;31(3):340–4.
7. Medvedev ZA, Medvedeva MN, Robson L. Tissue specificity and age changes for the pattern of the H1 group of histones in chromatin from mouse tissues. Gerontology. 1978;24(4):286–92.
8. Tas S, Tam CF, Walford RL. Disulfide bonds and the structure of the chromatin complex in relation to aging. Mech Ageing Dev. 1980;12(1):65–80.
9. Chaturvedi MM, Kanungo MS. Analysis of conformation and function of the chromatin of the brain of young and old rats. Mol Biol Rep. 1985;10(4):215–9.
10. Gravina S, Vijg J. Epigenetic factors in aging and longevity. Pflugers Archiv: European J Physiol. 2010;459(2):247–58.
11. Benayoun BA, Pollina EA, Brunet A. Epigenetic regulation of ageing: linking environmental inputs to genomic stability. Nat Rev Mol Cell Biol. 2015;16(10):593–610.
12. Lopez-Otin C, Blasco MA, Partridge L, Serrano M, Kroemer G. The hallmarks of aging. Cell. 2013;153(6):1194–217.
13. Jin B, Li Y, Robertson KD. DNA methylation: superior or subordinate in the epigenetic hierarchy? Genes Cancer. 2011;2(6):607–17.
14. Klose RJ, Bird AP. Genomic DNA methylation: the mark and its mediators. Trends Biochem Sci. 2006;31(2):89–97.
15. Pal S, Tyler JK. Epigenetics and aging. Sci Adv. 2016;2(7):e1600584.
16. Jung M, Pfeifer GP. Aging and DNA methylation. BMC Biol. 2015;13(7):1–8.
17. Watson JD, Baker TA, Bell SP, Gann A, Levine M, Losick R. Genome structure, chromatin, and the nucleosome Molecular Biology of the gene. 6th ed. Pearson: CSHL Press; 2008. p. 135–93.
18. Davey CA, Sargent DF, Luger K, Maeder AW, Richmond TJ. Solvent mediated interactions in the structure of the nucleosome core particle at 1.9 a resolution. J Mol Biol. 2002;319(5):1097–113.
19. Bannister AJ, Kouzarides T. Regulation of chromatin by histone modifications. Cell Res. 2011;21(3):381–95.
20. Dang L, White DW, Gross S, Bennett BD, Bittinger MA, Driggers EM, et al. Cancer-associated IDH1 mutations produce 2-hydroxyglutarate. Nature. 2009;462(7274):739–44.
21. Feser J, Truong D, Das C, Carson JJ, Kieft J, Harkness T, et al. Elevated histone expression promotes life span extension. Mol Cell. 2010;39(5):724–35.
22. Tsuchiya M, Dang N, Kerr EO, Hu D, Steffen KK, Oakes JA, et al. Sirtuin-independent effects of nicotinamide on lifespan extension from calorie restriction in yeast. Aging Cell. 2006;5(6):505–14.
23. Kanfi Y, Naiman S, Amir G, Peshti V, Zinman G, Nahum L, et al. The sirtuin SIRT6 regulates lifespan in male mice. Nature. 2012;483(7388):218–21.
24. Peleg S, Sananbenesi F, Zovoilis A, Burkhardt S, Bahari-Javan S, Agis-Balboa RC, et al. Altered histone acetylation is associated with age-dependent memory impairment in mice. Science. 2010;328(5979):753–6.
25. Larson K, Yan SJ, Tsurumi A, Liu J, Zhou J, Gaur K, et al. Heterochromatin formation promotes longevity and represses ribosomal RNA synthesis. PLoS Genet. 2012;8(1):e1002473.

26. Wood JG, Hillenmeyer S, Lawrence C, Chang C, Hosier S, Lightfoot W, et al. Chromatin remodeling in the aging genome of Drosophila. Aging Cell. 2010;9(6):971–8.
27. Liu B, Wang Z, Zhang L, Ghosh S, Zheng H, Zhou Z. Depleting the methyltransferase Suv39h1 improves DNA repair and extends lifespan in a progeria mouse model. Nat Commun. 2013;4:1868.
28. Wang CM, Tsai SN, Yew TW, Kwan YW, Ngai SM. Identification of histone methylation multiplicities patterns in the brain of senescence-accelerated prone mouse 8. Biogerontology. 2010;11(1):87–102.
29. Ni Z, Ebata A, Alipanahiramandi E, Lee SS. Two SET domain containing genes link epigenetic changes and aging in Caenorhabditis elegans. Aging Cell. 2012;11(2):315–25.
30. Sen P, Dang W, Donahue G, Dai J, Dorsey J, Cao X, et al. H3K36 methylation promotes longevity by enhancing transcriptional fidelity. Genes Dev. 2015;29(13):1362–76.
31. McCormick MA, Mason AG, Guyenet SJ, Dang W, Garza RM, Ting MK, et al. The SAGA histone deubiquitinase module controls yeast replicative lifespan via Sir2 interaction. Cell Rep. 2014;8(2):477–86.
32. Sharma R, Nakamura A, Takahashi R, Nakamoto H, Goto S. Carbonyl modification in rat liver histones: decrease with age and increase by dietary restriction. Free Radic Biol Med. 2006;40(7):1179–84.
33. Kosar M, Bartkova J, Hubackova S, Hodny Z, Lukas J, Bartek J. Senescence-associated heterochromatin foci are dispensable for cellular senescence, occur in a cell type- and insult-dependent manner and follow expression of p16(ink4a). Cell Cycle. 2011;10(3):457–68.
34. Cairns BR. Chromatin remodeling: insights and intrigue from single-molecule studies. Nat Struct Mol Biol. 2007;14(11):989–96.
35. Hargreaves DC, Crabtree GR. ATP-dependent chromatin remodeling: genetics, genomics and mechanisms. Cell Res. 2011;21(3):396–420.
36. Vaquero A, Loyola A, Reinberg D. The constantly changing face of chromatin. Sci Aging Knowledge Environ: SAGE KE. 2003;2003(14):RE4.
37. De Vaux V, Pfefferli C, Passannante M, Belhaj K, von Essen A, Sprecher SG, et al. The Caenorhabditis elegans LET-418/Mi2 plays a conserved role in lifespan regulation. Aging Cell. 2013;12(6):1012–20.
38. Dang W, Sutphin GL, Dorsey JA, Otte GL, Cao K, Perry RM, et al. Inactivation of yeast Isw2 chromatin remodeling enzyme mimics longevity effect of calorie restriction via induction of genotoxic stress response. Cell Metab. 2014;19(6):952–66.
39. Pegoraro G, Kubben N, Wickert U, Gohler H, Hoffmann K, Misteli T. Ageing-related chromatin defects through loss of the NURD complex. Nat Cell Biol. 2009;11(10):1261–7.
40. Henikoff S, Smith MM. Histone variants and epigenetics. Cold Spring Harb Perspect Biol. 2015;7(1):a019364.
41. Pina B, Suau P. Changes in histones H2A and H3 variant composition in differentiating and mature rat brain cortical neurons. Dev Biol. 1987;123(1):51–8.
42. Urban MK, Zweidler A. Changes in nucleosomal core histone variants during chicken development and maturation. Dev Biol. 1983;95(2):421–8.
43. Saade E, Pirozhkova I, Aimbetov R, Lipinski M, Ogryzko V. Molecular turnover, the H3.3 dilemma and organismal aging (hypothesis). Aging Cell. 2015;14(3):322–33.
44. Borghesan M, Fusilli C, Rappa F, Panebianco C, Rizzo G, Oben JA, et al. DNA Hypomethylation and histone variant macroH2A1 synergistically attenuate chemotherapy-induced senescence to promote hepatocellular carcinoma progression. Cancer Res. 2016;76(3):594–606.
45. Jeyapalan JC, Ferreira M, Sedivy JM, Herbig U. Accumulation of senescent cells in mitotic tissue of aging primates. Mech Ageing Dev. 2007;128(1):36–44.
46. Rinn JL, Chang HY. Genome regulation by long noncoding RNAs. Annu Rev Biochem. 2012;81:145–66.
47. Saka K, Ide S, Ganley AR, Kobayashi T. Cellular senescence in yeast is regulated by rDNA noncoding transcription. Curr Biol. 2013;23(18):1794–8.

48. Mori MA, Raghavan P, Thomou T, Boucher J, Robida-Stubbs S, Macotela Y, et al. Role of microRNA processing in adipose tissue in stress defense and longevity. Cell Metab. 2012;16(3):336–47.
49. Ibanez-Ventoso C, Yang M, Guo S, Robins H, Padgett RW, Driscoll M. Modulated microRNA expression during adult lifespan in Caenorhabditis elegans. Aging Cell. 2006;5(3):235–46.
50. Kato M, Chen X, Inukai S, Zhao H, Slack FJ. Age-associated changes in expression of small, noncoding RNAs, including microRNAs, in C. elegans. RNA. 2011;17(10):1804–20.
51. Boehm M, Slack F. A developmental timing microRNA and its target regulate life span in C. elegans. Science. 2005;310(5756):1954–7.
52. de Lencastre A, Pincus Z, Zhou K, Kato M, Lee SS, Slack FJ. MicroRNAs both promote and antagonize longevity in C. elegans. Curr Biol. 2010;20(24):2159–68.
53. Szafranski K, Abraham KJ, Mekhail K. Non-coding RNA in neural function, disease, and aging. Front Genet. 2015;6:87.
54. Liu N, Landreh M, Cao K, Abe M, Hendriks GJ, Kennerdell JR, et al. The microRNA miR-34 modulates ageing and neurodegeneration in Drosophila. Nature. 2012;482(7386):519–23.
55. Lee J, Padhye A, Sharma A, Song G, Miao J, Mo YY, et al. A pathway involving farnesoid X receptor and small heterodimer partner positively regulates hepatic sirtuin 1 levels via microRNA-34a inhibition. J Biol Chem. 2010;285(17):12604–11.
56. Jung HJ, Suh Y. MicroRNA in aging: from discovery to biology. Curr Genomics. 2012;13(7):548–57.
57. Meier I, Fellini L, Jakovcevski M, Schachner M, Morellini F. Expression of the snoRNA host gene gas5 in the hippocampus is upregulated by age and psychogenic stress and correlates with reduced novelty-induced behavior in C57BL/6 mice. Hippocampus. 2010;20(9):1027–36.
58. Castel SE, Martienssen RA. RNA interference in the nucleus: roles for small RNAs in transcription, epigenetics and beyond. Nat Rev Genet. 2013;14(2):100–12.
59. Allis CD, Bowen JK, Abraham GN, Glover CV, Gorovsky MA. Proteolytic processing of histone H3 in chromatin: a physiologically regulated event in Tetrahymena micronuclei. Cell. 1980;20(1):55–64.
60. Purohit JS, Chaturvedi MM, Panda P. Histone protease: the tale of tail clippers. Int J Integr Sci, Innov Technol. 2012;1(1):51–60.
61. Lin R, Cook RG, Allis CD. Proteolytic removal of core histone amino termini and dephosphorylation of histone H1 correlate with the formation of condensed chromatin and transcriptional silencing during Tetrahymena macronuclear development. Genes Dev. 1991;5(9):1601–10.
62. Satchidananda PJ, Mohan CM. Chromatin and aging. In: Rath PS PC, Sharma S, editors. Topics in biomedical gerontology. Singapore: Springer; 2017. p. 205.
63. Mahendra G, Gupta S, Kanungo MS. Effect of 17beta estradiol and progesterone on the conformation of the chromatin of the liver of female Japanese quail during aging. Arch Gerontol Geriatr. 1999;28(2):149–58.
64. Mahendra G, Kanungo MS. Age-related and steroid induced changes in the histones of the quail liver. Arch Gerontol Geriatr. 2000;30(2):109–14.
65. Panda P, Chaturvedi MM, Panda AK, Suar M, Purohit JS. Purification and characterization of a novel histone H2A specific protease (H2Asp) from chicken liver nuclear extract. Gene. 2013;512(1):47–54.
66. Purohit JS, Tomar RS, Panigrahi AK, Pandey SM, Singh D, Chaturvedi MM. Chicken liver glutamate dehydrogenase (GDH) demonstrates a histone H3 specific protease (H3ase) activity in vitro. Biochimie. 2013;95(11):1999–2009.
67. Vander Heiden MG, Cantley LC, Thompson CB. Understanding the Warburg effect: the metabolic requirements of cell proliferation. Science. 2009;324(5930):1029–33.
68. Barzilai N, Huffman DM, Muzumdar RH, Bartke A. The critical role of metabolic pathways in aging. Diabetes. 2012;61(6):1315–22.
69. Bratic I, Trifunovic A. Mitochondrial energy metabolism and ageing. Biochim Biophys Acta. 2010;1797(6–7):961–7.

70. Ward PS, Patel J, Wise DR, Abdel-Wahab O, Bennett BD, Coller HA, et al. The common feature of leukemia-associated IDH1 and IDH2 mutations is a neomorphic enzyme activity converting alpha-ketoglutarate to 2-hydroxyglutarate. Cancer Cell. 2010;17(3):225–34.
71. Pan M, Reid MA, Lowman XH, Kulkarni RP, Tran TQ, Liu X, et al. Regional glutamine deficiency in tumours promotes dedifferentiation through inhibition of histone demethylation. Nat Cell Biol. 2016;18(10):1090–101.
72. Chowdhury R, Yeoh KK, Tian YM, Hillringhaus L, Bagg EA, Rose NR, et al. The oncometabolite 2-hydroxyglutarate inhibits histone lysine demethylases. EMBO Rep. 2011;12(5):463–9.
73. Bowling AC, Schulz JB, Brown RH Jr, Beal MF. Superoxide dismutase activity, oxidative damage, and mitochondrial energy metabolism in familial and sporadic amyotrophic lateral sclerosis. J Neurochem. 1993;61(6):2322–5.
74. Hagen TM, Yowe DL, Bartholomew JC, Wehr CM, Do KL, Park JY, et al. Mitochondrial decay in hepatocytes from old rats: membrane potential declines, heterogeneity and oxidants increase. Proc Natl Acad Sci U S A. 1997;94(7):3064–9.
75. Wallace DC, Fan W, Procaccio V. Mitochondrial energetics and therapeutics. Annu Rev Pathol. 2010;5:297–348.
76. Koopman WJ, Willems PH, Smeitink JA. Monogenic mitochondrial disorders. N Engl J Med. 2012;366(12):1132–41.
77. Bratic A, Larsson NG. The role of mitochondria in aging. J Clin Invest. 2013;123(3):951–7.
78. Sun N, Youle RJ, Finkel T. The mitochondrial basis of aging. Mol Cell. 2016;61(5):654–66.
79. Kim J, Kundu M, Viollet B, Guan KL. AMPK and mTOR regulate autophagy through direct phosphorylation of Ulk1. Nat Cell Biol. 2011;13(2):132–41.
80. Ren J, Pulakat L, Whaley-Connell A, Sowers JR. Mitochondrial biogenesis in the metabolic syndrome and cardiovascular disease. J Mol Med. 2010;88(10):993–1001.
81. Turner N, Heilbronn LK. Is mitochondrial dysfunction a cause of insulin resistance? Trends Endocrinol Metab. 2008;19(9):324–30.
82. Su B, Wang X, Zheng L, Perry G, Smith MA, Zhu X. Abnormal mitochondrial dynamics and neurodegenerative diseases. Biochim Biophys Acta. 2010;1802(1):135–42.
83. Mammucari C, Rizzuto R. Signaling pathways in mitochondrial dysfunction and aging. Mech Ageing Dev. 2010;131(7–8):536–43.
84. Verdin E. NAD(+) in aging, metabolism, and neurodegeneration. Science. 2015;350(6265):1208–13.
85. Scheibye-Knudsen M, Mitchell SJ, Fang EF, Iyama T, Ward T, Wang J, et al. A high-fat diet and NAD(+) activate Sirt1 to rescue premature aging in cockayne syndrome. Cell Metab. 2014;20(5):840–55.
86. German NJ, Haigis MC. Sirtuins and the metabolic hurdles in Cancer. Curr Biol. 2015;25(13):R569–83.
87. Vassilopoulos A, Fritz KS, Petersen DR, Gius D. The human sirtuin family: evolutionary divergences and functions. Hum Genomics. 2011;5(5):485–96.
88. Madeo F, Zimmermann A, Maiuri MC, Kroemer G. Essential role for autophagy in life span extension. J Clin Invest. 2015;125(1):85–93.
89. Colman RJ, Anderson RM, Johnson SC, Kastman EK, Kosmatka KJ, Beasley TM, et al. Caloric restriction delays disease onset and mortality in rhesus monkeys. Science. 2009;325(5937):201–4.
90. Fontana L, Partridge L, Longo VD. Extending healthy life span--from yeast to humans. Science. 2010;328(5976):321–6.
91. Wang L, Karpac J, Jasper H. Promoting longevity by maintaining metabolic and proliferative homeostasis. J Exp Biol. 2014;217(Pt 1):109–18.
92. Lee CK, Allison DB, Brand J, Weindruch R, Prolla TA. Transcriptional profiles associated with aging and middle age-onset caloric restriction in mouse hearts. Proc Natl Acad Sci U S A. 2002;99(23):14988–93.
93. Higami Y, Pugh TD, Page GP, Allison DB, Prolla TA, Weindruch R. Adipose tissue energy metabolism: altered gene expression profile of mice subjected to long-term caloric restriction. FASEB J: Off Pub Fed Am Soc Exp Biol. 2004;18(2):415–7.

94. Higami Y, Barger JL, Page GP, Allison DB, Smith SR, Prolla TA, et al. Energy restriction lowers the expression of genes linked to inflammation, the cytoskeleton, the extracellular matrix, and angiogenesis in mouse adipose tissue. J Nutr. 2006;136(2):343–52.
95. Weindruch R, Walford RL, Fligiel S, Guthrie D. The retardation of aging in mice by dietary restriction: longevity, cancer, immunity and lifetime energy intake. J Nutr. 1986;116(4):641–54.
96. Anderson RM, Weindruch R. Metabolic reprogramming in dietary restriction. Interdiscip Top Gerontol. 2007;35:18–38.
97. Solon-Biet SM, Mitchell SJ, Coogan SC, Cogger VC, Gokarn R, McMahon AC, et al. Dietary protein to carbohydrate ratio and caloric restriction: comparing metabolic outcomes in mice. Cell Rep. 2015;11(10):1529–34.
98. Solon-Biet SM, McMahon AC, Ballard JW, Ruohonen K, Wu LE, Cogger VC, et al. The ratio of macronutrients, not caloric intake, dictates cardiometabolic health, aging, and longevity in ad libitum-fed mice. Cell Metab. 2014;19(3):418–30.
99. Charville GW, Rando TA. Stem cell ageing and non-random chromosome segregation. Philos Trans R Soc Lond Ser B Biol Sci. 2011;366(1561):85–93.
100. Kimura KD, Tissenbaum HA, Liu Y, Ruvkun G. Daf-2, an insulin receptor-like gene that regulates longevity and diapause in Caenorhabditis elegans. Science. 1997;277(5328):942–6.
101. Kapahi P, Zid BM, Harper T, Koslover D, Sapin V, Benzer S. Regulation of lifespan in Drosophila by modulation of genes in the TOR signaling pathway. Curr Biol. 2004;14(10):885–90.
102. Apfeld J, O'Connor G, McDonagh T, DiStefano PS, Curtis R. The AMP-activated protein kinase AAK-2 links energy levels and insulin-like signals to lifespan in C. elegans. Genes Dev. 2004;18(24):3004–9.
103. Fielenbach N, Antebi A. C. elegans dauer formation and the molecular basis of plasticity. Genes Dev. 2008;22(16):2149–65.
104. Sallon S, Solowey E, Cohen Y, Korchinsky R, Egli M, Woodhatch I, et al. Germination, genetics, and growth of an ancient date seed. Science. 2008;320(5882):1464.
105. Cano RJ, Borucki MK. Revival and identification of bacterial spores in 25- to 40-million-year-old Dominican amber. Science. 1995;268(5213):1060–4.
106. Rando TA, Chang HY. Aging, rejuvenation, and epigenetic reprogramming: resetting the aging clock. Cell. 2012;148(1–2):46–57.
107. Briggs R, King TJ. Transplantation of living nuclei from blastula cells into enucleated Frogs' eggs. Proc Natl Acad Sci U S A. 1952;38(5):455–63.
108. Gurdon JB. The developmental capacity of nuclei taken from intestinal epithelium cells of feeding tadpoles. J Embryol Exp Morphol. 1962;10:622–40.
109. Campbell KH, McWhir J, Ritchie WA, Wilmut I. Sheep cloned by nuclear transfer from a cultured cell line. Nature. 1996;380(6569):64–6.
110. Takahashi K, Yamanaka S. Induction of pluripotent stem cells from mouse embryonic and adult fibroblast cultures by defined factors. Cell. 2006;126(4):663–76.
111. Stadtfeld M, Hochedlinger K. Induced pluripotency: history, mechanisms, and applications. Genes Dev. 2010;24(20):2239–63.
112. Rossant J. Stem cells from the mammalian blastocyst. Stem Cells. 2001;19(6):477–82.
113. Loh KM, Lim B. Recreating pluripotency? Cell Stem Cell. 2010;7(2):137–9.
114. Bunster E, Meyer RK. An improved method of parabiosis. Anat Rec. 1933;57(4):339–43.
115. Conboy IM, Conboy MJ, Wagers AJ, Girma ER, Weissman IL, Rando TA. Rejuvenation of aged progenitor cells by exposure to a young systemic environment. Nature. 2005;433(7027):760–4.
116. Villeda SA, Luo J, Mosher KI, Zou B, Britschgi M, Bieri G, et al. The ageing systemic milieu negatively regulates neurogenesis and cognitive function. Nature. 2011;477(7362):90–4.
117. Adler AS, Sinha S, Kawahara TL, Zhang JY, Segal E, Chang HY. Motif module map reveals enforcement of aging by continual NF-kappaB activity. Genes Dev. 2007;21(24):3244–57.
118. Harrison DE, Strong R, Sharp ZD, Nelson JF, Astle CM, Flurkey K, et al. Rapamycin fed late in life extends lifespan in genetically heterogeneous mice. Nature. 2009;460(7253):392–5.

119. Chen C, Liu Y, Liu Y, Zheng P. mTOR regulation and therapeutic rejuvenation of aging hematopoietic stem cells. Sci Signaling. 2009;2(98):ra75.
120. Feng S, Jacobsen SE, Reik W. Epigenetic reprogramming in plant and animal development. Science. 2010;330(6004):622–7.
121. Meissner A. Epigenetic modifications in pluripotent and differentiated cells. Nat Biotechnol. 2010;28(10):1079–88.
122. Mikkelsen TS, Hanna J, Zhang X, Ku M, Wernig M, Schorderet P, et al. Dissecting direct reprogramming through integrative genomic analysis. Nature. 2008;454(7200):49–55.
123. Vastenhouw NL, Zhang Y, Woods IG, Imam F, Regev A, Liu XS, et al. Chromatin signature of embryonic pluripotency is established during genome activation. Nature. 2010;464(7290):922–6.
124. Creyghton MP, Cheng AW, Welstead GG, Kooistra T, Carey BW, Steine EJ, et al. Histone H3K27ac separates active from poised enhancers and predicts developmental state. Proc Natl Acad Sci U S A. 2010;107(50):21931–6.
125. Krizhanovsky V, Lowe SW. Stem cells: the promises and perils of p53. Nature. 2009;460(7259):1085–6.
126. Greer EL, Maures TJ, Ucar D, Hauswirth AG, Mancini E, Lim JP, et al. Transgenerational epigenetic inheritance of longevity in Caenorhabditis elegans. Nature. 2011;479(7373):365–71.
127. Narita M, Nunez S, Heard E, Narita M, Lin AW, Hearn SA, et al. Rb-mediated heterochromatin formation and silencing of E2F target genes during cellular senescence. Cell. 2003;113(6):703–16.
128. Van Den Bogaert A, De Zutter S, Heyrman L, Mendlewicz J, Adolfsson R, Van Broeckhoven C, et al. Response to Zhang et al (2005): loss-of-function mutation in tryptophan hydroxylase-2 identified in unipolar major Depression. Neuron 45, 11–16. Neuron. 2005;48(5):704; author reply 5-6.
129. Gao S, Chung YG, Parseghian MH, King GJ, Adashi EY, Latham KE. Rapid H1 linker histone transitions following fertilization or somatic cell nuclear transfer: evidence for a uniform developmental program in mice. Dev Biol. 2004;266(1):62–75.
130. Apostolou E, Hochedlinger K. Chromatin dynamics during cellular reprogramming. Nature. 2013;502(7472):462–71.
131. Lapasset L, Milhavet O, Prieur A, Besnard E, Babled A, Ait-Hamou N, et al. Rejuvenating senescent and centenarian human cells by reprogramming through the pluripotent state. Genes Dev. 2011;25(21):2248–53.
132. Li H, Collado M, Villasante A, Strati K, Ortega S, Canamero M, et al. The Ink4/Arf locus is a barrier for iPS cell reprogramming. Nature. 2009;460(7259):1136–9.
133. Mahmoudi S, Brunet A. Aging and reprogramming: a two-way street. Curr Opin Cell Biol. 2012;24(6):744–56.
134. Kim K, Doi A, Wen B, Ng K, Zhao R, Cahan P, et al. Epigenetic memory in induced pluripotent stem cells. Nature. 2010;467(7313):285–90.

Mitochondria as a Key Player in Aging

10

Rupa Banerjee and Pramod C. Rath

Introduction

Mitochondria are the organelles found in the cytoplasm of all eukaryotic cells, except for *Monocercomonoides* [1]. Mitochondria are commonly referred to as the 'powerhouse of cell', as they play a major role in providing cells with ATP for energy. Structurally, mitochondria comprise four distinct compartments: outer membrane, intermembrane space, inner membrane and matrix. The number of mitochondria per cell depends on the type of tissue or the cell in question. For instance, *Trypanosoma brucei* has been found to have a single mitochondrion per cell and is used as a model system to understand mitochondrial biogenesis/division [2]. On the other hand, the mammalian liver cells possess 1000–2000 mitochondria per cell. Mitochondria within a cell are found as a mitochondrial network, where individual mitochondria fuse together to form a tubular network. The advantage of these tubular networks over individual mitochondrion is the widespread distribution of mitochondria across the cell and generation of membrane potential [3].

Mitochondria are known to have a prokaryotic origin. They are believed to have evolved into today's cellular organelles as a result of an endosymbiotic relationship between an ancient prokaryote and a host cell [4, 5]. Mitochondria carry multiple copies of its own genome, which is distinct from the nuclear genome. Both nuclear and

---

R. Banerjee (✉)
Molecular Biology Laboratory, School of Life Sciences, Jawaharlal Nehru University, New Delhi, India

Biomedical Center Munich, Ludwig-Maximilians-Universität München, Planegg-Martinsried, Germany
e-mail: rupabanerjee17@gmail.com

P. C. Rath (✉)
Molecular Biology Laboratory, School of Life Sciences, Jawaharlal Nehru University, New Delhi, India
e-mail: pcrath@mail.jnu.ac.in

© Springer Nature Singapore Pte Ltd. 2020
P. C. Rath (ed.), *Models, Molecules and Mechanisms in Biogerontology*,
https://doi.org/10.1007/978-981-32-9005-1_10

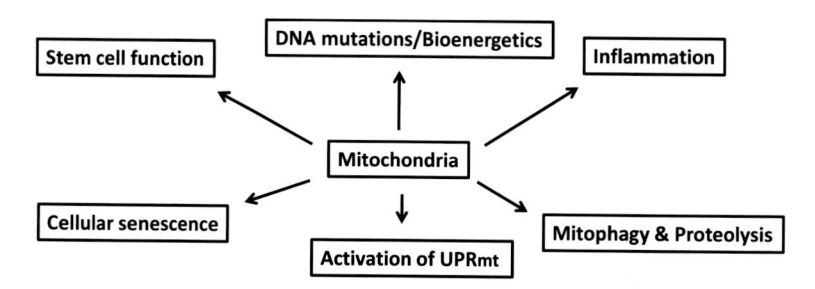

Fig. 10.1 Regulation of aging of organisms by mitochondria. (Sun N, Youle RJ, Finkel T (2016) The mitochondrial basis of aging. *Molecular Cell* 61: 654–666)

mitochondrial genomes encode for proteins that are found in mitochondria [6]. There are about 1000 proteins that are predicted to be mitochondrial proteins, of which the mitochondrial DNA (mtDNA) encodes for less than 1% of the total proteins. More than 99% of the proteins that are found in mitochondria are encoded by the nuclear DNA, are synthesized in the cytoplasm and are then translocated into their corresponding target mitochondrial compartment. The targeting of the proteins is carried out by various protein complexes, such as TOM complex at the outer membrane and TIM23 and TIM22 complexes at the inner membrane and so on. The targeting signal of the peptide is present in the sequence of the precursor proteins that has to be transported from the cytosol and is cleaved off as the protein enters its corresponding compartment. Translocation of proteins into the mitochondria takes place in its unfolded state, which is stabilized by various chaperones. Once the protein reaches its target compartment, it is folded into its native state with the help of various chaperone proteins [7, 8].

The major function of mitochondria in the cell is to generate ATP. In addition, mitochondria also play an important role in metabolism, cell signalling and cellular death. Since it plays such a central role in these processes, mitochondrial health and status is one of the key players in the process of aging. During the process of aging, the overall mitochondrial function has been observed to decline. It has been shown that the number of mitochondria and copies of mtDNA per mitochondria go down with age. The mitochondrial morphology is altered, and they become smaller and more round. Somatic mutations within mitochondrial genome increase with age [9–11]. In this chapter, we discuss how various pathways in mitochondria are implicated in the process of aging (Fig. 10.1).

Mitochondrial Cause of Aging

ROS Generation

Mitochondria generate ATP for the cellular functions by oxidizing nutrients. This oxidation is carried out by the process of oxidative phosphorylation. During this process, two electrons are derived from NADH, which are acquired by reducing NADH to NAD^+. NADH is obtained as an end product of many catabolic processes in the cell such as glycolysis and the Krebs cycle. These two electrons are then transported through a series of protein complexes residing in the inner membrane of mitochondria via a chain of redox reactions. These complexes are, namely, complex I (NADH: ubiquinone oxidoreductase), complex II (succinate dehydrogenase), complex III (coenzyme Q: cytochrome c oxidoreductase) and complex IV (cytochrome c oxidase), and together these form the electron transport chain (ETC). The electrons derived from NADH are transferred first to an electron carrier, ubiquinone, by the complex I. Furthermore, ubiquinone gets reduced as it transfers two electrons to two molecules of cytochrome c. This reduction of ubiquinone and oxidation of cytochrome c is mediated by complex III. Finally, a total of four electrons from four cytochrome c molecules are transferred to one oxygen molecule, which then reacts with four H^+ ions to form two water molecules, H_2O. Simultaneously, during these redox reactions, H^+ ions from the matrix are transported to the intermembrane space via complexes I, III and IV. As an end result, a proton gradient is generated across the mitochondrial inner membrane. This proton gradient is used as a form of energy by another inner-membrane protein complex, ATP synthase (F_oF_1 complex or complex V). ATP synthase phosphorylates ADP to give rise to a high-energy molecule, ATP. This thermodynamically unstable reaction is driven by the energy derived from the passage of H^+ moving down the electrochemical gradient, that is, from intermembrane space to mitochondrial matrix. This process is referred to as oxidative phosphorylation (OXPHOS) and is desirable for cell for the generation of energy [12–14]. In addition, complex II acts in parallel to complex I and funnels electrons to the ubiquinone in the respiratory chain by a FAD^+-dependent step in Krebs cycle. However, no reducing equivalents are transported across the membrane by this step. As a result, this step does not contribute in the membrane potential generation and produces less energy when compared with that of complex I [15].

The process of electron transport has an undesirable secondary effect as well. The transfer of electrons through the respiratory chain can lead to a premature escape of the electrons to generate highly reactive free radicals such as superoxide radicals, $O_2^{\cdot-}$. Furthermore, such superoxide radicals can react with H^+ ions within the cell to form reactive peroxide molecules. Such chemically reactive oxygen containing molecules are known as reactive oxygen species (ROS). ROS includes a number of moieties like peroxide, superoxide, hydroxyl radical, etc. The complex I and complex III are believed to be the main sites of the electron leakage. About 2–5% of oxygen molecules consumed in the mitochondria are converted to ROS [16, 17].

Due to its high reactivity, ROS can induce mitochondrial and cellular dysfunction by damaging protein, lipid and nucleic acid molecules. The accumulation of damaged molecules over a period of time causes gradual deterioration of mitochondrial and/or cellular function and is thought to contribute to aging. This is referred to as the mitochondrial free radical theory of aging [18]. This notion is supported by several studies that were carried out in *Saccharomyces cerevisiae*, in which protein molecules were oxidized by ROS. Several amino acids, especially cysteine, histidine and methionine, are highly susceptible to oxidative damage due to the presence of reactive functional groups in their side chains [19, 20]. Similarly, the ROS can oxidize lipid molecules, which can compromise the membrane integrity of the cell and the various organelles and cause organellar and cellular dysfunctions [21]. ROS can also react with nucleic acids, especially to the guanine residues to form 8-oxo-dG and thereby introducing mutations in both nuclear and mitochondrial genomes [22]. In addition, several enzymes can be inactivated by ROS by oxidation of their cofactors. A recent study suggests that activation of DNA damage response (DDR) can be mediated by the ROS, which leads to cell cycle arrest and causes cellular senescence [23].

Majority of ROS is eradicated from the cell by free radical scavenging enzymes and antioxidants. Two major classes of ROS scavenging enzymes are superoxide dismutase (SOD) and catalase. SOD catalyses the formation of peroxide from the oxygen-free radical. Peroxides are less damaging as compared to free radicals and are further broken down into water and oxygen by catalase. In humans, there are three different types of dismutases: SOD1 in cytosol, SOD2 in mitochondria and SOD3 in extracellular spaces. Additionally, in yeast, flies and mice, the deletion of mitochondrial SOD leads to a shorter lifespan and a high degree of oxidative stress [24–26]. There are several non-enzymatic antioxidants, such as glutathiones, vitamins and thiols, present in the cell that either inhibit or slow down the oxidation by ROS. However, their usage for medicinal purposes is still not clearly established [27].

Contrary to the free radical theory of aging, recent studies suggest that the ROS are not merely a by-product of the respiratory chain. The physiological levels of ROS play a significant role in the mitochondrial signalling processes. It has been shown in low-oxygen condition; H_2O_2 generated from mitochondria diffuses into the cell and stabilizes hypoxia-inducible factor-1α (HIF-1α), which induces the response pathway against hypoxia. Also, ROS have been shown to oxidize thiol groups in various enzymes that, in turn, are involved in other signalling pathways. Therefore, contrary to the free radical theory of aging, more recent studies suggest a hormetic response of cells towards ROS exposure. According to these, a low dosage of ROS is beneficial for cells and contributes to increased lifespan. On the other hand, abnormally high levels of ROS can be detrimental for cellular health and can lead to cell death [17, 28–30]. This is in agreement with the surprising observation in *Caenorhabditis elegans*, in which deletion of mitochondrial *sod-2* leads to an increased lifespan [31]. A recent study also reported that deletion of cytosolic *sod-1* and *sod-5* reduces the lifespan in a worm strain, suggesting that increase in only mitochondrial ROS increases the lifespan [32].

Mitochondrial DNA Damage

Human mitochondria comprise multiple copies of about 16.6 kbp large circular loop of double stranded DNA. They encode for 13 OXPHOS polypeptides, 2 rRNAs and 22 tRNAs. The remaining proteins are encoded by the nuclear DNA, which includes the remaining OXPHOS subunits. It is predicted that roughly 1000 proteins are targeted to mitochondria from the cytosol with a help of a targeting signal within the sequence of the precursor protein. Unlike nuclear DNA, the mtDNA is not known to be packaged by the histone proteins. Instead in humans, mtDNA is bound by a (HMG)-box containing protein, Tfam (transcription factor A, mitochondrial). Tfam plays an important role in the transcriptional activation of genes encoded by mitochondrial genome and in the maintenance and the organization of the mtDNA into a nucleoid [33, 34]. However, the protective function of Tfam to the mtDNA is not as robust as that of chromatin formation of the nuclear DNA by the histones. In addition, the DNA damage repair response for mtDNA is also limited. As a result, mtDNA is susceptible to undergo mutations, specifically generated by the ROS. A possible escape for mitochondria from this mutation is its capacity to have multiple copies of mtDNA within each mitochondrion. This may range from hundreds to thousands in number per mitochondrion depending on the tissue type. A gene encoded by the mutated mtDNA does not affect the mitochondrial function as the wild-type copy within the same mitochondrion can compensate for its function. This is referred to as heteroplasmy. However, over a period of time, the mutated mtDNA gets amplified by repeated replication and division. Deleterious mutations in mtDNA can hamper the respiratory chain by disrupting the OXPHOS complex organization. This, in turn, results in oxidative stress and accumulation of more ROS. Thereby, more mutations in mtDNA takes place until the mitochondrial function is completely disrupted. Dysfunctional mitochondria are recovered by an oxidative stress response or discarded by mitochondrial quality control. Inability of a cell to do so, gradually, leads to aging or manifestation of diseases [35].

Mitochondrial genome is replicated by DNA polymerase γ (Polg). Polg is a heterotrimer of a catalytic subunit, PolgA, and a dimeric accessory subunit, PolgB. PolgA subunit also carries out proofreading activity. Mice strains with mutations in the proofreading activity of Polg have been shown to have three to five times higher degree of random point mutations per mtDNA. These mice strains are referred to as the mutator mice strain and show severe defect in the respiratory chain and decreased lifespan, thereby, implicating mtDNA mutations in premature aging [36]. In *Saccharomyces cerevisiae*, deletion or loss of mtDNA gives rise to petite mutants (ρ^0). These mutants cannot survive in a non-fermentable (glucose deficient) medium in which cells are highly dependent on mitochondrial respiration and show slow growth in fermentable (glucose rich) medium [37]. These mutants are also under high oxidative stress, suggesting that mutations in mtDNA can contribute to various diseases and in the aging process. Interestingly, a study in conplastic mice strain suggests that variations in mtDNA can alter mitochondrial function and cellular response, implying that the interplay between mitochondrial and nuclear genome also influences the cellular health and lifespan [38].

Altered Gene Expression

Cells acquire ATP by supplying their nutrient metabolites into two different major pathways: the glycolysis and the oxidative phosphorylation. Glycolysis takes place in the cytosol in high nutrient conditions. The final products of glycolysis are ATP and NADH. Studies in *S. cerevisiae* have shown that when yeast cells are grown in a glucose-rich medium, the glucose is metabolized via glycolysis and the ratio of NADH/NAD$^+$ is relatively high. In contrast, when yeast cells are grown in low-glucose medium, they carry out oxidative phosphorylation to generate ATP. As a result, the mitochondrial pool of NADH is converted to NAD^+ and overall $NADH/NAD^+$ ratio gets low. Interestingly, this alteration in the ratio of $NADH/NAD^+$ in the cells can act as a sensor to determine the energy level of the cell and is, in turn, used by cell to regulate the mitochondrial biogenesis [39–42].

The first evolutionarily conserved protein to have been identified in this regard was a yeast protein, Silent information regulator 2 or Sir2p. Sir2p is an NAD^+-dependent histone deacetylase (HDAC) that regulates gene expression by activating chromatin compaction, specifically in the subtelomeric region and rDNA array. Sir2p has also been shown to have role in inheritance of undamaged protein and mitochondria in daughter cells by maintaining cell polarity. Overexpression of Sir2p has been implicated in extension of replicative lifespan of yeast, and downregulation has shown a decrease in lifespan [40].

In mammals, there are seven different types of non-redundant Sir2p-like proteins referred to as sirtuins: SIRT1, SIRT2, SIRT3, SIRT4, SIRT5, SIRT6 and SIRT7. SIRT1, SIRT6 and SIRT7 are found in the nucleus; SIRT2 in cytosol; and SIRT3, SIRT4 and SIRT5 are targeted to the mitochondrial matrix. Most of the sirtuins regulate the function of their target proteins by deacetylating the acetylated lysine residue at the ϵ-amino group. SIRT4, on the other hand, is not a deacetylase but mono-ADP-ribosyltransferase. Generally, the targets of the sirtuins comprise proteins involved in metabolism, stress response, DNA damage repair, etc. SIRT1 has been the most extensively studied homologue in mammals and has been shown to deacetylate proteins involved in a number of cellular processes such as histones, transcription factors and DNA repair proteins. Brain-specific upregulation of SIRT1 in mice has been shown to increase their lifespan. SIRT3 regulates mitochondrial function by deacetylating proteins involved in metabolism. In aged mice, the expression levels of SIRT3 in haematopoietic stem cells have been shown to decline [40].

Interestingly, during caloric restriction in worms, flies and mice, a number of proteins undergo deacetylation. The caloric restriction has been shown to activate the response pathway that prevents cells from the detrimental effects of ROS by activating ROS scavenging enzymes, preventing mtDNA against damage due to ROS and removing the damaged mitochondria by mitophagy. Since NAD^+ levels regulate the activation of sirtuins and are found in higher levels in a restricted dietary condition, it has been postulated that sirtuins mediate lifespan extension in caloric restriction. This also interlinks mitochondrial homeostasis with the DNA damage response (DDR) pathway as NAD^+ is a substrate for both sirtuins and poly (ADP-ribose) polymerase (PARP)-1, which is involved in DDR [39].

In yeast strains, replicative lifespan extension has been observed when yeasts are grown in medium with reduced glucose levels. However, whether it is mediated by Sir2p or not is unclear. In mammals, a tissue-specific involvement of SIRT1 during caloric restriction has not been unequivocally established. SIRT3, on the hand, has been shown to play a more defined role in caloric restriction. During caloric restriction, SIRT3 levels have been shown to increase in adipose tissue, skeletal muscle and liver. Also, in liver cells, deacetylation by SIRT3 activates SOD2 and IDH2, which prevents the accumulation of ROS in the cells [35, 40]. It would be interesting to see how sirtuins could be involved in lifespan extension in future studies. Interestingly, NAD⁺ is speculated to be of therapeutic importance for various cardiac and renal pathologies [43].

Maintenance of Mitochondrial Protein Homeostasis

It is quite clear that mitochondrial dysfunction can lead to aging and age-related diseases. For a mitochondrion to function properly, sustenance of the mitochondrial protein homeostasis (or proteostasis) is essential. Mitochondrial proteostasis is the maintenance of a functional balance between protein synthesis, folding and degradation in mitochondria. Proteome analysis of aging *C. elegans* suggests a decrease in the levels of cytosolic and mitochondrial ribosomal subunits, leading to a decreased protein synthesis and increased levels of proteasomal subunits for degradation of unwanted proteins [44]. When the capacity of a cell to maintain proteostasis is reduced, it leads to accumulation of protein aggregates with aging. As a result, a response is triggered by the cell in order to rescue itself from the deleterious effects of accumulation of misfolded/unfolded proteins. Interestingly, the activation of this response can also prolong the lifespan of an organism. This has been shown in the long-lived *daf-2* mutant of *C. elegans* that has enhanced proteome maintenance [44]. The recovery process, however, involves enhancement of the folding capacity of mitochondrial proteins and removal of the misfolded or aggregated proteins by degradation.

Mitochondrial Unfolded Protein Response (UPR^{mt})

The intricate pathway that is activated to ensure this mitochondrial proteostasis is known as mitochondrial unfolded protein response (UPR^{mt}). Unfolded protein responses (UPRs) can be activated in different cellular compartments to maintain subcellular proteostasis, and in case of each compartment, a different signalling pathway is followed. The unfolded protein response in cytosol is known as cytosolic heat shock response (HSR); and in endoplasmic reticulum (ER), it is known as unfolded protein response in ER (UPR^{ER}) [45–48]. Both HSR and UPR^{ER} have been extensively studied and will not be discussed in greater detail further as it is beyond the scope of this chapter.

A large number of studies on UPR^{mt} have been carried out in *Caenorhabditis elegans*. The advantages of using *C. elegans* as a model organism for the study of age-related proteostasis are as follows: (1) short lifespan of 2–3 weeks; (2) easy genetic manipulation; (3) subcellular, tissue-specific and organismal level study; and (4) established temperature-sensitive mutants and strains with marker (*hsp-6*) for study [49]. During aging, the overall mitochondrial function is impaired. In contrast, an RNAi screen in *C. elegans* had shown an increase in the lifespan of worms upon disruption of the mitochondrial function. These observations, however, are not in agreement with each other. But the contradictory results can be explained by the present studies that suggest that a moderate disruption in the mitochondrial function can activate UPR^{mt}. The activation of UPR^{mt} enhances the overall mitochondrial biogenesis and, thereby, causes an extension of lifespan. UPR^{mt} can be activated experimentally by several triggers, such as (1) accumulation of protein aggregates in the mitochondrial matrix; (2) imbalance in the ratio of nuclear and mitochondrial DNA-encoded proteins or mitonuclear protein imbalance; and (3) mitochondrial ROS and other compounds that interfere with ETC function. The activation of UPR^{mt} due to ROS is believed to be a result of protein damage in its vicinity rather than a direct effect of ROS itself. These result in a proteotoxic stress which in turn activates the UPR^{mt} and leads to an extended lifespan [46].

In *C. elegans*, an RNAi screen was done in which OXPHOS complex I, III, IV and V subunits encoded by nuclear or mitochondrial DNA were knocked down that resulted in extended lifespan in worms [50, 51]. In other studies, a mitochondrial ribosomal protein, *mrps-5*, was knocked down or inhibited with doxycycline, longevity in worms increased [52, 53]. Reduced expression of a mitoribosomal subunit *mrps-5* leads to a decreased translation of mtDNA-encoded proteins, and as a result, it impaired the mitonuclear balance, which in turn activated the UPR^{mt}. The activation of this response results in an increase in the transcriptional expression of different protein quality control genes such as chaperones and proteases leading to an increase in lifespan. Similarly, in a mice strain, knockdown of *Mrps-5* leads to an extended lifespan [52]. Interestingly, treatment in *C. elegans* and mice with NAD$^+$ boosters triggered the mitochondrial biogenesis by activating the sirtuin pathway, causing an activation of UPR^{mt} and extended lifespan [54, 55]. Also, in agreement with these findings, the knockdown of genes that are activated by UPR^{mt} leads to a shorter lifespan in worms, flies and mice [46].

The signalling molecule for UPR^{mt} activation in *C. elegans* has been shown to be a dually localizing target signal containing leucine zipper transcription factor, activating transcription factor associated with stress (ATFS)-1. ATFS-1 comprises a mitochondrial matrix targeting signal (MTS) and a nuclear localization signal (NLS). This protein is constitutively imported to the mitochondrial matrix. Within the mitochondrial matrix, this protein is degraded by the LON protease. During stress in *C. elegans*, the unfolded or misfolded proteins accumulate/aggregate in mitochondria or complex proteins that remain unbound in the mitochondrial matrix are cleaved off into smaller peptide fragments with a matrix localized proteolytic complex, CLPP-1. These peptide fragments are exported out of the mitochondria through an ATP-binding cassette protein transporter, HAF-1, embedded in the inner membrane into the cytosol. In the cytosol, these fragments have an inhibitory effect on the mitochondrial protein import machinery. The inhibitory mechanism of the peptides is not yet clearly understood. The inhibition in protein import prevents ATFS-1's entry into the mitochondria, and as a result, it is targeted into the nucleus. In the nucleus, ATFS-1 transcriptionally activates the expression of mitochondrial chaperones and several other genes that are involved in mitochondrial import, detoxification of ROS and glycolysis, which are necessary to resume the mitochondrial function in the cell [56]. Other than ATFS-1, a ubiquitin-like protein UBL-5 and a homeodomain-containing transcription factor DVE-1 are also believed to transcriptionally regulate UPR^{mt} [57]. These proteins assist the function of ATFS-1 and are speculated to play a role in regulation of the chromatin structure in the mtUPR. UPR^{mt} activation by ATFS-1 is also involved in maintenance of deleterious mitochondrial genome [58]. Furthermore, proteotoxic stress in mitochondria downregulates the protein translation in the cytosol. Downregulation of cytosolic protein translation reduces the additional burden of folding on mitochondrial chaperones. This is carried out by activation of general control non-repressed 2 (GCN-2) by ROS-mediated mitochondrial stress. Activated GCN-2 then inactivates the translation initiation factor, eIF2α, by phosphorylating it and preventing the cytosolic translation.

Interestingly, the first demonstration of the induction of UPR^{mt} was shown in mammalian cells, when aggregates of folding incompetent mutant ornithine transcarbamylase were targeted to mitochondrial matrix [59]. Although this was the first demonstration of a mitochondria-specific UPR, details of the process are not completely understood. Recently, it was observed that when mitochondrial proteostasis is disturbed in mammalian cells, there is an upregulation of the mitochondrial chaperones, proteases and import machinery upon induction of the response. Additionally, mammalian cells also show a reduced expression in a mitochondrial matrix ribonuclease, MRPP3, which is involved in pre-RNA processing, thereby reversibly inhibiting the protein translation in mitochondrial matrix and reducing the protein folding load in the matrix [60]. However, unlike *C. elegans*, the entire signalling pathway for such a response in mammalian cells is not clearly known. It has been shown that certain genes are transcriptionally activated by a heterodimer of leucine zipper transcription factors, CHOP and C/EBPβ. CHOP and C/EBPβ, on the other hand, have been shown to be activated by c-Jun N-terminal kinase (JNK)2 and dsRNA-activated protein kinase (PKR)-mediated c-Jun activation [56, 61]. CHOP has also been found to be activated and to play a role in UPR^{ER}. However, both CHOP and C/EBPβ comprise binding site for AP1 promoter site. The AP1 promoter site is believed to be essential for activation of the UPR^{mt} but not for the UPR^{ER} [46]; however, how the specificity of this pathway is maintained is poorly understood.

Recently, activation of an alternate pathway to UPR^{mt} has been shown to extend lifespan in worms when respiratory chain is disrupted. This involves activation of the p38 MAPK signalling pathway and is also under the promoter for ATFS-1 [62].

Mitochondrial Quality Control

The presence of functional mitochondria within a cell is a result of regulatory mechanisms that keep a check on mitochondrial quality. In order to achieve this, cells constantly generate functional mitochondria and degrade the damaged mitochondria to maintain organellar homoeostasis. This maintenance of homeostasis to keep a balance of functional mitochondria in the cell is referred to as mitochondrial quality control. An imbalance in mitochondrial proteostasis, such as protein aggregation, protein damage due to ROS, activates a stress response within mitochondria. The stress response, UPR^{mt}, enhances the protein folding capacity of the mitochondria and, thereby attempts to rescue mitochondrial function. However, mitochondria that are beyond recovery undergo a selective autophagy to remove the damaged mitochondria, referred to as mitophagy [35]. To this end, the stage at which this decision to undergo mitophagy is made by the cell is not clearly understood.

Fission and Fusion

Mitochondrial shape and organization within a cell is highly dynamic. Functional mitochondria constantly undergo fission, fusion and movement through the cellular cytoskeleton. This is necessary for the generation of new mitochondria, formation of mitochondrial network and distribution of mitochondria across the cell. A mitochondrion within a cell can exist as a single unit or as a network of a number of fused mitochondria. In yeast, there are three proteins belonging to the family of dynamin proteins that play a major role in the fusion and fission process. These three proteins are, namely, Mgm1, Fzo1 and Dnm1. Fzo1 and Mgm1 are localized in the mitochondrial outer membrane and inner membrane, respectively, and are responsible for mitochondrial outer- and inner-membrane fusion. Dnm1, on the other hand, is basally localized at the cytosol and is occasionally localized onto the surface of outer membrane. It is involved in mitochondrial fission. Other than these three proteins, there are several adaptor proteins that assist the process of fission and fusion [63]. In *C. elegans*/mammalian cells, the homologues of Mgm1, Fzo1 and Dnm1 are called EAT-3/Opa1, FZO-1/Mitofusin and DRP-1/Drp1, respectively. In case of mammalian Mitofusins, there are two variants, Mitofusin 1 (Mfn1) and Mitofusin 2 (Mfn2). In body wall skeletal muscle cells of *C. elegans*, the tubular mitochondrial network was found to be replaced with fragmented mitochondria with smaller volume upon aging [64, 65]. Similarly, in short-lived (*daf-16*) and long-lived (*clk-1*) mutants of *C. elegans*, the appearance of fragmented mitochondria during aging was faster and delayed, respectively [66]. Also, in yeast, when both fission and fusion processes are impaired by deleting Dnm1 and Mgm1, cells show a decreased replicative lifespan [67].

Mitochondria can undergo fusion followed by fission at the same cellular location. The process of fusion enables the formation of mitochondrial network that allows rapid distribution of ATP and dilutes the effect of the mtDNA mutations. The fission process, on the other hand, can give rise to a daughter mitochondrion with more concentrated subset of mutated mtDNA and damaged proteins. The process of fission requires energy. As a result of fission, a solitary mitochondrion with low membrane potential is acquired. In case this mitochondrion recovers, it is rejoined to the mitochondrial network. A mitochondrion with severely low membrane potential beyond repair is subject to elimination by mitophagy [64, 65, 68]. Interestingly, during the budding process in yeast, the mother cell has been found to inherit mitochondria with low membrane potential as compared to the daughter cell [69]. In human mammary stem cell division, asymmetric distribution of mitochondria takes place, such that the daughter cell inherits more of the newly synthesized and less damaged mitochondria and shows more stemness as compared to the mother cell [70].

Mitophagy-Damaged components

In the cell are removed by degradation in the lysozyme by the process of autophagy. Selective autophagy of damaged mitochondria is referred to as mitophagy. In worms, the accumulation of damaged mitochondria activates transcription factor,

SKN-1. SKN-1 regulates both mitochondrial biogenesis and mitophagy. The regulation of mitophagy by SKN-1 is mediated by DCT-1. Co-ordination between mitochondrial biogenesis and mitophagy leads to the maintenance of mitochondrial homeostasis. The deletion of *skn-1* and *dct-1* causes a very high oxidative stress and shortens the worm lifespan [71].

In a well-functioning mitochondrion in mammalian cells, PTEN-induced putative kinase 1 (PINK1) is targeted to the mitochondria and is degraded with the help of proteases. However, in a mitochondrion with impaired membrane potential, PINK1 accumulates in the outer mitochondrial membrane, where it does not undergo degradation. Accumulation of PINK1 on the mitochondrial surface leads to the recruitment of an ubiquitin ligase, Parkin, on the mitochondrial surface [72, 73]. To this end, how PINK1 recruits Parkin is not clearly understood. After the recruitment, Parkin ubiquitylates various mitochondrial outer-membrane target proteins, such as VDAC, TOM subunits and Mitofusins. Furthermore, PINK1 has been shown to phosphorylate the ubiquitin-tag at Ser65 position [72, 74, 75]. This is followed by recruitment of downstream cytosolic factors and adaptors, which eventually leads to recruitment of proteins involved in autophagy and proteasomal degradation. Cytosolic proteasomes degrade the ubiquitylated proteins, particularly, Mfn1 and Mfn2, which leads to the isolation of the impaired mitochondrion from the remaining healthy mitochondrial network in the cell [72, 73]. Furthermore, PINK1 has been shown to recruit downstream autophagy receptors, which eventually leads to mitophagy [76]. As a result, damaged mitochondria are eradicated from the cell. In case of a defect in the mitophagy pathway, the damaged mitochondria accumulate in the cell causing disruption in cellular function, which leads to the cell death. During the process of aging, dysfunction in mitochondrial quality control leads to accumulation of Krebs cycle intermediates. These intermediates act as a signalling molecule that changes the global chromatin landscape [77].

Manifestation of Age-Related Diseases

Excessive ROS production induces apoptosis via mitochondrial signalling. However, when there is a defect in the induction of UPR^{mt}, mitophagy or apoptosis, several deformities in cellular growth can be observed. The accumulation of such cells gives rise to several diseases. Pathologies due to mutations in genes involved in mitochondrial pathways are often overlapping and affecting mostly the nervous system, skeletal muscles, heart, liver, eyes and cochlea. This is because these tissues require more energy and are, hence, affected the most. These pathologies can be either due to mutations in the nuclear genome or mitochondrial genome [78]. These diseases can often be sporadic or familial. Mutations in the mitochondrial genome are maternally inherited. The diagnosis of mitochondrial diseases is often problematic as there is no clear clinical phenotype and often more than one organ system of the body are affected. Also, a number of mutations in a single gene have been implicated in various pathologies (Table 10.1). Some of the treatments of these diseases involve use of gene therapy and antioxidants [79, 80]. Recently, a study suggested

Table 10.1 Clinical pathologies due to mutations in different genes involved in various mitochondrial pathways

Function	Genes	Clinical Pathology
Complex assembly	NDUFS1, NDUFS2, NDUFS3, NDUFS4, NDUFS6, NDUFS7, NDUFS8, NDUFB3, NDUFV1, NDUFV2, NDUFA1, NDUFA2, NDUFA10, NDUFA11, SDH-A, SDH-B, SDH-C, SDH-D, UQCRB, UQCRQ, COX6B1, ATP5E, NDUFAF1 (CIA30), NDUFAF2 (B17.2L), NDUFAF3, NDUFAF4 (HRPAP20), C20orf7, NUBPL, FOXRED1, ACAD9, SDHAF1, SDHAF2, BCS1L, SURF1, SCO1, SCO2, COX10, COX15, LRPPRC, FASTKD2, TACO1, ATPAF2, TMEM70	Cardioencephalomyopathy, Leigh's syndrome, encephalopathy, cardiomyopathy, leukoencephalopathy, paraganglioma type 2, neonatal hepatic failure
Mitochondrial DNA assembly	POLG (PEOA1), POLG2 (PEOA4), ANT1 (PEOA2), MPV17, C10ORF2 (PEOA3), TYMP (ECGF1), DGUOK, RRM2B (PEOA5), SUCLA2, SUCLG1, TK2	Alpers syndrome, Alzheimer's disease, encephalomyopathy, hepatocerebral mtDNA syndrome
Mitochondrial protein synthesis	EFG1, YARS2, SARS2, DARS2, RARS2, MRPS16, MRPS22, TSFM, TUFM, MTTL1, MTTI, MTRNR1, MTTS1, MTTF, MTTV, MTTQ, MTTW, MTTK, MTTG, MTTH	Severe hepatoencephalopathy, myopathy, hyperuricaemia, leukoencephalopathy, pontocerebellar hypoplasia, cardiomyopathy, encephalomyopathy, leukodystrophy, lactic acidosis
Mitochondrial import and chaperones	DDP, DNAJC19, SPG7, HSPD1	Mohr–Tranebjaerg syndrome, cardiomyopathy, spastic paraplegia, leukodystrophy
mitochondrial integrity	OPA1, MFN2, DLP1, RMRP	Alzheimer's disease, Charcot–Marie–Tooth disease, microcephaly, Barth syndrome, metaphyseal chondrodysplasia
Iron homeostasis	FRDA, ABCB7, GLRX5, ISCU, BOLA3, NFU1	Friedreich ataxia, X-linked sideroblastic anaemia with ataxia, sideroblastic anaemia, myopathy, encephalomyopathy, lactic acidosis
Mitochondrial metabolism	PDHA1, ETHE1, PUS1	Leigh's syndrome, encephalopathy

Adopted from www.mitomap.org [82]

the use of hypoxia as a treatment for respiratory chain diseases. A mouse model with Leigh's syndrome showed alleviated neurometabolic dysfunction when subjected to chronic hypoxia conditions [81]. Other examples for treatment include different therapeutic methods such as medicine for diabetes mellitus, cochlear implants for hearing loss and surgeries like cataract surgery.

In the following text, some of the more common pathologies are discussed.

Neurodegeneration

Several maternally inherited homoplasmic mtDNA mutation diseases show a defect in the central nervous system (CNS). Similarly, several neurodegenerative disorders exhibit abnormal mitochondrial morphology and function [83, 84]. Neurodegenerative disorders can arise due to mitochondrial dysfunction arising due to mutation in nuclear DNA or mtDNA encoded genes. One such highly studied neurodegenerative disorder is Parkinson's disease (PD). It is primarily caused due to loss of dopaminergic neurons in the substantia nigra pars compacta, which causes loss of dopamine in striatum. It is characterized with unusual shaking of limbs, slow movements and rigid muscles. Defects in different aspects of mitochondrial processes have been linked to Parkinson's disease. In case of familial Parkinson's disease, mutations in a number of genes including α-synuclein, PINK1 and Parkin have been implicated. This suggests a defect in mitochondrial dynamics and mitophagy to play a role in pathogenesis of the disease. In agreement with this, in sporadic PD patient samples, misfolded and inactive S-nitrosylated Parkin have been observed. Additionally, in sporadic PD, patients show an impaired complex I function, which along with impaired mitophagy is thought to be a major cause of the disease [85].

Myopathy

Muscular tissues, such as muscles, heart and liver, are highly active and require ATP from mitochondria for their proper function. Therefore, defects in mitochondria can give rise to a number of diseases with muscular dysfunction that are referred to as mitochondrial myopathies. Mitochondrial myopathies can be caused by mutations in both the nuclear and mitochondrial DNA and also by the loss of mtDNA or mutations in mtDNA due to nuclear DNA mutation that exhibits impaired mtDNA maintenance. Mitochondrial myopathy could be affecting one or more tissues in humans. Also, several of these diseases are progressively severe during aging. One of the major reasons for this tissue specificity and age-related onset/progress is heteroplasmy in mtDNA. A mitochondrion comprises a number of mtDNA with or without deleterious mutations. A disproportionate distribution of mtDNA during mitochondrial division leads to accumulation of mitochondria consisting of mutant mtDNA in cells. As a result, over a period of time, manifestation of myopathy takes place. It is often characterized with muscle weakness, hearing loss, heart defects, impaired vision, etc. [86]. However, whether this occurs randomly or there is a genetic predisposition for such diseases is not clearly understood. Currently, the use of antioxidants and gene therapy as a treatment is being considered [79, 80].

Cancer

Warburg reported that the cancer cells generate ATP via the glycolysis pathway in aerobic manner. This gave rise to the idea that cancerous cells do not require functional mitochondria for their survival. However, studies with ρ^o cancerous cells show a defect in growth and reduced colony formation. In addition, nude mice with ρ^o show reduced tumour growth. These observations suggested that the functional mitochondria are prerequisite for the development of cancer. Our present understanding of mitochondrial function during cancer development suggests that the cells rewire themselves to undergo a metabolic shift, characterized by large number of mutations in the mtDNA. The cancerous cells switch to glycolysis for energy and adapt to hypoxia so as to shut down mitochondrial respiration. Mutations in both the nuclear- and mitochondrial-DNA-encoded mitochondrial genes have been implicated, especially the enzymes in the Krebs cycle, such as Succinate dehydrogenase and Isocitrate dehydrogenase [87, 88].

Conclusion

Mitochondria have a crucial role in cellular metabolism and death. As a result, they play a major role in the process of aging. Overall, a loss in mitochondrial function is observed during the process of aging. One of the major causes has been attributed to ROS generation in mitochondria. However, mitochondria are not the only source of ROS in the cell. Several other organelles, like peroxisome, also generate ROS [89]. There are also several other factors, such as hormones, that play a role in the process of aging. An interesting question that intrigues scientists is to determine if the mitochondrial dysfunction is the cause or the consequence of aging. However, considering how the different pathways in the cell are intricately connected with each other, at present there is no definite understanding of what the chronology is. Interestingly, a longitudinal study in killifish suggests that the transcriptome of the long- and short-lived organism varies remarkably in early adult stage. This suggests that the favourable conditions for an extended lifespan are acquired in the early adult stage of killifish [90]. The study also shows that inhibition of complex I with small molecules leads to an increase in lifespan by revamping the transcriptome of the organism at an early stage, suggesting complex I as the regulator of lifespan.

Acknowledgements We would like to apologize to all the colleagues whom we could not cite due to space limitation. Fellowship to RB and financial support from UGC-RNRC, -DRS and DST-FIST, -PURSE to SLS & PCR are acknowledged.

References

1. Karnkowska A, et al. A eukaryote without a mitochondrial organelle. Curr Biol. 2016;26(10):1274–84.
2. Matthews KR. The developmental cell biology of Trypanosoma brucei. J Cell Sci. 2005;118(Pt 2):283–90.

3. Hoitzing H, Johnston IG, Jones NS. What is the function of mitochondrial networks? A theoretical assessment of hypotheses and proposal for future research. BioEssays. 2015;37(6):687–700.
4. Gray MW. Mitochondrial evolution. Cold Spring Harb Perspect Biol. 2012;4(9):a011403.
5. Gray MW, Burger G, Lang BF. The origin and early evolution of mitochondria. Genome Biol. 2001;2(6):REVIEWS1018.
6. Chen XJ, Butow RA. The organization and inheritance of the mitochondrial genome. Nat Rev Genet. 2005;6(11):815–25.
7. Mokranjac D, Neupert W. Thirty years of protein translocation into mitochondria: unexpectedly complex and still puzzling. Biochim Biophys Acta. 2009;1793(1):33–41.
8. Neupert W, Herrmann JM. Translocation of proteins into mitochondria. Annu Rev Biochem. 2007;76:723–49.
9. Bratic A, Larsson NG. The role of mitochondria in aging. J Clin Invest. 2013;123(3):951–7.
10. Guarente L. Mitochondria--a nexus for aging, calorie restriction, and sirtuins? Cell. 2008;132(2):171–6.
11. Sun N, Youle RJ, Finkel T. The mitochondrial basis of aging. Mol Cell. 2016;61(5):654–66.
12. Chaban Y, Boekema EJ, Dudkina NV. Structures of mitochondrial oxidative phosphorylation supercomplexes and mechanisms for their stabilisation. Biochim Biophys Acta. 2014;1837(4):418–26.
13. Sazanov LA. A giant molecular proton pump: structure and mechanism of respiratory complex I. Nat Rev Mol Cell Biol. 2015;16(6):375–88.
14. van den Heuvel L, Smeitink J. The oxidative phosphorylation (OXPHOS) system: nuclear genes and human genetic diseases. BioEssays. 2001;23(6):518–25.
15. Cecchini G. Function and structure of complex II of the respiratory chain. Annu Rev Biochem. 2003;72:77–109.
16. Murphy MP. How mitochondria produce reactive oxygen species. Biochem J. 2009;417(1):1–13.
17. Marchi S, et al. Mitochondria-ros crosstalk in the control of cell death and aging. J Signal Transduct. 2012;2012:329635.
18. Harman D. Free radical theory of aging. Mutat Res. 1992;275(3–6):257–66.
19. Costa V, Quintanilha A, Moradas-Ferreira P. Protein oxidation, repair mechanisms and proteolysis in Saccharomyces cerevisiae. IUBMB Life. 2007;59(4–5):293–8.
20. Cabiscol E, et al. Oxidative stress promotes specific protein damage in Saccharomyces cerevisiae. J Biol Chem. 2000;275(35):27393–8.
21. Mylonas C, Kouretas D. Lipid peroxidation and tissue damage. In Vivo. 1999;13(3):295–309.
22. Boiteux S, Radicella JP. Base excision repair of 8-hydroxyguanine protects DNA from endogenous oxidative stress. Biochimie. 1999;81(1–2):59–67.
23. Correia-Melo C, et al. Mitochondria are required for pro-ageing features of the senescent phenotype. EMBO J. 2016;35(7):724–42.
24. Paul A, et al. Reduced mitochondrial SOD displays mortality characteristics reminiscent of natural aging. Mech Ageing Dev. 2007;128(11–12):706–16.
25. Elchuri S, et al. CuZnSOD deficiency leads to persistent and widespread oxidative damage and hepatocarcinogenesis later in life. Oncogene. 2005;24(3):367–80.
26. Wawryn J, et al. Deficiency in superoxide dismutases shortens life span of yeast cells. Acta Biochim Pol. 1999;46(2):249–53.
27. Fusco D, et al. Effects of antioxidant supplementation on the aging process. Clin Interv Aging. 2007;2(3):377–87.
28. Ristow M, Schmeisser K. Mitohormesis: promoting health and lifespan by increased levels of reactive oxygen species (ROS). Dose Response. 2014;12(2):288–341.
29. Finkel T. Signal transduction by reactive oxygen species. J Cell Biol. 2011;194(1):7–15.
30. Finkel T. Signal transduction by mitochondrial oxidants. J Biol Chem. 2012;287(7):4434–40.
31. Van Raamsdonk JM, Hekimi S. Deletion of the mitochondrial superoxide dismutase sod-2 extends lifespan in Caenorhabditis elegans. PLoS Genet. 2009;5(2):e1000361.
32. Schaar CE, et al. Mitochondrial and cytoplasmic ROS have opposing effects on lifespan. PLoS Genet. 2015;11(2):e1004972.

33. Alam TI, et al. Human mitochondrial DNA is packaged with TFAM. Nucleic Acids Res. 2003;31(6):1640–5.
34. Kukat C, et al. Cross-strand binding of TFAM to a single mtDNA molecule forms the mitochondrial nucleoid. Proc Natl Acad Sci U S A. 2015;112(36):11288–93.
35. Diot A, Morten K, Poulton J. Mitophagy plays a central role in mitochondrial ageing. Mamm Genome. 2016;27(7–8):381–95.
36. Trifunovic A, et al. Premature ageing in mice expressing defective mitochondrial DNA polymerase. Nature. 2004;429(6990):417–23.
37. Ferguson LR, Baguley BC. Induction of petite formation in Saccharomyces cerevisiae by experimental antitumour agents. Structure–activity relationships for 9-anilinoacridines. Mutat Res. 1981;90(4):411–23.
38. Latorre-Pellicer A, et al. Mitochondrial and nuclear DNA matching shapes metabolism and healthy ageing. Nature. 2016;535(7613):561–5.
39. Fang EF, et al. Nuclear DNA damage signalling to mitochondria in ageing. Nat Rev Mol Cell Biol. 2016;17(5):308–21.
40. Giblin W, Skinner ME, Lombard DB. Sirtuins: guardians of mammalian healthspan. Trends Genet. 2014;30(7):271–86.
41. Jasper H. Sirtuins: longevity focuses on NAD+. Nat Chem Biol. 2013;9(11):666–7.
42. Ryu D, et al. NAD+ repletion improves muscle function in muscular dystrophy and counters global PARylation. Sci Transl Med. 2016;8(361):361ra139.
43. Hershberger KA, Martin AS, Hirschey MD. Role of NAD+ and mitochondrial sirtuins in cardiac and renal diseases. Nat Rev Nephrol. 2017;13:213.
44. Walther DM, et al. Widespread proteome remodeling and aggregation in aging C. elegans. Cell. 2015;161(4):919–32.
45. Gariani K, et al. Eliciting the mitochondrial unfolded protein response by nicotinamide adenine dinucleotide repletion reverses fatty liver disease in mice. Hepatology. 2016;63(4):1190–204.
46. Jovaisaite V, Mouchiroud L, Auwerx J. The mitochondrial unfolded protein response, a conserved stress response pathway with implications in health and disease. J Exp Biol. 2014;217(Pt 1):137–43.
47. Mottis A, Jovaisaite V, Auwerx J. The mitochondrial unfolded protein response in mammalian physiology. Mamm Genome. 2014;25(9–10):424–33.
48. Taylor RC, Dillin A. Aging as an event of proteostasis collapse. Cold Spring Harb Perspect Biol. 2011;3(5)
49. Tissenbaum HA. Using C. elegans for aging research. Invertebr Reprod Dev. 2015;59(sup1):59–63.
50. Hamilton B, et al. A systematic RNAi screen for longevity genes in C. elegans. Genes Dev. 2005;19(13):1544–55.
51. Lee SS, et al. A systematic RNAi screen identifies a critical role for mitochondria in C. elegans longevity. Nat Genet. 2003;33(1):40–8.
52. Houtkooper RH, et al. Mitonuclear protein imbalance as a conserved longevity mechanism. Nature. 2013;497(7450):451–7.
53. Moullan N, et al. Tetracyclines disturb mitochondrial function across eukaryotic models: a call for caution in biomedical research. Cell Rep. 2015;10:1681.
54. Jensen MB, Jasper H. Mitochondrial proteostasis in the control of aging and longevity. Cell Metab. 2014;20(2):214–25.
55. Pena S, et al. The mitochondrial unfolded protein response protects against anoxia in Caenorhabditis elegans. PLoS One. 2016;11(7):e0159989.
56. Pellegrino MW, Nargund AM, Haynes CM. Signaling the mitochondrial unfolded protein response. Biochim Biophys Acta. 2013;1833(2):410–6.
57. Tian Y, et al. Mitochondrial stress induces chromatin reorganization to promote longevity and UPR(mt). Cell. 2016;165(5):1197–208.
58. Lin YF, et al. Maintenance and propagation of a deleterious mitochondrial genome by the mitochondrial unfolded protein response. Nature. 2016;533(7603):416–9.

59. Zhao Q, et al. A mitochondrial specific stress response in mammalian cells. EMBO J. 2002;21(17):4411–9.
60. Munch C, Harper JW. Mitochondrial unfolded protein response controls matrix pre-RNA processing and translation. Nature. 2016;534(7609):710–3.
61. Haynes CM, Ron D. The mitochondrial UPR – protecting organelle protein homeostasis. J Cell Sci. 2010;123(Pt 22):3849–55.
62. Munkacsy E, et al. DLK-1, SEK-3 and PMK-3 are required for the life extension induced by mitochondrial bioenergetic disruption in C. elegans. PLoS Genet. 2016;12(7):e1006133.
63. Westermann B. Molecular machinery of mitochondrial fusion and fission. J Biol Chem. 2008;283(20):13501–5.
64. van der Bliek AM, Shen Q, Kawajiri S. Mechanisms of mitochondrial fission and fusion. Cold Spring Harb Perspect Biol. 2013;5(6)
65. Chauhan A, Vera J, Wolkenhauer O. The systems biology of mitochondrial fission and fusion and implications for disease and aging. Biogerontology. 2014;15(1):1–12.
66. Regmi SG, Rolland SG, Conradt B. Age-dependent changes in mitochondrial morphology and volume are not predictors of lifespan. Aging (Albany NY). 2014;6:118.
67. Bernhardt D, et al. Simultaneous impairment of mitochondrial fission and fusion reduces mitophagy and shortens replicative lifespan. Sci Rep. 2015;5(5):7885.
68. Twig G, et al. Fission and selective fusion govern mitochondrial segregation and elimination by autophagy. EMBO J. 2008;27(2):433–46.
69. McFaline-Figueroa JR, et al. Mitochondrial quality control during inheritance is associated with lifespan and mother-daughter age asymmetry in budding yeast. Aging Cell. 2011;10(5):885–95.
70. Katajisto P, et al.. Stem cellsAsymmetric apportioning of aged mitochondria between daughter cells is required for stemness. Science. 2015;348(6232):340–3.
71. Palikaras K, Lionaki E, Tavernarakis N. Coordination of mitophagy and mitochondrial biogenesis during ageing in C. elegans. Nature. 2015;521(7553):525–8.
72. Jin SM, Youle RJ. PINK1- and Parkin-mediated mitophagy at a glance. J Cell Sci. 2012;125. (Pt 4:795–9.
73. Youle RJ, Narendra DP. Mechanisms of mitophagy. Nat Rev Mol Cell Biol. 2011;12(1):9–14.
74. Kazlauskaite A, et al. Binding to serine 65-phosphorylated ubiquitin primes Parkin for optimal PINK1-dependent phosphorylation and activation. EMBO Rep. 2015;16(8):939–54.
75. Kondapalli C, et al. PINK1 is activated by mitochondrial membrane potential depolarization and stimulates Parkin E3 ligase activity by phosphorylating Serine 65. Open Biol. 2012;2(5):120080.
76. Lazarou M, et al. The ubiquitin kinase PINK1 recruits autophagy receptors to induce mitophagy. Nature. 2015;524(7565):309–14.
77. Salminen A, et al. Krebs cycle dysfunction shapes epigenetic landscape of chromatin: novel insights into mitochondrial regulation of aging process. Cell Signal. 2014;26(7):1598–603.
78. Vafai SB, Mootha VK. Mitochondrial disorders as windows into an ancient organelle. Nature. 2012;491(7424):374–83.
79. Childers MK, et al. Gene therapy prolongs survival and restores function in murine and canine models of myotubular myopathy. Sci Transl Med. 2014;6(220):220ra10.
80. Dowling JJ, et al. Oxidative stress and successful antioxidant treatment in models of RYR1-related myopathy. Brain. 2012;135(Pt 4):1115–27.
81. Jain IH, et al. Hypoxia as a therapy for mitochondrial disease. Science. 2016;352(6281):54–61.
82. Ruiz-Pesini E, et al. An enhanced MITOMAP with a global mtDNA mutational phylogeny. Nucleic Acids Res. 2007;35(Database issue):D823–8.
83. Beck JS, Mufson EJ, Counts SE. Evidence for mitochondrial UPR gene activation in familial and sporadic Alzheimer's disease. Curr Alzheimer Res. 2016;13(6):610–4.
84. Liao C, et al. Dysregulated mitophagy and mitochondrial organization in optic atrophy due to OPA1 mutations. Neurology. 2017;88(2):131–42.
85. Exner N, et al. Mitochondrial dysfunction in Parkinson's disease: molecular mechanisms and pathophysiological consequences. EMBO J. 2012;31(14):3038–62.

86. Pfeffer G, Chinnery PF. Diagnosis and treatment of mitochondrial myopathies. Ann Med. 2013;45(1):4–16.
87. Singh KK, et al. Mitochondrial DNA determines the cellular response to cancer therapeutic agents. Oncogene. 1999;18(48):6641–6.
88. Wallace DC. Mitochondria and cancer. Nat Rev Cancer. 2012;12(10):685–98.
89. Lismont C, et al. Redox interplay between mitochondria and peroxisomes. Front Cell Dev Biol. 2015;(3):35.
90. Baumgart M, et al. Longitudinal RNA-Seq analysis of vertebrate aging identifies mitochondrial complex I as a small-molecule-sensitive modifier of lifespan. Cell Syst. 2016;2(2):122–32.

Aging, Free Radicals, and Reactive Oxygen Species: An Evolving Concept

11

Shyamal K. Goswami

Aging: An Overview

Aging is a natural process resulting in sequential changes in the physiology of an organism with the progression of time. It is an irreversible process caused by the environmental factors, diseases and innate programming called the "aging process." Aging enhances the chances of diseases and death [27]. As the quality of living in a particular population improves, the chances of death at an early age decrease. However, the aging population remains vulnerable to various debilitating diseases. Aging is thus a subject of intense research among the scientific community, policy makers, and health care professionals (Box 11.1).

Research on aging has been quite extensive and extended throughout the past century. However, with the progress of understanding the biology of metazoan organisms, the theories of aging has diversified and there is no universally accepted theory regarding the molecular and biochemical basis of aging. It is now believed that aging is caused by a number of factors which synergize with each other and such synergistic response increases with the progression of age.

Among various propositions on aging are (i) decreased immune responses followed by increased inflammation ("inflammaging", [13]); (ii) altered neuromodulation of brain functions [39]; (iii) alterations in the supramolecular compositions and structure of hierarchies of the organisms as governed by the thermodynamic principles [15]; (iv) genetic programming [11]; a drift in developmental programming [16].

However, the most comprehensive and widely accepted theory involves the nodal role of reactive species which causes a decline in plasticity of the genome, reduced mitochondrial function, compromised bioenergetic control, improper food utilization and metabolic homeostasis, decreased defense, and reproductive fidelity [32].

S. K. Goswami (✉)
School of Life Sciences, Jawaharlal Nehru University, New Delhi, India
e-mail: shyamal.goswami@gmail.com

Box 11.1
Free radicals are atoms or a group of atoms with one or more unpaired electron(s) in their outermost orbit. Commonly occurring free radicals in biological system are superoxide ($O_2\bullet-$), nitric oxide ($NO\bullet$), nitrogen dioxide ($NO_2\bullet$), hydroxyl- ($OH\bullet$), hydroperoxyl ($HO_2\bullet$), and peroxyl ($ROO\bullet$) radicals. Also, another group of molecules like hydrogen peroxide (H_2O_2), ozone (O_3), singlet oxygen ($^1O_2$), hypochlorous acid (HOCl), nitrous acid (HNO_2), peroxynitrite ($ONOO^-$), dinitrogen trioxide (N_2O_3), and lipid peroxide (LOOH) are highly reactive oxidants but not free radicals per se and they can generate free radicals in the cellular environment upon interaction with heavy metal, etc. The free radicals and nonradical oxidants are collectively called reactive oxygen/nitrogen species. According to this classification, hydrogen peroxide is not a free radical, but a reactive species while super oxide is both free radical and reactive oxygen species. Reactive oxygen/nitrogen species are formed endogenously by natural biological processes (like the activation of NADPH oxidases and during electron transport through Complex I and III in the mitochondrial electron transport chain). They are also generated upon stimulation by external agents such as UV radiation, genotoxic agents, etc. Since these free radicals and reactive oxygen species are damaging to cellular macromolecules like lipids, proteins and DNA, cells also have an extensive package of antioxidant molecules like vitamin E and C, and enzymes like glutathione peroxidase, thioredoxin peroxidase, etc. Cellular homeostasis is maintained by continuous generation of free radicals/reactive species and their immediate attenuation by the antioxidant system. In the past 20 years, extensive knowledge has accumulated on the various mechanisms of free radical/ROS generation, their intracellular sources, antioxidant defenses, and the chemistry of oxidative damage caused by the free radicals and reactive species [43].

The "Free Radical Theory of Aging": An Historical Perspective

The mechanisms of aging have been a subject interest for centuries [31, 40]. It is a manifestation of progressive accumulation of damaged cells, dysfunctional organs culminating to diseases and death. Although various theories of aging were proposed since the late nineteenth century, a discretely revolutionary concept attributing free radicals in the aging process was proposed by the gerontologist Denham Harman in 1955 when he published a report entitled "Aging: A theory based on free radical and radiation chemistry" in the University of California Radiation Laboratory report. He then published a full-length article in the *Journal of Gerontology* elaborating on his hypothesis [17]. Although his hypothesis was highly accretive in nature, especially in those days, it was also a technical challenge to prove or disprove it by experimentations (most of the present-day assays for free radicals and reactive species in biological samples were developed much later). In subsequent years, he substantiated his theory by extensive research on the role of free radicals in the life span and two age-related diseases, cancer, and atherosclerosis. Because of his extensive contribution to this area of research, he is considered as the "father of the free radical theory of aging" (https://en.wikipedia.org/wiki/Denham_Harman). Following are some of the major findings reported by him in support of his hypothesis:

1. Atherogenesis is caused by the oxidative polymerization of serum lipoproteins; anchoring of the oxidized materials in the arterial wall; and an inflammatory reaction induced in the arterial wall by these condensed products [18].
2. Serum mercaptan level decreases with age (in this article, he formally suggested the "Free radical theory of aging") [19].
3. Cancer and aging is caused by mutation induced by the free radicals [20].
4. Antioxidants can extend the life span and prevent cancer [21].
5. Serum copper level increases with aging [22].
6. Antioxidants increase the life span of mice [23].
7. Sustained oxygen consumption accumulates mitochondrial damage by the free radicals that ultimately cause death [24]. Based upon this study, more than 20 years later "Mitochondrial theories of aging" was proposed by others [34].
8. Enhanced level of lipid peroxidation adversely affects the CNS but does not increase the mortality rate. In this study, Harman showed the correlation between the degeneration of brain, an age-related disorder, with free radicals [28].
9. Antioxidants enhance the immune responses in mice, thus correlating the age, immune deficiency, and the free radicals [29].
10. The level of free radical–generating reactions increases with age, and the life span can be increased by 5–10 or more years by minimizing free radical reactions through lifestyle and diet control [25].

Also, in 1969, the discovery of superoxide dismutase (SOD) by McCord and Fridovich gave a major boost to the "Free radical theory of aging" [45]. It was generally accepted that since the free radicals are deleterious for health, nature has adopted SOD to destroy them.

More than 25 years after proposing the association between free radicals and aging, in 1984, Harman proposed the "Free radical theory of diseases" [26]. He hypothesized that free radicals cause damages that accumulate with age and it is common to all. However, due to genetic and environmental factors, patterns of changes differ from individual to individual. He thus perceived aging as a disease, although aging process is universal. Accordingly, he proposed that the chances of developing of age-related diseases can be reduced by lowering the level of free radicals by calorie restriction and antioxidants.

Emergence of the "Oxidative Stress Theory of Aging"

In 1970s and 1980s, while Harman was revising his "Free radical theory of aging" to "Free radical theory of diseases," it was becoming more and more evident that aging is also associated with the accumulation of oxidized DNA, lipids, and proteins. In certain age-related diseases like Parkinson's, cataract, cancer, and atherosclerosis; oxidative damages are extensive. Such observations thus led to the "Oxidative stress theory of aging," which hypothesized that the oxidation of cellular components contributes to the aging process (it is thus a variant of the "Free-radical theory of aging"). In agreement to this hypothesis, P. L. Larsen in 1996 demonstrated that mutation in two genes in *C. elegans* that increases the life span of the worm also increases the level of the antioxidant enzymes superoxide dismutase and catalase [38]. Later, Lithgow and coworkers demonstrated that the life span of *C. elegans* is also extended by the mimetic for superoxide dismutase and catalase [48]. In subsequent years, several other age-related parameters also emerged, and were incorporated into the Oxidative stress theory of aging. Some of those major hypotheses are given below.

1. In 1990, Bandy and Davison suggested that with aging, the mitochondrial DNA is damaged by the free radicals and oxidants generated in the mitochondria which in turn further increases the generation of oxidants and free radicals by mitochondria; accelerating the aging process [3].
2. In 1991, Pacifici and Davies proposed that with aging, levels of enzymes involved in repairing of oxidized biomolecules decline, contributing to the aging process [51].
3. In 2001, A. Terman proposed the "Garbage catastrophe theory of aging" which suggested that aging occurs due to the imperfect clearance of oxidatively damaged indigestible materials (e.g., lipofuscin pigment) [63]. This hypothesis was later revised as "The mitochondrial–lysosomal axis theory of aging," which proposed that with aging, due to the accumulation of lipofuscin, lysosomal function is impaired, that affects mitochondrial recycling. Aged cells thus unable to remove the oxidatively damaged mitochondria. Such damaged nonfunctional mitochondria gradually displace the normal ones and cumulative mitochondrial and lysosomal damage causes death of postmitotic cells [5].

Questioning the "Oxidative Stress Theory of Aging"

As described above, the "Free radical/Oxidative stress theory of aging" had a wide acceptance for about 4 decades since proposition in mid-1950s. Its acceptability was further consolidated by the genetic studies in model organisms like *Drosophila* and *C. elegans* [61, 62]. However, discrepancies in the hypothesis also emerged with more and more studies with these organisms. One such study was the demonstration that enhancing superoxide dismutase activity by its mimetic did not increase the life span of *C. elegans* [36]. The most incriminating evidences against this theory came from studies with transgenic and knockout mice. A comprehensive study by Arlan Richardson and coworkers on the effects of over- or underexpression of 18 genes encoding various antioxidant enzymes in mice showed that except one, that is, the superoxide dismutase gene, others had no effect on the life span [54]. In another study, it was demonstrated that the premature aging observed in mitochondrial DNA mutator mice was not due to oxidative stress but due to the respiratory chain dysfunction [65].

Signaling by Reactive Oxygen Species: A Paradigm Shift

During 1990s, while the tenability "Free radical theory of aging" was being questioned, a new concept on the role of reactive oxygen species in cellular physiology also emerged. It became apparent that reactive oxygen species are not necessarily deleterious to the cells and under certain circumstances, especially when generated in limited amount, they might have important regulatory functions. A seminal discovery in this regard was the observation that when human carcinoma cells treated with the epidermal growth factor (EGF), hydrogen peroxide is generated that inhibits protein tyrosine phosphatase activity, sustaining the tyrosine kinase signaling [2]. Another major discovery highlighting the role of ROS in cellular function was the presence of NADPH oxidases in nonphagocytic cells. NADPH oxidases are the multi-subunit enzymes involved in the generation of superoxide from molecular oxygen and NADPH. When it was first discovered in phagocytic cells, it was believed that the generation of superoxide is a highly specialized strategy to use ROS as the bactericidal agent. However, with the discovery of its isoforms in various other tissues led to the hypothesis that these enzymes are in fact the generator of superoxide (and hydrogen peroxide) for signaling purposes [58].

In the past 20 years, it became amply evident that they also play a role in normal physiological functions like cell growth, differentiation, embryonic development, angiogenesis, etc. It is now established that like kinases, reactive oxygen species can also covalently modify selective cysteine residues of various proteins, regulating their functions. In accordance, a concept of redox signaling had emerged in the past decade [60].

Under physiological conditions, any molecule mediating signal transduction must have substrate specificity and modifications induced by them on their targets should be reversible. Physiological oxidants like super oxide ($O_2^{\bullet-}$) and hydrogen peroxide (H_2O_2) have moderate oxidation potential and generally cause reversible modifications of cysteine residues eliciting specific biological responses (Box 11.2). On the other hand, certain highly reactive species, namely, hydroxyl- ($OH^\bullet$), hydroperoxyl- ($HO_2^\bullet$), and peroxyl- ($ROO^\bullet$) radicals, hypochlorous acid (HOCl) and peroxynitrite ($ONOO-$), etc., oxidize proteins and other biomolecules without specificity or preference. Oxidative modifications induced by these highly reactive species are irreversible, unlikely to have signaling functions. Besides cysteine, other amino acids like methionine, tryptophan, and tyrosine often undergo such irreversible oxidation under highly oxidizing environment, resulting in the loss of functions and diseases [42].

Box 11.2
Signaling by hydrogen peroxide selective cysteine oxidation. Hydrogen peroxide oxidizes cysteine thiols having lower pKa (~6.4, thus in the thiolate form) to sulfenic acid (RSOH). Sulfenic acid can react with the neighboring cysteine residues forming an intra- or intermolecular disulfide bond. It can also react with the reduced glutathione pool (S-glutathionylation) to form a mixed disulfide. These cysteine oxidations generally lead to altered biological activities of the target proteins. Oxidized cysteine residues are subsequently reduced thioredoxin and glutaredoxin, thus attenuating the redox signal. Hydrogen peroxide, especially when in higher concentrations, can also oxidize the cysteine residue sulfinic (RSOOH) and sulfonic (RSO3H) acids. The sulfinic acid residue can be reduced by another enzyme called sulfiredoxins (Srx). However, sulfonic acid formation is generally irreversible, nonspecific, and generally found in diseased tissue under oxidative stress [71].

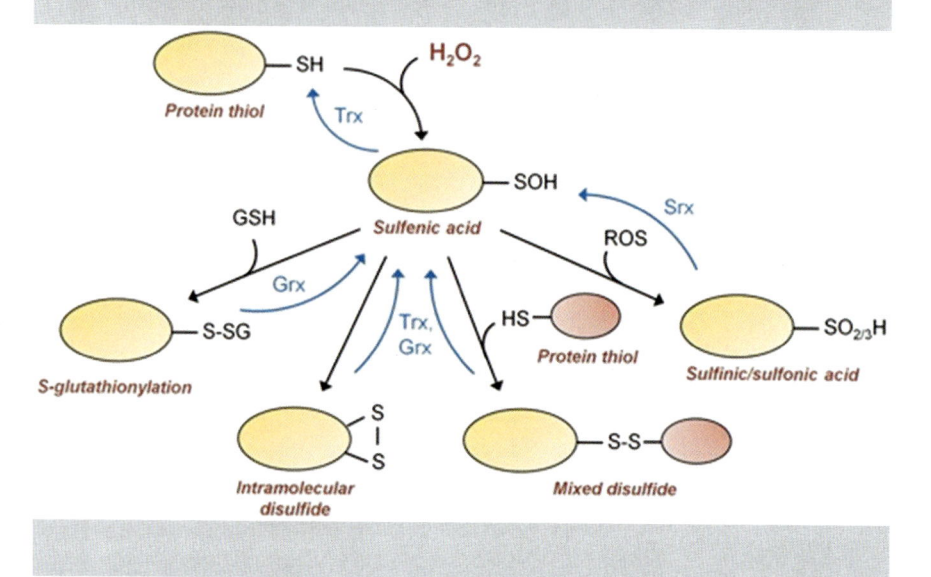

Numerous proteins contain cysteine residues that are highly conserved across species, suggesting functions related to oxidative signaling (and metal coordination). Quite often, they participate in metal ion coordination and are the targets of redox signaling as well (as in the case of nitric oxide-soluble guanylate cyclase signaling). There are several ways oxidative modification of a single cysteine (SOH) can affect the structure–function of a protein. While a protein can have many cysteine residues, few, especially those in the thiolate form (C–S–H) are capable of mediating redox signals. The pKa value of cysteines depends on the charge distribution of its adjacent amino acids in the 3D conformation of the target protein. Susceptibility to oxidative modification is governed by the concentration and reactivity of individual oxidants as well as the pKa of a cysteine thiol. Oxidative modifications of cysteines have been demonstrated in numerous proteins and its effects on the functions of a few have also been studied in details. Cysteine thiols thus serve as molecular switches, processing different redox-based signals into distinct functional responses [12, 70].

Role of Signaling by ROS in Aging Process

Although the intracellular sources of ROS and the mechanisms of ROS-mediated signaling are now well understood [55], our knowledge of the role of redox signaling in the aging process per se is quite limiting [8]. Since aging involves complex interactions between cellular and organismal processes, genetic tools have been used for delineating the role of ROS-mediated signaling in regulating the life span in organisms like yeast, *C. elegans*, *Drosophila*, and mice.

Calorie restriction (CR) decreases the level of oxidative stress and increases the life span in these organisms. It is thus used as an experimental model to study the role of oxidants in aging. In *S. cerevisiae*, two antioxidant proteins peroxiredoxin (*Tsa1*) and sulfiredoxin (*Srx1*) are required for the extension of replicative life span under CR. Under normal availability of nutrients, hydrogen peroxide is generated that inactivates Tsa1 by hyperoxidation of its catalytic Cys residues to sulfonic acid, accelerating the aging process. Under CR, expression of Srx1 is induced that reduces the sulfonic acid of Tsr1, restoring the function of Tsa1, decelerating the aging process. It has now been established that apart from being an antioxidant protein that decomposes hydrogen peroxide (Box 11.3), peroxiredoxin also acts as a sensor for hydrogen peroxide, regulating its intracellular flux [64]. However, the biochemical mechanisms of acceleration of aging process by the peroxiredoxin-sulfiredoxin system under normal nutritional condition and the extension of life span under CR is yet to be deciphered. Since peroxiredoxins are conserved in evolution, it is likely that they also contribute to the extension of life span under CR in higher eukaryotes. Caloric restriction in *S. cerevisiae* also reduces the target of rapamycin (TOR) signaling and upregulate the mitochondrial functions through "mitohormesis" (Box 11.4) that decelerates the aging process through ROS signaling. However, specific signaling pathways and the downstream effectors that are activated by CR-induced ROS production remain to be identified [52]. In *S. pombe*, low-glucose media reduces the signaling by TOR and protein kinase A (PKA), increases the generation

Box 11.3
Peroxiredoxin reduces hydrogen peroxide to water and in the process gets oxidized to sulfenic acid at the catalytic cysteine residue. Thioredoxin then regenerates it by the reduction of the oxidized cysteine. When it is hyperoxidized to sulfinic and sulfonic acids, thioredoxin cannot reduce those. However, another enzyme sulfiredoxin has the capacity to reduce the sulfinic and sulfonic acids to cysteine (see Box 11.2 for the details).

> **Box 11.4**
> While at a higher level ROS is detrimental to cellular function, when generated at a low concentration at a specific cellular location it also can act as signaling molecules. ROS generated in the mitochondria elicit various physiological responses including the extension of life span. Such beneficial response by a potentially harmful molecule at a lower concentration is named "hormesis" [7]. When the hermetic effect emanates from the ROS generated in the mitochondria, it is termed "mitochondrial hormesis" or "mitohormesis" [56]. A recent study has further extended this concept by showing that various organelle–organelle and organelle–cytosol communications regulate the chronological aging of *S. cerevisiae* through this mechanism [10].

of ROS and extends the chronological life span. Mitogen-activated protein kinase Sty1 is involved in this process [73]. So, as in other instances of regulation of cellular processes by ROS through the kinase network [57], it is likely that regulation of life span by ROS also involves the kinases activated by stimuli like CR and insulin. However, how kinase signaling cross-talk to the ROS signaling has to be mechanistically dissected [72].

In a recent study, it has been demonstrated that muscle mitochondrial injury in *Drosophila* impedes the age-dependent deterioration of muscle function and prolongs the life span. Further analysis suggested that mitochondrial ROS-dependent induction of genes regulating the mitochondrial unfolded protein response, insulin-like growth factor signaling, and mitophagy are involved in the process [50]. In another study, redox-proteomics were done for the cysteine residues in *Drosophila* during aging and fasting. It showed that the cysteine residues oxidized in the young flies did not become more oxidized with age. Also, fasting for 24 hours oxidized some of the cysteine residues that were in the reduced state under fed conditions, while few others which were oxidized under fed conditions were reduced by the fasting [49].

Genetic studies on *C. elegans* have also shed some light on the role of ROS signaling in aging. SOD and catalase are the two major (and well-investigated) antioxidant enzymes that are highly conserved in evolution. SOD reduces $O_2^{\cdot -}$ to H_2O_2 and catalase decomposes it to water and molecular oxygen. There are five *sod* genes in *C. elegans*, encoding for two mitochondrial (*sod-2* and *sod-3*); two cytosolic (*sod-1* and *sod-5*); and one extracellular (*sod-4*) enzymes. Also, *C. elegans* have three different catalases encoded by *ctl-1*, *ctl-2*, and *ctl-3* genes. According to the original postulations of the "Free radical theory of aging," deletion of genes encoding SOD or catalase would increase the level of ROS, resulting in decrease in life span. In contrast, overexpression of SOD and catalase would reduce the oxidative stress and extend the life span. In agreement, deletion of either mitochondrial or cytosolic *sod* genes in *Drosophila* (it has only two SOD encoding genes) shortens the life span [53]. Interestingly, deletion of individual *sod* genes in *C. elegans* not only failed to decrease the life span; rather deletion of *sod-2* even extended the life span along with increased levels of oxidative damage, strongly suggesting that (i) ROS toxicity does not play a major role in the reduction of life span and (ii) Increase in life span by ROS is presumably mediated by the redox signaling, and it is functionally separable from the deleterious effects of ROS. Further, worms devoid of two or three *sod* genes also lived as long as wild-type worms [66]. Therefore, ROS signaling rather than oxidative stress might be the nodal determining factor in the aging of this organism. Also, despite the fact that the mechanism of aging is likely to be conserved in evolution, subtle difference might exist between the pathways operating in *Drosophila* and *C. elegans*. More evidences favoring the role of ROS signaling in determining the life span in *C. elegans* also came from studies with the worms overexpressing *sod-1* and *sod-2* genes. Increased expression of both the proteins also enhanced the life span of the worm, but that of SOD-1 did not reduce the levels of lipid oxidation or glycation and rather increased the level of oxidized proteins [6]. Noticeably, increased level of SOD-1 also increased the steady-state level of H_2O_2, suggesting its role in delaying the aging process. However, simultaneous coexpression of catalase did not abrogate the life-span extension effect of *sod-1*. Since redox signaling is a compartmentalized process occurring in the cellular microdomain, it might be necessary to see whether catalase indeed neutralizes the H_2O_2 signaling [35].

Mitochondrial functions have long been associated with the aging process. The operational hypothesis is, sustained energy generation by mitochondria leads to the concurrent generation of ROS that damages the mitochondrial DNA and proteins; further affecting its function that leads to the generation of more ROS, accelerating the aging process. Various mitochondrial mutants of *C. elegans* have thus been studied for correlating the mitochondrial function, level of ROS generation and aging. In a seminal study, Hekimi and coworkers demonstrated that two such partial loss-of-function mutants namely, *nuo-6* (encodes a subunit of complex I) and *isp-1* (encodes an iron sulfur protein of complex III), have decreased electron transport and increased longevity. These mutants have a lower level of total ROS but a higher level of superoxide, suggesting its role in extending the life span through redox signaling [69]. Further, study with a number of other mutants also suggested that the pathway responsible for extending the life span in *isp-1* and *nuo-1* mutants through mitochondrial superoxide is different from those induced by insulin signaling, calorie restriction, hypoxia, and hormesis [56]. Another mutant *clk-1* (encoding a hydroxylase involved in the synthesis of coenzyme Q) has longer life span than their normal counterpart [37]. These worms have reduced rates of electron transport and oxidative phosphorylation, decreased levels of protein carbonylation and lipofuscin accumulation, the two prominent markers of oxidative stress [67, 68]. Upon deletion of superoxide dismutase (*sod*) genes in *clk-1* background, it was found that only the deletion of *sod-2* (primary mitochondrial SOD) resulted in further increase in the life span. On the contrary, deletion of either of the two genes for the cytoplasmic SOD, that is, *sod-1* and *sod-5*, significantly reduced the life span of the *clk-1* worms. Further, increasing the superoxide level through the treatment with paraquat boosted the life span in *clk-1-sod-1* double mutants. These results strongly suggest a compartment specific role of superoxide signaling in determining the life span of the *clk-1* worms. While cytoplasmic superoxide has a detrimental effect on the life span of the organism, its elevation in the mitochondria increases its life span [59].

The complexity of the ROS signaling in the aging process is further evident from the fact that treatment with paraquat increases the already extended life span of the clk-1mutant, but not of two other long-living mitochondrial mutants viz., *isp-1* and *nuo-6* [59].

Evidences in support of ROS signaling in the determination of life span in worms also came from the observation that when *C. elegans* are treated with a lower dose of paraquat, a compound that generates superoxide, its life span is extended [69]. Taken together, it appears that the regulation of life span in *C. elegans* by the oxidants is biphasic in nature. While at a lower dose, ROS elicits prosurvival signaling; at an increased level, it induces toxicity. Therefore, it is imperative that the role of ROS in aging and age-related diseases is far more nuanced than the original proposition that a unidirectional increase in the level of oxidants indiscriminately damage of macromolecules causing diseases and aging.

As it is evident from the studies with the model organisms like yeast, *Drosophila* and *C. elegans,* signaling by ROS plays a key role in physiological aging. When this homeostatic regulation fails, oxidative stress sets in, followed by the acceleration of aging and related disorders. Though it is expected that aging in higher metazoans like human and mice follow the same pathways, due to the complexity of their physiology, the mechanistic insights in general and the role of ROS in particular is largely obscure. In case of higher organisms, intracellular ROS generating organelles like mitochondria, peroxisome, and the endoplasmic reticulum are differently regulated in different tissues. Also, there are several ROS-generating enzymes like NADPH oxidases, cyclooxygenases, lipooxygenases, etc. in various tissues that are differentially modulated by the physiological stimuli. ROS generated from these organelles and enzymes regulates the intracellular signaling network of kinases, phosphatases, ion channels, transporters, etc., which then target the downstream gene regulatory modules of transcription factors and epigenetic regulators [1]. Any perturbation in such huge regulatory network either locally (cellular level) or globally (organismal level), would disturb the ROS homeostasis or vice versa [30]. Although the details of these networks in age-related diseases like cancer, Parkinson's, etc. are fast emerging [14], better understanding is needed to extrapolate this information to the aging process. Nevertheless, as per the present day knowledge, certain transcription factor like Nrf2, which elicits antioxidant responses, play a crucial role in the aging process. Nrf2 knockout mice have reduced protection against cancer under calorie restriction, suggesting that Nrf2 mediates the beneficial effects of calorie restriction on the carcinogenesis [44]. The sirtuins are a conserved family of protein deacylases of which the founding member SIR2 was first shown to extend the life span in *S. cerevisiae* [33]. There are seven mammalian orthologs named SIRT1–7 of which SIRT3, 4 and 5 are localized in the mitochondria. These mitochondrial proteins regulate almost each and every aspect of mitochondrial functions. As an example, SIRT3 has been shown to regulate mitochondrial reactive oxygen species homeostasis, ATP production, mitophagy, oxidation of metabolites, etc. SIRT3 knockout mice shows accelerated aging in association with various old age diseases like cancer, metabolic syndrome, neurodegenerative, and cardiovascular diseases [46]. The nuclear-localized SIRT1 also contributes to the extension of life span induced by CR. SIRT1 knockout mice do not have increased life span under CR and have decreased physical activity found in under CR [4]. Taken together, although the role of various ROS responsive pathways in increasing or decreasing life span in mammals have to be dissected, available information till date suggests a highly nuanced role these regulatory molecules as the major determinant of the aging process.

Concluding Remarks

Since the late nineteenth century, more than 300 theories have been put forward to explain the aging process, but none has been able to explain it well enough to be accepted universally [47]. This is not unusual, as there is not even a consensus on what aging is all about. Physiological aging involves structural wear and tear, tissue and organ degeneration, cellular senescence, and reduced activity at the organismal level. Aging process is thus likely to be multifactorial [41, 47]. Another major issue is the presumption that the mechanisms of aging are similar in all organisms and the same for a particular species. As an example, naked mole-rats live for more than 30 years despite having poor defense against oxidative damage comparable to that of mice which have much shorter life span. Even the aging phenotype in a single species is highly heterogeneous in terms of afflicting diseases, affected organs, molecular signatures, etc. So, the research on the mechanisms of aging at the cellular level may or may not fully reflect it at the organismal level. Thus, a better understanding of aging will need further investigations in the model organisms and then judiciously extrapolating the knowledge to others within the robust framework of biology and physiology [9].

In this context, the "Free radical theory of aging" as originally proposed, was definitely ahead of its time. Despite its ambiguities, especially at the later days, it has made an enormous contribution in understanding the aging process per se. It has motivated a large number of researchers to investigate the biochemical, physiological, and molecular basis of the aging process, generating a voluminous data. The late emergence of the role of ROS signaling is a paradigm shift from the original concept. It is thus expected that in coming years, this also will further expand the horizon of our understanding of the biochemical basis of aging.

References

1. B V, Poli G, Basaga H. Tumor suppressor genes and ROS: complex networks of interactions. Free Radic Biol Med. 2012;52(1):7–18.
2. Bae YS, Kang SW, Seo MS, Baines IC, Tekle E, Chock PB, Rhee SG. Role in EGF receptor-mediated tyrosine phosphorylation. J Biol Chem. 1997;272(1):217–21.
3. Bandy B, Davison AJ. Mitochondrial mutations may increase oxidative stress: implications for carcinogenesis and aging? Free Radic Biol Med. 1990;8(6):523–39.

4. Boily G, Seifert EL, Bevilacqua L, He XH, Sabourin G, Estey C, Moffat C, Crawford S, Saliba S, Jardine K, Xuan J, Evans M, Harper ME, McBurney MW. SirT1 regulates energy metabolism and response to caloric restriction in mice. PLoS One. 2008;3(3):e1759.
5. Brunk UT, Terman A. The mitochondrial-lysosomal axis theory of aging: accumulation of damaged mitochondria as a result of imperfect autophagocytosis. Eur J Biochem. 2002;269(8):1996–2002.
6. Cabreiro F, Ackerman D, Doonan R, Araiz C, Back P, Papp D, Braeckman BP, Gems D. Increased life span from overexpression of superoxide dismutase in *Caenorhabditis elegans* is not caused by decreased oxidative damage. Free Radic Biol Med. 2011;51(8):1575–82.
7. Calabrese EJ. Hormesis: a fundamental concept in biology. Microbial Cell. 1(5):145–9.
8. Chandrasekaran A, Idelchik M d PS, Melendez JA. Redox control of senescence and age-related disease. Redox Biol. 2017;11:91–102.
9. Alan A. Cohen, Aging across the tree of life: the importance of a comparative perspective for the use of animal models in aging, Biochim Biophys Acta 2017. pii: S0925-4439 (17)30219-3.
10. Dakik P, Titorenko VI. Communications between Mitochondria, The nucleus, vacuoles, peroxisomes, the endoplasmic reticulum, the plasma membrane, lipid droplets, and the cytosol during yeast chronological aging. Front Genet. 2016;7:–177.
11. de Grey AD. Do we have genes that exist to hasten aging? New data, new arguments, but the answer is still no. Curr Aging Sci. 2015;8(1):24–33.
12. Devarie-Baez NO, Silva Lopez EI, Furdui CM. Biological chemistry and functionality of protein Sulfenic acids and related thiol modifications. Free Radic Res. 2016;50(2):172–94.
13. Fulop T, Witkowski JM, Pawelec G, Alan C, Larbi A. On the immunological theory of aging. Interdiscip Top Gerontol. 2014;39:163–76.
14. Gào X, Schöttker B. Reduction-oxidation pathways involved in cancer development: a systematic review of literature reviews. Oncotarget. 2017;. [Epub ahead of print]
15. Gladyshev GP. Thermodynamics of the origin of life, evolution and aging. Adv Gerontol. 2014;27(2):225–8.
16. Gruber J, Yee Z, Tolwinski NS. Developmental drift and the role of Wnt signaling in aging. Cancers (Basel). 2016;8(8)
17. Harman D. Aging: a theory based on free radical and radiation chemistry. J Gerontol. 1956 Jul;11(3):298–300.
18. Harman D. Atherosclerosis: a hypothesis concerning the initiating steps on pathogenesis. J Gerontol. 1957;12:199–202.
19. Harman D. The free radical theory of aging: the effect of age on serum mercaptan levels. J Gerontol. 1960;15:38–40.
20. Harman D. Mutation, cancer, and ageing. Lancet. 1961;1(7170):200–1.
21. Harman D. Prolongation of the normal lifespan and inhibition of spontaneous cancer by antioxidants. J Gerontol. 1961;16:247–54.
22. Harman D. The free radical theory of aging: effect of age on serum copper levels. J Gerontol. 1965;20:151–3.
23. Harman D. Free radical theory of aging: effect of free radical reaction inhibitors on the mortality rate of male LAF mice. J Gerontol Oct. 1968;23(4):476–82.
24. Harman D. The biologic clock: the mitochondria? J Am Geriatr Soc. 1972;20(4):145–7.
25. Harman D. The aging process. Proc Natl Acad Sci U S A. 1981;78(11):7124–8.
26. Harman D. The aging process: major risk factor for disease and death. Proc Natl Acad Sci U S A. 1991;88(12):5360–3.
27. Harman D. Aging: overview. Ann N Y Acad Sci. 2001;928:1–21.
28. Harman D, Hendricks S, Eddy DE, Seibold J. Free radical theory of aging: effect of dietary fat on central nervous system function. J Am Geriatr Soc. 1976;24(7):301–7.
29. Harman D, Heidrick ML, Eddy DE. Free radical theory of aging: effect of free-radical-reaction inhibitors on the immune response. J Am Geriatr Soc. 1977;25(9):400–7.
30. Holmström KM, Finkel T. Cellular mechanisms and physiological consequences of redox-dependent signalling. Nat Rev Mol Cell Biol. 2014;15(6):411–21.
31. Jin K. Modern biological theories of aging. Aging Dis. 2010;1(2):72–4.

32. Jones DP. Redox theory of aging. Redox Biol. 2015;5:71–9.
33. Kaeberlein M, McVey M, Guarente L. The SIR2/3/4 complex and SIR2 alone promote longevity in Saccharomyces cerevisiae by two different mechanisms. Genes Dev. 1999;13(19):2570–80.
34. Kalous M, Drahota Z. The role of mitochondria in aging. Physiol Res. 1996;45(5):351–9.
35. Kaludercic N, Deshwal S, Di Lisa F. Reactive oxygen species and redox compartmentalization. Front Physiol. 2014;5:285.
36. Keaney M, Matthijssens F, Sharpe M, Vanfleteren J, Gems D. Superoxide dismutase mimetics elevate superoxide dismutase activity in vivo but do not retard aging in the nematode *Caenorhabditis elegans*. Free Radic Biol Med. 2004;37(2):239–50.
37. Lakowski B, Hekimi S. Determination of life-span in *Caenorhabditis elegans* by four clock genes. Science. 1996;272:1010–3.
38. Larsen PL. Aging and resistance to oxidative damage in *Caenorhabditis elegans*. Proc Natl Acad Sci U S A. 1993;90(19):8905–9.
39. Li SC, Rieckmann A. Neuromodulation and aging: implications of aging neuronal gain control on cognition. Curr Opin Neurobiol. 2014;29:148–58.
40. Ljubuncic P, Reznick AZ. The evolutionary theories of aging revisited - a mini-review. Gerontology. 2009;55(2):205–16.
41. López-Otín C, Blasco MA, Partridge L, Serrano M, Kroemer G. The hallmarks of aging. Cell. 2013;153:1194–217.
42. Lushchak VI. Free radicals, reactive oxygen species, oxidative stress and its classification. Chem Biol Interact. 2014;224:164–75.
43. Lushchak VI. Free radicals, reactive oxygen species, oxidative stress and its classification. Chem Biol Interact. 2014;224:164–75.
44. Martín-Montalvo A, Villalba JM, Navas P, de Cabo R. NRF2, cancer and calorie restriction. Oncogene. 2011;30(5):505–20.
45. McCord JM, Fridovich I, dismutase S. An enzymic function for erythrocuprein (hemocuprein). J Biol Chem. 1969;244(22):6049–55.
46. McDonnell E, Peterson BS, Bomze HM, Hirschey MD. SIRT3 regulates progression and development of diseases of aging. Trends Endocrinol Metab. 2015;26(9):486–92.
47. Medvedev ZA. An attempt at a rational classification of theories of aging. Biol Rev. 1990;65:375–98.
48. Melov S, Ravenscroft J, Malik S, Gill MS, Walker DW, Clayton PE, Wallace DC, Malfroy B, Doctrow SR, Lithgow GJ. Extension of life-span with superoxide dismutase/catalase mimetics. Science. 2000;289(5484):1567–9.
49. Menger KE, James AM, Cochemé HM, Harbour ME, Chouchani ET, Ding S, Fearnley IM, Partridge L, Murphy MP. Fasting, but not aging, dramatically alters the redox status of cysteine residues on proteins in *Drosophila melanogaster*. Cell Rep. 2015;11(12):1856–65.
50. Owusu-Ansah E, Song W, Perrimon N. Muscle mitohormesis promotes longevity via systemic repression of insulin signaling. Cell. 2013;155(3):699–712.
51. Pacifici RE, Davies KJ. Protein, lipid and DNA repair systems in oxidative stress: the free-radical theory of aging revisited. Gerontology. 1991;37(1–3):166–80.
52. Pan Y. Mitochondria, reactive oxygen species, and chronological aging: a message from yeast. Exp Gerontol. 2011;46(11):847–52.
53. Paul A, Belton A, Nag S, Martin I, Grotewiel MS, Duttaroy A. Reduced mitochondrial SOD displays mortality characteristics reminiscent of natural aging. Mech Ageing Dev. 2007;128(11–12):706–16.
54. Pérez VI, Bokov A, Van Remmen H, Mele J, Ran Q, Ikeno Y, Richardson A. Is the oxidative stress theory of aging dead? Biochim Biophys Acta. 2009;1790(10):1005–14.
55. Rhee SG, Woo HA, Kang D. The role of Peroxiredoxins in the transduction of H_2O_2 signals. Antioxid Redox Signal. 2017;. [Epub ahead of print]
56. Ristow M, Schmeisser S. Extending life span by increasing oxidative stress. Free Radic Biol Med. 2011;51(2):327–36.
57. Sanchis-Gomar F. Sestrins: novel antioxidant and AMPK-modulating functions regulated by exercise? J Cell Physiol. 2013;228(8):1647–50.

58. Saran M. To what end does nature produce superoxide? NADPH oxidase as an autocrine modifier of membrane phospholipids generating paracrine lipid messengers. Free Radic Res. 2003;37(10):1045–59.
59. Schaar CE, Dues DJ, Spielbauer KK, Machiela E, Cooper JF, Senchuk M, Hekimi S, Van Raamsdonk JM. Mitochondrial and cytoplasmic ROS have opposing effects on lifespan. PLoS Genet. 2015;11(2):e1004972.
60. Scialò F, Fernández-Ayala DJ, Sanz A. Role of mitochondrial reverse Electron transport in ROS signaling: potential roles in health and disease. Front Physiol. 2017;8:428.
61. Sohal RS. Role of oxidative stress and protein oxidation in the aging process. Free Radic Biol Med. 2002;33(1):37–44.
62. Sohal RS, Agarwal A, Agarwal S, Orr WC. Simultaneous overexpression of copper- and zinc-containing superoxide dismutase and catalase retards age-related oxidative damage and increases metabolic potential in Drosophila melanogaster. J Biol Chem. 1995;270(26):15671–4.
63. Terman A. Garbage catastrophe theory of aging: imperfect removal of oxidative damage? Redox Rep. 2001;6(1):15–26.
64. Tomalin LE, Day AM, Underwood ZE, Smith GR, Dalle Pezze P, Rallis C, Patel W, Dickinson BC, Bähler J, Brewer TF, Chang CJ, Shanley DP, Veal EA. Increasing extracellular H_2O_2 produces a bi-phasic response in intracellular H_2O_2, with peroxiredoxin hyperoxidation only triggered once the cellular H_2O_2-buffering capacity is overwhelmed. Free Radic Biol Med. 2016;95:333–48.
65. Trifunovic A, Hansson A, Wredenberg A, Rovio AT, Dufour E, Khvorostov I, Spelbrink JN, Wibom R, Jacobs HT, Larsson NG. Somatic mtDNA mutations cause aging phenotypes without affecting reactive oxygen species production. Proc Natl Acad Sci U S A. 2005;102(50):17993–8.
66. van Raamsdonk JM, Hekimi S. Deletion of the mitochondrial superoxide dismutase sod-2 extends lifespan in *Caenorhabditis elegans*. PLoS Genet. 2009;5(2):e1000361.
67. Van Raamsdonk JM, Hekimi S. Reactive oxygen species and aging in *Caenorhabditis elegans*: causal or casual relationship? Antioxid Redox Signal. 2010;13:1911–53.
68. Van Raamsdonk JM, Meng Y, Camp D, Yang W, Jia X, Bénard C, Hekimi S. Decreased energy metabolism extends life span in Caenorhabditis elegans without reducing oxidative damage. Genetics. 2010;185(2):559–71.
69. Yang W, Hekimi S. A mitochondrial superoxide signal triggers increased longevity in *Caenorhabditis elegans*. PLoS Biol. 2010;8(12):e1000556.
70. Yang J, Carroll KS, Liebler DC. The expanding landscape of the thiol redox proteome. Mol Cell Proteomics. 2016;15(1):1–11.
71. Yang J, Carroll KS, Liebler DC. The expanding landscape of the thiol redox proteome. Mol Cell Proteomics. 2016;15(1):1–11. https://doi.org/10.1074/mcp.O115.056051.. Epub 2015 Oct 30
72. Zhang J, Wang X, Vikash V, Ye Q, Wu D, Liu Y, Dong W. ROS and ROS-mediated cellular signaling. Oxidative Med Cell Longev. 2016;2016:4350965.
73. Zuin A, Carmona M, Morales-Ivorra I, Gabrielli N, Vivancos AP, Ayté J, Hidalgo E. Lifespan extension by calorie restriction relies on the Sty1 MAP kinase stress pathway. EMBO J. 2010;29(5):981–91.

Stem Cells and Aging

12

Jitendra Kumar Chaudhary and Pramod C. Rath

Introduction

Aging is the most important risk factor for various human pathologies, yet it is quite challenging to study owing to its multidimensional, multifactorial, and pervasive nature, affecting the entire organ systems of the body. The process of aging is driven by progressive loss of physiological, biochemical, and molecular integrity, leading to various diseases such as cardio-cerebro vascular and metabolic disorders, neurodegenerative diseases, diabetes, cancer, and eventual death [1]. The underlying causes include accumulation of subtle, irreversible cellular and molecular changes over an individual's lifespan, leading to progressive decline in the intrinsic regenerative and homeostatic potential. Such degenerative changes coupled with homeostatic alteration lead to stem cell exhaustion, genetic instability, cellular senescence, altered cellular communication, mitochondrial dysfunction, telomere attrition, multiple epigenome and transcriptome changes, loss of proteostasis, and deregulated nutrient sensing, among others [2]. However, the most important attributable factors among them are age-dependent changes and decline in tissue-/organ-specific stem cells. Such decline in stem cell population either precedes or followed by deterioration in niches surrounding them along with functional molecular cues that regulate their various biological activities. Repertoire of stem cells present in various organs steadily decline due to replicative aging that is preceded by global metabolic deterioration resulting from patho-physiological conditions which is, in turn, driven by

---

J. K. Chaudhary (✉)
Assistant Professor, Department of Zoology, Shivaji College, University of Delhi, New Delhi, India
e-mail: jnujitendra@gmail.com

P. C. Rath (✉)
Molecular Biology Laboratory, School of Life Sciences, Jawaharlal Nehru University, New Delhi, India
e-mail: pcrath@mail.jnu.ac.in

© Springer Nature Singapore Pte Ltd. 2020
P. C. Rath (ed.), *Models, Molecules and Mechanisms in Biogerontology*,
https://doi.org/10.1007/978-981-32-9005-1_12

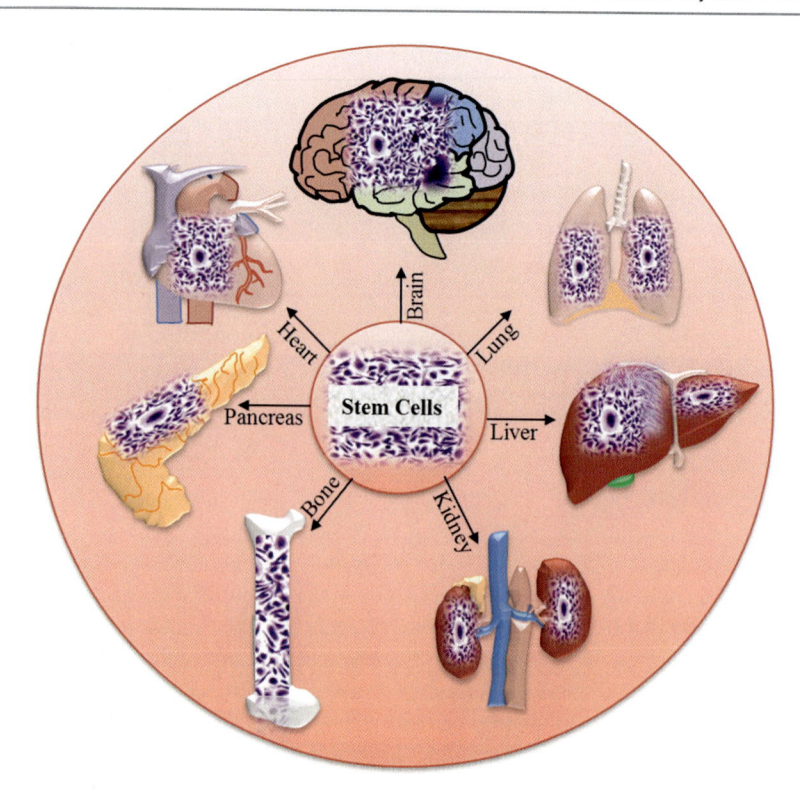

Fig. 12.1 Major locations of stem cells. Stem cells are located in almost every organs, including the brain, heart, lungs, liver, pancreas, major bones, and kidney, and help them remain functionally integrated and working throughout an individual's life. However, their associated functioning deteriorates as a function of aging, leading to physiological decline and onset of age-associated pathologies

aging process. Stem cells serve as cellular backup of various organs, continuously renewing them by supplying new daughter cells and requisite set of cytokines and growth factors required for regeneration and repair of tissues/organs so as to keep them integrated and functional throughout life span (Fig. 12.1). However, with the passage of time, stem cells also start showing multiple signs of replicative senescence and metabolic slowdown, thereby compromising on self-renewal and proliferation, cutting down the supply of new cells/secretory molecules and growth factors to the organs. These age-associated progressively deteriorating changes deprive the organ-specific niches/microenvironment of functional growth and molecules needed for mutually concerted functioning between stem cells and their respective organs [3]. The steady loss in the regenerative potential during aging is primarily attributed to telomere shortening/attrition, which flirts with the commitment of stem and non-stem cells [4], decrease in the ratio of DNA repair to damage, low level of ATP production owing to reactive oxygen species-induced (ROS) mitochondrial DNA damage, accumulation of damaged proteins and dysfunctional organelles, epigenetic and epitranscriptomic modifications, and consequent deregulated gene and protein expression, among others. The aforementioned biological changes affect stem cells as well as non-stem cell population on several accounts, which together contributes to development of age-related pathologies, and hence drive organismal aging. Such scientific correlation between stem cells and aging has been quite significant in developing our understanding to a certain extent, and hence becomes even more important to look at the multidimensional and multifactorial aging process from the stem cells perspective, which will be helpful in improving health span of organisms, including human beings. Considering all the findings in an integrated and holistic way would certainly help the ever-expanding aging society and address the issues regarding increase in age-related psychosomatic health problems confronting the human society worldwide.

Hallmarks of Stem Cell and Organismal Aging

Aging is associated with subtle but steady decline in organs' functions and structure at both microscopic and macroscopic levels, and hence substantially increases the risk factors for developing age-associated illness and diseases. The major underlying causes for such deterioration include stem cell exhaustion, genetic instability, cellular senescence, altered intercellular communication, telomere attrition, epigenetic and transcriptomic changes, mitochondrial dysfunction, immunosenescence, and loss of proteostasis (Fig. 12.2).

Stem Cell Pool, Self-Renewal, Quiescence, Terminal Differentiation, and Aging

Stem cells are very dynamic in nature, which helps them fulfill the growth, maintenance and regeneration demands of aging tissues undergoing slow but steady molecular, functional, and structural changes over an individual's lifetime. For instance, they divide very fast during fetal development so as to keep pace with tissue growth and development within evolutionary-allowed developmental timeframe. But this rapid cellular proliferation slows down considerably by stage of young adulthood, and later on variably ceases in mature mammalian tissues as they undergo quiescence, with intermittent division to maintain tissue homeostasis. In old adults, stem cells show enhanced tumor suppressor expression, possibly to avert tumorogenesis at the expense of tissue's intrinsic regeneration potential. With the advance in age, tissue repertoire of stem cells starts declining due to intrinsic and extrinsic factor-induced cellular exhaustion. Stem cell exhaustion is usually found to be age-dependent that is induced either by slow decline in self-renewal ability with age or progressive changes in the niches surrounding the pool of functional stem cells in various tissues. There have been various comprehensive studies showing age-dependent perturbed cell cycle regulation and depletion in stem cell abundance in a range of tissues, including muscles, brain, germline, liver, bone marrow,

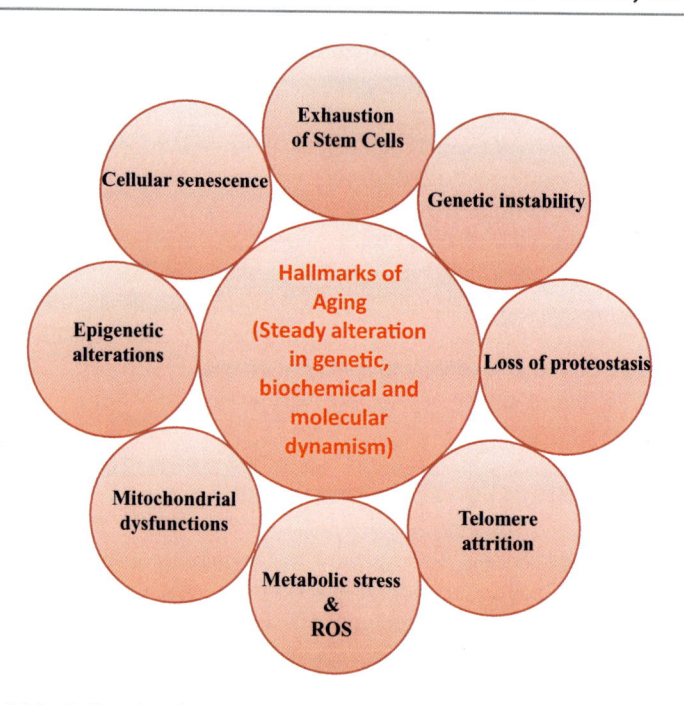

Fig. 12.2 Major hallmarks of aging. The diagram shows major hallmarks of aging, including exhaustion of stem cells, cellular senescence, genetic instability, epigenetic alterations, loss of proteostasis, mitochondrial dysfunctions, telomere attrition, metabolic stress, and reactive oxygen species (ROS)-induced changes

adipose tissues, etc. For instance, aged human brains have been found to possess significantly lower number, yet functional, of neuronal progenitor cells compared to young brains [5]. In addition, aged humans have lesser neurogenesis, but higher gliogenesis as opposed to their younger counterparts. However, there are few exceptions to the age-dependent decline of stem cells. For example, hematopoietic stem cell (HSC) populations have actually been found to increase in both number and frequency in aged mice, albeit, with reduced cell division and cell cycle progression, and higher accumulation of damaged cell cycle regulators such as p^{18} and p^{21} [6]. Furthermore, there has been empirical evidence suggesting subtle decrease in functionality of HSCs with each round of cell division [7], which may be further compounded by aging-associated DNA defects and resultant chromosomal lesions [8]. Apart from age-dependent decline in self-renewal ability, other underlying mechanisms may involve terminal differentiation, apoptosis, quiescence, differential niche-based selection pressure, and senescence of stem cells [9]. HSCs show age-dependent quiescence which, on the one hand, protects it from functional exhaustion and accumulation of damaged DNA and, on the other, promotes persistence of mutations as it allows the survival of cells with defective DNA. Surprisingly, cell cycle entry of damaged HSCs has been found to either help in DNA repair or getting the body rid of damaged and nonfunctional HSCs [10, 11]. Therefore, a precise balance between cell division and quiescence is of utmost importance for the proper functioning and maintenance of hematopoietic tissues in order to support hematopoiesis. Hematopoiesis depends on multiple intrinsic and extrinsic regulatory factors which, in turn, are being prevailed over by various pathophysiological parameters, including stress, immunity, and aging. For instance, under homeostatic conditions, hematopoiesis is usually maintained by short-term HSCs, also known as early hematopoietic progenitor cells (HPCs), while the same switches to long-term HSCs in response to stress [12]. Similarly, aged niche imposes differential selection pressure on various types of HPCs, favoring the monoclonality at the expense of natural polyclonality, which might be one of the underlying causes for higher incidence of age-associated blood-related diseases such as leukemia [13]. Besides, hematopoietic stem or progenitor cells (HSPCs) undergo age-dependent genetic alterations, including base-pair mutation, deletion, duplication, and other potentially harmful chromosomal anomalies. Aging-induced accumulation of unresolved DNA damage triggers the cell-intrinsic aged phenotype. For instance, humans under the age of 50 years show low frequency (0.2%–0.5%) of such chromosomal lesion, which drastically increases up to 2%–2.5% by the age of 80 years. Therefore, individuals over the age of 70–80 years have relatively higher risk of developing hematopoietic cancer [14]. In general, with advance in age, HSCs acquire lymphoid to myeloid lineage bias, reduced regenerative potential, and a dominant expansion of myeloid clones toward malignancies (Fig. 12.3).

Aging tends to variably destabilize the genomic integrity of almost all types of somatic and stem/progenitor cells irrespective of their location and functional specialization. Several comprehensive studies have shed light on how the parental age affects the offsprings, and their likelihood of developing genetic diseases quite early in the life. Spermatogonial stem cells or germline stem cells in aged male tend to have multiple molecular and genetic alterations which confer partisan advantage to mutant cells over their nonmutant counterparts. For example, mutations in Ras pathway cause one of parental age effect (PAE) diseases in the offspring. In *Drosophila*, there is age-dependent increase in frequency of stem cells with misaligned centrosome, contrary to centrosome orientation checkpoint, preventing such cells from division, and hence consequent decrease in sperm production. This leads to "selfish" proliferation and exponential increase of mutant spermatogonial cells over neutral mutation-carrying spermatogonial stem cells in the testes of aging men [15], which may have huge impact on the genetic make of the offspring and fertility.

Aging, slowly but steadily, tilts the balance between growth and atrophy in skeletal muscles that is empirically attributed to age-dependent loss of skeletal/muscle stem cells (SMSCs). Both young and adult maintain differential proportion of muscle stem cells with elder having less proportion owing to prevalence of lower asymmetric division. During muscle growth and development, the muscle precursor cells fuse with pre-formed muscle fibers, resulting into generation of new muscles and correspondingly increased muscle mass. Post-natal muscle development is followed by slow myonuclear turnover at around 15 years during adulthood. Human myoblast quiescence results from age-induced methylation-based alteration in sprouty 1 pathway, impairing the self-renewal potential of aged muscle stem cells. Moreover,

Fig. 12.3 Age-related changes in various stem cell populations. Aging has adverse effects on various functions of stem cells such as self-renewal, proliferation and differentiation

cellular senescence coupled with apoptosis has been proposed to be underlying mechanisms for age-based muscle stem cell loss as evidenced by higher susceptibility of aged murine muscle stem cells to in vitro apoptosis. Moreover, aged muscle stem cells tend to show considerably higher expression of myogenic differentiation markers such as Pitx2, MYH3, and MYL1, on the one hand, while having downregulation of sprouty1 and Pax7, markers of quiescent fate, on the other. Elderly muscle stem cells also possess inability to return to quiescence due to DNA methylation-induced suppression of quiescence pathways. Therefore, decreased asymmetric division/self-renewal in conjunction with higher likelihood of terminal myogenic differentiation may result into loss of reserved muscle stem cell pool in elderly and aged individuals [16]. Besides, there is an age-dependent changes in muscle stem cell niche which cause decline in their self-renewing ability due to excess proliferation in subset of satellite cells. For example, aged muscle stem cell niche, consisting of muscle fibers, expresses Fgf2, making subset of satellite cells, break quiescence, undergo proliferation-led depletion, and hence loss in long-term regeneration potential, whereas relatively dormant satellite cells have robust expression of Spry1, and that is why they do not undergo depletion, would persist even in aged muscle, and could be attributed to little regeneration occuring in aged individuals. Therefore, inhibition of Fgf2-mediated signaling by overexpressing *Spry1* has been found helpful in preserving satellite stem cell pool and stem cell functions in aged muscle [17].

Metabolic Stress, ROS Generation, Oxygen Sensitivity, and Mitochondrial Dysfunction

Long-lived tissues with minimal turnover are quite susceptible to the accumulation of oxidative, especially reactive oxygen-induced and nonoxidative damages, triggering a series of radical change in cellular phenomena, including cellular senescence, cell cycle arrest, decreased tissue-damage repair and regeneration, and eventual cell death. The damaging ROS is profoundly generated as a result of electron "leakage" during universally occurring biological process, oxidative phosphorylation, in mitochondria. The ROS-induced molecular damages can be understood in light of "free radical theory of aging" postulated by Harman in 1972 [18]. As per this theory, age-based accumulated cellular damage and compromised mitochondrial integrity cause elevated ROS production, thereby further damaging cellular macromolecules and already compromised mitochondrial oxidative phosphorylation, leading to eventual cellular decomposition and cell death. Reactive oxygen species such as superoxide (O_2^-) and hydroxyl radical ($^{\bullet}OH$) are highly reactive and consequently short-lived, damaging cellular DNA, proteins, and lipids either by direct or indirect chemical addition and/or modifications to the various functional groups present in them. ROS-induced oxidative modifications of biomolecules change their physicochemical properties, such as conformation, structure, solubility, reactivity, binding, proteolytic susceptibility, and enzyme activities. For example, 8-hydroxy-2-deoxyguanosine (8-Oxo-dG), an oxidized-derivative of deoxyguanosine (DNA), tends to show higher accumulation in aged tissues. Similarly, side chains of amino acid residues such as arginine, lysine, and proline undergo oxidative modification called protein carbonylation, a type of protein oxidation found to be highly accumulative, and reflective of cellular oxidative stress. Such chemical alteration and oxidation render these molecules nonfunctional and cause their accumulation, thereby decreasing the ratio of cell's overall functional-to-oxidized biomolecules, and consequently compromising the cellular functions.

The cellular ROS level may have different implications depending on the type of cells. For example, relatively increased ROS has been found to prolong the lifespan in *C. elegans* and yeast while it might have devastating effect on other types of eukaryotic cells. Under normal physiological conditions, ROS plays an important role in differentiation of hematopoietic stem/progenitor cells in *Drosophila* [19]. Generally, stem cells of various types and origins are differentially susceptible to damage due to elevated ROS level. Besides, experimental evidence has shown that stem cell ROS sensitivity also depends on the age of donor. For example, bone marrow and adipose tissue-derived mesenchymal stem cells (MSCs) isolated from aged donor show correspondingly increased susceptibility to ROS-induced oxidative damage [20, 21]. Moreover, there has been direct correlation between ROS level and aging in HSCs, which may damage the replicative potential and leads to their exhaustion. Therefore, containment of ROS level by overexpression of superoxide dismutase (SOD) in either stem cells or their supporting cells has been proven to prolong the stem cell functions. There are several intricately connected networks, involving forkhead box O (*FoxO*) transcription factors that are responsible for regulation of cellular oxidative stress in various stem and non-stem cells. *FoxO* transcription factors play very important role in global metabolism, proliferation, and oxidative stress by modulating the expression of a battery of genes encoding antioxidant enzymes and proteins. Therefore, they play very significant roles in maintaining appropriate oxidative state as the deletion of *FoxO1*, *FoxO3*, and *FoxO1* leads to increased ROS level in HSCs and other stem cells. This elevated ROS level, if not reduced by treatment with antioxidant such as N-acetyl-L-cysteine, depletes HSCs and neural stem cells [22]. Besides, there are several mechanisms and signaling pathways involved in regulating oxidative stress, such as polycomb family chromatin regulator and DNA damage signaling molecule ATM, and thereby maintains pool of stem cells in various tissues [23].

Apart from ROS, stem cells are also highly sensitive to cellular level of molecular oxygen (O_2) and abnormal mitochondrial functioning. The oxygen sensitivity in stem cells is accomplished by hypoxia-inducible factor 1α (*Hif1α*), a transcription factor that plays a very crucial role in stem cell function, maintenance, and aging. Under normoxic condition, the cellular level of Hif1α is kept low by its continuous E3 ubiquitin ligase, von Hippel Lindau (VHL)-mediated ubiquitination and proteosomal degradation. However, hypoxia causes stabilization of Hif1α which, in turn, activates transcription of a range of hypoxia survival genes such as, glucose transporter, heat shock protein (HSP), and glycolytic enzymes. Several HSCs and neural stem cells (NSCs), which are naturally and anatomically located in hypoxic microenvironment, have had relatively stable Hif1α for their maintenance and survival. That is why the dentate gyrus NSCs and HSCs, deficient for Hif1α, rapidly deplete during aging. Surprisingly, overstabilization of Hif1α does not rescue HSCs either, rather impedes their function, indicating the need of precise control of Hif1α level for stem cell maintenance and functions [24].

Mitochondrial dysfunctions can occur due to multiple factors, including error-prone mitochondrial DNA polymerase, abnormal constituent protein in mitochondria, increased ROS level, and impaired biogenesis. Mitochondrial DNA (mtDNA) is relatively more susceptible to mutation and deletion-based alteration as it lacks association of protective histones and lesser capability of DNA damage repair compared to nuclear DNA. Comprehensive analysis of single cell-derived mitochondria shows domination of state of homoplasmy over heteroplasmy, that is, simultaneous existence of both mutant and wild-type genomes within the same cell, as a function of age. This shift in homoplasmy-to-heteroplasmy status is dominated by mutant mitochondria over normal mitochondrial genome, considerably increasing the mutational load in aging cells. Careful consideration of available findings shows aging in cells is predominantly induced by erroneous mtDNA replication. There is a functional correlation between mitochondrial dysfunctioning and aging-associated phenotypes. However, mitochondria have also evolved certain protective and defensive mechanisms over a period of time to neutralize the oxidative damages to certain extent. For instance, to avert ROS-induced oxidative damage, mitochondria activate a range of ROS-detoxifying enzymes such as superoxide dismutase (SOD) and GPx1 with the help of peroxisome proliferator-activated receptor gamma co-activator 1 (PGC-1). In addition, PGC-1 promotes oxidative phosphorylation and mitochondrial biogenesis. This empirical evidence is further supported and strengthened by the fact that overexpression of PGC-1 delays the onset of age-related changes in intestine, increases tissue homeostasis and, more importantly, lengthens the lifespan in model organism like *Drosophila* [25].

Telomere Dysfunction

There is a direct correlation between extent of DNA damage and age, which, in turn, compromises genomic integrity and functionality. Though DNA damage can occur over any portion of chromosomes, however, some distinct regions on chromosomes are more susceptible compared to others. For instance, age-related chromosomal deterioration/attrition is more profound over the repetitive TTAGGG terminal ends of each chromosome, referred to as telomere. Telomeres protect genetic information and chromosome terminals from erosion and damage but they are shortened a bit each time cell divides as replicative DNA polymerases lack the capacity of replicating the chromosomal ends. Though nature has endowed each cell with a gene encoding the specialized reverse transcriptase, called telomerase for this function, but it expresses only in few cell types, including embryonic and adult stem cells. This leads to a steady and progressive loss of telomere-protective sequence located at either ends of each chromosome. The immediate question comes to mind, "why telomere shortening and/or breaking is not followed by the repair by DNA damage repair machinery? The reason being that each chromosome's telomeres are protectively occupied by a distinct multiprotein complex called as "shelterin." The main function of which is to deny DNA repair machinery access to telomeres. Otherwise, telomeres would be "repaired" immediately after DNA breaks leading to chromosome fusion and even more harmful consequences [26]. Heterozygous telomerase mutations potentially cause defect in organ regeneration and cancer development in humans. Telomere exhaustion underlies in vitro restricted proliferative potential that leads to replicative senescence or "Hayflick limit" [27]. Patients with telomerase deficiency or short telomeres are likely to develop aplastic anemia, dyskeratosis congentia, cirrhosis, and pulmonary fibrosis. Importantly, telomere shortening also precedes normal aging both in mice and in human.

Dynamic telomere length in cells is determined by ratio of functional telomerase to mitotic division-dependent telomere erosion. Unlike adult somatic cells, embryonic and adult stem cells have functional expression of telomerase that resist the telomere shortening as they keep elongating telomeres, albeit, to a certain and limited extent, making them better at balancing this ratio for longer duration. However, as a result of aging, telomeres shortening occurs even in the various stem cells, restricting their proliferation-based regenerative potential in aging individuals. For example, age-dependent telomere attrition occurs in HSCs and intestinal epithelium cells, which is further accelerated in case of chronic diseases. This is possibly linked to decreased regenerative potential, tissue malfunction, and increased incidence of diseases in older human population worldwide. Telomere shortening in HSCs results into accelerated aging-based imbalance in HSCs pool due to alteration in stem cell environment and differentiation-inducing checkpoints, which lead to the loss of self-renewing lymphoid-biased HPSCs [28]. Typically, HSCs aging is characterized by prevalence of myeloid-biased HPSCs over lymphoid-forming HPSCs. This age-dependent lineage biasness in HPSCs results in increased myeloid cells at the expense of lymphoid cells, and is likely to cause increased susceptibility to infections and other diseases (Fig. 12.3). In addition, telomere dysfunction induces defects in mRNA splicing which leads to a cascade of molecular events responsible for strong positive selection during human aging [29].

Epigenetic Alteration

Epigenetics refers to heritable trait which results from chromosomal changes without alterations in DNA sequences per se. Epigenetics determines the regulation of genes, and is involved in many cellular processes. Epigenetic-based gene regulation precisely controls the fate of cells to a great extent and is one of the several mechanisms responsible for cellular differentiation, and hence, formation of various types of cells such as neurons, liver cells, heart cells, pancreatic cells, and so on. Such alterations in the epigenome have been found to be age-dependent. DNA methylation, that adds methyl group to DNA, is one among several ways to bring about epigenetic changes. Comprehensive studies aimed at such chemical addition have shown hypermethylation both globally and at CpG islands in normally aging tissues. Upon enforced proliferation, HSCs show global hypomethylation, indicating the possibilities of either hypermethylation preceding HSCs aging or vice versa [30]. Aged mouse HSCs show unusual DNA methylation of genes involved in self-renewal and differentiation, imparting aging-characteristic phenotypes. Moreover, some distinct CpG islands undergo increased methylation as individual ages, which leads to myelodysplastic syndrome (MDS), and eventually to acute myeloid leukemia (AML) [31]. The alterations in epigenome are accomplished by multiple types of epigenetic regulators such as DNA methyl transferase (DNMT), Tet methylcytosine dioxygenase 2 (TET2), and additional sex combs-like 1 (ASXL1) among others. Meticulous analysis of reported works invariably shows mutations in the abovementioned epigenetic regulators underlie the earliest genetic changes in neoplastic progression. A recent study in mice showed enhanced self-renewal and impaired differentiation of HSCs following biallelic knockout of DNMT3a, a family member of DNMT, which predisposes them to hematological disorders such as MDS and AML [32]. This suggests that the presence of functional DNMT3a suppresses the set of self-renewal genes, including β-catenin and Runx1, by methylation and thereby regulates several cellular processes [33].

Aging of skeletal muscle is characterized by reduction in mass and strength of muscle owing to quantitative and qualitative deterioration and decline in contractile myofibers [34], and, therefore, substantial reduction in its intrinsic regeneration potential [35]. Under normal physiological conditions, new muscle tissue formation occurs following fusion of cellular progenies formed by asymmetric division in muscle stem cells (MuSCs), also popularly known as "satellite cells" owing to their sublaminar location and juxtaposed association with the plasma membrane of myofibers. The remaining half of progenies constitutes the pool of muscle precursor cells. However, aging causes diminution of such muscle stem cell pool. The reason for depletion in muscle stem cells pool encompasses age-dependent cellular senescence, impaired self-renewal, or death. These mechanisms, leading to MuSCs decline, depend on both intrinsic as well as extrinsic factors as evidenced from partial restoration of proliferation and differentiation capacities following exposure to young environment or to growth factors. In addition, it also suggests the likelihood of reversal of aging-induced cessation of self-renewal and differentiation potential [36]. During human myoblast quiescence, methylation suppresses sprouty1 pathways, involved in quiescence regulation. MuSCs, isolated from old mice, showed elevated repressive H3K27me3 marking on histone proteins genes, which would otherwise remain unmethylated in younger counterpart. A recent study has shown existence of different epigenetic stress response in satellite cells isolated from young and aged mice. Aged mice were found to have drastic induction of active chromatin marks both at site-specific and global locations, resulting in specific induction of Hoxa9 gene. The Hoxa9 gene, in turn, leads to activation of satellite stem cell function through activation of various pathways, including TGFβ, JAK/STAT, Wnt, and senescence signaling, indicating altered epigenetic stress response in activated MuSCs, and the consequent limited satellite stem cell-based muscle regeneration characteristic of aged muscle [37].

Age-Dependent Enhancement in Replication Stress in Stem Cells

Replication stress is highly complex nuclear phenomenon with wide range of effects on genome stability, cellular proliferation, and differentiation, resulting in multiple human diseases. Replication stress response could be triggered following multiple changes such as generation of single-stranded DNA containing aberrant replication fork structure, aneuploidy, chromosomal instability (CIN), genomic instability (GIN), and so on. Replication stress is predominantly mediated by the kinase ATM and Rad3-related (ATR) pathways. Adult stem cells are sensitive to replication-based stress, which is further compounded in wake of rapid demand for cell division. Therefore, replication stress and DNA damage triggered by multiple ways, including burst of oxidative stress, leads to accumulation of genetic alterations and exponential increase in genomic aberrations in aging stem cells. There are various underlying causes with varying implications and effects, including nicks, gaps, stretch of ssDNA, unrepaired DNA lesions, short hairpins, and DNA triplexes, among others. A recent study found increase in phosphorylated form of the variant histone H2AX (γH2AX) foci, indicative of DNA damage, as a function of time. Similarly, there is a direct correlation between age and DNA break as was shown using alkaline comet assay on the basis of experiments performed in purified HSCs from animals of different age groups [38]. Quiescent HSCs show high accumulation of all sorts of DNA damages compared to their proliferating counterparts owing to attenuation of DNA damage response and repair pathways. The underlying reason is aging-induced lower expression of mini-chromosome maintenance (MCM) helicase components. Furthermore, this leads to compromised HSC functions in response to various hematopoietic stresses, such as extensive blood loss and pathophysiological inflammation. Such stresses demand compensatory and dramatic increase in proliferation of quiescent or slow-proliferating HSCs, which is likely to result in increased DNA damage, and eventually, bone marrow failure [39]. One of the mechanisms leading to clonal dominance of mutant HSCs is thought to be replication stress, leading to higher susceptibility to developing diseases.

Aneuploidy has had several consequences on aging of stem cells. One such consequence is telomeric replication stress which, in turn, causes DNA damage at telomeres and consequent p53 activation. Telomerase-deficient mice show p53/RB-dependent depletion of hematopoietic stem cells. Contrary to this, endogenous telomerase expression in HSCs ensured alleviation of aneuploidy-induced replication stress and others. Therefore, telomerase plays a very crucial role in rescuing murine HSCs from aneuploidy-induced replication stress at telomeres and aneuploidy-induced senescence (AIS) and cell depletion, suggesting its suppressive role in telomere dysfunction-induced CIN, on the one hand, and ensuring replicative potential in aging stem and progenitor cells, on the other [40].

Age-Induced Shift in Proteostasis Equilibrium Drives Stem Cell Aging

Among multiple aging-inducing factors, impaired protein homeostasis, also known as proteostasis, has very important role to play as misfolded proteins can form toxic aggregates, disrupt membrane system, and thereby cause cell death and diseases. Under normal physiological conditions, protein synthesis occurs in a precisely controlled and concerted fashion through spatiotemporal control of ribosome biogenesis, recruitment, and loading, leveraging array of quality control molecular mechanisms. Proteostasis encompasses stabilization of correctly folded proteins, on the one hand, and removal of misfolded, damaged, aggregate, and unneeded ones through proteosome-based degradation, on the other [41]. However, with aging and age-induced extrinsic and intrinsic molecular alterations, there is an untoward shift in equilibrium leading to deranged proteostasis, which poses great risk of developing age-related diseases, such as Parkinson's disease, Alzheimer's disease, diabetes, cataracts, and Huntington's disease, among others [42, 43].

So, what does maintain proteostasis? There are two important global mechanisms employed by cells in order to maintain functional status of proteome in the cells, namely, chaperone-mediated protein folding and stability, and proteolytic system. Chaperones are protein molecules that provide assistance to proteins and other biomolecules at the various levels, including folding, unfolding, assembly, and disassembly, as well as prevent newly synthesized polypeptide chain to form nonfunctional aggregate with preassembled subunits. Heat-shock protein family is one type of cellular chaperone that responds to stress-induced protein denaturation. However, this weakens and impairs substantially during aging in various cells, including aged stem cells [44]. As a result, accumulation of damaged (carbonylation and glycation) and misfolded proteins increases in aging individual, also suggestive of decline in capacity to maintain protein homeostasis. Overexpression and upregulation of chaperone (heat shock proteins) and co-chaperone have been found to considerably lengthen the lifespan, indicating their potential involvement in proteostasis, hence conferring functional dynamism at both cell and organism levels [45]. Moreover, transcriptional activation of transcription factor HSF-1, master regulator of heat-shock response, has been found to increase the thermo-tolerance and longevity in several nematodes [46]. Similarly, in mammalian cells, transactivation of heat shock genes, including Hsp70, was found to be considerably enhanced following deacetylation of HSF-1 by SIRT1, and downregulation of SIRT1 expectedly reduces the response [47]. In addition, pharmacological induction of Hsp72 has been found to delay progression of dystrophic pathology and preserve muscle function in mouse model of muscular dystrophy, suggesting another approach to restore proteostasis by activating protein folding and stability [48].

Cellular protein quality control is accomplished through proteolytic system, which encompasses two components, namely, ubiquitin proteasome system and the autophagy–lysosomal system. However, the working of this system declines with aging. Treatment of human cultured cells with either proteasome activator or deubiquitylase inhibitors enhances the clearance and disposal of toxic proteins [49], and, in yeasts extends the replicative life. Similarly, in nematodes, epidermal growth factor (EGF) signaling-induced expression of ubiquitin–proteasome system increases the lifespan [50]. Autophagy plays very important role in maintaining protein homeostasis as evidenced from improved hepatic function in transgenic mice expressing an extra copy of the chaperone-mediated autophagy receptor LAMP2a. Similarly, induction of autophagy by regular administration of mTOR inhibitor, rapamycin, has lifespan-extending effects on yeast, flies, nematode, and mice [51]. Deletion of Atg7 or Fip200, involved in autophagy, causes rapid depletion of HSCs, indicating the crucial role of autophagy in HSCs maintenance [52]. FoxO transcription factor transcriptionally activates the expression of chaperone and thereby helps promote the longevity and stem cell functions. Therefore, all the above evidence suggests that perturbed proteostasis can further precipitate age-associated risk and pathology and, therefore, leads to several degenerative diseases, which can be controlled by developing deeper insight-based intervention strategy.

Nutrient Sensing and Changes in Nutrition Affect Stem Cell Functions

Ever-expanding research on interrelationship between nutrient and stem cell functions has shown lots of promising results. Caloric restriction, that is, substantial reduction in food intake without causing malnutrition, has been found to be helpful in delaying onset of age-associated degenerative diseases and extends lifespan, partly by influencing the function of stem cells. For example, in a rodent model study, caloric restriction was found to enhance proliferation of progenitor cells, increase survival of newly formed astrocytes and neurons, and thereby promote neurogenesis in the dentate gyrus [53]. Extending caloric restriction study in *Drosophila* has shown similar result, wherein reduction of age-associated germline stem cells was found to be reduced. The underlying mechanisms of age-lengthening effects of nutrients through stem cell-dependent functions are yet to be clearly elucidated. However, reasons underlying the nutrient-induced changes may be the expression of systemic factors which, in turn, regulate stem cell functions as is evidenced from loss of intestinal stem cells and male germline stem cells following protein starvation in *Drosophila*. Reduction in stem cells occurs because of decreased expression of insulin-like peptides in brain, which can be overcome by constitutive expression of active insulin receptor, indicating direct involvement of insulin in maintenance of germline stem cells [54].

Among multiple pathways underlying the calorie-induced beneficial effects, target of rapamycin (TOR) signaling plays a very crucial role. The downstream effect of TOR signaling is protein synthesis and cell growth. TOR, a conserved serine/threonine kinase, is activated by multiple factors, including amino acids, nutrients, growth factors, etc. [55]. Reduced TOR signaling may slow down aging and extends lifespan of organism, probably by increasing the proliferation and functioning of stem cells of various organs. However, further study would be needed to make conclusive statement in this regard.

Ex Vivo Stem Cell Aging

Aging of stem cells in vitro reflects the process of in vivo aging to a large extent, especially with respect to phenotypic features and molecular mechanisms. The mechanism underlying in vitro aging of stem cells shows a lot of species-specific and individual-specific variations. For instance, telomere shortening drives cellular senescence in cultured human cells, which is not reported in rodent cells following the trajectory of replicative senescence [56]. Almost all types of cells, including stem cells, irrespective of their source of origin, undergo aging during culture, called ex vivo aging. Expansion of cell through in vitro culture is limited to a certain number of cell division due to replicative senescence. Thereafter, cultured cells undergo cell cycle arrest, increases in size, and acquire "fried egg" morphology. Comprehensive molecular analysis of such cells shows aberrant alteration with respect to transcriptomics, epigenomics, and secretory profile, suggesting usefulness of replicative senescence for quality control of cell preparation for downstream therapeutic application [57]. Replicative senescence is, among others, induced by ever-changing DNA methylation landscape over the course of culture. However, it is not yet known as to what regulates age-specific methylation pattern, and how much significant role they play?

Mesenchymal stem cells show increased aging phenotype, such as decline in proliferation potential due to replicative senescence, downregulation of self-renewal-associated genes, such as Oct4, Sox2, and TERT, increased tendency for osteogenic differentiation, following repeated passage and long-term culture [58]. The actual age of culture is relied upon population doubling (PD). MSCs, derived from single-cell-based colony, can be expanded up to as many as 50 PD in about 10 weeks, which starts showing signs of senescence thereafter. The duration of PDs increases with increase in cell passage, suggesting decrease in proliferative potential as cell ages. The physical characteristics, reflective of cell aging such as cytoplasmic granules and debris, also increase with age of cultured cells. Telomere shortening usually occurs at the rate of around 50 bp over each passage which might differ according to type of cultured stem cells, medium, growth factors, etc. Telomere shortening destabilizes chromosome integrity, which affects expression dynamics of several genes. Cells at latter passages have also been found to have compromised lineage differentiation potential. In addition, the rate of senescence of stem cells during culture also depends on the age and health of donor. Aged in vitro cells also show substantial shift toward cancerous cellular morphology such as morphological transformation of elongated and adherent cells into round, nonadherent type. For instance, human adipose tissue-derived MSCs show spontaneous transformation into small, clustered aggregations, displaying chromosomal aberrations at the rate of around 50% [59]. In addition, various types of stem cells, including bone marrow–derived endothelial progenitor cells (EPCS), show decrease in clonogenicity and increased tendency to acquire round-shape morphology with aging [60]. There are various ways of measuring cellular senescence such as β-galactosidase assay, measuring level of senescence-associated genes, p21 and p16 among others. Although, these methods reveal and help quantify cellular senescence but they fail to unveil the underlying cellular mechanisms, leading to cell cycle arrest and cell senescence. Therefore, considering all the findings, ex vivo cell culture model could be tremendously useful in deepening our understanding of molecular pathways, and designing appropriate aging-lowering clinical intervention and reducing age-associated risk factors and pathologies.

Exercise Induces Stem Cell Functions and Slows Down Aging Process

Regular physical exercise has been proven to have many beneficial effects on functioning of tissues and organs by promoting the activities of resident stem cells. For example, neural stem cells were found to increase and so does the cognitive parameter, including learning and memory, following exercise in an experimental study involving mice and human [61]. In addition, an experimental group of voluntarily running mice showed stimulation of cell proliferation and consequent neurogenesis in the hippocampal region of the brain. Moreover, exercise also induces the number of neural stem cells apart from neurogenesis in subventricular zone of forebrain. This physical exercise-induced stem cell stimulation, proliferation, and functions are mediated by multiple factors and corresponding signaling pathways. In a rodent study, expression of insulin-like growth factor-1 (IGF1) and growth hormone were substantially activated following exercise, which upon binding to their respective receptor may induce the beneficial neurogenesis activity as the same effect was not found in growth hormone-deficient mice [62]. IGF1, mainly produced by liver, is actively taken up by specific group of neurons, resulting into their proliferation and adaptive responses, which can be blocked by administration of antibody against IGF1. Even subcutaneous administration of IGF1 is found to be sufficient for neurogenesis in dentate gyrus [63]. Considering aforementioned empirical evidences, physical exercise may play very important role in activating the stem cell functions and thereby slow down aging process.

Role of p53 in Aging of Stem Cells

Among tumor suppressor genes, p53 has been known to play very important role in a range of cellular activities, including cell cycle arrest and apoptosis. Throughout life of organisms, there is an accumulation of various forms of DNA damage owing to generation of reactive oxygen species (ROS), exposure to a range of potential mutagens, error-prone DNA replication, etc. These damages, if left unrepaired, may lead to several diseases such as cancer. However, our cells are endowed with a range of reparative potential which is mediated through p53-dependent mechanisms. Depending upon extent of DNA damage, p53 triggers different responses which would either repair the damage or lead the cell on the path of senescence and cell death, and that is how it helps preserve the genomic integrity and resist the development of diseases. In addition, p53 is involved in a range of very crucial cellular activities, such as cell cycle regulation, maintaining conducive cellular redox state, and various metabolic processes. The significance of DNA repair process can be observed in case of segmental progeria syndrome—rare human disease characterized by premature aging phenotypes such as skin atrophy, cataracts, osteoporosis, heart diseases, cerebellum degeneration, hair graying, immunodeficiency, cancers, and consequent reduced lifespan. This is triggered off by impaired DNA repair processes owing to loss of function in DNA damage signaling protein ATM and RecQ DNA helicase WRN [64]. Furthermore, loss of DNA damage signaling protein, ATM has been implicated in depletion of HSCs and heightened loss of melanocyte stem cells following low dose radiation. There has been similar and consistent progeroid phenotypic display and accelerated aging in mice carrying mutated genes associated with human progeroid syndromes [65], further reinforcing and consolidating the empirical evidences, which suggest unrepaired DNA damage as a fundamental cause of aging. Moreover, mutation in p53 is a very commonly observed phenomenon in various types of cancer, indicating its pervasive role in growth, development, and diseases [66].

Apart from being involved in normal cells, p53 also plays a very crucial role in promoting stem cell-based tissue regeneration, repair, and homeostasis by maintaining functional genomic integrity. For example, p53 has been found to regulate cell division, differentiation, and chromosomal stability in mouse olfactory bulb stem cells [67]. In addition, p53 has been found to regulate cell division polarity by restoring asymmetric division and, thus, the self-renewing potential of mammary stem cells. Similarly, there are other types of stem cell and progenitor cell populations whose proliferation and differentiation are regulated in a similar manner [68–70]. Therefore, p53-mediated DNA repair and other regulatory pathways help restore DNA damage, maintain stem cell repertoire, and thereby delay cellular aging along with substantial reduction in the rate of organismal aging (Fig. 12.4).

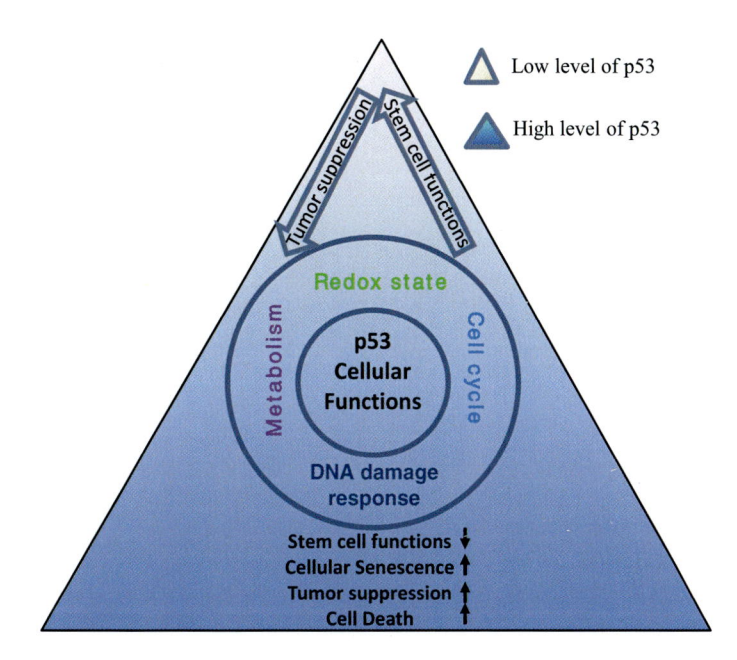

Fig. 12.4 Cellular expression of p53 and its effect on various stem cell functions. P53 plays a very crucial role in many vital cellular functions such as cell cycle regulation, metabolism, maintaining conducive redox state, and DNA damage response, among others. Elevated level of p53 (bottom of pyramid) represses stem cell proliferative functions while inducing cellular senescence, tumor suppression, and cell death. On the other hand, its lower expression level promotes stem cell proliferation and associated functions at the cost of increased incidence of tumorogenesis (top of pyramid)

Discussion and Conclusions

Aging is characterized by a gradual decline and deterioration in tissue homeostasis, which is attributed to age-associated impairments in tissue function, intrinsic regeneration, and developmental potential. Generally, tissue homeostasis is maintained by balancing the ratio of tissue damage to tissue repair, leading to continuous renewal of structural and functional aspects of organs over time. Under normal physiological conditions, adult stem cells, residing in various tissues and organs, carry out such renewal and repair processes as and when need arises. Although, there is no doubt whether age-dependent decline in tissue regenerative potential occurs, but an important question arises—is it because of intrinsic aging of stem cells or impairment of stem cell function in aged microenvironment or both? Answer seems both ways considering all the empirical evidences. Unraveling the underlying molecular mechanisms and their functional integration will be critical in designing stem cell-based therapeutic applications with regard to aging intervention, tissue injury, and age-associated degenerative diseases. Though there are a lot of commonalities regarding molecular pathways and their regulators between stem and non-stem cells, many of them are unique to former only. This suggests that underlying aging mechanisms in various cells is similar on many accounts and some are unique to stem cells only. For instance, common hallmarks of aging phenotypes both in stem cells and in non-stem cells include ROS production, DNA damage, telomere attrition, proteotoxicity, and aberrant changes in epigenetic landscape, among others.

Over the past decades or so, there have been tremendous progresses in understanding of mechanisms underlying the molecular control of stem cell functions during normal and pathophysiological conditions. Furthermore, emerging empirical evidences from range of stem cells distinctly unravel prevalence of many overlapping mechanisms and their differential regulations in different microenvironments during the aging. This ever-expanding multidimensional knowledge on interrelationship of stem cells and aging will help us in strengthening the currently existing regime of clinical intervention in aging as well as designing new treatment which would help improve the stem cell-based tissue homeostasis and regeneration and thereby reduce the incidence of age-associated diseases in the ever-expanding aged human population worldwide.

Acknowledgments This work was supported by UGC-RNRC, -DRS; DST-FIST and -PURSE grants to SLS & PCR and ICMR-JRF & SRF to JKC.

Conflict of Interest The authors declare no conflict of interest.

References

1. Niccoli T, Partridge L. Ageing as a risk factor for disease. Curr Biol. 2012;22:R741–52.
2. López-Otín C, Blasco MA, Partridge L, Serrano M, Kroemer G. The hallmarks of aging. Cell. 2013;153(6):1194–217.

3. Justin V, Leanne Jones D. Stem cells and the niche: a dynamic duo. Cell Stem Cell. 2010;6(2):103–15.
4. Allsopp R. Short Telomeres Flirt with Stem Cell Commitment. Cell Stem Cell. 2013;12(4):383–4.
5. Macas J, Nern C, Plate KH, Momma S. Increased generation of neuronal progenitors after ischemic injury in the aged adult human forebrain. J Neurosci. 2006;26:13114–9.
6. Noda S, Ichikawa H, Miyoshi H. Hematopoietic stem cell aging is associated with functional decline and delayed cell cycle progression. Biochem Biophys Res Commun. 2009;383:210–5.
7. Beerman I, Bock C, Garrison BS, Smith ZD, Gu H, Meissner A, Rossi DJ. Proliferation-dependent alterations of the DNA methylation landscape underlie hematopoietic stem cell aging. Cell Stem Cell. 2013;12:413–25.
8. Flach J, Bakker ST, Mohrin M, Conroy PC, Pietras EM, Reynaud D, Alvarez S, Diolaiti ME, Ugarte F, Forsberg EC, et al. Replication stress is a potent driver of functional decline in ageing haematopoietic stem cells. Nature. 2014;512:198–202.
9. Juhyun O, Lee YD, Wagers AJ. Stem cell aging: mechanisms, regulators and therapeutic opportunities. Nat Med. 2014;20(8):870–80.
10. Beerman I, Seita J, Inlay MA, Weissman IL, Rossi DJ. Quiescent hematopoietic stem cells accumulate DNA damage during aging that is repaired upon entry into cell cycle. Cell Stem Cell. 2014;15:37–50.
11. Walter D, Lier A, Geiselhart A, Thalheimer FB, Huntscha S, Sobotta MC, Moehrle B, Brocks D, Bayindir I, Kaschutnig P, et al. Exit from dormancy provokes DNA-damage-induced attrition in haematopoietic stem cells. Nature. 2015;520:549–52.
12. Busch K, Klapproth K, Barile M, Flossdorf M, Holland-Letz T, Schlenner SM, Reth M, Höfer T, Rodewald HR. Fundamental properties of unperturbed haematopoiesis from stem cells in vivo. Nature. 2015;518:542–6.
13. Vas V, Senger K, Dörr K, Niebel A, Geiger H. Aging of the microenvironment influences clonality in hematopoiesis. PLoS One. 2012;7(8):e42080.
14. Jacobs KB, Yeager M, Zhou W, Wacholder S, Wang Z, Rodriguez-Santiago B, Hutchinson A, Deng X, Liu C, Horner MJ, et al. Detectable clonal mosaicism and its relationship to aging and cancer. Nat Genet. 2012;44:651–8.
15. Goriely A, Wilkie AO. Paternal age effect mutations and selfish spermatogonial selection: causes and consequences for human disease. Am J Hum Genet. 2012;90:175–200.
16. Bigot A, Duddy WJ, Ouandaogo ZG, Negroni E, Mariot V, Ghimbovschi S, et al. Age-associated methylation suppresses SPRY1, leading to a failure of re-quiescence and loss of the reserve stem cell pool in elderly muscle. Cell Rep. 2015;13(6):1172–82.
17. Chakkalakal JV, Jones KM, Basson MA, Brack AS. The aged niche disrupts muscle stem cell quiescence. Nature. 2012;490:355–60.
18. Harman D. Free radical theory of aging: dietary implications. Am J Clin Nutr. 1972;25:839–43.
19. Owusu-Ansah E, Banerjee U. Reactive oxygen species prime Drosophila haematopoietic progenitors for differentiation. Nature. 2009;461:537–41.
20. De Barros S, Dehez S, Arnaud E, Barreau C, Cazavet A, Perez G, Galinier A, Casteilla L, Planat Benard V. Agingrelated decrease of human ASC angiogenic potential is reversed by hypoxia preconditioning through ROS production. Mol Ther. 2013;2013(21):399–408.
21. Stolzing A, Jones E, McGonagle D, Scutt A. Agerelated changes in human bone marrow derived mesenchymal stem cells: consequences for cell therapies. Mech Aging Dev. 2008;129:163–73.
22. Renault VM, Rafalski VA, Morgan AA, Salih DA, Brett JO, Webb AE, Villeda SA, Thekkat PU, Guillerey C, Denko NC, et al. FoxO3 regulates neural stem cell homeostasis. Cell Stem Cell. 2009;5:527–39.
23. Maryanovich M, Oberkovitz G, Niv H, Vorobiyov L, Zaltsman Y, Brenner O, Lapidot T, Jung S, Gross A. The ATM-BID pathway regulates quiescence and survival of haematopoietic stem cells. Nat Cell Biol. 2012;14:535–41.

24. Takubo K, Goda N, Yamada W, Iriuchishima H, Ikeda E, Kubota Y, Shima H, Johnson RS, Hirao A, Suematsu M, Suda T. Regulation of the HIF-1alpha level is essential for hematopoietic stem cells. Cell Stem Cell. 2010;7:391–402.
25. Rera M, Bahadorani S, Cho J, Koehler CL, Ulgherait M, Hur JH, Ansari WS, Lo T Jr, Jones DL, Walker DW. Modulation of longevity and tissue homeostasis by the Drosophila PGC-1 homolog. Cell Metab. 2011;14:623–34.
26. Palm W, de Lange T. How shelterin protects mammalian telomeres. Annu Rev Genet. 2008;42:301–34.
27. Hayflick L, Moorhead PS. The serial cultivation of human diploid cell strains. Exp Cell Res. 1961;25:585–621.
28. Wang J, Sun Q, Morita Y, Jiang H, Gross A, Lechel A, Hildner K, Guachalla LM, Gompf A, Hartmann D, et al. A differentiation checkpoint limits hematopoietic stem cell self-renewal in response to DNA damage. Cell. 2012;148:1001–14.
29. Colla S, Ong DS, Ogoti Y, Marchesini M, Mistry NA, Clise-Dwyer K, Ang SA, Storti P, Viale A, Giuliani N, et al. Telomere dysfunction drives aberrant hematopoietic differentiation and myelodysplastic syndrome. Cancer Cell. 2015;27:644–57.
30. Beerman I, Bock C, Garrison BS, Smith ZD, Gu H, Meissner A, Rossi DJ. Proliferation-dependent alterations of the DNA methylation landscape underlie hematopoietic stem cell aging. Cell Stem Cell. 2013;12:413–25.
31. Maegawa S, Gough SM, Watanabe-Okochi N, Lu Y, Zhang N, Castoro RJ, Estecio MR, Jelinek J, Liang S, Kitamura T, et al. Age-related epigenetic drift in the pathogenesis of MDS and AML. Genome Res. 2014;24:580–91.
32. Mayle A, Yang L, Rodriguez B, Zhou T, Chang E, Curry CV, Challen GA, Li W, Wheeler D, Rebel VI, Goodell MA. Dnmt3a loss predisposes murine hematopoietic stem cells to malignant transformation. Blood. 2015;125:629–38.
33. Challen GA, Sun D, Mayle A, Jeong M, Luo M, Rodriguez B, Mallaney C, Celik H, Yang L, Xia Z, et al. Dnmt3a and Dnmt3b have overlapping and distinct functions in hematopoietic stem cells. Cell Stem Cell. 2014;15:350–64.
34. Nilwik R, Snijders T, Leenders M, Groen BBL, van Kranenburg J, Verdijk LB, van Loon LJC. The decline in skeletal muscle mass with aging is mainly attributed to a reduction in type II muscle fiber size. Exp Gerontol. 2013;48:492–8.
35. Pollina EA, Brunet A. Epigenetic regulation of aging stem cells. Oncogene. 2011;30:3105–26.
36. Collins CA, Zammit PS, Ruiz AP, Morgan JE, Partridge TA. A population of myogenic stem cells that survives skeletal muscle aging. Stem Cells. 2007;25:885–94.
37. Schworer S, et al. Epigenetic stress responses induce muscle stem-cell ageing by *Hoxa9* developmental signals. Nature. 2016;540:428–32.
38. Beerman I, Seita J, Inlay MA, Weissman IL, Rossi DJ. Quiescent hematopoietic stem cells accumulate DNA damage during aging that is repaired upon entry into cell cycle. Cell Stem Cell. 2014;15:37–50.
39. Walter D, Lier A, Geiselhart A, Thalheimer FB, Huntscha S, Sobotta MC, Moehrle B, Brocks D, Bayindir I, Kaschutnig P, et al. Exit from dormancy provokes DNA-damage-induced attrition in haematopoietic stem cells. Nature. 2015;520:549–52.
40. Meena JK, Cerutti A, Beichler C, et al. Telomerase abrogates aneuploidy-induced telomere replication stress, senescence and cell depletion. EMBO J. 2015;34(10):1371–84.
41. Koga H, Kaushik S, Cuervo AM. Protein homeostasis and aging: the importance of exquisite quality control. Ageing Res Rev. 2011;10:205–15.
42. Ross CA, Poirier MA. Protein aggregation and neurodegenerative disease. Nat Med Suppl. 2004;10:S10–7.
43. Powers ET, Morimoto RI, Dillin A, Kelly JW, Balch WE. Biological and chemical approaches to diseases of proteostasis deficiency. Annu Rev Biochem. 2009;78:959–91.
44. Calderwood SK, Murshid A, Prince T. The shock of aging: molecularchaperones and the heat shock response in longevity and aging—amini-review. Gerontology. 2009;55:550–8.

45. Swindell WR, Masternak MM, Kopchick JJ, Conover CA, Bartke A, Miller RA. Endocrine regulation of heat shock protein mRNA levels in long-lived dwarf mice. Mech Ageing Dev. 2009;130:393–400.
46. Chiang WC, Ching TT, Lee HC, Mousigian C, Hsu AL. HSF-1 regulators DDL-1/2 link insulin-like signaling to heat-shock responses and modulation of longevity. Cell. 2012;148:322–34.
47. Westerheide SD, Anckar J, Stevens SM Jr, Sistonen L, Morimoto RI. Stress-inducible regulation of heat shock factor 1 by the deacetylase SIRT1. Science. 2009;323:1063–6.
48. Gehrig SM, van der Poel C, Sayer TA, Schertzer JD, Henstridge DC, Church JE, Lamon S, Russell AP, Davies KE, Febbraio MA, Lynch GS. Hsp72 preserves muscle function and slows progression of severe muscular dystrophy. Nature. 2012;484:394–8.
49. Lee BH, Lee MJ, Park S, Oh DC, Elsasser S, Chen PC, Gartner C, Dimova N, Hanna J, Gygi SP, et al. Enhancement of proteasome activity by a small-molecule inhibitor of USP14. Nature. 2010;467:179–84.
50. Liu G, Rogers J, Murphy CT, Rongo C. EGF signalling activates the ubiquitin proteasome system to modulate C. elegans lifespan. EMBO J. 2011a;30:2990–3003.
51. Blagosklonny MV. Rapamycin-induced glucose intolerance: hunger or starvation diabetes. Cell Cycle. 2011;10:4217–24.
52. Mortensen M, Soilleux EJ, Djordjevic G, Tripp R, Lutteropp M, Sadighi-Akha E, Stranks AJ, Glanville J, Knight S, Jacobsen SE, et al. The autophagy protein Atg7 is essential for hematopoietic stem cell maintenance. J Exp Med. 2011;208:455–67.
53. Bondolfi L, Ermini F, Long JM, Ingram DK, Jucker M. Impact of age and caloric restriction on neurogenesis in the dentate gyrus of C57BL/6 mice. Neurobiol Aging. 2004;25:333–40.
54. McLeod CJ, Wang L, Wong C, Jones DL. Stem cell dynamics in response to nutrient availability. Curr Biol. 2010;20:2100–5.
55. Laplante M, Sabatini DM. mTOR signaling in growth control and disease. Cell. 2012;149:274–93.
56. Campisi J. From cells to organisms: can we learn about aging from cells in culture? Exp Gerontol. 2001;36:607–18.
57. Kuilman T, Michaloglou C, Mooi WJ, Peeper DS. The essence of senescence. Genes Dev. 2010;24:2463–79.
58. Li Z, Liu C, Xie Z, Song P, Zhao RC, Guo L, Liu Z, Wu Y. Epigenetic dysregulation in mesenchymal stem cell aging and spontaneous differentiation. PLoS One. 2011;6:e20526.
59. Rubio D, et al. Spontaneous human adult stem cell transformation. Cancer Res. 2005;65(8):3035–9.
60. Hassanpour M, et al. A reversal of age dependent proliferative capacity of endothelial progenitorcells from different species origin in in vitro condition. J Cardiovasc Thorac Res. 2016;8(3):102–6.
61. Hillman CH, Erickson KI, Kramer AF. Be smart, exercise your heart: exercise effects on brain and cognition. Nat Rev. 2008;9:58–65.
62. Blackmore DG, Golmohammadi MG, Large B, Waters MJ, Rietze RL. Exercise increases neural stem cell number in a growth hormone-dependent manner, augmenting the regenerative response in aged mice. Stem Cells. 2009;27:2044–52.
63. Carro E, Nunez A, Busiguina S, Torres-Aleman I. Circulating insulin-like growth factor I mediates effects of exercise on the brain. J Neurosci. 2000;20:2926–33.
64. Martin GM. Genetic modulation of senescent phenotypes in Homo sapiens. Cell. 2005;120:523–32.
65. Chang S, Multani AS, Cabrera NG, Naylor ML, Laud P, Lombard D, Pathak S, Guarente L, DePinho RA. Essential role of limiting telomeres in the pathogenesis of Werner syndrome. Nat Genet. 2004;36:877–82.
66. Kandoth C, McLellan MD, Vandin F, Ye K, Niu B, Lu C, Xie M, Zhang Q, McMichael JF, Wyczalkowski MA, et al. Mutational landscape and significance across 12 major cancer types. Nature. 2013;502:333–9.

67. Armesilla-Diaz A, Bragado P, Del Valle I, Cuevas E, Lazaro I, Martin C, Cigudosa JC, Silva A. p53 regulates the self-renewal and differentiation of neural precursors. Neuroscience. 2009;158:1378–89.
68. Cicalese A, Bonizzi G, Pasi CE, Faretta M, Ronzoni S, Giulini B, Brisken C, Minucci S, Di Fiore PP, Pelicci PG. The tumor suppressor p53 regulates polarity of self-renewing divisions in mammary stem cells. Cell. 2009;138:1083–95.
69. Tosoni D, Zecchini S, Coazzoli M, Colaluca I, Mazzarol G, Rubio A, Caccia M, Villa E, Zilian O, Di Fiore PP, Pece S. The numb/p53 circuitry couples replicative self-renewal and tumor suppression in mammary epithelial cells. J Cell Biol. 2015;211:845–62.
70. McConnell AM, Yao C, Yeckes AR, Wang Y, Selvaggio AS, Tang J, Kirsch DG, Stripp BR. p53 regulates progenitor cell quiescence and differentiation in the airway. Cell Rep. 2016;17(9):2173–82.

Can Autophagy Stop the Clock: Unravelling the Mystery in *Dictyostelium discoideum*

13

Priyanka Sharma, Punita Jain, Anju Shrivastava, and Shweta Saran

Aging is a fundamental process followed by all organisms, but some organisms like tortoise lays eggs at the age of 100, while whales live for 200 years, and few clamps probably make it past 400 years. Way back in the late 1800s, August Weismann suggested that aging was a part of life's program as the old needed to remove itself to make space for the next generation sustaining turnover, which is essential for evolution [1]. It was thus evident that the purpose of aging was known, but the mechanism was far from being understood. Many theories have been put forward to understand this phenomenon.

It is still not clear why some animals live longer than the others? We find diverse forms of aging but are not sure if they follow the same mechanism. By studying and comparing the aging process of other animals to humans may possibly help in understanding this mechanism(s). The prokaryotic unicellular microorganisms, for instance, the bacteria, reproduce by cell division where the mother cell divides into two equal and undistinguishable daughter cells suggests that these organisms do not age. However, *Caulobacter crescentus*, a bacterium, shows asymmetric division and a decrease in reproduction with age [2]. *Escherichia coli* cells show growth arrest upon starvation and lose their ability to reproduce suggesting aging. Bacteria till recently was thought to be immortal, but functional asymmetric division of *E. coli*

Authors Priyanka Sharma and Punita Jain have equally contributed to this chapter

P. Sharma · S. Saran (✉)
School of Life Sciences, Jawaharlal Nehru University, New Delhi, India
e-mail: ssaran@mail.jnu.ac.in

P. Jain
School of Life Sciences, Jawaharlal Nehru University, New Delhi, India

Department of Zoology, University of Delhi, New Delhi, India

A. Shrivastava
Department of Zoology, University of Delhi, New Delhi, India

© Springer Nature Singapore Pte Ltd. 2020
P. C. Rath (ed.), *Models, Molecules and Mechanisms in Biogerontology*,
https://doi.org/10.1007/978-981-32-9005-1_13

serves as an evidence of aging in this species. Besides having similar basic cellular functions, it is very difficult to compare aging between bacteria and humans.

Asexually reproducing organisms show clonal or replicative aging where they stop dividing after certain numbers of division. *Paramecium* and *Tetrahymena* are good model systems to analyze aging as they reproduce asexually to ultimately reach clonal extinction [3].

Do plants age and die? It is very difficult to answer this question and aging is difficult to define in them. Many plant species show a slow aging rate, for example fruit trees like apple, orange, etc., where the function(s) declines with age. Some other plants, like the citrus fruits, can rejuvenate by vegetative reproduction [4]. Many hormones produced by roots or growing tips influence this process. On the other hand, plants like bamboos show clear signs of aging as they reproduce and then die indicating programmed mechanisms.

One of the favorite fungus model is *Saccharomyces cerevisiae*, a yeast that reproduce by budding. Both replicative and chronological forms of senescence are well documented. Another model is *Caenorhabditis elegans*, a worm, which is largely composed of post-mitotic cells. Many mechanisms governing aging have been understood from these model systems.

Evolution has optimized to select the most successful offspring. Thus, after the reproduction of an individual takes place, evolution selects them to play a role in supporting its offspring. Given limited resource for a species in a microenvironment, the offspring does better only if the older generation does not compete for the same resource.

Over the years, many theories have emerged to understand the mechanism of aging [5]. Testing these theories is difficult as one cannot test them on humans, and we lack many model systems which may follow the same principles. Therefore, any consensus on causes of aging has not yet been reached. It is still not possible to measure the rate of aging, the effect of environment, and other factors on aging. In complex organisms, aging may be brought about by changes due to the intrinsic cellular mechanisms or changes in one of the predominant tissues, for example, brain [6].

To develop a rational theory of aging, it is important to separate the causes from the effects, thus making it difficult to predict the correct theory of aging. It is possible to discriminate between causes from effects by taking a system-level approach, where perturbation of each component of the given pathway under study is carried and the results obtained are then integrated to formulate new hypothesis. Perturbations generally refer to genetic manipulations and very less on other socio-economic factors, which may be a large concern while understanding human aging.

What Is Aging and Longevity?

Senescence (to grow old) or biological aging is a gradual deterioration of functions that are characteristic of complex life forms and at the level of the organism increases the mortality after maturation [7]. It can refer to both cellular and organismal senescence.

Over a period of time, the organism accumulates damage, which ultimately leads to death. Environmental hazards like radiation, chemicals, and internal error in copying lead to general wear and tear of DNA, which may lead to disruptions in cell division, misfolding of proteins ultimately leading to a gradual deterioration of tissues, or cause disease or maybe a total shutdown of the system.

Longevity is known as a by-product of genes that are selected for additional help to the organism to survive till the age of reproduction, whereas aging echoes random events that take place during early reproductive age and is not a product of evolution. It is believed that the incidence of non-disease disorders could be kept under check if the older population followed a recommended nutrition style. Therefore, it is important for a nutrition researcher to know the difference between aging and longevity and then focus on alterations in the expression of specific genes by specific nutrients.

Longevity in Different Model Systems

Selection of widely acceptable and justified model systems to represent the best start in understanding the mechanism of aging/longevity helps to overcome many other aspects like studying human subjects, which is extremely difficult and impractical, as well as long lifespan. Aging is well studied in multicellular organisms having a defined lifespan, whereas longevity is best understood in unicellular as well as smaller multicellular organisms. Valuable understanding of the genetics and cellular biology of aging has been put forth by the invertebrate models like nematodes and fruit flies, but they also have their own limitations. Therefore, to further enhance our knowledge in this area, many more model organisms are required which could help understand one or other mechanisms of longevity or aging.

Exploitation of model animals in aging/longevity studies has provided beneficial information regarding the mechanisms of human aging and longevity. Animal models are vital in deducing the pathways that modulate human aging. Model organisms are elementary for aging research due to significant shortcomings of using human subjects, like lifespan, genetic heterogeneity, and differences in the environmental impacts. There exists a general similarity in aging between humans and model animals, but still there are some significant differences [8]. For example, overexpression of *Sir2* gene in yeast [9], nematodes [10], and flies [11] elicits longevity. Resveratrol, a small molecule that can activate Sir2 protein and its mammalian ortholog SIRT1 [12], plays role in lifespan extension in mice that were fed on high-fat diet. The results obtained with different invertebrate model organisms do not imply that we would expect the same results in humans. But it is true that researchers are still deciphering human biology of aging with the help of various model systems as each of them offers its own positive and negative points. Below are some widely used animal models.

Saccharomyces cerevisiae (yeast): Yeast has helped to identify the major conserved aging/longevity pathways that are shared among a large number of species [13]. Despite being unicellular organisms and also having significant differences with humans, they prove to be advantageous as a model organism. The replicative lifespan assay involving asymmetric division into mother and daughter cells helped to identify various factors involved in aging/longevity. The extra-ribosomal DNA circle that is inherited by the mother cells contributes largely to senescence [14].

Caenorhabditis elegans (worm): *Caenorhabditis elegans*, a nematode, serves as an important model for studying aging [15], and the major understanding came from the dauer larva. Since they are evolutionarily distant from humans, they have few drawbacks in serving as a model system, such as their inability to regenerate their tissues. In fact, three major aging pathways, including insulin/IGF-1 signaling [16], TOR pathway [17], and SIR2 pathway [10], were identified in *C. elegans*.

Drosophila melanogaster (fruit fly): Fruit flies serve as a powerful model system for aging studies as they have specialized tissues and proliferating stem cell populations in their guts [18].

Hydra: Hydras are well known for their negligible senescence and thus prove to be desirable for understanding longevity. Absence of senescence in asexually reproducing hydras and aging observed after sexual reproduction makes them a good model system [19].

Danio rerio (zebrafish): Zebrafish has an exceptional ability to regenerate its tissues, which proves to be advantageous for decoding the mechanisms of tissue regeneration and longevity [20].

Rodents: Mice are irreplaceable in aging studies as they share nearly 99% of human orthologs compared to the invertebrate organisms. Interestingly, results obtained with rats are not always useful in understanding the biology of human aging, for example treatment with metformin could extend lifespan in mice, but not in F344 rats [21].

Cellular Senescence: Since cells are the fundamental units of our bodies, it is but obvious to assume that changes in them would contribute to the process of aging. Thus, studying cellular aging in vitro (e.g., replicative senescence) may provide clues in understanding longevity.

There are some cellular and molecular hallmarks of aging, which if fulfilled can allow the cells to be used as a model organism for studying longevity. Few hallmarks possibly could be the fixed number of times a normal cell can divide [7]. Genomic instability is due to aggregation of genetic damages [22] accumulated due to various hazards they face like reactive oxygen species, replication errors, etc. Also, DNA damages like chromosomal aneuploidies, copy number variations, etc., are involved in aging [23]. Deletion mutations in mitochondrial DNA are considered as a hallmark of aging. Telomere shortening is reported during normal aging in many different model systems. Histone modifications are also considered as a hallmark of aging. Stress resistance is an important factor in aging, and lifespan extension is associated with increased stress resistance [24] and shows an association with organismal aging. Nutrient signaling and sensing systems are important hallmarks of longevity as they induce autophagy-mediated longevity [25].

Presently we focus on autophagy-mediated longevity in a eukaryotic model system *Dictyostelium discoideum*.

Autophagy and Longevity

Autophagy predominantly is protective rather than self-destructive process where small fractions of cytoplasm is sequestered within cytosolic double-membrane vesicles and delivered to the lysosome/vacuole for degradation and further reutilized for survival [26]. Upon continued starvation, autophagy process is triggered in such a way that the cell starts to consume its important organelles for survival, ultimately leading to programmed cell death type II or the autophagic cell death (ACD). It is an evolutionary conserved pathway in eukaryotic cells, which plays an essential role in stress response and also degradation of damaged cytosolic components. Various studies have demonstrated that autophagy genes are important for longevity [27], wherein the cells enter the process of autophagy for its survival.

There are various mechanisms by which the cytosolic materials are transported to the lysosomes for degradation: microautophagy, chaperone-mediated autophagy (CMA), and macroautophagy [17]. In case of microautophagy, there is invagination of the lysosomal membrane, which sequesters cytosolic materials, whereas CMA allows translocation of unfolded substrates across lysosomal membrane. It selectively degrades proteins having a particular consensus motif KFERQ. The process of macroautophagy starts with the formation of phagophore to the completion of autophagosome, which subsequently fuses with the lysosome to form autophagolysosome or autolysosome that degrades the enclosed material by acid hydrolases. Autophagy (from now on means macroautophagy) is an evolutionarily conserved process required for recycling of the cellular components to maintain metabolism during nutrient starvation and also prevent the accumulation of damaged proteins and organelles. It is a dynamic process (Fig. 13.1) having sequential stages, namely initiation, nucleation, elongation, and maturation, which are controlled by various autophagic genes (*Atg* genes) [28]. Autophagy initiates by the formation of a phagophore, which requires the class III phosphoinositide 3-kinase (PI3K) Vps34 and few autophagy proteins like Atg6 (Beclin1), Atg14, and Vps15. Other proteins that are involved in the early autophagy stages are Atg5, Atg12, Atg16, and the focal adhesion kinase (FAK) family-interacting protein that interacts with Atg1 (Ulk1) and Atg13. During elongation, association with two ubiquitination-like reactions takes place. In the first reaction, Atg12 is conjugated to Atg5 with the help of Atg7 (similar to E1 ubiquitin-activating enzyme) and Atg10 (similar to E2 ubiquitin-conjugating enzyme). This complex of Atg5 and Atg12 conjugates with Atg16, which is important for phagophore formation, but later it gets dissociated after the complex is formed. In the second reaction, LC3/Atg8 is conjugated to the lipid phosphatidylethanolamine (PE) by Atg7 and Atg3. Many such autophagosomes are formed in the cytoplasm, which are trafficked on the microtubules to reach the lysosome. The autophagosome–lysosome fusion takes place with the help of SNARE and VAMP proteins [29].

The mammalian target of rapamycin (mTOR) is upstream of Atg proteins and plays a critical role. It is a serine/threonine protein kinase, which helps to converge many upstream pathways that regulate cell proliferation, cell growth, cell survival,

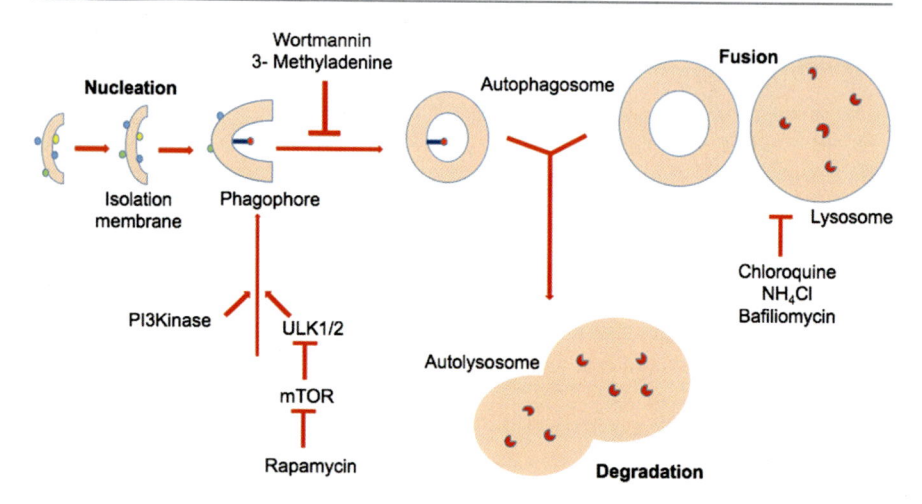

Fig. 13.1 Diagrammatic representation of the autophagic process. Three main steps comprising initiation, nucleation, and elongation in autophagosome formation are illustrated. During nucleation, the formation of isolation membrane takes place which invaginates leading to sequestration of cytoplasmic components. Fusion of outer membrane of the autophagosome to the lysosome forms autolysosome, where the cargo is degraded and later recycled. Different modulators involved in the regulation of autophagosome formation are rapamycin that attenuates at the step of phagophore formation, wortmannin and 3-methyladenine block the formation of autophagosome and chloroquine, NH$_4$Cl, bafilomycin inhibit the fusion of autophagosome to the lysosome

cell motility, protein synthesis, translation, and autophagy [30]. Inhibition of mTOR induces autophagy. mTOR is a known negative regulator of autophagy from yeast to humans and is inhibited under starvation conditions. This can be mimicked by rapamycin. One of the most important pathways regulating mTOR is the insulin-like growth factor, which binds to its receptor and activates Akt and this inhibits TSC1/2, a negative regulator of autophagy. mTOR can also be regulated by GTPase. AMP-activated protein kinase (AMPK), the energy sensor, can directly activate autophagy under starvation conditions. IP3 also regulates autophagy by binding to IP3 receptors, which release calcium from ER stores resulting in the activation of proteases that inhibit autophagy [31].

Increased Autophagy Delays Aging and Increases Lifespan

Calorie restriction is a physiological inducer of autophagy [32], and its inhibition prevents anti-aging effects in almost all organisms. Autophagy in this case is via the AMPK and Sirtuin 1 (Sirt1). Results obtained from calorie restriction and anti-aging show that they both probably follow the same mechanisms. Rapamycin induces autophagy, but if the *Atg* genes are knocked out/down, then lifespan extension is abolished. mTOR inhibits translation by hypophosphorylating its substrates S6K (kinase activity is lost upon dephosphorylation) or the 4EBP (hypophosphorylation becomes active as a translational repressor). Overexpression of S6K does not allow rapamycin to extend lifespan, and on the other hand, impaired S6K activity extends lifespan [33]. Sirtuins and its orthologs from other systems reduce aging and prolong longevity if they are overexpressed, though their mechanism of action is still not understood. Resveratrol, an activator of Sirtuin, helps to prolong lifespan.

Apart from mobilizing endogenous macromolecules for reutilization, autophagy also maintains organelle homeostasis and proteotoxic stress, and avoids age-associated processes. There are ample numbers of evidences showing nutritional, pharmacological, and genetic manipulations stimulate autophagy to increase longevity in variety of organisms.

In general, autophagy is regarded as a mechanism for cellular 'cleaning' and can be correlated to cell survival. During metabolic stress, like nutrient depletion, hypoxia, etc., autophagy increases cellular survival and in turn increases cellular fitness, whereas its inhibition leads to bioenergetic failure and cell death [34]. It is observed that autophagy is induced under low-insulin conditions during fasting, whereas high-insulin levels are suppressive [35]. Furthermore, prolonged expression of *Atg8* gene in *Drosophila* extends lifespan by 50%. Similarly, knock down of *Atg7* and *Atg12* genes partially suppresses the lifespan extension in *daf2* mutants, encoding IGF-1 receptor in *C. elegans* [36]. These results suggest that lifespan extension is mediated via autophagy and is regulated by insulin signaling pathway. Sirtuins are known to extend lifespan upon calorie restriction via autophagy, but still it is not sure whether TOR has any role to play. Unraveling the molecular mechanism underlying autophagic regulation may help to maneuver autophagy for therapeutic uses.

How Longevity Is Studied in Different Model Systems

A salient breakthrough was observed when lifespan extensions with improved health in various organisms occurred due to basic environmental and genetic involvements [37]. A conclusion drawn from these studies was that lifespan extension was correlated to aging process and was evolutionarily conserved. It was strongly attributed to nutritional status or mechanisms that could sense nutrients. Way back in the 1930s, dietary restriction (DR) was identified to increase lifespan of rats that were subjected to low nutritional uptake. Effects of DR on aging were also studied in *S. cerevisiae*, *C. elegans*, and *Drosophila*, as well as in rhesus monkeys [38]. Single-gene mutations that could extend lifespan were first identified in *C. elegans*, which happened to be the insulin/insulin-like growth factor 1 (IGF-1) signaling (IIS) pathway [37]. The IIS pathway cross-talks with the target of rapamycin (TOR) kinase, an intracellular nutrient-sensing network like energy sensing, amino acid starvation, etc. It is activated by nutrients and stimulates protein synthesis and aging in a variety of animals and probably has evolved with the beginning of multicellularity [39]. It is essential in directing responses in various parts of the organism toward nutrition. It was observed that the long-lived mutant mice showed enhanced health. In case of humans, genetic variants associated with genes encoding different components of the insulin/IGF-1 pathway are generally associated with survival during aging. Recently, mutants of various components of the TOR pathway were shown to be involved in lifespan extension in organisms like *S. cerevisiae, C. elegans, Drosophila*, and mice [40]. Secondly, we could use drugs in laboratory models to alter signaling network(s) that would ultimately provide health benefits and extend lifespan. Rapamycin is an immune suppressant approved for human use during cancer chemotherapy and also inhibits the activity of TOR kinase. It shows substantial effects on both aging and aging-related diseases. For example, the drug can extend lifespan in both *Drosophila* and mice. Middle-aged mice treated with rapamycin extend the lifespan.

Mutant yeast cells defective in autophagy-related genes show decreased chronological lifespan (CLS) in minimal medium as compared to control strains [41]. Autophagy is shown to be key process for the extension of yeast CLS induced by calorie restriction [42], but not for CR-mediated yeast replicative lifespan (RLS) extension [43]. Many other studies signify the interconnection between autophagy, calorie restriction and longevity in yeast.

Drugs That Influence Longevity

Interestingly, anti-aging drugs like spermidine, resveratrol, and rapamycin are autophagic regulators/activators [44]. Rapamycin stimulates autophagy in a TOR-dependent manner, whereas resveratrol and spermidine evokes a TOR-independent autophagy by different pathways. Thus, there are great deals of evidences that support autophagy and aging to be modulated by different signaling pathways.

Both resveratrol and rapamycin target the conserved pathways for longevity and mimics dietary restriction mechanism and help increase lifespan in many species. Clinical trials of resveratrol for anticancer treatment are underway [45]. It is a small polyphenolic compound that can activate the Sirtuin family of proteins. Overexpression of Sirtuin increases lifespan [9]. Recent studies have shown AMPK to be another target of resveratrol, which help maintain the energy balance. AMPK can activate the enzyme Sirtuin or in a way resveratrol can activate Sirtuin via AMPK to increase longevity [46]. On the other hand, rapamycin increases lifespan by reducing the TOR activity. Supplementation of rapamycin to mice increases lifespan. Both these aspects are well documented.

Resveratrol It is a polyphenol found in red wine and is an activator of Sirtuin 1 that helps in lifespan extension in yeast, worms, and flies. It has been shown that resveratrol increases lifespan of *C. elegans* through Sirtuin 1-dependent autophagy. Recent studies show that resveratrol may not directly activate Sirtuin 1 [47], but may function on various targets like enzymes, receptors, and transporters. Resveratrol is known to exert its effect in dosage-dependent manner. Studies show the dual role of resveratrol in inhibiting self-renewal of cells and thereby inducing apoptosis in human cancer stem cells. Sirtuins deacetylate and destabilize p53 activity which results in lifespan extension/cell survival because of delayed apoptosis.

Spermidine It is a polyamine that induces longevity by promoting autophagy. Spermidine (spd) is present in all living cells whether it is a prokaryote, eukaryote, plants, or animals. Spd is an organic cation that is protonated at physiological pH and therefore is a trivalent [48]. It is a low-molecular-weight aliphatic polycation important for cell proliferation and differentiation [49]. When cell faces depletion of spermidine, it ceases to proliferate, contrarily when the cell is provided with spermidine, it restores normal proliferation. Spermidine is essential for life and modulates diverse cellular processes including autophagy. Among the different polyamines, spermidine is considered to be maximally involved in regulating cell proliferation and differentiation, alteration of signal pathways, control of cellular Ca^{2+} homeostasis, and intracellular antioxidant defense. Moreover, spermidine is known to have a property of prolonging lifespan of several model organisms and decreases age-related oxidative protein damage in mice, depicting that it may act as an anti-aging drug.

Using the chronological aging experiments, the $\Delta spe1$ yeast mutant cells showed shortened lifespan, which could be restored by the addition of exogenous spermidine or its precursor putrescine [50]. The wild-type yeast cells could increase the lifespan upon addition of spermidine, which correlated with reduced acetylation of many lysine residues present in the N-terminal tail of histone H3. The intracellular level of individual polyamines diminishes as aging takes place, showing the anti-aging property of spermidine. Prolongation of lifespan is because of reduced acetylation of many lysine residues located in the N-terminal tail of histone H3 (i.e., Lys 9, Lys 14, Lys 8). Thus, it can be concluded that spermidine inhibits histone acetylases. Restriction to dietary supplement slows down aging and is mediated by target of rapamycin (TOR) kinase inhibition. TOR signaling represses autophagy and thus spermidine blocks autophagy, leading to reduction of lifespan prolongation in combination with dietary restriction and TOR inhibition [51].

External administration of spermidine extends lifespan in yeast, worms, and flies in an autophagy-dependent fashion. Spermidine causes global hypoacetylation of histone H3 and the *ATG7* promoter [50] leading to its upregulation. Induction of autophagy is observed by spermidine-mediated transcription of autophagy-relevant genes. Sirtuin 1 affects autophagy by deacetylation of important autophagy-relevant proteins, whereas spermidine probably induces autophagy through a combination of nuclear (transcriptional) and cytoplasmic (transcription-independent) mechanisms. It is probably attributed to the inhibition of protein acetyltransferases (by spermidine) and activation of deacetylases (by resveratrol) that may converge on similar regulators or effectors of autophagy. Sirtuins also known as anti-aging proteins in various model systems like yeast, *Drosophila*, *C. elegans*, and mouse are regulated by 'calorie restriction' and are shown to be involved in metabolic activity of cells as they are NAD-dependent enzymes [52]. Resveratrol activates autophagy through Sirtuins and most probably autophagy is required for lifespan extension via calorie restriction. Several studies prove the role(s) of calorie restriction, Sirtuin activation, TOR inhibition, and autophagy requirements during longevity. Calorie restriction (CR) is known as the most consistent and non-pharmacological way of increasing lifespan and provides protection against deterioration of biological functions. A decrease in diet by 20–40% in calorie intake allows the rats to live longer. Similar results have been reported in organisms like yeast, worms, flies, fish, etc. [53].

SIRT1 homolog SIR2.1 of *C. elegans* regulates lifespan through DAF16 (dauer formation protein 16), a homolog of FOXO protein which is the target of insulin/IGF pathway. This homolog associates with scaffolding proteins, 14-3-3, which help in interaction of SIRT1 with DAF16 in an IGF-independent manner. An extra copy of *dSir2* gene in *D. melanogaster* leads to lifespan extension (approximately 18–29% longer) by regulating insulin IGF-1 pathway [11]. So, IGF signaling, a nutrient-activated pathway, plays a prominent role in aging. Organisms that show CR-mediated longevity have increased insulin sensitivity and reduced blood glucose and insulin levels. SIRT1 regulates gluconeogenesis in liver that is mediated either via the FOXO protein or PGC1-α. During conditions of stress, SIRT1 deacetylates the nuclear FOXO1 to promote transcription of gluconeogenic genes, thus allowing glucose metabolism [54]. It was observed that knock down or inhibition of SIRT1 impairs insulin signaling, whereas its overexpression increases insulin resistance and impaired glucose transport. These genetic studies gave a different view regarding the shift in the metabolic strategy in cells with reference to calorie restriction, which favored longevity. During Sirtuin-mediated longevity, the cells have high calories and the carbon flow through glycolysis is high, thus resulting in the glycolytic enzyme glyceraldehyde 3-phosphate dehydrogenase (GAPDH) to use NAD and the resulting NADH is recycled by electron transport chain. But when there is anaerobic condition, then it recycles to acetyl CoA to generate fermentation products. Thus, only a substantial portion of NAD pool remains available to high flow of carbon through glycolysis. On the contrary, when calories are low, the flow of carbon through glycolysis is low and there is more availability of NAD for other NAD-binding proteins like Sirtuins, which leads to increase in lifespan.

How *D. discoideum* as a Model Organism Overcomes These Limitations

Ever since the discovery of *D. discoideum* is introduced as the model organism, the multicellular morphogenesis of this simple eukaryote offers an eminent chance to deduce the different fundamental principles of cell biology. Fulfilment of primarily at least two criteria by any model organism is necessary for it to be proved as a model system. Firstly, the model organism should offer experimental advantages, which are not offered by other competing systems. Secondly, wider relevance discoveries should be made. *D. discoideum* is one of the simplest present-day unicellular eukaryotes currently showing developmental cell death. It offers the following advantages as a model system for the study of the various aspects of biology.

It is a connecting link between unicellular and multicellular transitions. *D. discoideum* is instrumented to study chemotaxis and phagocytosis and behaves similarly to the macrophages of mammalian cells. It is a completely sequenced haploid genome and promotes genetic studies. The functional analysis of genes can be studied due to its genetically docile system. Propagation of developmental mutants that behave as conditional mutants can be achieved under vegetative conditions, as both vegetative and developmental programs are independent of each other [55]. *D. discoideum* proves to be advantageous over bacteria and yeast for the heterologous expression of genes and recombinant proteins required for the post-translational modifications, which are properly folded and remain active [56]. There are orthologs of nearly 33 disease-related genes from humans [57], and thus, can be used in understanding of human pathological conditions as well as the biology of the disease [58]. Database is available at http://www.dictybase.org, which is updated regularly [59].

Similar to yeast, *Dictyostelium* culture can be propagated eternally, but individual cells age and eventually die. Thus, it can also be used to study longevity in this organism. The benefit of this organism is that it shares homology to humans, is a eukaryotic system, and follows programmed developmental cell death by inducing autophagy. Since there are no caspases present and autophagic cell death is the only mechanism of programmed cell death, one can study autophagy-mediated longevity in this organism. Other parameters also help this organism to be better than yeast, as cell biology, biochemistry, and molecular biology tools are very well developed.

Autophagy is conserved throughout the eukaryotes and has been well studied in yeast [60]. Recent studies show that despite its early evolutionary divergence from the higher eukaryotes (i.e., metazoans), the lower eukaryote, *D. discoideum*, possesses an autophagic pathway, which is more similar to mammalian cells as compared to *S. cerevisiae*. As during the course of evolution, *S. cerevisiae* has undergone various genome duplications and gene loss making it exquisitely specialized but more divergent from metazoans. Although studies carried with yeast makes it an invaluable model system for study of highly conserved autophagy pathway similar to mammals, there are substantial differences in various elements of its autophagic machinery making it more divergent from mammals. Few differences existing between mammalian and *S. cerevisiae* autophagy are as follows:

1. Presence of Lysosomes: There is a single degradative vacuole present in yeast, as opposed to a numerous acidic lysosomal vesicles in mammalian cells. *Dictyostelium* cells also have a classical lysosomal compartment consisting of numerous acidified and proteolytic vesicles [61].
2. Location of Autophagosome Formation: Both in case of mammals and *Dictyostelium*, nascent autophagosomes originate in the cytoplasm from multiple origins. They are formed by regions of the endoplasmic reticulum (ER) or mitochondrial membranes into phagophore nucleation sites and produce many autophagosomes simultaneously [62, 63]. In contrast, in *S. cerevisiae*, phagophore membrane initiates and expands from a single specialized structure known as the phagophore assembly site (PAS), which is disconnected from other cellular organelles [64].
3. Presence of Macroautophagy Proteins: Besides the conserved core machinery of autophagy genes present in yeast and other eukaryotic systems, there are some other essential conserved proteins required for macroautophagy, which are absent in organism like *S. cerevisiae*. Homologs of vacuolar membrane protein 1

(VMP1) and etoposide-induced protein 24 (EI24) are present from non-metazoans to metazoans. Reports suggest that these genes were present in diverse organisms such as *Trypanosoma brucei* and plants during their early eukaryotic evolution but were missing in *S. cerevisiae*. The presence of conserved role of VMP1 in autophagy in *Dictyostelium* suggests the possible role of this organism in evolutionary background [65].

Autophagy being a fundamental process, vital for cellular health and survival, is highly conserved mechanism, extensively studied in yeast, but diversity in eukaryotes, suggests that there could not be one generic model for all studies because of its limitations. Thus, the need of the hour is to study the mechanism of autophagy in other evolutionary conserved organism such as *D. discoideum* to get in-depth picture of the autophagy process.

Life Cycle of *D. discoideum*

Dictyostelium discoideum, a cellular slime mold, is haploid, unicellular eukaryote isolated from the forest soil by Raper [66]. The amoebae proliferate by mitotic division in the presence of nutrients, which usually are bacteria. Upon depletion of food, approximately 100,000 amoebae signal each other by releasing the chemoattractant, cAMP [67], and initiate aggregation at common collecting points to give rise to multicellular structures. Multicellularity in this case is not by the repeated division of the 'zygote' (as in case of metazoans), but by coming together of the spatially segregated cell types, which undergo morphogenetic movements to ultimately culminate into fruiting bodies consisting of two terminally differentiated cell types, namely spores (viable) and stalk cells (dead vacuolated). Anytime during its development, about 15–20% of the total population is destined to form the stalk and the rest (80–85%) form spores. This organism has been widely used for understanding cellular motility, signal transduction, differential cell sorting, pattern formation, phagocytosis, and chemotaxis.

It undergoes transition from single-celled amoeba to multicellular organism as a natural part of its life cycle. The life cycle (Fig. 13.2) of *D. discoideum* begins with the spores released from mature fruiting bodies. Each spore germinates into an amoeba under favorable (nutrient-rich and moist) conditions. During vegetative stage, the amoebae phagocytose bacteria, but can also pinocytose liquid axenic medium. These organisms show conditional multicellularity with cell-type differentiation.

The amoebae move toward the bacteria, attracted by the folic acid secreted by them. Actively dividing amoebae continuously secrete the autocrine factor PSF (*pre-starvation factor*, a glycoprotein with a molecular mass of 68 kDa), which accumulates in the media in proportion to the cell density, which helps in quorum sensing [68]. The presence of bacteria as a food source inhibits the response of the cells to PSF. This enables the cells to sense their own density in context to bacteria. High extracellular PSF during late exponential growth induces several early developmental genes essential for aggregation and intracellular signaling [69]. Upon starvation,

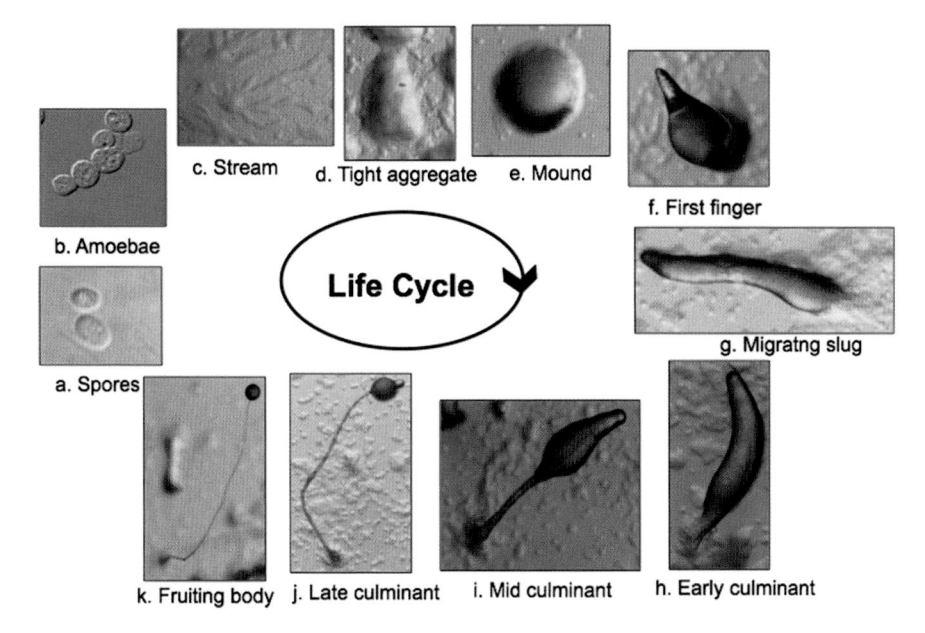

Fig. 13.2 **Asexual life cycle of *Dictyostelium discoideum*.** Amoebae stay in vegetative cycle when there is abundant nutrition. Starvation elicits the developmental cycle leading to sexual or social depending upon the availability of appropriate mating types. Upon starvation, amoebae form aggregates which undergoes morphogenetic movements to finally culminate into a fruiting body (**c–k**)

condition medium factor (CMF) protein is secreted by the amoeba into the media [70], which help regulate cAMP signal transduction, a prerequisite for the cells to aggregate. Inactivation of CMF leads to direct inhibition of aggregate formation. An important component in this signal system also involves the gene *srsA* [71] whose expression is induced within few minutes after the onset of starvation and repressed within 2 h. *srsA* appears to function independent of *yakA* as the expression of *srsA* does not get altered in *yakA* null cells. Disruption or overexpression of *srsA* resulted in reduced expression of *acaA* and *carA*, resulting in a delay in development along with the formation of aberrant structures, suggesting that certain critical level of the protein is crucial for the regulation of early development [71]. *yakA* induces the expression of aggregation-specific genes at the onset of development. Transition from growth to development is under the tight regulation of *yakA*, as overexpression of *yakA* could induce growth arrest and rapid development. *yakA*-deficient strain shows a short cell cycle duration and does not undergo normal development. In addition, *yakA* levels are modulated by PSF and observed to be essential for aggregation, cell-type differentiation, and spore germination. Disruption of *smlA* gene leads to the formation of abnormally small-sized aggregates, suggesting that *D. discoideum* uses a cell-counting mechanism to regulate the size of its aggregates [72]. Counting factor (CF) is a secreted complex of polypeptides with a molecular mass of approximately 450 kDa [73] that help determine stream break-up and aggregate size. When any one of the counting factor components, that is *countin*, *cf451*, *cf50*, or *cf60*, is knocked down or knocked out, it results in fewer but larger aggregates and aberrant fruiting bodies [73]. Oversecretion of CF results from the disruption of *smlA* gene. When wild-type cells are exposed to CF or to mutants where they are overexpressed, the streams break-up into large numbers of small-sized aggregates, which ultimately form small fruiting bodies.

cAMP facilitates two distinct responses in nutrition-deprived amoebae: (i) produces and relays the cAMP signal and (ii) responds to cAMP (i.e., chemotaxis). An anterior–posterior gradient of cAMP in multicellular structures possibly determines the prespore and prestalk distribution.

Apart from cAMP, the differentiation-inducing factor (DIF), a lipophilic morphogen induces prestalk differentiation [74]. There are many DIFs identified, among which DIF-1 [1-(3,5-dichloro-2,6-dihydroxy-4methoxyphenyl) hexan-1-one] is the most potent one. DIF-3 is one of the breakdown products of DIF-1 and shows only 3.5% of the activity of DIF-1. DIF-2 which is neither a precursor nor a breakdown product of DIF-1 shows ~40% activity.

Dictyostelium discoideum is commonly used as a model organism, as it can be easily observed at various organismic, cellular, and molecular levels because of its restricted number of cell types, behaviors, and rapid growth. It is closely related to higher metazoans, and many of the genes are homologous to human genes. Many processes and aspects of development such as cell differentiation, programmed cell death, autophagy, and motility are either absent or too difficult to view in other model organisms, making this organism a good model system.

Starvation activates the free-living amoebae to initiate multicellular development leading to the formation of two terminally differentiated cell types (the stalk cells) showing a vacuolated form of programmed cell death and viable spores. Thus, question arises whether autophagic cell death could be the process for the developmental cell death? There are few evidences which strongly support the study of the process of developmental cell death mainly by autophagy in *D. discoideum* [75]: Presence of acidic vacuoles in the prestalk/stalk cells, caspase-independent vacuolated developmental cell death programme [76] and isolated prestalk cells showing higher levels of intracellular calcium as compared to the isolated prespore cells [77]. Calcium is also reported to regulate mTORC1 signaling in HeLa cells. ROS has been observed as a vital player in the early stages of development in *D. discoideum* [78].

Autophagy in *D. discoideum*

Dictyostelium discoideum follows a developmental cell death program which is caspase independent [79]. It is seen that approximately 15–20% (i.e. the stalk cells) of the total cell population die to ensure the survival of the remaining 80–85% of the cells that comprise the spores.

An important aspect of this model system is that autophagy can be analyzed independent of apoptosis as there are no caspases or meta/paracaspase found in this organism [80]. Autophagic cell death is classified as type II programmed cell death. Autophagy is characterized by enormous degradation of cellular components that may also include essential organelles via the lysosomal pathway.

Dictyostelium discoideum shows vacuolated developmental cell death as stalk cells are highly vacuolated and they fail to regrow in nutrient-rich medium [80]. Differentiation-inducing factor (DIF) triggers cell death only when cells are starved for a certain time period, therefore it is easy to separate autophagy from autophagic cell death (ACD). Under experimental conditions (in monolayer cultures), *Dictyostelium* follow two types of cell death: (i) autophagic cell death, which is vacuolar and probably identical to the death of stalk cells during normal development and (ii) necrotic cell death is observed in the mutant *Atg1* cells. Triggering of autophagic cell death requires two signals: the first being starvation/cAMP that leads to autophagy and a second being DIF-1 that leads to ACD.

Dictyostelium shows striking similarities with higher eukaryotes in many biological aspects including chemotaxis, developmental signaling pathways, response to bacterial infections, and programmed cell death. *Dictyostelium* genome has evolutionary conserved autophagic genes, which aid in the formation of autophagosome. A serine/threonine kinase Atg1 activity is required for autophagy in different model systems including *Dictyostelium* [81]. The nutrient sensor TOR receives a wide variety of intracellular and extracellular signals like nutrients, energy, growth factors, calcium, and amino acids. TOR in association with other proteins forms a TORC1 complex, which is involved in autophagy. TORC1 regulates cell growth, metabolism, and is upstream of Atg1 complex which plays a central role in the regulation of autophagy [65]. There are many autophagy proteins present in *D. discoideum* which are involved in the transport and recycling of components from the autophagosome. Several autophagy proteins like Atg2, Atg9, and Atg18 are recognized in *S. cerevisiae*, *Dictyostelium*, and human genome. Atg9 in *Dictyostelium* is located in small vesicles that travel from cells periphery to the microtubule-organizing center and its deletion leads to autophagic defects [82]. In summary, *Dictyostelium* genome has the components of basic machinery which regulates autophagy. Moreover, the strong similarity and presence of conserved proteins in *Dictyostelium* and humans emphasize the role of autophagy in *Dictyostelium* and use of this model system to study autophagy (Fig. 13.3).

A large number of proteins are involved in autophagosome formation and its maturation. The Atg1 complex is required for recruiting other Atg proteins to the autophagosome assembly site, thereby initiating autophagosme formation in the form of isolation membrane. The expansion and formation of autophagic vesicle is dependent on two ubiquitin-like conjugation systems involving Atg8 and Atg12 (Atg5–Atg12–Atg16). Membrane trafficking and biogenesis of autophagosome is dependent on Atg9 and Atg18.

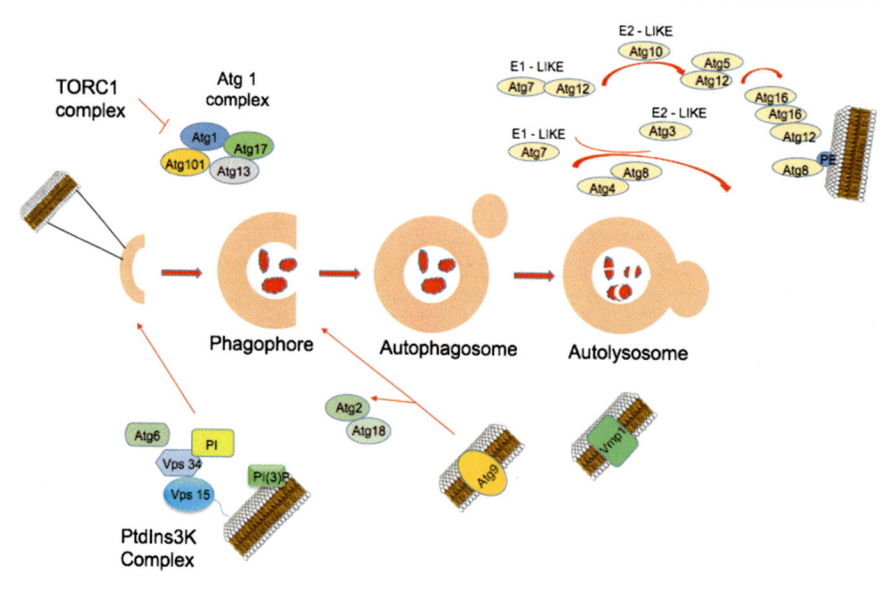

Fig. 13.3 Autophagosome formation and signaling pathways in *Dictyostelium discoideum*. The phagophore is a double membrane which enlarges and engulfs parts of the cytoplasm. Later it fuses with lysosomes where the macromolecules are degraded and recycled. Autophagic proteins have been arranged in hypothetical functional complexes similar to yeast *Saccharomyces cerevisiae* and mammalian cells. Various protein complexes involved in membrane trafficking during autophagosome formation are shown. TORC1 inhibits Atg1 complex which weakens the phagophore formation. Transmembrane proteins Vmp1 and Atg9 play roles in membrane trafficking during autophagosome formation

Signaling Pathways Involved in Autophagy in *D. discoideum*

Signaling pathways like insulin/IGF-1, TOR kinase, protein kinase, CR, mitochondrial respiration, etc. are important for longevity probably via autophagy induction wherein damaged proteins are removed for cell survival [83]. An important function of autophagy in multicellular organisms is the clearance of cytosolic ubiquitinated substrates and/or protein aggregate. Below are the two main cellular functions of autophagic degradation:

(i) It acts as an alternate energy source for fueling the cells during nutrient depletion by providing macromolecular components, the products of lysosomal degradation, including amino acids and fatty acids.
(ii) Autophagy contributes to the cellular clearance by eliminating damaged macromolecules and organelles.

There are nearly 31 autophagy-related genes (Atg) identified so far and the majority of them function in the autophagosome formation. Based on their function(s), the Atg proteins are classified into different groups as shown below [84]:

(i) Atg1 protein kinase along with Atg2 and Atg18 are involved in the induction of autophagy and cycling of integral membrane proteins.
(ii) Atg9 and Atg14 are part of phosphatidylinositol 3-kinase complex required for vesicle nucleation.
(iii) Atg8 and Atg12 are component of ubiquitin-like conjugation systems that are involved in autophagosome formation.

Some of the signaling pathways (Fig. 13.4) leading to autophagy are as follows:

(i) *TOR complex I*: Rapamycin stimulates autophagy by inactivating TORC1 suggesting that TOR inhibits autophagy [85].
(ii) *The Ras/PKA pathway*: A common mechanism found in different organisms regarding glucose sensing is via the Ras/cAMP-dependent protein kinase A (PKA) signaling pathway. Inhibition of autophagy by Ras/PKA probably is mediated by Atg1, which is identified as a phosphorylation substrate of PKA [86].
(iii) *Insulin/insulin-like growth factor pathways:* Insulin and insulin-like growth factors regulate mTOR through the class I PtdIns3K [35].
(iv) *Energy sensing*: Reduced cellular energy (ATP) levels are sensed by AMPK, which phosphorylates and activates TSC1/2 complex that inhibits mTOR activity via Rheb. This results in increased ATP levels via recycling of nutrients [87].

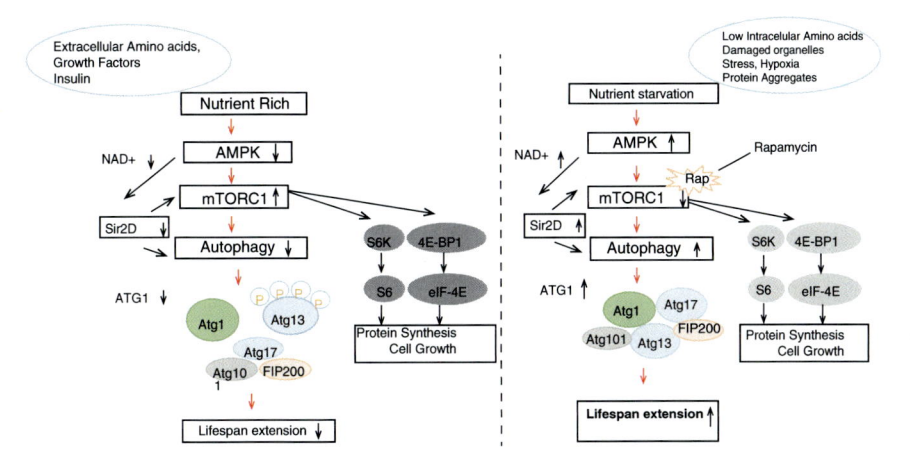

Fig. 13.4 Nutrient-dependent regulation of autophagy-mediated lifespan extension in *Dictyostelium discoideum*. In nutrient-rich condition autophagy proceeds at low basal rate and is induced to high levels in response to stress, hypoxia, starvation, or low growth factors. The TOR signaling plays a central role in these responses. Downstream of TOR, protein kinases Atg1, S6K, and 4EBP-1 inhibit TOR signaling through negative feedback loops. Starvation activates AMPK in response to AMP/ATP levels and inhibits TOR which modulates Sir2D to induce autophagy-mediated lifespan extension by forming complex of autophagy initiation proteins, viz. Atg1, Atg13, Atg101, Atg17, and FIP200. [Rap: Rapamycin inhibitor of TOR; AMPK: AMP-activated protein kinases; S6K: p70 S6 kinase; 4EBP: 4E-binding protein 1; Atg: Autophagy genes; Sir2D: NAD-dependent deacetylases, a homolog of Sirt1 of humans]

(v) *Stress response*: There are various kinds of stress faced by the organisms, which generally induce autophagy for survival of the organism. There are enough reports regarding stress-induced autophagy.
(vi) *Oxidative stress*: Reactive oxygen species is one of the major stresses faced by *Dictyostelium* cells upon starvation which leads to autophagy. The major source of ROS generation is the mitochondria, which ultimately damage the organelles.
(vii) *Pathogen infection*: Elimination of pathogens is mediated via autophagy and is observed to be TOR independent [88].

D. discoideum to Study Longevity

We have analyzed longevity in case of *Dictyostelium* using chronological lifespan mechanism like that studied in yeast. Lifespan extension in yeast is mainly studied by two different mechanisms: replicative and chronological. Yeast replicative lifespan (RLS) is defined as number of times a single cell divides before its senescence. This method is useful to study aging process of mitotically active cells in multicellular organisms [13]. Chronological lifespan (CLS) is a suitable model for post-mitotic cells in multicellular organisms as it determines the length of time a cell can survive under non-proliferative conditions, that is, number of days cells remain viable in stationary phase culture.

Similar to yeast chronological aging/longevity model, chronological lifespan determination was performed in wild-type (Ax2) cells grown in axenic media containing standard amount of glucose. Briefly, fresh cells were grown for 72 h to enter into stationary phase. The midpoint of stationary culture, that is 3 days after inoculation, was considered as the beginning of the chronological aging or 0 h. Subsequently viability of non-proliferative cells in stationary culture was determined at different time points and regrowth assay was performed. We have shown that cells overexpressing Sirtuin2D (DdSir2D) extends lifespan by inducing autophagy [89, 90].

Monitoring Autophagic Flux in *D. discoideum*

Autophagy is a process which ultimately is completed when the degradation of the cytosolic components takes place in the lysosomes. It is thus essential that autophagic flux (AF; the fusion of autophagosomes with lysosomes and subsequent degradation of cargo) be monitored as formation of autophagosomes only does not allow one to get conclusive answers. In case of *D. discoideum*, AF is monitored essentially by two following methods:

Quantification of Atg8 Puncta Per Cell by Confocal Microscopy Autophagic flux is assessed using RFP-GFP-Atg8 (an autophagosome marker expressed tandemly with reporter genes). The strain is used for confocal microscopy for tracing the presence of red puncta, as RFP fluorescence is more resistant to the acidic and protease-rich conditions of the lysosome, whereas GFP fluorescence is rapidly quenched.

The total number of Atg8 puncta per cell in *act15 (RFP-GFP-ATG8)/sir2D⁻* are significantly less as compared to wild type, thereby suggesting the crucial role played by *Ddsir2D in* regulating autophagy pathway [90].

Proteolytic Cleavage of GFP-Tkt-1 Autophagic flux in *Dictyostelium* is monitored in the presence of proteolytic fragments derived from autophagic degradation by the expressed fusion proteins. Lysosomotropic agent such as NH_4Cl is used for the measurement of these cleaved fragments [90, 91]. Tkt-1 cytosolic protein under *actin 15* promoter in pTX-GFP vector helped detect the presence of free GFP fragments from the fusion protein *GFP-Tkt-1*. Further, western blotting analysis is checked in the control GFP-TKT-1 and the treated GFP-TKT-1 (treatment with 100 mM NH_4Cl) to quantitate the cleaved GFP product.

In order to understand the role of DdSir2D in autophagy, we examined the semi-quantitative mRNA expression patterns of different autophagy-related genes (*Atg1, 8, 5, 16, 18, 9*) in wild-type, overexpressor and knock-out strains of DdSir2D. Significant reduction in expression pattern of *Dictyostelium* homolog of yeast Atg1 that is involved in the induction of autophagy, in knock-out strains, suggests a link between Sirtuin and autophagy in *D. discoideum*. To further examine its function in the process of autophagy, different *Atg* genes like *Atg8* (autophagosome marker), *Atg5* and *Atg16* (forms ubiquitin-like conjugation system), and *Atg9* and *Atg18* (involved in membrane trafficking) were studied. Increase in above-mentioned autophagy-related genes in overexpressors and subsequent decrease in knockout strain suggest its role in autophagy. Furthermore, role of DdSir2D in basal and induced autophagy was examined by transforming randomly tagged RFP-GFP-Atg8 plasmid in wild-type and knock-out strains and autophagic fluxes were monitored. As reported by Calvo-Garrido et al. [91], RFP fluorescence is resistant to acidic pH of the lysosome compared to GFP fluorescence. Thus, the red–green puncta marks the early autophagosomes and the red puncta that lack green fluorescence marks the fusion of autophagosomes with lysosomes. On induction with known inducers of autophagy, that is, starvation, rapamycin and NH_4Cl, a high autophagy flux was observed in wild-type cells as compared to knockout cells.

To confirm our hypothesis that DdSir2D plays role in autophagy, proteolytic cleavage of GFP-Tkt-1 was performed and decrease in the free GFP levels in knockout cells after NH_4Cl treatment was observed similar to the wild-type cells [90]. Starvation is a condition known to stimulate autophagic flux, similarly calorie restriction or starvation stimulates Sirtuin activity.

Other Molecules Involved in Autophagy-Mediated Longevity in *D. discoideum*

We have measured autophagy-mediated longevity in case of *D. discoideum* by resveratrol, spermidine, and rapamycin (unpublished). We found Sir2D of *D. discoideum* to be involved in autophagy-mediated lifespan extensions. Treatment with resveratrol, an activator of Sirtuins, also increases lifespan in a dose-dependent manner. Rapamycin, a positive regulator of autophagy, inhibits TOR activity and shows a dose-dependent increase in autophagy [92]. Spermidine is a physiological inducer of autophagy in many systems studied. We found higher levels of spermidine in isolated prespore cells as compared to isolated prestalk cells [93], suggesting higher levels were required for survival. A dose-dependent increase in autophagy was observed in this organism, which also increased lifespan extension. Conversely, 2-(difluoromethyl) ornithine (DFMO) could decrease the spermidine levels and also autophagic flux. All these show that spermidine can induce autophagy in *D. discoideum* and is required for its survival.

Conclusions

Lifespan can be extended both by calorie extension and pharmacological agents that can mimic the same. There is no doubt that autophagy has antiaging potential. Major function of autophagy is to protect cells during stress conditions, remove aggregate-prone proteins, etc. In the present study, we consider *D. discoideum* as one efficient model system to delineate the pathways involved in autophagy and find the mechanism of action of anti-aging agents. Additionally, this organism helps to study autophagy in the absence of caspases. Since it holds an important phylogenetic position and shows much homology to human as compared to other lower model organisms, we consider that the results obtained would solve few underlying processes involved in longevity. Since nutrition plays an important role in its life cycle, we think that autophagy via the TOR pathway could be delineated well in this model organism. The results obtained with this organism would help understand the process of cell survival/longevity. Studies that could utilize genetic, nutritional, and pharmacological manipulations to address the longevity issues could use *D. discoideum* as a model system to expand our knowledge.

Acknowledgements SS would like to thanks Indian Council of Medical Research, India, (54/41/CFP/GER/2011-NCD-II) and University Grants Commission, India (42-181/2013-SR), for the research grants. PS thanks UGC for JRF and PJ thanks DST for the INSPIRE fellowship.

References

1. Gavrilov LA, Gavrilova NS. Evolutionary theories of aging and longevity. Sci World J. 2002;2:339–56.
2. Ackermann M, Stearns SC, Jenal U. Senescence in a bacterium with asymmetric division. Science. 2003;300:1920.
3. Nanney DL. Aging and long-term temporal regulation in ciliated protozoa. A critical review. Mech Ageing Dev. 1974;3:81–105.
4. Finch CE. Longevity, senescence, and the genome. The comparative biology of senescence. Chicago/London: The University of Chicago Press; 1990.
5. Medvedev ZA. On the immortality of the germ line: genetic and biochemical mechanism – a review. Mech Ageing Dev. 1981;17:331–59.

6. Mattson MP, Duan W, Maswood N. How does the brain control lifespan? Ageing Res Rev. 2002;1:155–65.
7. Lopez-Otin C, Blasco MA, Partridge L, Serrano M, Kroemer G. The hallmarks of aging. Cell. 2013;153:1194–217.
8. Mitchell SJ, Scheibye-Knudsen M, Longo DL, de Cabo R. Animal models of aging research: implications for human aging and age-related diseases. Annu Rev Anim Biosci. 2015;3:283–303.
9. Kaeberlein M, McVey M, Guarente L. The SIR2/3/4 complex and SIR2 alone promote longevity in *Saccharomyces cerevisiae* by two different mechanisms. Genes Dev. 1999;13:2570–80.
10. Tissenbaum HA, Guarente L. Increased dosage of a *sir-2* gene extends lifespan in *Caenorhabditis elegans*. Nature. 2001;410:227–30.
11. Rogina B, Helfand SL. Sir2 mediates longevity in the fly through a pathway related to calorie restriction. Proc Natl Acad Sci. 2004;101:15998–6003.
12. Howitz KT, Bitterman KB, Cohen HY, Lamming DW, Lavu S, Wood JG, et al. Small molecule activators of Sirtuins extend *Saccharomyces cerevisiae* lifespan. Nature. 2003;425:191–6.
13. Wasko BM, Kaeberlein M. Yeast replicative aging: a paradigm for defining conserved longevity interventions. FEMS Yeast Res. 2014;14:148–59.
14. Kaeberlein M. Cell biology: a molecular age barrier. Nature. 2008;454:709–10.
15. Tissenbaum HA. Using *C. elegans* for aging research. Invertebr Reprod Dev. 2015;59:59–63.
16. Kenyon C, Chang J, Gensch E, Rudner A, Tablang R. A *C. elegans* mutant lives twice as long as wild type. Nature. 1993;366:461–4.
17. Vellai T, Vellai KT, Sass M, Klionsky DJ. The regulation of aging: does autophagy underlie longevity? Trends Cell Biol. 2009;19:487–94.
18. Wang L, Jones DL. The effects of aging on stem cell behavior in *Drosophila*. Exp Gerontol. 2011;46:340–4.
19. Yoshida K, Fujisawa T, Hwang JS, Ikeo K, Takashi. Degeneration after sexual differentiation in hydra and its relevance to the evolution of aging. Gene. 2006;385:64–70.
20. Lepilina A, Coon AN, Kikuchi K, Holdway JE, Roberts RW, Burns CG, Poss KD. A dynamic epicardial injury response supports progenitor cell activity during zebrafish heart regeneration. Cell. 2006;127:607–19.
21. Martin-Montalvo A, Mercken EM, Mitchell SJ, Palacios HH, Mote PL, Scheibye-Knudsen M. Metformin improves healthspan and lifespan in mice. Nat Commun. 2013;4:2192.
22. Moskalev AA, Shaposhnikov MV, Plyusnina EN, Zhavoronkov A, Budovsky A, Yanai H, Fraifeld VE. The role of DNA damage and repair in aging through the prism of Koch-like criteria. Ageing Res Rev. 2013;12:661–84.
23. Faggioli F, Wang T, Vijg J, Montagna C. Chromosome-specific accumulation of aneuploidy in the aging mouse brain. Hum Mol Genet. 2012;21:5246–53.
24. Longo VD, Liou LL, Valentine JS, Gralla EB. Mitochondrial superoxide decreases yeast survival in stationary phase. Arch Biochem Biophys. 1999;365:131–42.
25. Houtkooper RH, Williams RW, Auwerx J. Metabolic networks of longevity. Cell. 2010;142:9–14.
26. Ohsumi Y. Yoshinori Ohsumi: autophagy from beginning to end. Interview metabolic networks of longevity by Caitlin Sedwick. J Cell Biol. 2012;197:164–5.
27. Madeo F, Tavernarakis N, Kroemer G. Can autophagy promote longevity? Nat Cell Biol. 2010;12:842–6.
28. Pyo JO, Nah J, Jung YK. Molecules and their functions in autophagy. Exp Mol Med. 2012;44:73–80.
29. Fraldi A, Annunziata F, Lombardi A, Kaiser HJ, Medina DL, Spampanato C. Lysosomal fusion and SNARE function are impaired by cholesterol accumulation in lysosomal storage disorders. EMBO J. 2010;29:3607–20.
30. Rubinsztein DC, Mariño G, Kroemer G. Autophagy and aging. Cell. 2011;146:682–95.
31. Decuypere JP, Monaco G, Bultynck G, Missiaen L, De Smedt H, Parys JB. The IP(3) receptor-mitochondria connection in apoptosis and autophagy. Biochim Biophys Acta. 2011;13:1003–13.

32. Levine B, Kroemer G. Autophagy in the pathogenesis of disease. Cell. 2008;132:27–42.
33. Bjedov I, Toivonen JM, Kerr F, Slack C, Jacobson J, Foley A, Partridge L. Mechanisms of lifespan extension by rapamycin in the fruit fly *Drosophila melanogaster*. Cell Metab. 2010;11:35–46.
34. Kroemer G. Autophagy: a druggable process that is deregulated in aging and human disease. J Clin Invest. 2015;125:1–4.
35. Mortimore GE, Poso AR, Lardeux BR. Mechanism and regulation of protein degradation in liver. Diabetes Metab Rev. 1989;5:49–70.
36. Hars ES, Qi H, Ryazanov AG, Jin S, Cai L, Hu C, Liu LF. Autophagy regulates ageing in *C. elegans*. Autophagy. 2007;3:93–5.
37. Kenyon C. A pathway that links reproductive status to lifespan in *Caenorhabditis elegans*. Ann N Y Acad Sci. 2010;1204:156–62.
38. Mattison JA, Colman RJ, Beasley TM, Allison DB, Kemnitz JW, Roth GS. Caloric restriction improves health and survival of rhesus monkeys. Nat Commun. 2017;8:14063.
39. Morris JZ, Tissenbaum HA, Ruvkun G. A phosphatidylinositol-3-OH kinase family member regulating longevity and diapause in *Caenorhabditis elegans*. Nature. 1996;382:536–9.
40. Kaeberlein M, Powers RW III, Steffen KK, Westman EA, Hu D, Dang N. Regulation of yeast replicative lifespan by TOR and Sch9 in response to nutrients. Science. 2005;310:1193–6.
41. Alvers AL, Fishwick LK, Wood MS, Hu D, Chung HS, Dunn WA Jr, Aris JP. Autophagy and amino acid homeostasis are required for chronological longevity in *Saccharomyces cerevisiae*. Aging Cell. 2009;8:353–69.
42. Aris JP, Alvers AL, Ferraiuolo RA, Fishwick LK, Hanvivatpong A, Hu D. Autophagy and leucine promote chronological longevity and respiration proficiency during calorie restriction in yeast. Exp Gerontol. 2013;48:1107–19.
43. Tang F, Watkins JW, Bermudez M, et al. A life-span extending form of autophagy employs the vacuole-vacuole fusion machinery. Autophagy. 2008;4:874–86.
44. Morselli E, Maiuri MC, Markaki M, Megalou E, Pasparaki A, Palikaras K. Caloric restriction and resveratrol promote longevity through the Sirtuin-1-dependent induction of autophagy. Cell Death. 2010;1:e10.
45. Carter LG, D'Orazio JA, Pearson KJ. Resveratrol and cancer: focus on in vivo evidence. Endocr Relat Cancer. 2014;21:R209–25.
46. Price NL, Gomes AP, Ling AJ, Duarte FV, Martin-Montalvo A, North BJ, et al. SIRT1 is required for AMPK activation and the beneficial effects of resveratrol on mitochondrial function. Cell Metab. 2012;15:675–90.
47. Pacholec M, Bleasdale JE, Chrunyk B, Cunningham D, Flynn D, Garofalo RS, et al. SRT1720, SRT2183, SRT1460, and resveratrol are not direct activators of SIRT1. J Biol Chem. 2010;285:8340–51.
48. Wallace HM. Polyamines: specific metabolic regulators or multifunctional polycations? Biochem Soc Trans. 1998;26:569–71.
49. Hougaard DM, Bolund L, Fujiwara K, Larsson LI. Endogenous polyamines are intimately associated with highly condensed chromatin *in vivo*. A fluorescence cytochemical and immunocytochemical study of spermine and spermidine during the cell cycle and in reactivated nuclei. Eur J Cell Biol. 1987;44:151–5.
50. Eisenberg T, Knauer H, Schauer A, Büttner S, Ruckenstuhl C, Carmona-Gutierrez D, et al. Induction of autophagy by spermidine promotes longevity. Nat Cell Biol. 2009;11:1305–14.
51. Colman RJ, Anderson RM, Johnson SC, Kastman EK, Kosmatka KJ, Beasley TM, et al. Caloric restriction delays disease onset and mortality in rhesus monkeys. Science. 2009;325:201–4.
52. Haigis MC, Mostoslavsky R, Haigis KM, Fahie K, Christodoulou DC, et al. SIRT4 inhibits glutamate dehydrogenase and opposes the effects of calorie restriction in pancreatic beta cells. Cell. 2006;126:941–54.
53. Koubova J, Guarente L. How does calorie restriction work? Genes Dev. 2003;17:313–21.
54. Frescas D, Valenti L, Accili D. Nuclear trapping of the forkhead transcription factor FoxO1 via Sirt-dependent deacetylation promotes expression of glucogenetic genes. J Biol Chem. 2005;280:20589–95.

55. Loomis WF. Genetic tools for *Dictyostelium discoideum*. Methods Cell Biol. 1987;28:31–65.
56. Yighua L, Xiaoxia W, Zhinan X, Qingbiao L, Xu D. Advances in *Dictyostelium discoideum* as an expression system. Chem Mag. 2004;6:58–63.
57. Eichinger L, Pachebat J, Glöckner G, Rajandream MA, Sucqang R, Berriman M, et al. The genome of the social amoeba *Dictyostelium discoideum*. Nature. 2005;435:43–57.
58. Williams JG. Transcriptional regulation of *Dictyostelium* pattern formation. EMBO Rep. 2006;7:694–8.
59. Gaudet P, Fey P, Basu S, Bushmanova YA, Dodson R, Sheppard KA, et al. Functionality and the initial steps towards a genome portal for the Amoebozoa. Nucleic Acids Res. 2011;39:D620–4.
60. Tsukada M, Ohsumi Y. Isolation and characterization of autophagy-defective mutants of *Saccharomyces cerevisiae*. FEBS Lett. 1993;333:169–74.
61. Korolchuk VI, Saiki S, Lichtenberg M, Siddiqi FH, Roberts EA, Imarisio S, et al. Lysosomal positioning coordinates cellular nutrient responses. Nat Cell Biol. 2011;13:453–60.
62. Suzuki K, Kirisako T, Kamada Y, Mizushima N, Noda T, Ohsumi Y. The pre-autophagosomal structure organized by concerted functions of APG genes is essential for autophagosome formation. EMBO J. 2001;20:5971–81.
63. Axe EL, Walker SA, Manifava M, Chandra P, Roderick HL, Habermann A, et al. Autophagosome formation from membrane compartments enriched in phosphatidylinositol 3-phosphate and dynamically connected to the endoplasmic reticulum. J Cell Biol. 2008;182:685–701.
64. Hailey DW, Rambold AS, Satpute-Krishnan P, Mitra K, Sougrat R, Kim PK, et al. Mitochondria supply membranes for autophagosome biogenesis during starvation. Cell. 2010;141:656–67.
65. Calvo-Garrido J, Escalante R. Autophagy dysfunction and ubiquitin-positive protein aggregates in *Dictyostelium* cells lacking Vmp1. Autophagy. 2010;6:100–9.
66. Raper KB. *Dictyostelium discoideum*: a new species of slime mold from decaying forest leaves. J Agic Res. 1935;50:135–47.
67. Konijn TM, Raper KB. Cell aggregation in *Dictyostelium discoideum*. Dev Biol. 1961;3:725–56.
68. Clarke M, Yang J, Kayman S. Analysis of the pre-starvation response in growing cells of *Dictyostelium discoideum*. Dev Genet. 1988;9:315–26.
69. Rathi A, Clarke M. Expression of early developmental genes in *Dictyostelium discoideum* is initiated during exponential growth by an autocrine dependent mechanism. Mech Dev. 1992;36:173–82.
70. Clarke M, Gomer RH. PSF and CMF, autocrine factors that regulate gene expression during growth and early development of *Dictyostelium*. Experientia. 1995;51:1124–34.
71. Sasaki K, Chae SC, Loomis WF, Iranfar N, Amagai A, Maeda Y. An immediate-early gene, *srsA*: its involvement in the starvation response that initiates differentiation of *Dictyostelium* cells. Differentiation. 2008;76:1093–103.
72. Brown JM, Firtel RA. Just the right size: cell counting in *Dictyostelium*. Trends Genet. 2000;16:191–3.
73. Brock DA, Gomer RH. A cell-counting factor regulating structure size in *Dictyostelium*. Genes Dev. 1999;13:1960–9.
74. Kay RR, Berks M, Traynor D. Morphogen hunting in *Dictyostelium*. Dev Suppl. 1989;107:81–90.
75. Mesquita A, Cardenal-Muñoz E, Dominguez E, Muñoz-Braceras S, Nuñez-Corcuera B, Phillips BA, et al. Autophagy in *Dictyostelium*: mechanisms regulation and disease in a simple biomedical model. Autophagy. 2017;23:1324–40.
76. Cornillon S, Foa C, Davoust J, Buonavista N, Gross JD, Golstein P. Programmed cell death in *Dictyostelium*. J Cell Sci. 1994;107:2691–704.
77. Saran S, Nakao H, Tasaka M, Iida H, Tsuji FI, Nanjundiah V, Takeuchi I. Intracellular free calcium level and its response to cAMP stimulation in developing *Dictyostelium* cells transformed with jellyfish apoaequorin cDNA. FEBS Lett. 1994;337:43–7.
78. Bloomfield G, Pears C. Superoxide signalling required for multicellular development of *Dictyostelium*. J Cell Sci. 2003;116:3387–97.

79. Roisin-Bouffay C, Luciani MF, Klein G, Levraud JP, Adam M, Golstein P. Developmental cell death in *Dictyostelium* does not require paracaspase. J Biol Chem. 2004;279:11489–94.
80. Giusti C, Tresse E, Luciani MF, Golstein P. Autophagic cell death: analysis in *Dictyostelium*. Biochim Biophys Acta. 2009;1793:1422–31.
81. Tekinay T, Wu MY, Otto GP, Anderson OR, Kessin RH. Function of the *Dictyostelium discoideum*Atg1 kinase during autophagy and development. Eukaryot Cell. 2006;5:1797–806.
82. Tung SM, Unal C, Ley A, Pena C, Tunggal B, Noegel AA, et al. Loss of *Dictyostelium*ATG9 results in a pleiotropic phenotype affecting growth, development, phagocytosis and clearance and replication of *Legionella pneumophila*. Cell Microbiol. 2010;12:765–80.
83. Vellai T. Autophagy genes and ageing. Cell Death Differ. 2009;16:94–102.
84. Xie Z, Nair U, Klionsky DJ. Atg8 controls phagophore expansion during autophagosome formation. Mol Biol Cell. 2008;19:3290–8.
85. Heitman J, Movva NR, Hall MN. Targets for cell cycle arrest by the immunosuppressant rapamycin in yeast. Science. 1991;253:905–9.
86. Yen W-L, Klionsky DJ. How to live long and prosper: autophagy mitochondria, and aging. Physiology. 2008;23:248–62.
87. Smith EM, Finn SG, Tee AR, Browne GJ, Proud CG. The tuberous sclerosis protein TSC2 is not required for the regulation of the mammalian target of rapamycin by amino acids and certain cellular stresses. J Biol Chem. 2005;280:18717–27.
88. Colombo MI. Autophagy: a pathogen driven process. IUBMB Life. 2007;59:238–42.
89. Jain P, Sharma P, Shrivastava A, Saran S. *Dictyostelium discoideum*: a model system to study autophagy-mediated life extension. In: Rath PC, Sharma R, Prasad S, editors. Topics in biomedical gerontology. Puchong: Springer; 2016. p. 35–55.
90. Lohia R, Jain P, Jain M, Burma PK, Shrivastava A, Saran S. *Dictyostelium discoideum* Sir2D protein modulates cell type specific gene expression and is involved in autophagy. Int J Dev Biol. 2017;61:95–104.
91. Calvo-Garrido J, Carilla-Latorre S, Mesquita A, Escalante R. A proteolytic cleavage assay to monitor autophagy in *Dictyostelium discoideum*. Autophagy. 2011;7:1063–8.
92. Swer PB, Lohia R, Saran S. Analysis of rapamycin induced autophagy in *Dictyostelium discoideum*. Indian J Exp Biol. 2014;52:295–304.
93. Saran S. Changes in endogenous polyamine levels are associated with differentiation in *Dictyostelium discoideum*. Cell Biol Int. 1998;22:575–80.

14. Aging: Reading, Reasoning, and Resolving Using *Drosophila* as a Model System

Nisha, Kritika Raj, Pragati, Shweta Tandon, Soram Idiyasan Chanu, and Surajit Sarkar

Introduction

Aging in contemporary biology could be described as a collection of gradual senescence processes which operates at both physiological and cellular levels. In broadest sense, aging imitates all the changes that happen over the entire course of life. Interestingly, the evolutionary biologists define aging as an age-dependent decline in essential physiological and cellular functions, leading to decline in reproductive capability and an increase in age-specific mortality rate [67]. Therefore, the aging could be best defined as a persistent decline in the age-specific fitness components of an organism due to internal physiological deterioration [226].

Aging is a complex process, involving both the genetic and environmental factors [250]. The phenomenon of aging is represented by some most prominent characteristics such as a progressive decrease in physiological capacity, reduced ability to respond adaptively to environmental stimuli, increased susceptibility to infection and complex diseases, and increased mortality [4, 97, 261]. Some irretrievable series of biological changes which occur during the advanced stages of aging inevitably result in the death of the organism. Although the exact cause of these changes is still unsolved and almost unrelated in different cases entailing no common mechanism, yet they often indicate some shared elements of descent. Several genes and many biological processes have been found to be associated with the phenomenon of aging; however, numerous unsolved questions remain to be deliberated. This is also largely due to absence of a large number of molecular markers which could be used to measure the aging process in a tissue-specific manner. Some of the critical questions include (i) How does the aging progress? (ii) Which biological processes lead to the age-dependent cellular and physiological dysfunction? (iii) Is it possible to target the molecular pathways to combat or restrict the aging phenomenon and

Nisha · K. Raj · Pragati · S. Tandon · S. I. Chanu · S. Sarkar (✉)
Department of Genetics, University of Delhi, New Delhi, India
e-mail: sarkar@south.du.ac.in

associated impairments? (iv) Can we identify some genetic modifiers with the ability to recapitulate the after-aging effects? The researchers are still trying to disentangle the various aspects of aging phenomenon in different model organisms. Therefore, even after almost an era of attentiveness to the human race since the establishment of documented history, aging remained as a most enigmatic field of biomedical research.

The last decades have shown remarkable improvement in the genetic analysis of aging, with a greater prominence near interpretation of molecular mechanisms, pathways, and physiological processes associated with longevity. Since limitations associated with human genetics do not permit comprehensive analysis on the functional and mechanistic aspects of the candidate gene(s) in greater details, and with the fact that the basic biological processes remain largely conserved in various organisms, therefore utilization of model organisms to decipher the different aspects of aging phenomenon and modifier screening has emerged as a prime approach to study the in-depths of aging process. Extensive research has been performed utilizing several model systems such as *C. elegans*, *Drosophila*, and mice to elucidate the essential genetic/cellular pathways of aging. Subsequently, classical model systems such as *Caenorhabditis elegans* and *Drosophila melanogaster* have emerged as one of the prime organisms to elucidate the essential genetic/cellular pathways of human aging. *Drosophila*, particularly, holds tremendous promise for identifying genes and also to deduce other possible mechanisms which stimulate age-associated functional declines. Some of the most important features of *Drosophila* for aging studies have been discussed below:

Drosophila melanogaster as a Model Organism for Aging Research

Drosophila is one of the oldest and the most versatile model organisms to study a diverse range of biological processes including genetics, development, learning, behavior, and aging. For the first time, Thomas H. Morgan used the tiny invertebrate *Drosophila melanogaster* for his studies on the "chromosomal theory of inheritance" and this marks the beginning of an era of revolutionary research utilizing this humble organism in his "fly room" at the University of Columbia, USA. Subsequently, the researcher has traveled a long way, and *Drosophila* emerged as an excellent model system for aging studies due to a number of advantages, that is, short lifespan (50–70) days, high fecundity (female lay up to 100 eggs per day), availability of powerful genetic tools, accessibility of stocks with many different alterations, knowledge of the complete genomic sequence, and large homogeneous populations. In addition, ease of culturing, low maintenance costs, and affordability of maintaining large population within the confines of laboratory further make flies a remarkable model organism (Fig. 14.1). Further, the absence of meiotic recombination in male flies and availability of balancer chromosomes allow populations of flies carrying heterozygous mutations to be maintained without undergoing any constant screening for the mutations. Moreover, completely sequenced and annotated

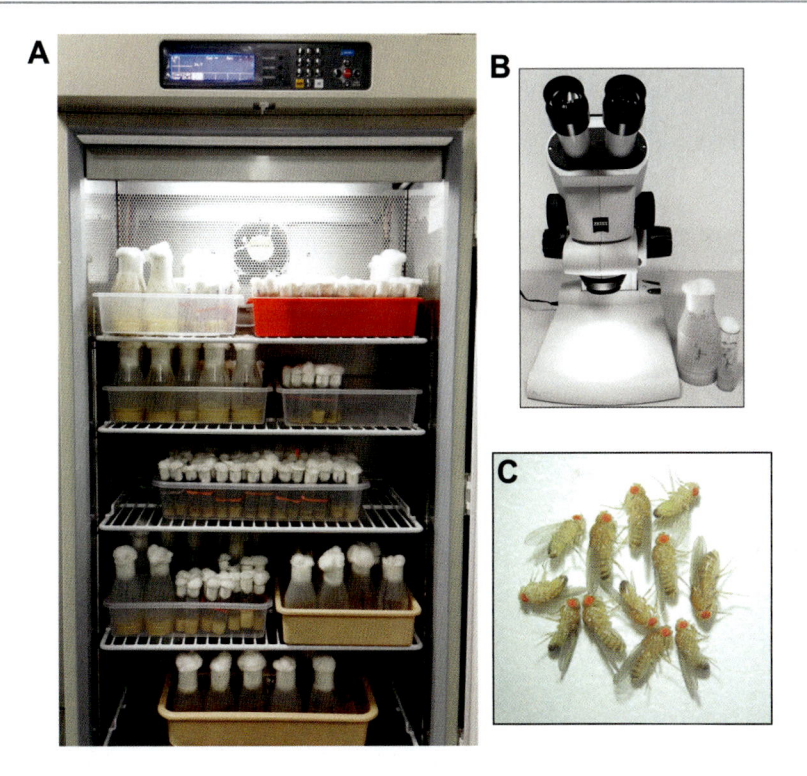

Fig. 14.1 Maintenance and handling of *Drosophila* in laboratory: (**a**) BOD incubator showing rearing of *Drosophila* at 25 °C in culture bottles and vials. (**b**) Stereo zoom binocular microscope used for routine fly work. (**c**) Stereo zoom binocular microscopic view of mixed population (male and female) of wild-type (*Oregon*$^{R+}$) *Drosophila* (Images not on scale)

genome distributed on four chromosomes make *Drosophila* a well-acceptable system to perform large-scale genetic screens for identification of potential modifiers of aging and disease-related phenotype(s). Due to existence of morphologically distinct developmental stages in *Drosophila* which include embryonic, larval, pupal, and adult phase (Fig. 14.2), it is easy to distinguish the sexually matured "aging" adults in the developing population. In several model organisms, it is not so convenient to visually distinguish the mature aging adults from immature or juvenile stage. *Drosophila* life cycle varies with temperature, and in the laboratories, it is generally maintained at 22 ± 2 °C. Since morphological features and developmental processes of *Drosophila* have been well documented, environmental and genetic manipulations which modulate the aging dynamics and lifespan could be easily performed and scored. Besides, availability of the large number of mutants and transgenic lines at several *Drosophila* stock centers further makes it a popular model organism [53, 231].

Interestingly, it has been estimated that more than 50% of the *Drosophila* genes have homologs in humans [2, 187], and nearly 75% of known human disease genes have functional homologs in the fly [221]. This striking similarity makes *Drosophila*

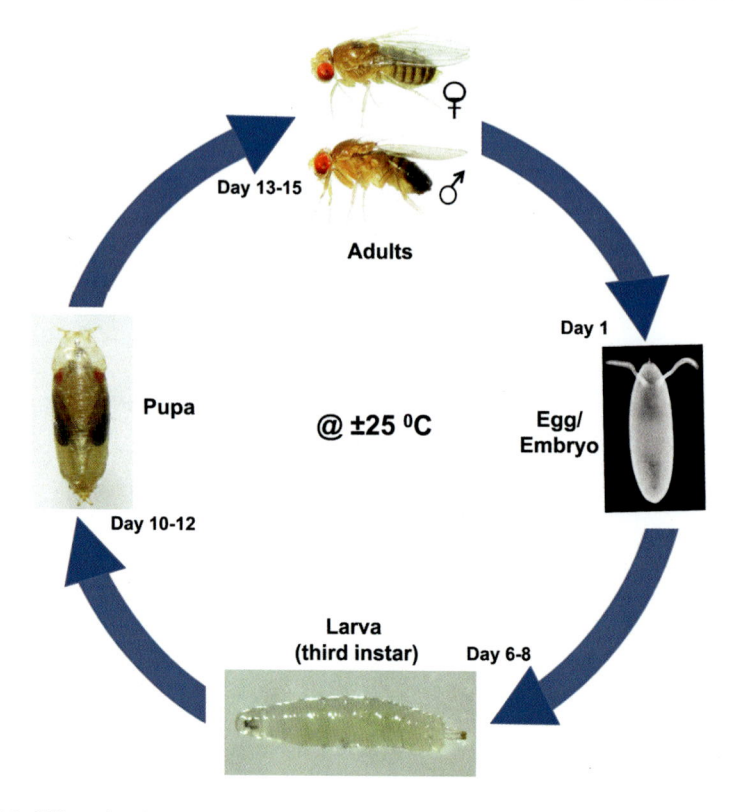

Fig. 14.2 Life cycle of *Drosophila melanogaster* at 25 °C

as the model organism of choice for several human-related studies such as aging and longevity. The adult fly harbors a well-coordinated sophisticated brain and nervous system, which makes it capable of displaying complex behaviors such as learning and memory, much like the human brain [123]. Disruption of this synchronized motor behaviors results in neuronal death and dysfunction. The aging-related characteristic phenotypes, such as locomotor and sensory impairments, learning disabilities, and sleep-like behavior, are well manifested in *Drosophila* [81]. *Drosophila* lacks a functional blood–brain barrier which could otherwise prevent access of drugs to the neuronal cells of central nervous system; and therefore, *Drosophila* has emerged extremely useful for pharmacological screening for identification of novel therapeutic drug targets [123]. In this context, it is interesting to note that response toward many drugs that has shown effects within the *Drosophila* CNS is reasonably comparable to the mammalian systems [201, 277].

Subsequent to above, *Drosophila* provides great genetic tools which facilitate manipulation of gene expression in a tissue-specific manner during various stages of life cycle. The *UAS-Gal4* system is a frequently used genetic tool to achieve ectopic expression of a gene of interest or to suppress it by *UAS-RNAi* transgene [27]. Furthermore, FLP-FRT (Flippase - Flippase Recognition Target) system, a site-directed recombination technique, has been progressively used to manipulate the fly genome *in vivo* in somatic and germ cells [257]. Using this genetic tool, loss-of-function of a lethal gene could be easily studied in target organ or tissues [248]. The effect of altered gene expression can also be studied over time, by using an inducible promoter to trigger the recombination activity late in development. This prevents the genetic alteration from affecting overall development of the organ, and also allows single-cell comparison of the one lacking the gene to normal neighboring cells in the same environment [282].

Drosophila also offers some additional advantages for aging studies. The presence of almost fully differentiated post-mitotic cells throughout the adult stage represents synchronized aging [7]. The initiation of adulthood in *Drosophila* has been proposed to occur only after the pupal eclosion and the fly becomes sexually mature and competent to reproduce [237]. This provides a great advantage over other model systems where it is often difficult to exactly determine when the organism has attained the maturity [102, 103]. Further, the rarely dividing neurons of the *Drosophila* brain makes it an excellent model to study various aspects of human brain aging and neurodegenerative disorders [104]. Aging-mediated cellular and structural changes could be convincingly inferred by examining synchronously aging neuronal cells in adult *Drosophila* brain. Due to the lack of blood vessels in the fly brain, the pathological complexities due to blood vasculature can be excluded. Taken together, in view of above-noted several advantages, *Drosophila* has been extensively used to decipher various aspects of aging. A brief account of the *Drosophila* aging research has been provided below.

Drosophila in Aging Research: An Overview

Loeb and Northrop in 1916 reported the first use of *Drosophila* as a model organism for aging-related studies [148]. They demonstrated the effects of temperature and food on the lifespan and concluded that longevity of flies as poikilothermic organisms depends on the temperature of the environment [148]. In addition, they also studied the effect of starvation and sugar-rich diet on fly longevity [149]. Subsequently, it was demonstrated that longevity in flies is heritable [204, 205]. Consistent with the findings of Pearl and Parker, the role of genetic influence in regulation of lifespan of adult flies was validated further [45]. By utilizing *Drosophila* as a model system, several small compounds such as biotin, pyridoxine, and pantothenic acid were identified which extend the lifespan upon regulated feeding [71]. The relationship between reproduction and fly longevity was for the first time studied by J Maynard Smith in 1958 [241]. It was found that longevity could be considerably extended when female flies were selected late for fertility [153, 154, 227, 228].

The role of reproductive behavior on aging has been a topic of aging research since middle of the twentieth century when it was reported that longevity of *Drosophila* could be influenced by reproductive behavior [241]. This established *Drosophila* as an excellent model system to study the fitness trade-offs and lifespan [241].

Thereafter, studies on to establish the mechanistic correlation between reproduction and longevity has been a topic of immense interest in the aging research. Consistently, the plasticity behaviors between fly longevity and reproductive output was further confirmed by the selection experiments, which demonstrated that lifespan could be significantly extended when female flies were selected for late-life fertility [153, 154, 227, 228]. Michael Rose has reviewed the history of laboratory-based evolution experiments and the use of different genomic technologies to unravel the genetics of longevity in *Drosophila* [227, 228]. Interestingly, random mutagenesis approach during the end of twentieth century led to identification of two independent life-extending genes in *Drosophila*. Interestingly, *methuselah* (*mth*) was the first such gene in which P-element insertion-mediated downregulation was found to increase the lifespan [143]. In another study, it was found that five independent P-element insertional mutation in gene *I'm not dead yet* (*indy*) resulted in a near doubling of the average adult lifespan without making any negative impact on the fertility or physical activity of the flies [230]. *Drosophila* has further been used as a model to study the role of immune senescence and inflammatory responses in aging [87, 142]. Major milestone in aging research using *Drosophila* can be summarized as shown in Fig. 14.3.

In contemporary aging research, various approaches and strategies are being adopted to decipher the mechanistic insights of aging and longevity. Some of the widespread genetic approaches include random mutagenesis followed by forward genetic analysis, selective breeding, biochemical, cellular and molecular assays, and QTL analysis. These methods collectively have allowed identification of numerous genes those are involved in diverse cellular functions including aging and longevity in *Drosophila*. Table 14.1 provides a brief account of some genes and their

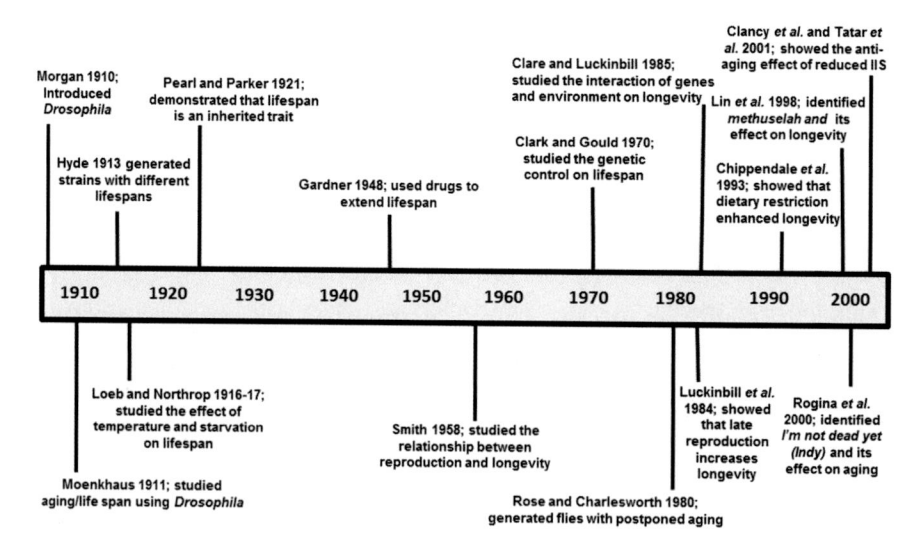

Fig. 14.3 Timeline representing historical overview of some major findings in *Drosophila* aging research

Table 14.1 A brief collection of some genes found to extend lifespan in *Drosophila melanogaster*

Gene	Endogenous function	Positive effect on longevity due to	References
Indy	Succinate and citrate transmembrane transporter	Knockdown	[225]
dsir2	Histone and nonhistone, NAD-dependent deacetylase	Overexpression	[14, 224]
gsh	Antioxidant enzyme involved in formation of reduced glutathione	Overexpression	[174]
mth	G-protein-coupled receptor	Knockdown	[143]
SOD	Antioxidant enzyme involved in partitioning of superoxide radicals to molecular oxygen	Overexpression	[202, 249]
Hep	JNK kinase	Overexpression	[270]
Puc	Inhibits JNK by specific JNK phosphatase activity in JNK signaling	Knockdown	[270, 287]
Hsp22	Molecular chaperons involved in stress response	Overexpression	[179]
Hsp23	Molecular chaperons involved in stress response	Overexpression	[177]
Hsp26	Molecular chaperons involved in stress response	Overexpression	[141]
Hsp27	Molecular chaperons involved in stress response	Overexpression	[141]
Hsp68	Molecular chaperons involved in stress response	Overexpression	[270]
Hsp70	Molecular chaperons involved in stress response	Overexpression	[253]
14-3-3ε	Antagonist to *dFoxo*	Knockdown	[192]
chico	Insulin receptor substrate in *Drosophila*	Knockdown	[44]
dFoxo	Forkhead transcription factor in *Drosophila*	Overexpression	[75]
dilps	Insulin-like peptides in *Drosophila*	Knockdown	[80]
InR	Insulin receptor in *Drosophila*	Knockdown	[254]
dPTEN	*Drosophila* phosphatase and tensin homolog control cell growth and proliferation by negatively regulating insulin signaling	Overexpression	[111]
dS6K	Important downstream kinase involved in TOR pathway	Knockdown	[124]
dTOR	It is a serine/threonine protein kinase which regulates cellular growth, proliferation, survival, transcription, etc.	Knockdown	[124]
dTsc1, *dTsc2*	Act synergistically to inhibit TOR	Overexpression	[124]

potential function(s) which have been found to modulate the aging and longevity in *Drosophila*. A brief overview of various techniques and approaches which have been adopted in *Drosophila* aging research is provided below.

Evaluating Aging in *Drosophila*

The rate of aging in *Drosophila* is affected by a combination of environmental and genetic factors. Thus, various approaches have been used to evaluate aging in *Drosophila*. Some of these approaches are briefly described below.

Environmental and Physiological Approaches

Environmental parameters which significantly influence lifespan in *Drosophila* include diet, oxidative stress, and conditions causing inflammation [99]. The following means can be exploited in order to analyze aging using physiological approaches.

Analyzing Demographics

This assay is based on the calculation of survivorship and mortality curves. A typical survivorship curve remains relatively flat for the early period of life and starts to decline at older ages, which corresponds to a period of low mortality followed by a period of an exponential increase in mortality (Fig. 14.4a). A stressful environment will usually manifest as an excess of early death in the population and an abnormal dip in the survivorship curve. Assuming that the shortening or lengthening of lifespan of an organism is the result of relative aging, comparative analysis among mean, median, and maximum lifespan of different populations under conditions could be treated as one of the factors to measure the aging process [145].

Dietary Restriction

Dietary restriction (DR) refers to a moderate reduction in food intake that leads to extension of lifespan beyond that of normal healthy individuals. DR in *Drosophila* usually involves reduction of the yeast and sugar components of the diet [203], and interestingly yeast appears to account for the majority of the DR effect on lifespan [42, 158]. It has been found that *Drosophila* fed on low yeast/low sugar diet has the highest lifespan [167]. DR impacts the physiology of flies in two major ways: extension of lifespan and reduction in reproductive ability [203]. It has been found that DR-mediated lifespan extension is controlled by metabolic pathways such as insulin/IGF-1 signaling and TOR pathway [203], as described in detail in the later sections. Two major methods have been generally used to measure the food intake in *Drosophila*. One method includes direct measurement of the amount of liquid food consumed by flies using a capillary feeder (CAFE) [116], and the second indirect method deals with the estimation of the food intake by measuring the uptake of a dye or radioactive tracer added in the food [278].

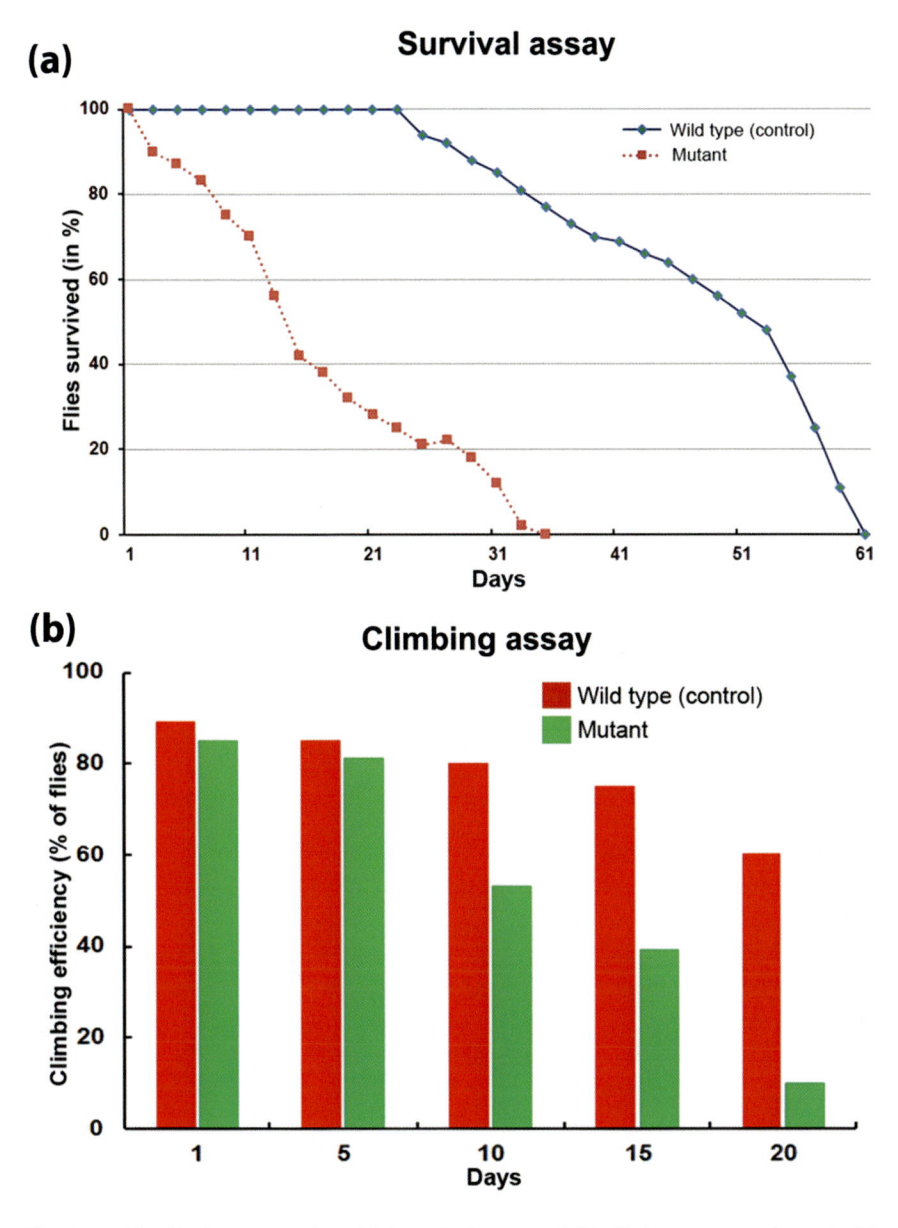

Fig. 14.4 Graphical representation of (**a**) survival assay and (**b**) climbing assay performed with wild type and a mutant line of *Drosophila*. In comparison to wild type, mutant population show increase in the mortality rate, and decline in the climbing efficiency with aging

Stress Resistance

Some amount of resistance could be observed when flies are exposed to the various physiological stresses such as oxidative stress, starvation, heat/cold shock, and desiccation. Level of such resistance has been found to be positively correlating with the longevity [88, 130]. Selection for increased resistance to starvation and desiccation in *Drosophila* has been shown to increase longevity [229]. The relative responses against desiccation and starvation and subsequent selection of lines with increased lifespan are partly independent of each other, indicating a multiplicity of physiological mechanisms involved in aging and longevity [229]. Oxidative stress resistance is usually measured in *Drosophila* by feeding paraquat (N,N′-dimethyl-4, 4′-bipyridinium dichloride), which produces ROS upon ingestion and consequently induces oxidative damage [8]. Starvation assay, on the other hand, is typically performed by measuring the survival of adult flies fed solely on water [109].

Reproductive Output

A negative correlation between reproductive output and lifespan was observed and termed as "cost of reproduction" [252]. Late reproduction in *Drosophila* has been found to increase lifespan [110], and decrease in early reproduction corresponds to long-lived flies [292]. Moreover, sterile flies tend to live longer than their fertile controls [13], and long-lived mutants flies have been shown to exhibit reduced fecundity [66]. The methods used to assess reproductive output in flies include measuring lifetime egg production in once-mated females or calculating progeny number from a pair mating of female and male. Other environmental factors such as temperature, humidity, and circadian light rhythm also affect longevity in flies, and thus can be used to study the process of aging like other environmental approaches.

Behavioral Approaches

Changes in behavioral pattern can also be used to study the cognitive functions during various stages of aging. Relationship between behavioral pattern and aging has been well established; for instance, *Drosophila* experiences a decline in sleep time with aging [132]. One such behavioral aspect is the locomotory activities. In general, there is a gradual fall in the locomotory activities with aging [114]. Two methods are commonly employed in *Drosophila* aging studies for assessment of locomotory behaviors.

Rapid Iterative Negative Geotaxis (RING) Assay

Based on the inherent negative geotaxis behavior of flies, this assay measures an innate escape response during which flies ascend the wall of a cylinder after being tapped to its bottom [72]. Climbing ability of *Drosophila* has been found to be deteriorating during the process of aging (Fig. 14.4b). In RING assays, digital photography could be used to document negative geotaxis in multiple groups of animals simultaneously [72].

Drosophila Activity Monitoring (DAM) System
In this method, flies are kept in sealed activity tubes individually and are placed in the DAM system [210]. Fly activity is measured by the frequency of an "activity event," which is recorded each time when a fly breaks an infrared light beam across the middle of the activity tube [210].

In addition to the above-noted two relatively simple systems, sophisticated video tracking systems have been developed to analyze various fly behaviors, such as movement pattern and courting, which can be potentially used to measure lifetime behavioral changes and locomotory activity-related health-span parameters [28, 82].

Genetic Approaches
Genetic approaches remain an invaluable method for identification of casual genetic factors which modulate the aging process. Single gene mutations and the resultant phenotypes help in determining the complex pathways of aging and longevity. Genetic approaches could also be adopted to confirm the already available hypotheses using candidate gene approaches, or to explore for novel ones, using random single-gene alteration approaches [102, 103]. Genetic approaches have also emerged as an important strategy for screening of the genetic interacting partners and modifiers. Several genetic tools are available in *Drosophila* to investigate the in-depths of aging and longevity. Some of the widely utilized genetic approaches being used in *Drosophila* aging research include selective breeding, quantitative trait loci (QTL) mapping [216], achieving ectopic expression of gene by *UAS-Gal4* system, inducible gene expression by gene-switch Gal4 (*GSG-UAS*) system, and gene knockdown by RNA interference (RNAi) strategy [250].

Cellular, Molecular, Biochemical, and Other Approaches
In addition to the above-mentioned strategies, various contemporary techniques are widely utilized to investigate the cellular, molecular, and biochemical aspects of aging-mediated changes in *Drosophila*. Several techniques such as reverse transcription PCR, real-time PCR, microarray, whole mount *in situ* hybridization, immunohistochemistry-based staining techniques, western blot analysis, co-immunoprecipitation, and microscopy techniques can be used appropriately to investigate different facets of aging in *Drosophila*. Examination of the systemic biomarkers of aging-like protein carbonylation, lipid peroxidation, protein aggregation, and accumulation of advanced glycation end products (AGEs) has also emerged as an important area to establish the age-related changes and cellular dysfunctions [99]. Further, status of gut homoeostasis and various organs such as heart, muscle, brain, etc. could be investigated in *Drosophila* to study the effect of aging-mediated changes on vital organs [99].

Cellular Pathways Affecting Aging in *Drosophila*

Aging is an inevitable phenomenon which affects all cell types in every hierarchy of organism. It is a natural process in which cells along with divisions accumulate certain defects leading to activation of some specific signaling pathways/stress responses that leads to senescence and finally cell death. There can be many stimuli or damages which can trigger the aging processes. Some of the key aging-related hallmarks are described below.

Genomic Instability

One of the common causes of aging is the progressive accumulation of genetic damages during the lifespan of an organism [182]. The veracity and stability of genomic DNA is being continuously affected by various exogenous stresses such as physical–chemical agents, and endogenous stimuli like DNA replication errors, spontaneous hydrolytic reactions, and ROS levels [105]. These stresses could cause damages encompassing point mutations, translocations, chromosomal gains, and losses. Here it is interesting to note that premature aging diseases such as Werner and Bloom syndromes are caused by accumulation of increased DNA damage [32]. However, the DNA repair systems of the organism collectively function with various cellular systems to minimize the nuclear damages [152]. Also, defects in nuclear architecture can be a cause of genomic instability which results in premature aging syndrome [280].

Defects in Nuclear Architecture

In addition to the genomic damage, defects in nucleus structure, that is, nuclear lamina, also contribute to genomic instability [52]. Nuclear lamins are the major components of nuclear lamina and they play an important role in maintaining the genomic stability by providing a scaffold for tethering chromatin and protein complexes, which are important for the maintenance of genomic stability [77, 146]. Lamins got attention in aging research recently after the discovery that mutation in lamins causes accelerated aging syndrome such as Hutchinson–Gilford (HGPS) and Néster-Guillermo progeria syndrome [33, 51, 58]. Accumulation of Progerin, an aberrant pre-lamin isoform, has been found during normal human aging [219, 233]. Interestingly, it has been reported that altered telomere function promotes production of Progerin in human fibroblast upon prolonged *in vitro* culture [34]. Aberration in nuclear lamina triggers various stress pathways such as activation of p53 [264], deregulation of somatotrophic axis [159], and attrition of adult stem cells [59, 233]. The role of nuclear lamins in aging is supported by the observation that decreased cellular level of pre-lamin A delays the onset of progeroid symptoms and extends lifespan in mouse models of HGPS [197]. An interesting approach using induced pluripotent stem cells (iPSCs) derived from HGPS patients has been developed to correct the lamin A/C (*LMNA*) mutations by homologous recombination-based strategy, which could be utilized in future cell therapies [147].

Telomere Abrasion

Telomeres are repetitive ends of chromosome whose length gets shortened by every cell division cycle. Progressive shortening of telomeric region has been observed during normal aging in human and other model systems [21]. Therefore, the length of the telomeres is quite heterogeneous from chromosome to chromosome and from cell to cell within a population. The telomeric ends of chromosomes are repetitive in nature and they are replicated by special DNA polymerase known as telomerase [79]. Since the normal mammalian somatic cells do not express telomerase, this leads to progressive shortening of the chromosomal ends, whereas in germline cells and many immortalized cell lines and cancers, telomere length is maintained by the enzyme telomerase [238]. Telomere abrasion explains the reason of limited proliferative capacity of *in vitro* cultured cells, this is called replicative senescence or Hayflick limit [98, 194]. Interestingly, ectopic expression of telomerase is sufficient to confer immortality to otherwise mortal cells [22].

In telomerase-positive cells, telomeres are maintained to a stable length resulting in the bypass of senescence and cellular immortalization. Both telomere protection and the regulation of telomere length are mediated by a stably associated complex, called shelterin. Mammalian shelterin masks the chromosome ends and makes them inaccessible for the telomerase and the DNA repair machinery [200]. Therefore, shelterins covered telomeres continuously accumulate exogenous DNA damage during cell divisions. Abnormalities in telomeres have been found to be associated with premature manifestation of many diseases in humans, such as pulmonary fibrosis, dyskeratosis congenita, and aplastic anemia [9]. Moreover, mutations in shelterins have been reported in some cases of aplastic anemia and dyskeratosis congenita [200]. In mice, shortened or lengthened telomeres exhibit decreased and increased lifespan, respectively [10, 230, 259]. Interestingly, human meta-analysis also suggests a strong correlation between short telomeres and mortality risk, particularly at younger age [24].

Nuclear–Mitochondrial (NM) Signaling in Aging

The mitochondria are double membrane-bound cytoplasmic organelles found in all eukaryotic organisms, which function as an energy production site of a cell. In addition, mitochondria are also involved in several other processes such as signaling, growth and differentiation, cell cycle, and apoptosis [6]. Mitochondria constantly maintain their morphology and function in response to changing microenvironment by multiple processes including fusion and fission, DNA repair, and mitophagy (clearance of damaged mitochondria). Impairment of any of these processes leads to mitochondrial dysfunction, which can be a causal for many mitochondrial diseases, neurodegenerative disorders, cancer, diabetes, heart disease, immunodeficiency, and early aging [160, 220, 269, 285]. For instance, impairment of mitophagy has been reported in Parkinson's disease, Alzheimer's disease, and pathological aging [62, 166, 199]. Such dysfunctioning in mitochondria was suggested to be due to impairments in mitochondrial proteins and aberrant nuclear-to-mitochondrial signaling [63].

In view of the mitochondrial involvement in several cellular processes as noted above, proper functioning of a mitochondrion is important for the maintenance of the cellular homeostasis. Indeed, there are several evidences which show that damaged mitochondria accumulate with age from unicellular organisms to humans [47, 62, 183, 209]. Also, some recent reports suggest that compromised nuclear-to-mitochondrial signaling is a key component of mammalian aging, which initiates because of the nuclear damage accumulated over time due to progressive aging [63]. Number of factors which senses DNA damage includes poly(ADP ribose) polymerase 1 (PARP1), ataxia telangiectasia mutated (ATM), and transcription factor p53 [63]. Thereafter, NAD-dependent protein deacetylase sirtuin 1 (SIRT1) and AMP-activated protein kinase (AMPK) trigger chromatin remodeling through post-translation modification of histones and peroxisome proliferator-activated receptor-γ co-activator 1α (PGC1α); and subsequently, some other proteins assist in propagating the nuclear-to-mitochondrial signaling ahead [63]. Therefore, progressively accumulating DNA damages leads to downstream changes in cellular transcriptome, epigenome, metabolome, and in bioenergetics, which ultimately contribute to aging and its associated disorders.

Oxidative Stress

Almost about a century ago, it was observed that animals with higher metabolic activity generally have shorter lifespan, and this observation leads to the emergence of "the rate of living hypothesis" of aging [65]. Interestingly, in contrast to this theory, some species, that is, birds and primates, do not show such inverse correlation. Also, Denham Harman proposed a "Free radical theory," which postulates that ROS generated inside a cell results in oxidative damages to the cells, which in turn accelerates aging [92, 283]. Almost about a decade later, this theory was supported by the identification of superoxide dismutase (SOD) enzyme, which solely degenerate the superoxide anions [163]. Later, this theory was modified to oxidative stress theory which emerged as most convincing theory of aging [208]. Various studies attempted to substantiate this theory, however, the results were inconsistent and sometime challenging as well [136]. However, findings in several model organisms including *Drosophila* substantiate that a decrease in level of ROS is directly correlated with an increase in lifespan [23]. Hence, it appears that an intricate balance between ROS production and the ability of the cells to counteract it drives the progression of a cell toward aging.

D. melanogaster has been widely used to investigate mechanistic correlation between ROS levels and aging. In fly, the relationship between oxidative stress and longevity was examined by modulating the expression levels of various antioxidant gene(s) by mutagenesis approaches. The driving hypothesis in finding this correlation postulates that factors which reduce the cellular levels of ROS should have beneficial effects against aging, and thus would increase the life expectancy. Supporting this claim, a positive correlation between decrease in the levels of ROS and increase in lifespan has been found in *Drosophila* [55]. Such *Drosophila* strains with extended lifespan either have low levels of ROS or have an increased level of antioxidant enzymes [55, 94]. For instance, P element insertion-mediated reduction in the level of *Methuselah* (a G-protein-coupled receptor) leads to ~35% increase in lifespan [143]. In addition to this, reduced level of *Methuselah* also increases the resistance of the fly toward various stresses such as high temperature, dietary paraquat (generates free radicals), and starvation [143]. Further, overexpression of an antioxidant gene glutathione reductase (GSH) leads to increased lifespan in hyperoxic conditions but no effect was evident under normoxic conditions [174]. Also, decrease in the expression of an antioxidant enzyme superoxide dismutase (SOD) and catalase (involved in H_2O_2 eradication) reduces the lifespan, which suggests a positive correlation between reduced level of cellular ROS and increased lifespan [129, 171, 211, 212]. In this context, it is important to note that since such mutation(s) are prevalent during the development of the fly, decrease in the lifespan could also be due to the cellular damages accumulated over time and not solely due to oxidative stress. However, studies conducted in the *Drosophila* lines in which antioxidant genes, that is, SOD and catalase, were overexpressed or SOD was overexpressed showed oxidative stress resistance and increase in the life expectancy [196, 202, 249]. Interestingly, the transgenic flies expressing human SOD gene in their motor neurons exhibited 40–50% increase in their lifespan [202]. These studies suggest specific role(s) of SOD and catalase in modulating the lifespan by regulating the cellular level of ROS. Here it is also important to note that some studies with several other antioxidant genes reported only slight or insignificant increase in the level of oxidative stress tolerance and life expectancy [195, 236]. Taken together, the above studies convincingly suggest a direct correlation between cellular level of ROS and progression of aging.

What Does Oxidative Stress Do?

Mitochondria produce ATP *via* the process of oxidative phosphorylation which involves consumption of oxygen, and therefore, surplus availability of oxygen in the mitochondria renders it to be site of ROS production. The major types of ROS found in living animals include superoxide anion ($O2^{•−}$), hydrogen peroxide (H_2O_2), and hydroxyl radical (•OH). In mitochondria, ROS is produced when (a) it is not generating ATP and thus has a high proton-motive force and a reduced level of coenzyme Q; and (b) there are high levels of NADH/NAD$^+$ in matrix [186]. Under normal physiological conditions, approximately 2% of the electrons leak from the electron transport chain (ETC) and account for ROS production [37]. When an electron encounters an oxygen atom, $O_2^{•−}$ is formed due to a reduction reaction. $O_2^{•−}$ is considered to be the most important oxygen-free radical and the source of other ROS molecules. $O_2^{•−}$ is readily converted into H_2O_2 by superoxide dismutase (SOD) enzyme which is in turn converted into •OH in the presence of ferrous (Fe^{2+}) or cuprous (Cu^{2+}) ions [54]. While •OH is highly reactive, H_2O_2 is more stable and membrane permeant. Although there are around eight known sites in the mitochondria which possess the ability to produce $O_2^{•-}$, however, the two major sites in ETC include complex I (NADH dehydrogenase) and complex III (Ubiquinone–cytochrome c reductase) [26, 262]. In normal conditions, complex III is the major site of ROS production [37]. Non-ETC sources of mitochondrial ROS production include monoamine oxidase which locates in the outer mitochondrial membrane and produces H_2O_2 as a byproduct of oxidative deamination. Further, under the elevated level of NADPH/NADP+ ratio and calcium, glycerol-3-phosphate dehydrogenase (GPDH) and α-ketoglutarate dehydrogenase (α-KGDH) in the mitochondrial matrix produces $O_2^{•-}$, and the both $O_2^{•-}$ and H_2O_2 respectively. In view of above, it is increasingly clear now that cellular level of ROS primarily depends upon the metabolic status of an individual. Intriguingly, ROS exhibit beneficial as well as deleterious effects. Since ROS are highly reactive in nature, excessive production or accumulation of ROS most often proves to be detrimental for the cell. The enhanced level of oxidative stress may show its effect on cells by damaging cellular components and/or by modulating some signaling cascades. A brief overview of the impact of ROS on cellular functioning is discussed below.

Effect of Oxidative Stress on Cellular Components

An increased level of ROS leads to oxidative damage to all the macromolecules such as nucleic acids, proteins, and lipids present in a cell, which in turn causes imbalance in the cellular homeostasis and instigates the aging process [138]. Interestingly, mitochondrion, in spite of being the major source of ROS, also becomes the key target of oxidants. Also, due to the close vicinity of mitochondrial elements to ROS production site, they are more susceptible to the damage by ROS. Lack of histone protection and repair mechanism in mitochondrial DNA further aggravates the susceptibility to ROS-mediated damages. Collectively, these factors add to the risk of mitochondrial dysfunction which has been greatly linked to manifestation of aging process [243, 268]. Several studies have been carried out in *Drosophila* which correlate age-associated changes with the structure and functions of mitochondria, which is indicative of the notion that gradual mitochondrial dysfunctioning is concomitant with aging [268]. For instance, one such study examining the effect of aging on *Drosophila* flight muscles demonstrated a specific "swirl"-like rearrangement of mitochondrial cristae under oxidative stress, with aging [266]. Interestingly, rapid and widespread accrual of similar pathological condition was perceived even in young flies under severe oxidative stress condition. Correlating with the functional aspects of this pathological condition, cristae with swirling pattern were found to have reduced enzymatic activity of cytochrome c (COX) or complex IV, which is an important enzyme complex in ETC involved in ATP production. Moreover, existence of swirls is accompanied by modifications in the structural conformation of cytochrome c, and extensive apoptosis of the flight muscles in *Drosophila* [43, 268].

ETC in mitochondria is associated with energy production in the cell, which is the most vital process required for the maintenance of cellular homeostasis. Studies in *Drosophila* reported an overall decrease in various aspects of ETC with aging [64]. Interestingly, compared to the several other mitochondrial ETC enzymes that were examined, age-associated reduction was predominantly found in the activity of COX [64]. Correspondingly, drug-mediated impairment of COX in young flies results in enhanced ROS production in mitochondria [50]. These observations suggest that ROS-induced mitochondrial impairment results in further enhanced production of ROS which exaggerates the mitochondrial damages, forming a "vicious cycle" and thereby acting as driving force in aging and age-associated impairments [161].

Signaling Cascades Activated by ROS

Several stress pathways like the extracellular signal-regulated kinase (ERK), c-Jun amino-terminal kinase (JNK), p38 mitogen-activated protein kinase (MAPK) signaling cascades, the phosphoinositide 3-kinase [PI(3)K]/Akt pathway, the nuclear factor (NF)-kappaB signaling system, p53 activation, heat shock response, etc. get activated as a mechanism to combat oxidative stress. Here it is important to note that in addition to stress response, these pathways also play essential role(s) during normal growth and metabolism [65]. Among them, the JNK pathway has been identified as an evolutionarily conserved cascade which can potentially increase the lifespan in flies by activating a set of protective genes to mitigate the toxic effects of oxidative stress [270, 271].

Compared to JNK pathway in vertebrates which is relatively complicated due to involvement of huge gene families, JNK signaling in *Drosophila* is significantly less complicated, which makes the genetic analysis much simpler than other model organisms [113, 122]. The JNK signaling pathway in *Drosophila* constitutes various JNK kinase kinases (JNKKK) such as TGF-β-activated kinase1 (TAK1), mixed lineage protein kinase 2/slipper (MLK), MEK kinase 1 (MEKK1), apoptotic signal regulating kinase 1 (ASK1), two JNK kinase (JNKK; Hemipterous and dMKK4), and one JNK [Basket (Bsk)] [19, 25, 40, 74]. Oxidative stress results in activation of transcription factors AP-1 and dFoxo (*Drosophila* forkhead transcription factor) by Bsk phosphorylation, which in turn triggers stress-specific cellular responses by activating several response genes. JNK pathway is negatively regulated by a key target of AP-1 known as puckered (*puc*), which reduces JNK signaling because of JNK-specific phosphatase activity [19, 271]. Genetic manipulations of genes dosage in *Drosophila*, that is, downregulation and overexpression of the *puc* and JNKK/Hep, respectively, enhances the basal JNK signaling levels and leads to enhanced JNK signaling, which in turn result in improved oxidative stress tolerance and increased lifespan [270, 287]. On the contrary, mutant flies for JNKK/Hep gene displayed higher sensitivity toward oxidative stress and were observed incapable of eliciting JNK signaling-dependent transcriptional factor-induced stress response (Fig. 14.5) [275].

It has been found that adequate availability of dFoxo is essential to accomplish JNK signaling-mediated increased longevity in *Drosophila*. This suggests an

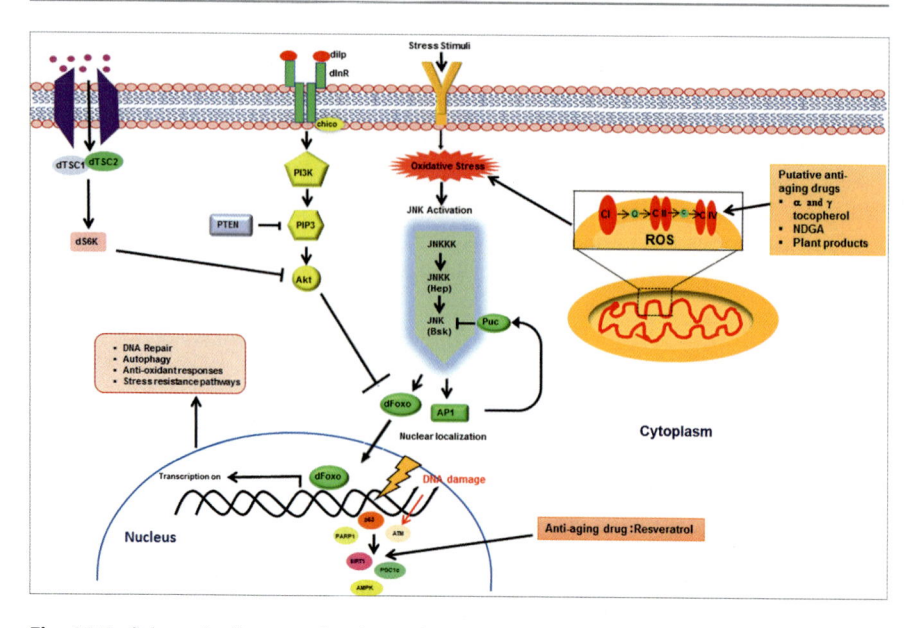

Fig. 14.5 Schematic diagram of various stimuli, participating signaling cascades, and putative drug targets which could modulate progression of aging and longevity in *Drosophila*

antagonistic relationship between JNK and insulin/insulin-like growth factor (IGF)-like signaling (IIS) pathways [271]. In *Drosophila*, JNK inhibits IIS pathway autonomously and/or systemically (endocrine mechanism) to regulate the life expectancy. While functioning cell autonomously, JNK inhibits insulin signaling by promoting nuclear localization of dFoxo, which subsequently activates transcription of genes involved in stress response, damage repair, and growth control [19, 106, 271]. In addition, JNK also inhibits insulin signaling systemically by repressing the expression of its ligand, *Drosophila* insulin-like peptide2 (dilp2) in insulin-producing neuroendocrine cells present in the fly brain [125, 271]. Therefore, antagonistic relationship between JNK and insulin pathway has emerged as an important aspect to combat oxidative stress during progressing of aging.

Proteostasis Loss During Aging

Proteostasis or protein homeostasis deals with the quality control process which regulates the complex signaling pathways to control biogenesis, folding, signaling, and degradation of proteins inside and outside the cell. The process of proteostasis involves various mechanisms such as stabilization of correctly folded protein, refolding of denatured protein, and/or degradation of misfolded proteins by proteasome or lysosomal pathways to remove them from the cell [95, 131, 173]. Moreover, there are regulators such as MOAG-4 which deal with age-related proteotoxicity and act *via* a different pathway from that of molecular chaperones and proteases

[151]. Aging and some age-related diseases impair the cellular protein homeostasis, and defect in these pathways result in progressive accumulation of abnormally folded or misfolded proteins in cellular compartments and triggers pathogenesis of several neurodegenerative disorders [218]. Molecular chaperones represent a major group of proteins which regulate the cellular proteostasis. A brief overview of the molecular chaperones and their involvement in regulation of cellular homeostasis and aging has been discussed below.

Molecular Chaperones Facilitated Protein Folding

It is increasingly clear now that aging is driven by both the genetic and nongenetic factors. In addition to the *in vivo* factors, several environmental stimuli such as cytokines, UV radiation, chemotherapeutic agents, hyperthermia, and some growth factors can also lead to enhanced production of ROS, which may potentially disturb the balance between normal redox levels and oxidative stress [65]. Molecular chaperones are ubiquitous and highly conserved protein families which utilize cycles of ATP-driven conformational changes to either stabilize the nascent and/or stresses-mediated unfolded proteins, or unfold them to translocate across membranes, or mark them for degradation [232]. Molecular chaperones are also regarded as Heat shock proteins (Hsps) or stress proteins because of their induced expression during stress condition(s). With the discovery of the molecular chaperones for the first time in *Drosophila* [222, 223], functions of these proteins and their correlation with aging and longevity have emerged as a prime area of research. Based on their amino acid sequence homologies, molecular weight, and functional aspects, Hsps have been divided into 5 major families: Hsp100 (100–104 kDa), Hsp90 (82–90 kDa), Hsp70 (68–75 kDa), Hsp 60 (58–65 kDa), and small Hsps (15–30 kDa) [232]. As in other model organisms, fly also has homologs of Hsp families, for example, Hsp83 (Hsp90 family), Hsp/Hsc 70 complex (Hsp 70 family), Hsp60, Hsp40, and small Hsps [177]. Expression of the Hsps is facilitated by binding of Heat Shock Factor (HSF) to the heat-shock response elements localized at promoter region of genes, and induction of their high-level transcription [265]. Since the basic property of Hsps involves refolding of protein denatured due to stressors, enhanced expression of Hsps could be observed due to increased level of cellular ROS [175, 185]. It has been reported that subsets of Hsps are also induced by oxidative stress through dFoxo transcription factor and the JNK pathway [271].

Interestingly, constitutive expression of Hsps exhibits well-regulated and stage-specific expression pattern during development; however, enhanced expression of several Hsps could be seen upon exposure to environmental stresses like heat [177]. Some Hsps such as Hsp70 do not express during normal physiological conditions and demonstrate stress-induced expression pattern. The definite involvement of Hsps in aging and increased sensitivity of the aged flies to environmental stimuli has originated from the comparative analysis of the stress response between young and old flies [68]. Subsequently, studies in different animal models support the idea of causative impact of chaperones decline in longevity. For example, transgenic *C.*

elegans and *Drosophila* with increased level of molecular chaperones show relatively longer lifespan [179, 267]. Also, mutant mice with reduced level of co-chaperones show enhanced rate of aging and manifestation of age-associated phenotypes [168].

Comparative analysis of old and young flies showed a greater abundance of damaged proteins in the old flies. Induction of the identical set of proteins in young *Drosophila* fed with canavanine (an arginine analog used to mimic accumulation of damaged proteins which were otherwise present only in old flies) suggests an increased sensitivity due to accumulation of aging-mediated damaged proteins [68, 191]. Moreover, tissue-specific enhanced expression patterns of several Hsps have been reported during normal fly aging [177]. For instance, enhanced expression of *hsp22* and *hsp70* at both RNA and protein level and upregulation of *hsp23* at RNA level could be observed during *Drosophila* aging [177]. Further, with the aim to elucidate transcriptional dynamics due to aging, genome-wide gene expression profiling in *Drosophila* has also revealed age-associated upregulation of several Hsps [49, 215, 290]. Interestingly, in addition to the notable upregulation of subsets of Hsps including Hsp70 and sHsps, enhanced expression of the genes for innate immune response has also been reported in old flies [49, 215, 290]. On the contrary, downregulation of the genes involved in energy synthesis and mitochondrial ETC was found in the same set of flies [49, 215, 290]. In this context, it also important to note that an extensive overlap between the gene expression profile of aged flies and the young flies exposed to oxidative stress further establishes the potential relationship between aging and oxidative stress [290].

The beneficiary effects of Hsps on longevity was also confirmed by the "Hormesis" in *Drosophila*, in which mild dosage of stressors are used to activate stress response without causing cellular damages [169]. Exposure to sublethal levels of stress induces hormetic effect which in turn modulates the heat shock response and this helps the organism to survive longer by reducing the negative effects generated due to aging [169]. Also, *Drosophila* strains with improved lifespan also show intrinsic increased level of cellular sHsps, which further correlates enhanced expression of Hsps with aging [135]. Among multiple sHsp in *Drosophila*, Hsp27, Hsp26, Hsp23, and Hsp22 have been established to increase the lifespan significantly upon tissue-specific overexpression [177, 260, 270]. Enhanced expression of Hsp22 in motor neurons has been found to increase the lifespan by 30%, and these flies also exhibited improved stress tolerance and locomotor activity [179]. In agreement to the above observation, mutation in *hsp70* or *hsp22* has been found to be associated with decreased lifespan and increased sensitivity to stress. The beneficial effect of Hsps in longevity was further demonstrated by the fact that histone deacetylases (HDAC) inhibitors-mediated enhanced expression of Hsp70 and sHsps increases the lifespan of adult flies [288]. Independent studies have revealed decreased survival rate of *Drosophila* when *hsp22* or all six copies of *hsp70* [76, 178] were mutated and exposed to heat and/or other stresses. In addition, it has been found that *hsp83* mutant *Drosophila* becomes more sensitive to the toxic effects of stresses such as sleep deprivation [237]. Unlike sHsps, major Hsps such as Hsp70 and Hsp60 have failed to make any notable effect on longevity, except causing reduced mortality rates upon mild stress, improved heat tolerance, and an insignificant increase in overall lifespan [170, 253]. Therefore, due to ubiquitous nature of Hsps and their crucial involvement in a variety of cellular processes by interacting with various cellular proteins, it can be concluded that the prevalent outcome of aging could be the consequence of the associated chaperone failure, and therefore, molecular chaperones itself represent one of the vital inherent regulators of aging and longevity.

Impact of Epigenetic Changes on Aging

Age-dependent changes in the chromatin configuration and subsequent amendment in gene expression is primarily regulated by various epigenetic modifications. However, establishing a direct correlation between aging and epigenetic modification is complex. The aging-induced epigenetic changes are largely mediated via methylation of the regulatory regions of genes, modification of the core histone proteins, and by controlled expression of several regulatory noncoding RNAs [60]. Different epigenetic changes affect all types of cells in an organism throughout the life [251]. In agreement to this, epigenetic changes in the genome also affect the expression of the genes involved in aging and longevity [89, 133], which subsequently leads to various molecular and physiological changes during the aging process [258]. Some epigenetic modifications like increased H4K16 acetylation, H4K20 or H3K4 trimethylation, and decreased H3K9 methylation or H3K27 trimethylation constitute the hallmark of aging-mediated epigenetic changes [70, 89]. The enzymes involved in generation and maintenance of such epigenetic hallmarks include DNA methyltransferases (Dnmts), histone deacetylases (HDACs) and acetylases, histone methylases and demethylases, as well as some other proteins involved in chromatin modifications. A brief overview of various epigenetic changes and its impact on aging has been provided below.

DNA Methylation

The degree of DNA methylation in genome is inversely proportional to the number of activated genes. Although establishing one-to-one correlation between methylation status and aging is complex, however, some reports have revealed locus-specific hypermethylation of various tumor suppressor genes and polycomb target genes, with advancing age [156]. With aging, the number of methylated cells and the extent of methylation in the CpGs of various promoters increases, which in turn cause reduced gene expression [189]. A connection between chronological age and 5-methylcytosine DNA methylation has been observed in humans [18, 91, 172], and therefore, methylation status can serve as an "aging clock" for determining the chronological age of an individual.

It is increasingly accepted now that DNA methylation regulates the process of aging [17]. Several studies have reported that DR, which is a major risk factor for aging, causes changes in the DNA methylation pattern at specific loci of some cancer-causing genes such as increased methylation of proto-oncogene ras [96].

Also, a study demonstrates that normal cells subjected to 4 weeks of glucose restriction shows increased methylation of tumor repressor $p16^{INK4a}$ [140]. Interestingly, it was found that gene expression of Dnmt 1 (prime methyltransferase under normal conditions) and Dnmt3a significantly declines in aged ells, which is paradoxical to the finding that widespread hypermethylation occurs during aging [35]. This finding was explained based on transcriptional upregulation of the third methyltransferase Dnmt3b which could be responsible for causing increased methylation with advancing age [35].

Histone Modification with Aging
The extent of expression of a gene is also regulated by histone modifications carried out by various acetyl transferases, deacetylases, methyltransferases, and demethylases [133, 181]. Additionally, histone demethylases can modulate lifespan by targeting key components of several pathways affecting longevity [121]. The ADP ribosyltranferases and sirtuin family of NAD-dependent deacetylases have been extensively studied as potential antiaging factors due to their role in chromatin remodeling. This was further supported by the fact that overexpression of dSir2 in *Drosophila* also extends the lifespan [224]. However, these findings came into question with a report showing that perplexed genetic background was the leading factor to cause the increase lifespan, and not only the dSir2 overexpression [30]. Several of the mammalian sirtuins have been reported to delay various factors of aging in mice [107, 235]. Among them, the mitochondrial-located SIRT3 has been shown to mediate some of the beneficial roles of DR in longevity [245]. Interestingly, SIRT3 has also been reported to converse the regenerative capacity of aged hematopoietic stem cells [30].

Levels of some chromosomal enzymes such as heterochromatin protein 1α (HP1α), and chromatin remodeling factors like Polycomb group of proteins, or the NuRD complex get diminished both in normal and pathologically aged cells [206, 217]. This was also supported by the finding that loss of function of HP1α in flies leads to early death, and on the contrary, its overexpression increases longevity and delays muscular deterioration (a characteristic of old age) [137].

Noncoding RNAs
Noncoding genes have emerged as important regulators of the aging-associated epigenetic changes. A variety of noncoding RNAs such as microRNAs (miRNAs), siRNAs, piwi interacting RNAs, QDE-2 interacting RNAs (qiRNAs), and long noncoding RNAs have been found to regulate the epigenetic aspects of aging. They affect various biological processes by regulating the gene expression and also assist in maintenance of the integrity of genome.

Reduced expression of argonaute-like gene-1 (alg-1) affects the lifespan in *C. elegans* [126]. Adequate expression of alg-1 is required for the processing and functioning of miRNAs in *C. elegans* [126]. In *Drosophila*, human homolog of miR-200, i.e., miR-8 affects the aging process by inhibiting PI3K kinase of the insulin-signaling pathway [112]. The miR-8 knockout flies were smaller in size which is also an indicative of defective insulin signaling pathway [111, 112]. The above reports suggest the involvement of miR-8 and miR-200 in regulation of aging in *Drosophila* and human aging, respectively [112].

Long noncoding RNAs have emerged as one of the important regulators of the gene expression. Several noncoding RNAs have been found to control various aging-associated cellular activities such as proliferation, differentiation, quiescence, and stress response [78, 120]. Long noncoding RNAs such as Telomerase RNA component (TERC), Telomeric repeat-containing RNA (TERRA) control telomere length during aging; *Airn*, *PTENpg1-AS*, and *H19* regulate the epigenetic changes; *lncRNA-p21* deals with proteostasis, and *MALAT1*, *ANRIL*, *eRNAs*, and 7SL control cell division; Kcnq1ot1, NeST, and ANRASSF1 regulate histone modifications, and linc-RoR, ES1, ES2, and ES3 have been found to regulate the stem cell behavior [78, 128].

Dietary Restriction

Diet is one of the major factors affecting the quality and duration of life in various living organisms. DR refers to the reduction in energy intake without being malnourished. More explicitly, during DR, calorie intake is restricted by about 30–40% in comparison to controls fed *ad libitum* [115]. Several studies have demonstrated lifespan extension in response to DR [256]. The first report of this kind was published in 1935 in which it was demonstrated that rats subjected to DR displayed increased mean as well as average lifespan [162]. Several hypotheses have been put forward to explain this effect of DR on longevity. One of such hypothesis states that DR slows down the metabolism which in turn restricts the production of reactive oxygen species (ROS), thereby decelerating the aging process [84], whereas the other asserts that DR extends lifespan by delaying the onset of age-related disorders [247]. The latter has been elucidated in David Sinclair's unified theory of aging which perceives DR as a highly complex yet conserved stress response that increases an organism's likelihood of surviving adversity by modulating key cellular processes like cell protection, repair mechanisms, and metabolism [239]. DR has also been shown to prevent muscle damage [165] and inhibit aging cardiomyopathy [12, 284].

The phenomenon of lifespan extension by DR is conserved between species; however, the mechanisms underlying such prolongation may not be conserved which intensifies the problem of comparative DR studies [213]. In case of *Drosophila*, DR effect is brought about by using a new medium for feeding that has controlled concentration of nutrients, which allows determination of specific nutrients that are essential for the organism's response to DR [214]. Since flies visit the food several times each hour to eat, and since restricting the access to food by intermittent starvation leads to considerable deaths [193], diluting the food proved to be more practical and effective [203]. Flies subjected to DR exhibit increased lifespan as compared to their control counterparts [39]. In fact, when flies were shifted from DR food to control food, they adopted an increased mortality rate when compared with the flies subjected to continued DR diet [157]. This suggests that death-causing damage accumulates in both DR and control flies at the same rate, but a high-nutrient diet increases the risk of death. Moreover, it has been suggested that the protein/carbohydrate ratio also plays a role in modulating longevity in *Drosophila* [244]. A higher ratio tends to shorten the lifespan and *vice versa*. Furthermore, it has been demonstrated that within the protein component of the diet also, specific amino acids have crucial roles to play. For example, restriction of methionine extends average and maximal lifespan in *Drosophila* [1]. Additionally, whole genome transcript profiling in *Drosophila* has shown that DR is capable of reverting aging-specific transcriptional changes and limits cell growth, metabolism, and reproduction [215].

DR-mediated effects on longevity have been shown to be brought about by a set of molecular effectors including FOXO transcription factor, AMP kinase (AMPK), sirtuins, Heat shock factor-1 (HSF1), and NRF-2 transcription factor [69]. Inhibition of Akt due to DR activates the transcription factor FOXO which is involved in the upregulation of several pathways such as DNA repair, autophagy, antioxidant responses, stress resistance, and cell proliferation, which in turn promote longevity [272–274]. Similarly, ectopic overexpression of a few sirtuins like SIRT1, SIRT3, and SIRT6 reduces NF-κB signaling, increases genomic stability, and improves metabolic homeostasis via histone deacetylation [83]. Combined activation of SIRT1 and AMPK activates PGC-1α, which is a major transcriptional regulator of mitochondrial function and antioxidant defense [281]. Further, DR-mediated upregulation of HSP70 and p62 activates the transcription factors HSF-1 and NRF-2, which are also involved in enhancing antioxidant responses, preventing age-dependent impairment of proteostasis and promoting maintenance of cell structure and metabolism [3]. Thus, multiple yet parallel processes contribute to DR-mediated lifespan extension.

Insulin Signaling/mTOR Network

The **I**nsulin/**I**nsulin-like Growth Factor (IGF) **S**ignaling (IIS) pathway is one of the major pathways involved in cellular metabolism and growth and differentiation of somatic cells, whereas the mTOR pathway is vital for nutrient/energy/redox sensing and control of protein synthesis in the cell. DR has been suggested to exert its modulation on lifespan mostly via the IIS/mTOR network [56]. DR reduces plasma insulin/IGF levels in humans [275], and evidences comply that compromised insulin signaling results in increased lifespan in various model organisms [73]. Partridge and coworkers reported the first IIS mutation that extends lifespan which was present in the *Drosophila* homolog of the insulin receptor substrate CHICO [44]. The *chico* null flies were found to exhibit up to 48% increased median lifespan in homozygous females, 31% in heterozygous females, and 13% in homozygous males. Subsequently, it was also found that a hypomorphic mutation in the *Drosophila* insulin receptor (dINR) also affects the longevity positively [254]. Interestingly, these mutants also display increased triglyceride content and super oxide dismutase (SOD) activity pointing toward enhanced stress response, and thus increased lifespan. Similarly, reduced expression of the *Drosophila* insulin-like peptides (dilps), the ligands for dINR, also extends lifespan [80]. In fact, ablation of dilp-producing cells and median neurosecretory cells (MNCs) in the late-staged larval brain also produces an analogous effect [29], as does the deletion of dilp-encoding genes [80]. Furthermore, enhanced expression of dPTEN (*Drosophila* phosphatase and tensin homolog), a negative regulator of IIS signaling, has been shown to bring about lifespan extension by antagonizing the action of phosphatidylinositol-3-kinase (PI3K) and promoting nuclear localization of FOXO [111]. This, in turn, causes an upsurge in transcriptional activity of FOXO. Since inhibition of the activity of *Drosophila* FOXO homolog *daf-16* leads to decreased lifespan [127], it is affirmatory that the key molecular effector of IIS in context of aging is dFOXO. Inevitably, downregulation of 14-3-3ε, the negative regulator of FOXO, also tends to extend lifespan [192].

IIS has been suggested to exert its effects, in part, by the mTOR pathway. It has been observed that systemic overexpression of dTSC1 and dTSC2, antagonists of TOR activity, increases lifespan [124]. Similarly, expression of a dominant negative form of TOR or mutating the major downstream effector of this pathway, S6 kinase (S6K), also extends lifespan [124]. It was also demonstrated that reduced TOR activity exhibits ~20% increase in median lifespan without any associated stress resistance, as compared to the controls [156]. Moreover, rapamycin-mediated inhibition of mTORC1, the chief signaling complex of TOR pathway, also increases lifespan [20]. Inhibition of mTORC1 enhances processes like proteostasis, autophagy, and stem cell functions [69]. Since autophagic processes are activated in response to damaged or malfunctioning of proteins and/or organelles, they play a vital role in eliminating the damaged macromolecules and/organelles that contribute to intensifying the aging process. In fact, it has been demonstrated that inhibition of autophagic processes makes positive impact on longevity [20].

Another important cellular pathway involved in stress response against DR and inducing longevity is the Jun-N-terminal kinase (JNK) pathway. Although it acts as an independent pathway in the cell, it ultimately converges at the same molecular effectors as the IIS and mTOR pathway. JNK primarily antagonizes IIS and causes FOXO to localize to the nucleus and activate its downstream gene targets [271]. Taken together, it is increasingly clear now that several pathways act synergistically in the cells to bring about lifespan extension without making any adverse effects or fitness cost.

Aging-Associated Diseases

The risk of developing several diseases such as diabetes type 2, heart diseases, obesity, cancer, arthritis, kidney, and neurodegenerative disorders such as Parkinson's disease (PD) and Alzheimer's disorder (AD) increases with aging. With a rapidly growing aging population, these disorders have become a prodigious economic burden on the society. Therefore, due to absence of effective therapies, it has become even essential to find effective strategies for the benefit of the aging population. As discussed earlier, manifestation of aging-associated impairments could be minimized to certain extent by genetic, dietary, and pharmacological interventions, which generally target different molecular pathways involved in aging, because these diseases have shown interference with age-related molecular mechanisms. Advancing mechanistic understanding of aging-associated diseases might be a clue for development of new therapeutic strategies.

Interestingly, age comes out as a critical factor for the onset of several human neurodegenerative disorders. Neuronal loss, shrinkage of cell bodies and axons of neuronal cells, and loss of synapse collectively lead to reduced brain volume and weight in aging individuals, who are cognitively normal [221]. Neurofibrillary tangles and senile plaques that show sparse distribution are neuropathological hallmark of AD, which have been found to accumulate in cortical region and adversely affect the cognitive function of the individual [86]. Similarly, common pathology of polyglutamine [poly(Q)]-mediated neurotoxicity in a variety of poly(Q) disorders is presented by degeneration of neuronal cell bodies, axons, synapse, and specific parts of the nervous system [61]. Moreover, it is still enigmatic whether both aging and disease-associated proteins act synergistically to extend neuronal dysfunctions, or only aging-related changes are accountable for driving the neuronal pathology.

It appears relatively coherent to hypothesize that disease-related proteins enhance disease toxicity by accelerating the aging process. For instance, in *C. elegans*, mutation that increases longevity in poly(Q) disease divulges age-dependent reduction in protein aggregate formation and toxicity, subsequently affirming the effect of aging in poly(Q)-mediated cellular dysfunction [176]. Several reports including our own findings demonstrate gradually aggravating poly(Q)-mediated neurotoxicity in an age-dependent manner [240]. Targeted expression of SCA-78(Q) in *Drosophila* eye causes manifestation of poly(Q) disease in form of cellular degeneration, retinal depigmentation, and neurotoxicity [240]. Our studies on flies expressing SCA3-78(Q) transgene during aging suggest that the extent of retinal depigmentation and cellular toxicity gradually increases with age. Similarly, in *Drosophila* human neuronal tauopathy models, tissue-specific expression of human tau (h-tau) transgene causes severe degradation of neuronal tissue [38]. Figure 14.6 depicts extensive degeneration of mushroom body upon pan neuronal expression of h-tau transgene in 3-day-old *Drosophila* adult brain. Mushroom body of *Drosophila* is a specialized structure which functions as a center of associative learning and also regulates a wide range of behaviors including habituation, olfactory learning, temperature preference, and sleep [164]. Contribution of common signaling networks in longevity and alleviation of neurodegenerative disorders further suggests that slowing down the aging process may act as a neuroprotective measure. Therefore, in order to develop novel strategies to obstruct onset and progression of such deadly disorders, it will be interesting to walk around how aging dysfunction and neuropathology are intertwined, and how they act together during disease pathogenesis.

As stated earlier, all eukaryotic life forms have well-regulated protein quality control system which includes chaperone network, ubiquitin–proteasome system, and lysosome-mediated autophagy. Proper functioning of this system is essential to achieve post-translational modifications, protein folding, stress response, and clearance/translocation of damaged proteins [11, 246]. It has been found that process of aging deteriorates the functional capacity of the cellular protein folding machinery,

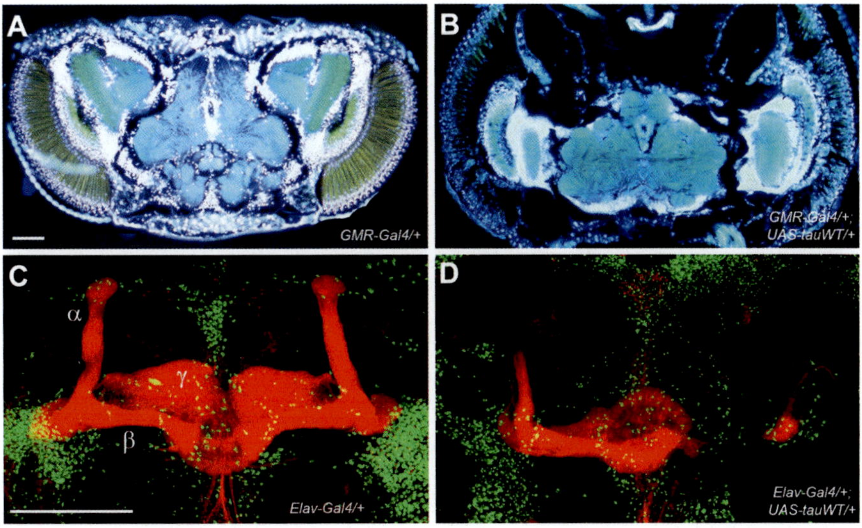

Fig. 14.6 Paraffin sections of a 3-day-old adult head across the midbrain stained with DAPI. In comparison to wild-type (**a**), eye-specific expression of human tau (h-tau) transgene results in severe tissue degeneration (arrows in **b**). Anti-Fasciclin II (FasII) staining shows that compared to the wild-type mushroom body with distinct presence of α, β, and γ lobes (**c**), pan neuronal expression of h-tau transgene results in notable degeneration of mushroom body (**d**) as distinctly seen in α, β, and γ lobes (arrowhead in **d**) (Scale **a**, **b** = 100 μm; **c**, **d** = 100 μm)

proteosome activity and the stress response; therefore, the post-mitotic neurons become susceptible to toxic protein aggregates and ultimately lead to cell death [184]. Several studies have been performed using *Drosophila* to illustrate the potential role(s) of molecular chaperones in suppression of neurodegenerative disorders. Not surprisingly, tissue-specific upregulation of molecular chaperones ameliorates the disease toxicity and also minimizes age-related cellular impairments. Targeted upregulation of Hsp70 along with Hsp40/DnaJ (HJD1) suppresses neurodegenerative phenotypes and also improves the lifespan in *Drosophila* Machado–Joseph disease (MJD) and Huntington disease (HD) model [184]. Further, role of Hsp70 and Hsp40 in regulation of poly(Q) aggregation and cellular toxicity has been further validated in *S. cerevisiae*, *C. elegans*, and mouse [48, 185].

In order to explain the progressive decline of Hsps in neurodegenerative diseases, several mechanisms have been hypothesized, including transcriptional deficit of *hsps* expression via toxic misfolded protein, and sequestration of cellular soluble Hsps along with the toxic aggregates to form inclusion bodies [90]. Moreover, evidences like CBP-induced transcriptional impairment of Hsp70 in *Drosophila* via reduction of Hsf-1 activity further support the transcriptional deficit hypothesis [90]. It appears that misregulation of molecular pathways and several factors those are responsible for regulation of protein quality control mechanism at cellular level might be the risk factor for disease occurrence. Therefore, novel therapeutic strategies could be provided by rejuvenating the protein quality control machinery for restoration of cellular homeostasis and to delay the aging onset of diseases.

Focusing on the disease-associated stress condition, it was fascinating to find a correlation between insulin/IGF-1 signaling in protein aggregation and toxicity, as aging is the key factor for disease onset. For the first time, studies on the *C. elegans* provided a direct link between insulin/IGF-1 signaling and protein aggregation. It was demonstrated that reduced level of insulin/IGF-1 neutralizes the poly(Q) aggregation and protects the worms from motility impairment and neurotoxicity [255]. The above finding suggests that lowered level of insulin/IGF-1 signaling pathway restricts the neurodegenerative disease phenotypes by modulating the aging process. Subsequently, studies in other model systems also suggest neuroprotective properties of insulin/IGF-1 signaling, which is primarily achieved by modulating the aging processes. Therefore, insulin/IGF-1 signaling can be considered as a novel target to combat aging-mediated impairment.

Antiaging Drugs and Natural Products

As discussed earlier, aging is a complex process due to involvement of multiple factors which influence this phenomenon. Such factors include genetic components, environment, metabolism, as well as reproduction. These multiple factors generate logistical difficulties in the development and evaluation of antiaging compounds. Therefore, studies focused on relatively simpler model organisms such as *Drosophila* and *C. elegans* have emerged as excellent systems for screening of genetic modifiers and antiaging drug molecules. In these model organisms, longevity can be altered and scored by genetic manipulations, and potential drugs which can increase the lifespan. Such antiaging compounds could be identified and categorized based on their functioning and mode of action. In this context, it is also important to note that several physiological and biological pathways are conserved in humans and *Drosophila*. Several antiaging molecules such as anticonvulsants (ethosuximide), antidepressants (mianserin), antioxidants, and others such as inhibitors of histone deacetylase, and resveratrol, a sir2 activator, have been identified and characterized in these model organisms [134]. A brief collection of antiaging molecules which modulate aging in *Drosophila* and other model systems has been provided in Table 14.2.

It was reported for the first time in *Drosophila* that resveratrol extends lifespan by activating sirtuins, without making any negative impact on fecundity [279]. This finding was further supported by the fact that feeding of resveratrol and rapamycin to 1-year-old mice improves the lifespan and heath, respectively [15, 93]. In view of above findings, it was postulated that resveratrol-induced increase in lifespan was sirtuin-dependent, and functions through pathways related to caloric intake [279]. However, studies based on biochemical assays with native substrates suggest that resveratrol does not activate SIRT1 directly [198]. Therefore, it appears that pharmacologic–genetic interplay should be taken into account while investigating the antiaging compounds and their operating mechanism(s). Moreover, this could also facilitate screening of additional genes and pathways which influence the aging process.

Table 14.2 A brief collection of some drugs and natural products found to delay aging and extend lifespan

S.no.	Drug/antiaging compound	Molecular target(s)	Generalized function	References
1.	Metformin	Mitochondrial respiratory complex I, AKT/TOR signaling modulation	Treatment of type 2 diabetes, antitumor; extends life span and inhibit age-related centrosome amplification in *Drosophila*	[5, 188]
2.	NSAID (celecoxib)	3′-Phosphoisositide-dependent kinase-1 (PDK-1) component of insulin/IGF-1 signaling cascade	Increases life span in *C. elegance*	[41]
3.	NSAID (ibuprofen)	Inhibits the tryptophan permease Tat2p, a component of Pkh2-ypk1-lem3-tat2 signaling pathway	Increases life span in *S. cerevisiae*, *C. elegance*, and *Drosophila*	[101]
4.	Sc-560, trans resveratrol, and Valdecoxib Aspirin, and NS-398, APHS, valeryl salicylate	Inhibition of COX and reduction in the production of ROS	Increases life span in *Drosophila*	[50]
5.	Ethosuximide	Inhibitor of T-type calcium channel, anti-convulsant, inhibits the function of specific chemosensory neurons	Delay age-related changes and extend life span of *C. elegance*	[46]
6.	Lithium	Inhibits GSK3 and activates NRF-2	Extends life span of *Drosophila*	[36]
7.	Spermidine (natural polyamine)	Activates autophagic machinery	Extends lifespan of *S. cerevisiae*, *C. elegance*, and *Drosophila*	[57]
8.	Sodium butyrate	HDAC inhibitor	Promotes longevity in *Drosophila*	[263]
9.	Cranberry plant extract	Minimizes oxidative stress, activates ERK/MAPK signaling and AKT pathway	Promote longevity in *Drosophila* and *C. elegance*	[85, 272, 273]

(continued)

Table 14.2 (continued)

S.no.	Drug/antiaging compound	Molecular target(s)	Generalized function	References
10.	Blueberry plant extract	Upregulates superoxide dismutase (SOD), catalase (CAT), and Rpn11, and downregulates methuselah (*mts*)	Promotes longevity in *Drosophila*	[207]
11	Extract of *Rhodiola rosea*	Acts against oxidative stress and decreases the production of ROS	Extends lifespan in *Drosophila*, *C. elegance* and *S. cerevisiae*	[16, 118, 276]
12.	Extract of *Rosa amascena*	Acts against oxidative stress	Extends lifespan in *Drosophila*	[119]
13.	Curcumin	Activates TOR pathway	Extends lifespans in *Drosophila*	[139, 242]
14.	Extract of *Ludwigia octovalvis*	Activates AMP-activated protein kinase (AMPK) pathway	Extends lifespan in *Drosophila*	[144]

The genetic approach in *Drosophila* generally follows the effects of antiaging drugs or compounds on pathways those are potentially involved in the aging process; there are still several pharmacologic compounds that are supposed to exert an impact on aging; however, their mechanisms are yet to be determined. Such putative antiaging compounds may facilitate discovery of novel antiaging compounds and also help in unraveling additional insights into the antiaging pathways. Several studies were focused on to examine the antioxidant effects of some selected compounds such as tocopherol-p-chloro-phenoxy acetate, nordihydroguaiaretic acid (NDGA), and Mg-TCA, and α and γ-tocopherol in the *Drosophila* [286, 291]. Interestingly, Jafari and coworkers have identified several antiaging pharmaceutical and botanical agents using *Drosophila* as a model organism [117]. A few of such antiaging agents include extracts of the plants *Rhodiola rosea*, *Rosa damascene*, cinnamon, green tea, and antidiabetic drug pioglitazone [117, 150, 234]. Regulated dosage of above compounds has been suggested to decrease the mortality rate in male and female flies without making any significant negative impact [117, 150, 234]. Further, extracts from cranberry plant has been found to contain antiaging and anti-inflammatory bioactive compounds and found to extend the lifespan in *C. elegans* and *Drosophila* significantly [85, 100, 108, 190, 272, 273]. Age-related functional decline of pancreatic β-cells in rats has been found to be delayed by using cranberry extract [289]. Interestingly, extracts from cranberry plant have been suggested to induce some epigenetic changes in chromatin, which in turn alter the dynamics of the aging-related signaling pathways and minimize the cellular damages [250].

As discussed earlier, chronic inflammation is associated with development of several aging-related diseases, and therefore, pharmacological inhibition of inflammatory processes using certain drugs has emerged as an effective antiaging strategy. A large number of nonsteroidal anti-inflammatory drugs (NSAIDs) such as

CAY10404, aspirin, APHS, SC-560, NS-398, SC-58125, valeroyl salicylate, transresveratrol, valdecoxib, and licofelone have been screened using *Drosophila* [50]. It was deduced that regulated feeding of anti-inflammatory drugs to *Drosophila* results in extended lifespan, delayed age-dependent decline of locomotor activities, and increased stress resistance [50].

Although numerous pharmacological drugs and natural compounds have been studied using various model organisms to target the aging phenomenon individually, however, the major challenge remains that once an antiaging compound shows antiaging activity deprived of any impact on physiological processes, or any undesirable effects on health span, it may require further evaluation in other genetic backgrounds or in additional model organisms. Further, deciphering the mechanism of action of such compounds could be worthwhile for identifying added antiaging agents or assessing combination therapies. However, there are still quite a lot of challenges for studies in this field, which is a major prerequisite to overcome.

Concluding Remarks

Aging research has perceived a notable acceleration due to the inclusion of several model organisms and development of contemporary tools that permit rapid screening of genetic modifiers and novel drug molecules, and analysis of the genome, transcriptome, epigenome, proteome, and metabolome of aging cells and tissues. This information is now being utilized for development of possible therapies to minimize the deleterious aspects of aging. It is increasingly clear now that aging is not an irreversible process, and senescence is not the inevitable fate of all organisms and it could be significantly delayed without any significant fitness cost. However, in spite of a considerable advancement in aging research, several questions related to molecular and neurological aspects of aging remain to be answered. Besides, most of the life-extension molecules/mechanisms have been observed in simpler model organisms, and these have still to be verified as viable antiaging therapies in humans. Here, it is also worth considering that a number of genetic manipulations which extend lifespan in *Drosophila* and other species show ex-specific inclination. The histrionic progress made in recent years established the feasibility to disentangle the mysteries of aging and to reach to a logical and decisive conclusion.

Acknowledgments Research programs in the laboratory have been supported by grants from the Department of Science and Technology (DST), Department of Biotechnology (DBT), Council of Scientific and Industrial Research (CSIR), Government of India, New Delhi, and DU/DST-PURSE scheme to SS. Nisha, KR, Pragati, ST, and SIC are supported by DBT-SRF, UGC-SRF, CSIR-JRF, UGC-JRF, and UGC-SRF fellowships, respectively.

References

1. Ables GP, Brown-Borg HM, Buffenstein R, Church CD, Elshorbagy AK, Gladyshev VN, Huang TH, Miller RA, et al. The first international mini-symposium on methionine restriction and lifespan. Front Genet. 2014;5:122.
2. Adams MD, Celniker SE, Holt RA, Evans CA, Gocayne JD, Amanatides PG, Scherer SE, Li PW, et al. The genome sequence of *Drosophila melanogaster*. Science. 2000;287:2185–95.
3. Akerfelt M, Morimoto RI, Sistonen L. Heat shock factors: integrators of cell stress, development and lifespan. Nat Rev Mol Cell Biol. 2010;11:545–55.
4. Anantharaju A, Feller A, Chedid A. Aging liver. A review. Gerontology. 2002;48:343–53.
5. Anisimov VN. Metformin: do we finally have an anti-aging drug? Cell Cycle. 2013;12:3483–9.
6. Antico Arciuch VG, Elguero ME, Poderoso JJ, Carreras MC. Mitochondrial regulation of cell cycle and proliferation. Antioxid Redox Signal. 2012;16:1150–80.
7. Arking R. Biology of aging: observations and principles. 3rd ed. Prentice Hall: Englewood Cliffs; 1991.
8. Arking R, Buck S, Berrios A, Dwyer S, Baker GT 3rd. Elevated paraquat resistance can be used as a bioassay for longevity in a genetically based long-lived strain of *Drosophila*. Dev Genet. 1991;12:362–70.
9. Armanios M, Blackburn EH. The telomere syndromes. Nat Rev Genet. 2012;13:693–704.
10. Armanios M, Alder JK, Parry EM, Karim B, Strong MA, Greider CW. Short telomeres are sufficient to cause the degenerative defects associated with aging. Am J Hum Genet. 2009;85:823–32.
11. Arslan MA, Csermely P, Soti C. Protein homeostasis and molecular chaperones in aging. Biogerontology. 2006;7:383–9.
12. Bales CW, Kraus WE. Caloric restriction: implications for human cardiometabolic health. J Cardiopulm Rehabil Prev. 2013;33:201–8.
13. Barnes AI, Wigby S, Boone JM, Partridge L, Chapman T. Feeding, fecundity and lifespan in female *Drosophila melanogaster*. Proc Biol Sci. 2008;275:1675–83.
14. Bauer JH, Morris SN, Chang C, Flatt T, Wood JG, Helfand SL. dSir2 and Dmp53 interact to mediate aspects of CR-dependent lifespan extension in *D. melanogaster*. Aging (Albany NY). 2009;1:38–48.
15. Baur JA, Pearson KJ, Price NL, Jamieson HA, Lerin C, Kalra A, Prabhu VV, Allard JS, et al. Resveratrol improves health and survival of mice on a high-calorie diet. Nature. 2006;444:337–42.
16. Bayliak MM, Lushchak VI. The golden root, *Rhodiola rosea*, prolongs lifespan but decreases oxidative stress resistance in yeast Saccharomyces cerevisiae. Phytomedicine. 2011;18:1262–8.
17. Ben-Avraham D, Muzumdar RH, Atzmon G. Epigenetic genome-wide association methylation in aging and longevity. Epigenomics. 2012;4:503–9.
18. Benayoun BA, Pollina EA, Brunet A. Epigenetic regulation of ageing: linking environmental inputs to genomic stability. Nat Rev Mol Cell Biol. 2015;16:593–610.
19. Biteau B, Karpac J, Hwangbo D, Jasper H. Regulation of *Drosophila* lifespan by JNK signaling. Exp Gerontol. 2011;46:349–54.
20. Bjedov I, Toivonen JM, Kerr F, Slack C, Jacobson J, Foley A, Partridge L. Mechanisms of lifespan extension by rapamycin in the fruit fly *Drosophila melanogaster*. Cell Metab. 2010;11:35–46.
21. Blasco MA. Telomere length, stem cells and aging. Nat Chem Biol. 2007;3:640–9.
22. Bodnar AG, Ouellette M, Frolkis M, Holt SE, Chiu CP, Morin GB, Harley CB, Shay JW. Extension of life-span by introduction of telomerase into normal human cells. Science. 1998;279:349–52.
23. Bokov A, Chaudhuri A, Richardson A. The role of oxidative damage and stress in aging. Mech Ageing Dev. 2004;125:811–26.

24. Boonekamp JJ, Simons MJ, Hemerik L, Verhulst S. Telomere length behaves as biomarker of somatic redundancy rather than biological age. Aging Cell. 2013;12:330–2.
25. Boutros M, Agaisse H, Perrimon N. Sequential activation of signaling pathways during innate immune responses in *Drosophila*. Dev Cell. 2002;3:711–22.
26. Brand MD. The sites and topology of mitochondrial superoxide production. Exp Gerontol. 2010;45:466–72.
27. Brand AH, Perrimon N. Targeted gene expression as a means of altering cell fates and generating dominant phenotypes. Development. 1993;118:401–15.
28. Branson K, Robie AA, Bender J, Perona P, Dickinson MH. High-throughput ethomics in large groups of *Drosophila*. Nat Methods. 2009;6:451–7.
29. Broughton SJ, Piper MD, Ikeya T, Bass TM, Jacobson J, Driege Y, Martinez P, Hafen E, et al. Longer lifespan, altered metabolism, and stress resistance in *Drosophila* from ablation of cells making insulin-like ligands. Proc Natl Acad Sci U S A. 2005;102:3105–10.
30. Brown K, Xie S, Qiu X, Mohrin M, Shin J, Liu Y, Zhang D, Scadden DT, et al. SIRT3 reverses aging-associated degeneration. Cell Rep. 2013;3:319–27.
31. Burnett C, Valentini S, Cabreiro F, Goss M, Somogyvári M, Piper MD, Hoddinott M, Sutphin GL, et al. Absence of effects of Sir2 overexpression on lifespan in *C. elegans* and *Drosophila*. Nature. 2011;477:482–5.
32. Burtner CR, Kennedy BK. Progeria syndromes and ageing: what is the connection? Nat Rev Mol Cell Biol. 2010;11:567–78.
33. Cabanillas R, Cadinanos J, Villameytide JA, Perez M, Longo J, Richard JM, Alvarez R, Duran NS, et al. Nestor-Guillermo progeria syndrome: a novel premature aging condition with early onset and chronic development caused by BANF1 mutations. Am J Med Genet A. 2011;155A:2617–25.
34. Cao K, Blair CD, Faddah DA, Kieckhaefer JE, Olive M, Erdos MR, Nabel EG, Collins FS. Progerin and telomere dysfunction collaborate to trigger cellular senescence in normal human fibroblasts. J Clin Invest. 2011;121:2833–44.
35. Casillas MA Jr, Lopatina N, Andrews LG, Tollefsbol TO. Transcriptional control of the DNA methyltransferases is altered in aging and neoplastically transformed human fibroblasts. Mol Cell Biochem. 2003;252:33–43.
36. Castillo-Quan JI, Li L, Kinghorn KJ, Ivanov DK, Tain LS, Slack C, Kerr F, Nespital T, et al. Lithium promotes longevity through GSK3/NRF2-dependent hormesis. Cell Rep. 2016;15:638–50.
37. Chance B, Sies H, Boveris A. Hydroperoxide metabolism in mammalian organs. Physiol Rev. 1979;59:527–605.
38. Chanu SI, Sarkar S. Targeted downregulation of dMyc suppresses pathogenesis of human neuronal tauopathies in *Drosophila* by limiting heterochromatin relaxation and Tau hyperphosphorylation. Mol Neurobiol. 2017;54:2706–19.
39. Chapman T, Partridge L. Female fitness in *Drosophila melanogaster*: an interaction between the effect of nutrition and of encounter rate with males. Proc R Soc Lond Ser B. 1996;263:755–9.
40. Chen W, White MA, Cobb MH. Stimulus-specific requirements for MAP3 kinases in activating the JNK pathway. J Biol Chem. 2002;277:49105–10.
41. Ching TT, Chiang WC, Chen CS, Hsu AL. Celecoxib extends *C. elegans* lifespan via inhibition of insulin-like signaling but not cyclooxygenase-2 activity. Aging Cell. 2011;10:506–19.
42. Chippindale AK, Leroi AM, Kim SB, Rose MR. Phenotypic plasticity and selection in *Drosophila* life-history evolution. I. Nutrition and the cost of reproduction. J Evol Biol. 1993;6:171–93.
43. Cho J, Hur JH, Walker DW. The role of mitochondria in *Drosophila* aging. Exp Gerontol. 2011;46:331–4.
44. Clancy DJ, Gems D, Harshman LG, Oldham S, Stocker H, Hafen E, Leevers SJ, Partridge L. Extension of life-span by loss of CHICO, a *Drosophila* insulin receptor substrate protein. Science. 2001;292:104–6.

45. Clark AM, Gould AB. Genetic control of adult lifespan in *Drosophila melanogaster*. Exp Gerontol. 1970;15:157–62.
46. Collins JJ, Evason K, Pickett CL, Schneider DL, Kornfeld K. The anticonvulsant ethosuximide disrupts sensory function to extend *C. elegans* lifespan. PLoS Genet. 2008;4:e1000230.
47. Conley KE, Jubrias SA, Esselman PC. Oxidative capacity and ageing in human muscle. J Physiol. 2000;526:203–10.
48. Cummings CJ, Sun Y, Opal P, Antalffy B, Mestril R, Orr HT, Dillmann WH, Zoghbi HY. Overexpression of inducible HSP70 chaperone suppresses neuropathology and improves motor function in SCA1 mice. Hum Mol Genet. 2001;10:1511–8.
49. Curtis C, Landis GN, Folk D, Wehr NB, Hoe N, Waskar M, Abdueva D, Skvortsov D, et al. Transcriptional profiling of MnSOD-mediated lifespan extension in *Drosophila* reveals a species-general network of aging and metabolic genes. Genome Biol. 2007;8:R262.
50. Danilov A, Shaposhnikov M, Shevchenko O, Zemskaya N, Zhavoronkov A, Moskalev A. Influence of non-steroidal anti-inflammatory drugs on *Drosophila melanogaster* longevity. Oncotarget. 2015;6:19428–44.
51. De Sandre-Giovannoli A, Bernard R, Cau P, Navarro C, Amiel J, Boccaccio I, Lyonnet S, Stewart CL, et al. Lamin a truncation in Hutchinson-Gilford progeria. Science. 2003;300:2055.
52. Dechat T, Pfleghaar K, Sengupta K, Shimi T, Shumaker DK, Solimando L, Goldman RD. Nuclear lamins: major factors in the structural organization and function of the nucleus and chromatin. Genes Dev. 2008;22:832–53.
53. Dietzl G, Chen D, Schnorrer F, Su KC, Barinova Y, Fellner M, Gasser B, Kinsey K, et al. A genome-wide transgenic RNAi library for conditional gene inactivation in *Drosophila*. Nature. 2007;448:151–6.
54. Dröge W. Aging-related changes in the thiol/disulfide redox state: implications for the use of thiol antioxidants. Exp Gerontol. 2002;371:333–1345.
55. Dudas SP, Arking R. A coordinate upregulation of antioxidant gene activities is associated with the delayed onset of senescence in a long-lived strain of *Drosophila*. J Gerontol A Biol Sci Med Sci. 1995;50:B117–27.
56. Efeyan A, Zoncu R, Sabatini DM. Amino acids and mTORC1: from lysosomes to disease. Trends Mol Med. 2012;18:524–33.
57. Eisenberg T, Knauer H, Schauer A, Beuttner S, Ruckenstuhl C, Carmona-Gutierrez D, Ring J, Schroeder S, et al. Induction of autophagy by spermidine promotes longevity. Nat Cell Biol. 2009;11:1305–14.
58. Eriksson M, Brown WT, Gordon LB, Glynn MW, Singer J, Scott L, Erdos MR, Robbins CM, et al. Recurrent de novo point mutations in lamin A cause Hutchinson-Gilford progeria syndrome. Nature. 2003;423:293–8.
59. Espada J, Varela I, Flores I, Ugalde AP, Cadinanos J, Pendas AM, Stewart CL, Tryggvason K, et al. Nuclear envelope defects cause stem cell dysfunction in premature-aging mice. J Cell Biol. 2008;181:27–35.
60. Esteller M. CpG island hypermethylation and tumor suppressor genes: a booming present, a brighter future. Oncogene. 2002;21:5427–40.
61. Fan HC, Ho LI, Chi CS, Chen SJ, Peng GS, Chan TM, Lin SZ, Harn HJ. Polyglutamine (PolyQ) diseases: genetics to treatments. Cell Transplant. 2014;23:441–8.
62. Fang EF, Scheibye-Knudsen M, Brace LE, Kassahun H, SenGupta T, Nilsen H, Mitchell JR, Croteau DL, et al. Defective mitophagy in XPA via PARP-1 hyperactivation and NAD(+)/SIRT1 reduction. Cell. 2014;157:882–96.
63. Fang EF, Scheibye-Knudsen M, Chua KF, Mattson MP, Croteau DL, Bohr VA. Nuclear DNA damage signalling to mitochondria in ageing. Nat Rev Mol Cell Biol. 2016;17:308–21.
64. Ferguson M, Mockett RJ, Shen Y, Orr WC, Sohal RS. Age-associated decline in mitochondrial respiration and electron transport in *Drosophila melanogaster*. Biochem J. 2005;390:501–11.
65. Finkel T, Holbrook NJ. Oxidants, oxidative stress and biology of ageing. Nature. 2000;408:239–47.
66. Flatt T. Survival costs of reproduction in *Drosophila*. Exp Gerontol. 2011;46:369–75.
67. Flatt T. A new definition of aging? Front Genet. 2012;3:148.

68. Fleming JE, Walton JK, Dubitsky R, Bensch KG. Aging results in an unusual expression of *Drosophila* heat shock proteins. Proc Natl Acad Sci U S A. 1988;85:4099–103.
69. Fontana L, Partridge L. Promoting health and longevity through diet: from model organisms to humans. Cell. 2015;161:106–18.
70. Fraga MF, Esteller M. Epigenetics and aging: the targets and the marks. Trends Genet. 2007;23:413–8.
71. Gardner TS. The use of *Drosophila melanogaster* as a screening agent for longevity factors II. The effects of biotin, pyridoxine, sodium yeast nucleate, and pantothenic acid on the lifespan of the fruit fly. J Gerontol. 1948;3:9–13.
72. Gargano JW, Martin I, Bhandari P, Grotewiel MS. Rapid iterative negative geotaxis (RING): a new method for assessing age-related locomotor decline in *Drosophila*. Exp Gerontol. 2005;40:386–95.
73. Gems D, Partridge L. Insulin/IGF signalling and ageing: seeing the bigger picture. Curr Opin Genet Dev. 2001;11:287–92.
74. Geuking P, Narasimamurthy R, Lemaitre B, Basler K, Leulier F. A non-redundant role for *Drosophila* Mkk4 and hemipterous/Mkk7 in TAK1-mediated activation of JNK. PLoS One. 2009;4:e7709.
75. Giannakou ME, Goss M, Jünger MA, Hafen E, Leevers SJ, Partridge L. Long-lived *Drosophila* with overexpressed dFOXO in adult fat body. Science. 2004;305:361.
76. Gong WJ, Golic KG. Loss of Hsp70 in *Drosophila* is pleiotropic, with effects on thermotolerance, recovery from heat shock and neurodegeneration. Genetics. 2006;172:275–86.
77. Gonzalez-Suarez I, Redwood AB, Perkins SM, Vermolen B, Lichtensztejin D, Grotsky DA, MorgadoPalacin L, Gapud EJ, et al. Novel roles for A-type lamins in telomere biology and the DNA damage response pathway. EMBO J. 2009;28:2414–27.
78. Grammatikakis I, Panda AC, Abdelmohsen K, Gorospe M. Long noncoding RNAs(lncRNAs) and the molecular hallmarks of aging. Aging. 2014;6:992–1009.
79. Greider CW, Blackburn EH. Identification of a specific telomere terminal transferase activity in *Tetrahymena* extracts. Cell. 1985;43:405–13.
80. Gronke S, Clarke DF, Broughton S, Andrews TD, Partridge L. Molecular evolution and functional characterization of *Drosophila* insulin-like peptides. PLoS Genet. 2010;6:e1000857.
81. Grotewiel MS, Martin I, Bhandari P, Cook-Wiens E. Functional senescence in *Drosophila melanogaster*. Aging Res Rev. 2005;4:372–97.
82. Grover D, Yang J, Ford D, Tavaré S, Tower J. Simultaneous tracking of movement and gene expression in multiple *Drosophila melanogaster* flies using GFP and DsRED fluorescent reporter transgenes. BMC Res Notes. 2009;2:58.
83. Guarente L. Calorie restriction and sirtuins revisited. Genes Dev. 2013;27:2072–85.
84. Guarente L, Kenyon C. Genetic pathways that regulate ageing in model organisms. Nature. 2000;408:255–62.
85. Guha S, Cao M, Kane RM, Savino AM, Zou S, Dong Y. The longevity effect of cranberry extract in *Caenorhabditis elegans* is modulated by daf-16 and osr-1. Age (Dordr). 2013;35:1559–74.
86. Guillozet AL, Weintraub S, Mash DC, Mesulam MM. Neurofibrillary tangles, amyloid, and memory in aging and mild cognitive impairment. Arch Neurol. 2003;60:729–36.
87. Guo L, Karpac J, Tran SL, Jasper H. PGRP-SC2 promotes gut immune homeostasis to limit commensal dysbiosis and extend lifespan. Cell. 2014;156:109–22.
88. Haigis MC, Yankner BA. The aging stress response. Mol Cell. 2010;40:333–44.
89. Han S, Brunet A. Histone methylation makes its mark on longevity. Trends Cell Biol. 2012;22:42–9.
90. Hands S, Sinadinos C, Wyttenbach A. Polyglutamine gene function and dysfunction in the ageing brain. Biochim Biophys Acta. 2008;1779:507–21.
91. Hannum G, Guinney J, Zhao L, Zhang L, Hughes G, Sadda S, Klotzle B, Bibikova M, et al. Genome-wide methylation profiles reveal quantitative views of human aging rates. Mol Cell. 2013;49:359–67.

92. Harman D. Aging: a theory based on free radical and radiation chemistry. J Gerontol. 1956;11:298–300.
93. Harrison DE, Strong R, Sharp ZD, Nelson JF, Astle CM, Flurkey K, Nadon NL, Wilkinson JE, et al. Rapamycin fed late in life extends lifespan in genetically heterogeneous mice. Nature. 2009;460:392–5.
94. Harshman LG, Haberer BA. Oxidative stress resistance: a robust correlated response to selection in extended longevity lines of Drosophila melanogaster. J Gerontol A Biol Sci Med Sci. 2000;55:B415–7.
95. Hartl FU, Bracher A, Hayer-Hartl M. Molecular chaperones in protein folding and proteostasis. Nature. 2011;475:324–32.
96. Hass BS, Hart RW, Lu MH, Lyn-Cook BD. Effects of caloric restriction in animals on cellular function, oncogene expression, and DNA methylation in vitro. Mutat Res. 1993;295:281–9.
97. Hayflick L. The biology of human aging. Plast Reconstr Surg. 1981;67:536–50.
98. Hayflick L, Moorhead PS. The serial cultivation of human diploid cell strains. Exp Cell Res. 1961;25:585–621.
99. He Y, Jasper H. Studying aging in *Drosophila*. Methods. 2014;68:129–33.
100. He X, Liu RH. Cranberry phytochemicals: isolation, structure elucidation, and their antiproliferative and antioxidant activities. J Agric Food Chem. 2006;54:7069–74.
101. He C, Tsuchiyama SK, Nguyen QT, Plyusnina EN, Terrill SR, Sahibzada S, Patel B, Faulkner AR, et al. Enhanced longevity by ibuprofen, conserved in multiple species, occurs in yeast through inhibition of tryptophan import. PLoS Genet. 2014;10:e1004860.
102. Helfand SL, Rogina B. Genetics of aging in the fruit fly, *Drosophila melanogaster*. Annu Rev Genet. 2003a;37:329–48.
103. Helfand SL, Rogina B. From genes to aging in *Drosophila*. Adv Genet. 2003b;49:67–109.
104. Herman MM, Miquel J, Johnson M. Insect brain as a model for study of aging. Age-related changes in *Drosophila melanogaster*. Acta Neuropathol. 1971;19:167–83.
105. Hoeijmakers JH. DNA damage, aging, and cancer. N Engl J Med. 2009;361:1475–85.
106. Hotamisligil GS. Inflammation and metabolic disorders. Nature. 2006;444:860–7.
107. Houtkooper RH, Pirinen E, Auwerx J. Sirtuins as regulators of metabolism and healthspan. Nat Rev Mol Cell Biol. 2012;13:225–38.
108. Howell AB. Bioactive compounds in cranberries and their role in prevention of urinary tract infections. Mol Nutr Food Res. 2007;51:732–7.
109. Huey RB, Suess J, Hamilton H, Gilchrist GW. Starvation resistance in *Drosophila melanogaster*: testing for a possible 'cannibalism' bias. Funct Ecol. 2004;18:952–4.
110. Hutchinson EW, Shaw AJ, Rose MR. Quantitative genetics of postponed aging in *Drosophila melanogaster*. II Analysis of selected lines. Genetics. 1991;127:729–37.
111. Hwangbo DS, Gershman B, Tu MP, Palmer M, Tatar M. *Drosophila* dFOXO controls lifespan and regulates insulin signalling in brain and fat body. Nature. 2004;429:562–6.
112. Hyun S, Lee JH, Jin H, Nam J, Namkoong B, Lee G, Chung J, Kim VN. Conserved microRNA miR-8/miR-200 and its target USH/FOG2 control growth by regulating PI3K. Cell. 2009;139:1096–108.
113. Igaki T. Correcting developmental errors by apoptosis: lessons from *Drosophila* JNK signaling. Apoptosis. 2009;14:1021–8.
114. Iliadi KG, Boulianne GL. Age-related behavioral changes in *Drosophila*. Ann N Y Acad Sci. 2010;1197:9–18.
115. Ingram DK, Zhu M, Mamczarz J, Zou S, Lane MA, Roth GS, deCabo R. Calorie restriction mimetics: an emerging research field. Aging Cell. 2006;5:97–108.
116. Ja WW, Carvalho GB, Mak EB, de la Rosa NN, Fang AY, Liong JC, Brummel T, Benzer S. Prandiology of *Drosophila* and the CAFE assay. Proc Natl Acad Sci U S A. 2007;104:8253–6.
117. Jafari M. *Drosophila melanogaster* as a model system for the evaluation of anti-aging compounds. Fly (Austin). 2010;4:253–7.
118. Jafari M, Felgner JS, Bussel II, Hutchili T, Khodayari B, Rose MR, Vince-Cruz C, Mueller LD. Rhodiola: a promising anti-aging Chinese herb. Rejuvenation Res. 2007;10:587–602.

119. Jafari M, Zarban A, Pham S, Wang T. *Rosa damascena* decreased mortality in adult *Drosophila*. J Med Food. 2008;11:9–13.
120. Jain S, Thakkar N, Chhatai J, Bhadra MP, Bhadra U. Long non-coding RNA: functional agent for disease traits. RNA Biol. 2016;16:1–14.
121. Jin C, Li J, Green CD, Yu X, Tang X, Han D, Xian B, Wang D, et al. Histone demethylase UTX-1 regulates *C. elegans* lifespan by targeting the insulin/IGF-1 signaling pathway. Cell Metab. 2011;14:161–72.
122. Johnson GL, Nakamura K. The c-jun kinase/stress-activated pathway: regulation, function and role in human disease. Biochim Biophys Acta. 2007;1773:1341–8.
123. Jones MA, Grotewiel M. *Drosophila* as a model for age-related impairment in locomotor and other behaviors. Exp Gerontol. 2011;46:320–5.
124. Kapahi P, Zid BM, Harper T, Koslover D, Sapin V, Benzer S. Regulation of lifespan in *Drosophila* by modulation of genes in the TOR signaling pathway. Curr Biol. 2004;14:885–90.
125. Karpac J, Jasper H. Insulin and JNK: optimizing metabolic homeostasis and lifespan. Trends Endocrinol Metab. 2009;20:100–6.
126. Kato M, Chen X, Inukai S, Zhao H, Slack FJ. Age-associated changes in expression of small, noncoding RNAs, including microRNAs, in *C. elegans*. RNA. 2011;17:1804–20.
127. Kenyon CJ. The genetics of ageing. Nature. 2010;464:504–12.
128. Kim J, Kim KM, Noh JH, Yoon JH, Abdelmohsen K, Gorospe M. Long noncoding RNAs in diseases of aging. Biochim Biophys Acta. 2016;1859:209–21.
129. Kirby K, Hu J, Hilliker AJ, Phillips JP. RNA interference-mediated silencing of Sod2 in *Drosophila* leads to early adult-onset mortality and elevated endogenous oxidative stress. Proc Natl Acad Sci U S A. 2002;99:16162–7.
130. Kirkwood TB, Austad SN. Why do we age? Nature. 2000;408:233–8.
131. Koga H, Kaushik S, Cuervo AM. Protein homeostasis and aging: the importance of exquisite quality control. Ageing Res Rev. 2011;10:205–15.
132. Koh K, Evans JM, Hendricks JC, Sehgal A. A *Drosophila* model for age-associated changes in sleep: wake cycles. Proc Natl Acad Sci U S A. 2006;103:13843–7.
133. Kouzarides T. Chromatin modifications and their function. Cell. 2007;128:693–705.
134. Kuno A, Horio Y. Anti-aging drugs. Nihon Rinsho. 2009;67:1384–8.
135. Kurapati R, Passananti HB, Rose MR, Tower J. Increased hsp22 RNA levels in *Drosophila* lines genetically selected for increased longevity. J Gerontol A Biol Sci Med Sci. 2000;55:B552–9.
136. Lapointe J, Hekimi S. When a theory of ageing ages badly. Cell Mol Life Sci. 2010;67:1–8.
137. Larson K, Yan SJ, Tsurumi A, Liu J, Zhou J, Gaur K, Guo D, Eickbush TH, et al. Heterochromatin formation promotes longevity and represses ribosomal RNA synthesis. PLoS Genet. 2012;8:e1002473.
138. Le Bourg E. Oxidative stress, aging and longevity in *Drosophila melanogaster*. FEBS Lett. 2001;498:183–6.
139. Lee KS, Lee BS, Semnani S, Avanesian A, Um CY, Jeon HJ, Seong KM, Yu K, et al. Curcumin extends lifespan, improves health span, and modulates the expression of age-associated aging genes in *Drosophila melanogaster*. Rejuvenation Res. 2010;13:561–70.
140. Li Y, Liu L, Tollefsbol TO. Glucose restriction can extend normal cell lifespan and impair precancerous cell growth through epigenetic control of hTERT and p16 expression. FASEB J. 2010;24:1442–53.
141. Liao PC, Lin HY, Yuh CH, Yu LK, Wang HD. The effect of neuronal expression of heat shock proteins 26 and 27 on lifespan, neurodegeneration, and apoptosis in Drosophila. Biochem Biophys Res Commun. 2008;376:637–41.
142. Libert S, Chao Y, Chu X, Pletcher SD. Trade-offs between longevity and pathogen resistance in *Drosophila melanogaster* are mediated by NFkappaB signaling. Aging Cell. 2006;5:533–43.
143. Lin YJ, Seroude L, Benzer S. Extended life-span and stress resistance in the *Drosophila* mutant methuselah. Science. 1998;282:943–6.

144. Lin WS, Chen JY, Wang JC, Chen LY, Lin CH, Hsieh TR, Wang MF, Fu TF, et al. The anti-aging effects of *Ludwigia octovalvis* on *Drosophila melanogaster* and SAMP8 mice. Age. 2014;36:689–703.
145. Linford NJ, Bilgir C, Ro J, Pletcher SD. Measurement of lifespan in *Drosophila melanogaster*. J Vis Exp. 2013;71(71):pii 50068.
146. Liu B, Wang J, Chan KM, Tjia WM, Deng W, Guan X, Huang JD, Li KM, et al. Genomic instability in laminopathy-based premature aging. Nat Med. 2005;11:780–5.
147. Liu GH, Suzuki K, Qu J, Sancho-Martinez I, Yi F, Li M, Kumar S, Nivet E, et al. Targeted gene correction of laminopathy-associated LMNA mutations in patient-specific iPSCs. Cell Stem Cell. 2011;8:688–94.
148. Loeb J, Northrop JH. Is there a temperature coefficient for the duration of life? Proc Natl Acad Sci U S A. 1916;2:456–7.
149. Loeb J, Northrop JH. On the influence of food and temperature upon the duration of life. J Biol Chem. 1917;32:103–21.
150. Lopez T, Schriner SE, Okoro M, Lu D, Chiang BT, Huey J, Jafari M. Green tea polyphenols extend the lifespan of male *Drosophila melanogaster* while impairing reproductive fitness. J Med Food. 2014;17:1314–21.
151. López-Otín C, Blasco MA, Partridge L, Serrano M, Kroemer G. The hallmarks of aging. Cell. 2013;153:1194–217.
152. Lord CJ, Ashworth A. The DNA damage response and cancer therapy. Nature. 2012;481:287–94.
153. Luckinbill L, Clare M. Selection for lifespan in *Drosophila melanogaster*. Heredity. 1985;55:9–18.
154. Luckinbill L, Arking R, Clare MJ, Cirocco WC, Buck S. Selection for delayed senescence in *Drosophila melanogaster*. Evolution. 1984;38:996–1003.
155. Luong N, Davies CR, Wessells RJ, Graham SM, King MT, Veech R, Bodmer R, Oldham SM. Activated FOXO-mediated insulin resistance is blocked by reduction of TOR activity. Cell Metab. 2006;4:133–42.
156. Maegawa S, Hinkal G, Kim HS, Shen L, Zhang L, Zhang J, Zhang N, Liang S, et al. Widespread and tissue specific age-related DNA methylation changes in mice. Genome Res. 2010;20:332–40.
157. Mair W, Goymer P, Pletcher SD, Partridge L. Demography of dietary restriction and death in *Drosophila*. Science. 2003;301:1731–3.
158. Mair W, Piper MD, Partridge L. Calories do not explain extension of lifespan by dietary restriction in *Drosophila*. PLoS Biol. 2005;7:e223.
159. Marino G, Ugalde AP, Fernandez AF, Osorio FG, Fueyo A, Freije JM, Lopez-Otin C. Insulin-like growth factor 1 treatment extends longevity in a mouse model of human premature aging by restoring somatotroph axis function. Proc Natl Acad Sci U S A. 2010;107:16268–73.
160. Maynard S, Fang EF, Scheibye-Knudsen M, Croteau DL, Bohr VA. DNA damage, DNA repair, aging, and neurodegeneration. Cold Spring Harb Perspect Med. 2015;5:a025130.
161. McCarroll SA, Murphy CT, Zou S, Pletcher SD, Chin CS, Jan YN, Kenyon C, Bargmann CI, et al. Comparing genomic expression patterns across species identifies shared transcriptional profile in aging. Nat Genet. 2004;36:197–204.
162. McCay CM, Crowell MF, Maynard LA. The effect of retarded growth upon the length of lifespan and upon the ultimate body size. 1935 Nutrition. 1989;5:155–71.
163. McCord JM, Fridovich I. Superoxide dismutase. An enzymatic function for erythrocuperin (hemocuperin). J Biol Chem. 1969;244:6049–605.
164. McGuire SE, Le PT, Davis RL. The role of *Drosophila* mushroom body signaling in olfactory memory. Science. 2001;293:1330–3.
165. McKiernan SH, Colman RJ, Lopez M, Beasley TM, Aiken JM, Anderson RM, Weindruch R. Caloric restriction delays aging-induced cellular phenotypes in rhesus monkey skeletal muscle. Exp Gerontol. 2011;46:23–9.

166. Menzies FM, Fleming A, Rubinsztein DC. Compromised autophagy and neurodegenerative diseases. Nat Rev Neurosci. 2015;16:345–57.
167. Min KJ, Flatt T, Kulaots I, Tatar M. Counting calories in *Drosophila* diet restriction. Exp Gerontol. 2007;42:247–51.
168. Min JN, Whaley RA, Sharpless NE, Lockyer P, Portbury AL, Patterson C. CHIP deficiency decreases longevity, with accelerated aging phenotypes accompanied by altered protein quality control. Mol Cell Biol. 2008;28:4018–25.
169. Minois N. Longevity and aging: beneficial effects of exposure to mild stress. Biogerontology. 2000;1:15–29.
170. Minois N, Khazaeli AA, Curtsinger JW. Locomotor activity as a function of age and lifespan in *Drosophila melanogaster* overexpressing hsp70. Exp Gerontol. 2001;36:1137–53.
171. Missirlis F, Phillips JP, Jackle H. Cooperative action of antioxidant defense systems in *Drosophila*. Curr Biol. 2001;11:1272–7.
172. Mitteldorf J. An epigenetic clock controls aging. Biogerontology. 2016;17:257–65.
173. Mizushima N, Levine B, Cuervo AM, Klionsky DJ. Autophagy fights disease through cellular self digestion. Nature. 2008;451:1069–75.
174. Mockett RJ, Sohal RS, Orr WC. Overexpression of glutathione reductase extends survival in transgenic *Drosophila melanogaster* under hyperoxia but not normoxia. FASEB J. 1999;13:1733–42.
175. Morimoto RI. Proteotoxic stress and inducible chaperone networks in neurodegenerative 1002 disease and aging. Genes Dev. 2008;22:1427–38.
176. Morley JF, Brignull HR, Weyers JJ, Morimoto RI. The threshold for polyglutamine-expansion protein aggregation and cellular toxicity is dynamic and influenced by aging in *Caenorhabditis elegans*. Proc Natl Acad Sci U S A. 2002;99:10417–22.
177. Morrow G, Tanguay RM. Heat shock proteins and aging in *Drosophila melanogaster*. Semin Cell Dev Biol. 2003;14:291–9.
178. Morrow G, Battistini S, Zhang P, Tanguay RM. Decreased lifespan in the absence of expression of the mitochondrial small heat shock protein Hsp22 in *Drosophila*. J Biol Chem. 2004a;279:43382–5.
179. Morrow G, Samson M, Michaud S, Tanguay RM. Overexpression of the small mitochondrial Hsp22 extends *Drosophila* life span and increases resistance to oxidative stress. FASEB J. 2004b;18:598–9.
180. Morrow G, Heikkila JJ, Tanguay RM. Differences in the chaperone-like activities of the four main small heat shock proteins of *Drosophila melanogaster*. Cell Stress Chaperones. 2006;11:51–60.
181. Mosammaparast N, Shi Y. Reversal of histone methylation: biochemical and molecular mechanisms of histone demethylases. Annu Rev Biochem. 2010;79:155–79.
182. Moskalev AA, Shaposhnikov MV, Plyusnina EN, Zhavoronkov A, Budovsky A, Yanai H, Fraifeld VE. The role of DNA damage and repair in aging through the prism of Koch-like criteria. Ageing Res Rev. 2012;12:661–84.
183. Mouchiroud L, Houtkooper RH, Moullan N, Katsyuba E, Ryu D, Cantó C, Mottis A, Jo YS, et al. The NAD(+)/Sirtuin pathway modulates longevity through activation of mitochondrial UPR and FOXO signaling. Cell. 2013;154:430–41.
184. Muchowski PJ, Wacker JL. Modulation of neurodegeneration by molecular chaperones. Nat Rev Neurosci. 2005;6:11–22.
185. Muchowski PJ, Schaffar G, Sittler A, Wanker EE, Hayer-Hartl MK, Hartl FU. Hsp70 and hsp40 chaperones can inhibit self-assembly of polyglutamine proteins into amyloid-like fibrils. Proc Natl Acad Sci U S A. 2000;97:7841–6.
186. Murphy MP. How mitochondria produces reactive oxygen species. Biochem J. 2009;417:1–13.
187. Myers EW, Sutton GG, Delcher AL, Dew IM, Fasulo DP, Flanigan MJ, Kravitz SA, Mobarry CM, et al. A whole-genome assembly of *Drosophila*. Science. 2000;287:2196–204.

188. Na H, Park J, Pyo J, Jeon H, Kim Y, Arking R, Yoo M. Metformin inhibits age-related centrosome amplification in *Drosophila* midgut stem cells through AKT/TOR pathway. Mech Aging Dev. 2015;149:8–18.
189. Nakagawa H, Nuovo GJ, Zervos EE, Martin EW Jr, Salovaara R, Aaltonen LA, de la Chapelle A. Age-related hypermethylation of the 5′ region of MLH1 in normal colonic mucosa is associated with microsatellite-unstable colorectal cancer development. Cancer Res. 2001;61:6991–5.
190. Neto CC. Cranberry and its phytochemicals: a review of in vitro anticancer studies. J Nutr. 2007;137:186S–93S.
191. Niedzwiecki A, Kongpachith AM, Fleming JE. Aging affects expression of 70-kDa heat shock proteins in *Drosophila*. J Biol Chem. 1991;266:9332–8.
192. Nielsen MD, Luo X, Biteau B, Syverson K, Jasper H. 14-3-3 Epsilon antagonizes FoxO to control growth, apoptosis and longevity in *Drosophila*. Aging Cell. 2008;7:688–99.
193. Oishi K, Shiota M, Sakamoto K, Kasamatsu M, Ishida N. Feeding is not a more potent Zeitgeber than the light-dark cycle in *Drosophila*. Neuroreport. 2004;15:739–43.
194. Olovnikov AM. Telomeres, telomerase, and aging: origin of the theory. Exp Gerontol. 1996;31:443–8.
195. Orr WC, Sohal RS. Effects of Cu-Zn superoxide dismutase overexpression on lifespan and 1034 resistance to oxidative stress in transgeneic *D. melanogaster*. Arch Biochem Biophys. 1993;301:34–40.
196. Orr WC, Sohal RS. Extension of life-span by overexpression of superoxide dismutase and 1037 catalase in *Drosophila melanogaster*. Science. 1994;263:1128–30.
197. Osorio FG, Navarro CL, Cadinanos J, Lopez-Mejia IC, Quiros PM, Bartoli C, Rivera J, Tazi J, et al. Splicing-directed therapy in a new mouse model of human accelerated aging. Sci Transl Med. 2011;3:106ra107.
198. Pacholec M, Bleasdale JE, Chrunyk B, Cunningham D, Flynn D, Garofalo RS, Griffith D, Griffor M, et al. SRT1720, SRT2183, SRT1460, and resveratrol are not direct activators of SIRT1. J Biol Chem. 2010;285:8340–51.
199. Palikaras K, Lionaki E, Tavernarakis N. Coordination of mitophagy and mitochondrial biogenesis during ageing in *C. elegans*. Nature. 2015;521:525–8.
200. Palm W, de Lange T. How shelterin protects mammalian telomeres. Annu Rev Genet. 2008;42:301–34.
201. Pandey UB, Nichols CD. Human disease models in *Drosophila melanogaster* and the role of the fly in therapeutic drug discovery. Pharmacol Rev. 2011;63:411–36.
202. Parkes TL, Elia AJ, Dickinson D, Hilliker AJ, Phillips JP, Boulianne GL. Extension of *Drosophila* lifespan by overexpression of human SOD1 in motor neurons. Nat Genet. 1998;19:171–4.
203. Partridge L, Piper MD, Mair W. Dietary restriction in *Drosophila*. Mech Ageing Dev. 2005;126:938–50.
204. Pearl R, Parker SL. Experimental studies on duration of life I. Introductory discussion of the duration of life in of *Drosophila*. Am Nat. 1921a;60:481–509.
205. Pearl R, Parker SL. Experimental studies on duration of life II. Hereditary differences in duration of life in line-bread strains of *Drosophila*. Am Nat. 1921b;56:174.
206. Pegoraro G, Kubben N, Wickert U, Göhler H, Hoffmann K, Misteli T. Ageing-related chromatin defects through loss of the NURD complex. Nat Cell Biol. 2009;11:1261–7.
207. Peng C, Zuo Y, Kwan KM, Liang Y, Ma KY, Chan HY, Huang Y, Yu H, et al. Blueberry extract prolongs lifespan of *Drosophila melanogaster*. Exp Gerontol. 2012;47:170–8.
208. Pérez VI, Bokov A, Van Remmen H, Mele J, Ran Q, Ikeno Y, Richardson A. Is the oxidative stress theory of aging dead? Biochim Biophys Acta. 2009;1790:1005–14.
209. Petersen KF, Befroy D, Dufour S, Dziura J, Ariyan C, Rothman DL, DiPietro L, Cline GW, et al. Mitochondrial dysfunction in the elderly: possible role in insulin resistance. Science. 2003;300:1140–2.

210. Pfeiffenberger C, Lear BC, Keegan KP, Allada R. Locomotor activity level monitoring using the *Drosophila* activity monitoring (DAM) system. Cold Spring Harb Protoc. 2010;2010(11):pdb.prot5518.
211. Phillips JP, Hilliker AJ. Genetic analysis of oxygen defense mechanisms in *Drosophila melanogaster*. Adv Genet. 1990;28:43–71.
212. Phillips JP, Campbell SD, Michaud D, Charbonneau M, Hilliker AJ. Null mutation of copper/zinc superoxide dismutase in *Drosophila* confers hypersensitivity to paraquat and reduced longevity. Proc Natl Acad Sci U S A. 1989;86:2761–5.
213. Piper MD, Partridge L. Dietary restriction in *Drosophila*: delayed aging or experimental artefact? PLoS Genet. 2007;3:e57.
214. Piper MD, Blanc E, Leitao-Goncalves R, Yang M, He X, Linford NJ, Hoddinott MP, Hopfen C, et al. A holidic medium for *Drosophila melanogaster*. Nat Methods. 2014;11:100–5.
215. Pletcher SD, Macdonald SJ, Marguerie R, Certa U, Stearns SC, Goldstein DB, Partridge L. Genomewide transcript profiles in aging and calorically restricted *Drosophila melanogaster*. Curr Biol. 2002;12:712–23.
216. Poirier L, Seroude L. Genetic approaches to study aging in *Drosophila melanogaster*. Age. 2005;27:165–82.
217. Pollina EA, Brunet A. Epigenetic regulation of aging stem cells. Oncogene. 2011;30:3105–26.
218. Powers ET, Morimoto RI, Dillin A, Kelly JW, Balch WE. Biological and chemical approaches to diseases of proteostasis deficiency. Annu Rev Biochem. 2009;78:959–91.
219. Ragnauth CD, Warren DT, Liu Y, McNair R, Tajsic T, Figg N, Shroff R, Skepper J, et al. Prelamin A acts to accelerate smooth muscle cell senescence and is a novel biomarker of human vascular aging. Circulation. 2010;121:2200–10.
220. Randow F, Youle RJ. Self and nonself: how autophagy targets mitochondria and bacteria. Cell Host Microbe. 2014;15:403–11.
221. Reiter LT, Potocki L, Chien S, Gribskov M, Bier E. A systematic analysis of human disease-associated gene sequences in *Drosophila melanogaster*. Genome Res. 2001;11:1114–25.
222. Ritossa F. A new puffing pattern induced by temperature shock and DNP in *Drosophila*. Experientia. 1962;18:571–3.
223. Ritossa F. Discovery of the heat shock response. Cell Stress Chaperones. 1996;1:97–8.
224. Rogina B, Helfand SL. Sir2 mediates longevity in the fly through a pathway related to calorie restriction. Proc Natl Acad Sci U S A. 2004;101:15998–6003.
225. Rogina B, Reenan RA, Nilsen SP, Helfand SL. Extended life-span conferred by co transporter gene mutations in *Drosophila*. Science. 2000;290:2137–40.
226. Rose MR. Evolutionary biology of aging. New York: Oxford University Press; 1991.
227. Rose M, Charlesworth B. A test of evolutionary theories of senescence. Nature. 1980;287:141–2.
228. Rose MR, Charlesworth B. Genetics of life history in *Drosophila melanogaster*. II. Exploratory selection experiments. Genetics. 1981;97:187–96.
229. Rose MR, Vu LN, Park SU, Graves JL Jr. Selection on stress resistance increases longevity in *Drosophila melanogaster*. Exp Gerontol. 1992;27:241–50.
230. Rudolph KL, Chang S, Lee HW, Blasco M, Gottlieb GJ, Greider C, DePinho RA. Longevity, stress response, and cancer in aging telomerase-deficient mice. Cell. 1999;96:701–12.
231. Ryder E, Ashburner M, Bautista-Llacer R, Drummond J, Webster J, Johnson G, Morley T, Chan YS, et al. The DrosDel deletion collection: a *Drosophila* genomewide chromosomal deficiency resource. Genetics. 2007;177:615–62.
232. Sarkar S, Singh MD, Yadav R, Arunkumar KP, Pitman GW. Heat shock proteins: molecules with assorted functions. Front Biol. 2011;6:312–67.
233. Scaffidi P, Misteli T. Lamin A-dependent misregulation of adult stem cells associated with accelerated ageing. Nat Cell Biol. 2008;10:452–9.
234. Schriner SE, Kuramada S, Lopez TE, Truong S, Pham A, Jafari M. Extension of *Drosophila* lifespan by cinnamon through a sex-specific dependence on the insulin receptor substrate Chico. Exp Gerontol. 2014;60:220–30.

235. Sebastián C, Satterstrom FK, Haigis MC, Mostoslavsky R. From sirtuin biology to human diseases: an update. J Biol Chem. 2012;287:42444–52.
236. Seto NO, Hayashi S, Tener GM. Overexpression of Cu-Zn superoxide dismutase in *Drosophila* does not affect life-span. Proc Natl Acad Sci U S A. 1990;87:4270–4.
237. Shaw P, Ocorr K, Bodmer R, Oldham S. *Drosophila* aging 2006/2007. Exp Gerontol. 2008;43:5–10.
238. Shay JW, Bacchetti S. A survey of telomerase activity in human cancer. Eur J Cancer. 1997;33:787–91.
239. Sinclair DA. Toward a unified theory of caloric restriction and longevity regulation. Mech Ageing Dev. 2005;126:987–1002.
240. Singh MD, Raj K, Sarkar S. *Drosophila Myc*, a novel modifier suppresses the poly(Q) toxicity by modulating the level of CREB binding protein and histone acetylation. Neurobiol Dis. 2013;63:48–61.
241. Smith JM. The effect of temperature and egg-laying on the longevity of *Drosophila subobscura*. J Exp Biol. 1958;35:832–42.
242. Soh JW, Marowsky N, Nichols TJ, Rahman AM, Miah T, Sarao P, Khasawneh R, Unnikrishnan A, et al. Curcumin is an early-acting stage-specific inducer of extended functional longevity in *Drosophila*. Exp Gerontol. 2013;48:229–39.
243. Sohal RS. Oxidative stress hypothesis of aging. Free Radic Biol Med. 2002;33:573–4.
244. Solon-Biet SM, McMahon AC, Ballard JW, Ruohonen K, Wu LE, Cogger VC, Warren A, Huang X, et al. The ratio of macronutrients, not caloric intake, dictates cardiometabolic health, aging, and longevity in *ad libitum*-fed mice. Cell Metab. 2014;19:418–30.
245. Someya S, Yu W, Hallows WC, Xu J, Vann JM, Leeuwenburgh C, Tanokura M, Denu JM, et al. Sirt3 mediates reduction of oxidative damage and prevention of age-related hearing loss under caloric restriction. Cell. 2010;143:802–12.
246. Soti C, Csermely P. Aging and molecular chaperones. Exp Gerontol. 2003;38:1037–40.
247. Speakman JR, Mitchell SE. Caloric restriction. Mol Asp Med. 2011;32:159–221.
248. Stowers RS, Schwarz TL. A genetic method for generating *Drosophila* eyes composed exclusively of mitotic clones of a single genotype. Genetics. 1999;152:1631–9.
249. Sun J, Tower J. FLP recombinase-mediated induction of Cu/Zn-superoxide dismutase transgene expression can extend the lifespan of adult *Drosophila melanogaster* flies. Mol Cell Biol. 1999;19:216–28.
250. Sun Y, Yolitz J, Wang C, Spangler E, Zhan M, Zou S. Aging studies in *Drosophila melanogaster*. Methods Mol Biol. 2013;1048:77–93.
251. Talens RP, Christensen K, Putter H, Willemsen G, Christiansen L, Kremer D, Suchiman HE, Slagboom PE, et al. Epigenetic variation during the adult lifespan: cross-sectional and longitudinal data on monozygotic twin pairs. Aging Cell. 2012;11:694–703.
252. Tatar M. Reproductive aging in invertebrate genetic models. Ann N Y Acad Sci. 2010;1204:149–55.
253. Tatar M, Khazaeli AA, Curtsinger JW. Chaperoning extended life. Nature. 1997;390:30.
254. Tatar M, Kopelman A, Epstein D, Tu MP, Yin CM, Garofalo RS. A mutant *Drosophila* insulin receptor homolog that extends life-span and impairs neuroendocrine function. Science. 2001;292:107–10.
255. Teixeira-Castro A, Ailion M, Jalles A, Brignull HR, Vilaça JL, Dias N, Rodrigues P, Oliveira JF, et al. Neuron-specific proteotoxicity of mutant ataxin-3 in *C. elegans*: rescue by the DAF-16 and HSF-1 pathways. Hum Mol Genet. 2011;20:2996–3009.
256. Testa G, Biasi F, Poli G, Chiarpotto E. Calorie restriction and dietary restriction mimetics: a strategy for improving healthy aging and longevity. Curr Pharm Des. 2014;20:2950–77.
257. Theodosiou NA, Xu T. Use of FLP/FRT system to study *Drosophila* development. Methods. 1998;14:355–65.
258. Thompson RF, Atzmon G, Gheorghe C, Liang HQ, Lowes C, Greally JM, Barzilai N. Tissue-specific dysregulation of DNA methylation in aging. Aging Cell. 2010;9:506–18.

259. Tomas-Loba A, Flores I, Fernandez-Marcos PJ, Cayuela ML, Maraver A, Tejera A, Borras C, Matheu A, et al. Telomerase reverse transcriptase delays aging in cancer-resistant mice. Cell. 2008;135:609–22.
260. Tower J. Transgenic methods for increasing *Drosophila* lifespan. Mech Aging Dev. 2000;118:1–14.
261. Troen BR. The biology of aging. Mt Sinai J Med. 2003;70:3–22.
262. Turrens JF. Superoxide production by the mitochondrial respiratory chain. Biosci Rep. 1997;17:3–8.
263. Vaĭserman AM, Koliada AK, Koshel NM, Simonenko AV, Pasiukova EG. Effect of the histone deacetylase inhibitor sodium butyrate on the viability and lifespan in *Drosophila melanogaster*. Adv Gerontol. 2012;25:126–31.
264. Varela I, Cadinanos J, Pendas AM, Gutierrez-Fernandez A, Folgueras AR, Sanchez LM, Zhou Z, Rodriguez FJ, et al. Accelerated ageing in mice deficient in Zmpste24 protease is linked to p53 signalling activation. Nature. 2005;437:564–8.
265. Voellmy R. On mechanisms that control heat shock transcription factor activity in metazoan cells. Cell Stress Chaperones. 2004;9:122–33.
266. Walker DW, Benzer S. Mitochondrial "swirls" induced by oxygen stress and in the *Drosophila* mutant hyperswirl. Proc Natl Acad Sci U S A. 2004;101:10290–5.
267. Walker GA, Lithgow GJ. Lifespan extension in *C. elegans* by a molecular chaperone dependent upon insulin-like signals. Aging Cell. 2003;2:131–9.
268. Wallace DC. A mitochondrial paradigm of metabolic and degenerative diseases, aging, and cancer: a dawn for evolutionary medicine. Annu Rev Genet. 2005;39:359–407.
269. Wallace DC. Mitochondrial DNA variation in human radiation and disease. Cell. 2015;163:33–8.
270. Wang MC, Bohmann D, Jasper H. JNK signaling confers tolerance to oxidative stress and extends lifespan in *Drosophila*. Dev Cell. 2003;5:811–6.
271. Wang MC, Bohmann D, Jasper H. JNK extends lifespan and limits growth by antagonizing cellular and organism-wide responses to insulin signaling. Cell. 2005;121:115–25.
272. Wang C, Yolitz J, Alberico T, Laslo M, Sun Y, Wheeler CT, Sun X, Zou S. Cranberry interacts with dietary macronutrients to promote healthy aging in *Drosophila*. J Gerontol A Biol Sci. 2014a;69:945–54.
273. Wang L, Karpac J, Jasper H. Promoting longevity by maintaining metabolic and proliferative homeostasis. J Exp Biol. 2014b;217:109–18.
274. Webb AE, Brunet A. FOXO transcription factors: key regulators of cellular quality control. Trends Biochem Sci. 2014;39:159–69.
275. Weiss EP, Racette SB, Villareal DT, Fontana L, Steger-May K, Schechtman KB, Klein S, Holloszy JO. Improvements in glucose tolerance and insulin action induced by increasing energy expenditure or decreasing energy intake: a randomized controlled trial. Am J Clin Nutr. 2006;84:1033–42.
276. Wiegant FA, Surinova S, Ytsma E, Langelaar-Makkinje M, Wikman G, Post JA. Plant adaptogens increase lifespan and stress resistance in *C. elegans*. Biogerontology. 2009;10:27–42.
277. Wolf FW, Heberlein U. Invertebrate models of drug abuse. J Neurobiol. 2003;54:161–78.
278. Wong R, Piper MD, Wertheim B, Partridge L. Quantification of food intake in *Drosophila*. PLoS One. 2009;4:e6063.
279. Wood JG, Rogina B, Lavu S, Howitz K, Helfand SL, Tatar M, Sinclair D. Sirtuin activators mimic caloric restriction and delay ageing in metazoans. Nature. 2004;430:686–9.
280. Worman HJ. Nuclear lamins and laminopathies. J Pathol. 2012;226:316–25.
281. Wu Z, Puigserver P, Andersson U, Zhang C, Adelmant G, Mootha V, Troy A, Cinti S, et al. Mechanisms controlling mitochondrial biogenesis and respiration through the thermogenic coactivator PGC-1. Cell. 1999;98:115–24.
282. Xu T, Rubin GM. Analysis of genetic mosaics in developing and adult *Drosophila* tissues. Development. 1993;117:1223–37.

283. Yadav R, Kundu S, Sarkar S. *Drosophila* glob1 expresses dynamically and is required for development and oxidative stress response. Genesis. 2015;53:719–37.
284. Yan L, Gao S, Ho D, Park M, Ge H, Wang C, Tian Y, Lai L, et al. Calorie restriction can reverse, as well as prevent, aging cardiomyopathy. Age (Dordr). 2013;35:2177–82.
285. Youle RJ, van der Bliek AM. Mitochondrial fission, fusion, and stress. Science. 2012;337:1062–5.
286. Yu BP. Approaches to anti-aging intervention: the promises and the uncertainties. Mech Ageing Dev. 1999;111:73–87.
287. Zeitlinger J, Bohmann D. Thorax closure in *Drosophila*: involvement of Fos and the JNK pathway. Development. 1999;126:3947–56.
288. Zhao Y, Sun H, Lu J, Li X, Chen X, Tao D, Huang W, Huang B. Lifespan extension and elevated hsp gene expression in *Drosophila* caused by histone deacetylase inhibitors. J Exp Biol. 2005;208:697–705.
289. Zhu M, Hu J, Perez E, Phillips D, Kim W, Ghaedian R, Napora JK, Zou S. Effects of long-term cranberry supplementation on endocrine pancreas in aging rats. J Gerontol A Biol Sci Med Sci. 2011;66:1139–51.
290. Zou S, Meadows S, Sharp L, Jan LY, Jan YN. Genome-wide study of aging and oxidative stress response in *Drosophila melanogaster*. Proc Natl Acad Sci U S A. 2000;97:13726–31.
291. Zou S, Sinclair J, Wilson MA, Carey JR, Liedo P, Oropeza A, Kalra A, de Cabo R, et al. Comparative approaches to facilitate the discovery of prolongevity interventions: effects of tocopherols on lifespan of three invertebrate species. Mech Ageing Dev. 2007;128:222–6.
292. Zwaan B, Bijlsma R, Hoekstra RE. Direct selection on life-span in *Drosophila melanogaster*. Evolution. 1995;49:649–59.

Nothobranchius furzeri as a New Model System for Ageing Studies

15

Eva Terzibasi Tozzini

Ageing is the age-dependent reduction in the performance of basically every physiological function leading to increased disease susceptibility and mortality. Ageing results from age-dependent accumulation of molecular and cellular damages, it is largely inevitable and irreversible, but it can be influenced by intrinsic factors (both heritable and random variation) and extrinsic factors, such as nutrition [1–4] and pharmacological treatments [5, 6]. Disentangling the contributions of these factors is challenging, but it is a necessary step to develop strategies for reducing the incidence of ageing-related disabilities, a pressing need of our society.

For practical reasons, the current molecular understanding of the ageing process derives almost exclusively from the study of random or targeted single-gene mutations in highly inbred laboratory species, mostly invertebrates. Little information is available as to the genetic mechanisms responsible for natural lifespan variation and the evolution of longevity, especially in vertebrates. Yet, natural variability in lifespan across vertebrate species greatly exceeds the magnitude of life extension that has been obtained by single-gene manipulations. The genetics of naturally evolved lifespan differences, a highly relevant biological question in itself, also represents an untapped source of possible targets for modulating ageing.

Natural ageing is a complex trait whose inheritance is controlled by the interaction of multiple loci, and their identification using conventional model organisms is challenging. In addition, different aspects of natural ageing (or different age-dependent diseases) may have distinct genetic architectures [7, 8], thereby complicating the genetic dissection of "ageing". Finally, phenotypic differences due to non-heritable (random) variation may also influence lifespan [3]. Investigating the nature of this non-heritable source of phenotypic variability observed during the

---

E. T. Tozzini (✉)
Laboratorio di Biologia BIO@SNS, Scuola Normale Superiore, Pisa, Italy
e-mail: eva.terzibasi@sns.it

© Springer Nature Singapore Pte Ltd. 2020
P. C. Rath (ed.), *Models, Molecules and Mechanisms in Biogerontology*,
https://doi.org/10.1007/978-981-32-9005-1_15

growth and maturation phases of vertebrate lifespan may provide novel molecular insights into the control of longevity.

One of the main limitations to investigate natural ageing variation is the lack of a group of related species that are good laboratory organisms, are genetically tractable and at the same time show naturally evolved large-scale differences in lifespan. Comparative genomics can identify genes bearing signatures of adaptive evolution in long-living mammals, but the phenotypic consequences of these molecular changes cannot be tested easily. Mouse populations, on the other hand, do not show large natural variation in lifespan, and the vast majority of mouse studies are based on domesticated strains adapted to captive life (e.g. [9]).

Annual fishes of the genus *Nothobranchius* evolved short lifespan as a trade-off for rapid growth and maturation that represent primary adaptations to survive in an ephemeral habitat (Fig. 15.1). These species provide a system to investigate the genetic basis for lifespan evolution. At the same time, they are laboratory organisms that can be genetically manipulated and represent a unique model bridging evolutionary biology and biomedical research (reviewed in [10]). They indeed are specifically suitable for many types of experimental analyses in the ageing field, such as:

(i) To assess the consequences on organismal ageing of adaptive molecular changes associated with evolution of short lifespan

Fig. 15.1 Nothobranchius life cycle – When the rainy season starts, *N. furzeri* larvae hatch and develop rapidly, reaching sexual maturity into 4–5 weeks and beginning laying eggs. Depending on the availability of water, a fraction of the embryos complete development and hatch in the same season. Most of the embryos, however, will enter diapause at any of the three stages indicated (dispersed phase, somite stage, hatching stage). The arrest in diapause will last for several months, until the next rainy season starts again (Source: FLI/© Alexander Schmidt, Atelier Symbiota)

(ii) To identify early molecular signatures of individual lifespan
(iii) To dissect different biological components of ageing and their corresponding genetic architecture

Model Organisms in Ageing Research

Genetic studies in invertebrate model organisms, such as nematode worms and the fruit fly, lead to the identification of *conserved* genetic pathways that influence ageing across metazoan, such as the IGF/insulin signalling pathway [11], the target of rapamycin (TOR) pathway [12] and the mitonuclear balance system [9]. **Invertebrate models**, however, **have some intrinsic biological limitations**: their anatomical organization is fundamentally different from mammals; they lack adaptive immunity and bones, have limited stem cell populations and lack some genes that are highly relevant for human ageing. Among these, for example, the *INK4* locus (coding for the cyclin kinases inhibitors CDKN2A and CDKN2B), in genome-wide association studies (GWAS), was associated with a large number of age-associated diseases (including glaucoma, Alzheimer's, cancer, coronary artery disease, type 2 diabetes, etc.) [13].

Vertebrate models are indispensable to investigate pathophysiological mechanisms of age-dependent organ functional decline. Although a number of long-living mouse mutants were identified, their characterization is space-, funding- and time-intensive and genetic **research into mouse ageing faces serious practical limitations**. Teleost fishes are established and widely accepted as alternative vertebrate models in many fields of biomedical research because of their small size, high fecundity and ease of genetic manipulation while sharing many anatomical and genetic similarities with humans [14–16]. Yet, **the lifespan of both zebrafish and medaka is 3–5 years** [17, 18] preventing their widespread application to ageing research.

Nothobranchius furzeri as an Experimental Model

In 2003, the short-lived annual fish *Nothobranchius furzeri* has been proposed as an alternative model for ageing [19]. In the last decade, this species – originally received as a zoological curiosity – gradually gained recognition as an innovative model organism that could become a game changer in ageing research [20–24].

***N. furzeri* is the shortest-lived vertebrate that can be cultured in captivity**. Its short lifespan of few months is characterized by expression of a large number of typical vertebrate ageing markers at behavioural [6, 19], histopathological [7, 25, 26] and cellular/molecular level, [27–29]. Most importantly, RNA-seq analysis revealed evolutionary conservation of age-dependent patterns of transcript regulation between *N. furzeri* and other vertebrates, including humans [30–32].

This species is amenable to genetic manipulations: efficient random insertion of constructs in the genome can be obtained by Tol2-mediated transposition [33–35], and techniques for CRISPR/Cas9 genome editing, including homologous recombination enabling single-codon substitutions, were established [36].

Finally, *N. furzeri* lifespan and ageing phenotypes can be improved by environmental, dietary and pharmacological interventions [6, 32, 37, 38]. It is highly relevant for the present proposal that modulation of ageing-related phenotypes by genetic manipulations were recently reported [35, 36]. Therefore, *N. furzeri* **represents a platform to rapidly assess the impact of genetic manipulations on vertebrate ageing and ageing-related pathologies** [10, 36].

Nothobranchius furzeri as Experimental Ageing Model

N. furzeri is a small (6 cm) annual fish which is adapted to the alternation of wet and rainy season. It inhabits ephemeral habitats that last a few months, thereby setting the upper limit of its natural lifespan [49]. This short lifespan is retained under captive conditions (3–7 months depending on genetic background, Fig. 15.2-a), making *N. furzeri* **the shortest-lived vertebrate that can be cultured in captivity**. These animals develop from fry to a sexually mature adult in as short as 3 weeks (Fig. 15.2-b).

In the last decade, the research work of several scientific groups was largely dedicated to producing compelling evidence that **short lifespan is a consequence of rapid ageing and is experimentally malleable** (for a systematic review, see [10]). At a behavioural level, there are reduced locomotor activity and impairment in learning paradigms [6]. Histopathological examinations have revealed accumulation of the fluorescent age pigment lipofuscin [7, 26] and age-dependent lesions in the heart, liver and kidney that are similar to those observed in other small teleosts [25]. At a cellular/molecular level, ageing of *N. furzeri* is associated with increased apoptosis [25, 7], telomere erosion [27], reduced mitochondrial function [28], expression of cellular senescence markers in vivo and age-dependent impairment of adult neuronal stem cells [29].

At a global level, RNA-seq analysis in several organs highlighted the **evolutionary conservation of age-dependent gene expression between *N. furzeri* and other vertebrates, including humans** [30–32]. For example, in key terms previously identified by meta-analysis in vertebrates, such as lysosome, inflammation and ribosome, TCA cycle and collagen are also regulated in *N. furzeri*. Further, unbiased analysis of *N. furzeri* transcriptome led to the identification of novel conserved and age-regulated genes that are expressed in neuronal stem cells [31]. Notably, **this species is amenable to genetic manipulations**: random insertion of constructs in the genome can be obtained by **Tol2- mediated transgenesis** and a number of promoters for ubiquitous or tissue-specific expression were identified [33, 34, 35, 40, 41]. Most relevant for the present proposal was the demonstration that **CRISPR/Cas9genome**

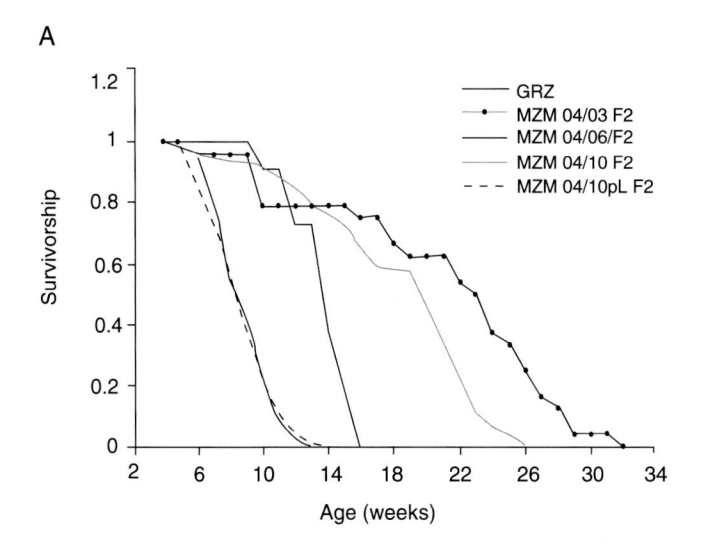

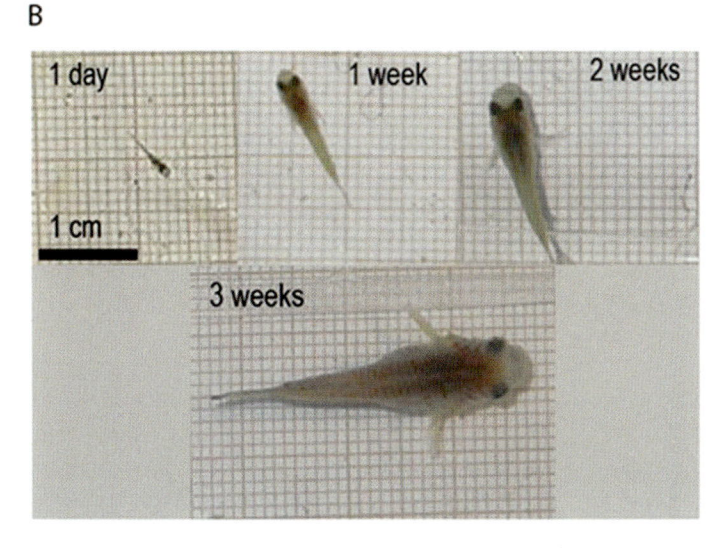

Fig. 15.2 Survival and growth of *Nothobranchius furzeri* laboratory strains. (a) Survivorship of the F2 generation of MZM-04/03 (n = 24), MZM-04/06 (n = 11), MZM-04/10G (n = 47) and MZM-04/10P (n = 90) and of the inbred line GRZ (n = 93). (b) Growth record of *N. furzeri* larva in the first 3 weeks of live

editing is feasible [36] and can **introduce specific point mutations** via homologous recombination [36].

N. furzeri lifespan and ageing phenotypes can be improved by environmental, dietary and pharmacological interventions [6, 32, 37, 38], and **modulations of ageing-related phenotypes by genetic manipulations were recently reported** [35, 36]. For all the above-mentioned reasons, *N. furzeri* **can be with full rights** considered as the ideal model to investigate the effects of genetic manipulations on lifespan and ageing-related phenotypes.

Nothobranchius furzeri as a Model for Evolutionary Genetics of Ageing

The genus *Nothobranchius* evolved from a non-annual (therefore long-living) ancestor: the non-annual sister genus (*Aphyosemion*) is clearly identified and the two taxa provide a sharp phenotypic contrast. Duration of the habitat (aridity) strictly limits natural lifespan. We specifically tested whether **differences in habitat duration led to the evolution of a different rate of senescence in Nothobranchius** populations from Southern and Central Mozambique, a region characterized by a major gradient in aridity. Two **independent evolutionary lineages of Nothobranchius** are found in this area (Fig. 15.3): *N. furzeri* and *N. kuhntae* belong to one lineage, while *N. rachovii* and *N. pienaari* belong to another lineage. For each lineage, one species originates from semi-arid habitat (*N. furzeri* and *N. pienaari*, respectively) and another species from the humid habitat (*N. kuhntae* and *N. rachovii*, respectively). In both species pairs, the **species from more humid habitat showed shortened lifespan and accelerated expression of ageing markers** [39], thereby providing a clear example of parallel evolution.

In 2015, has been sequenced and assembled the genome of *N. furzeri* as well as the transcriptomes of *N. pienaari*, *N. rachovii*, *N. kuhntae* and *N. kadleci* (the sister species of *N. furzeri*) together with *N. korthausae* (a long-living *Nothobranchius*, lifespan 18 months) and *Aphyosemion striatum* (lifespan >3 years) as a representative of the non-annual sister genus [42]. This allows the identification of genes bearing signatures of *positive selection*. Positive selection is the process by which a mutation becomes fixed in a population because it confers an advantage by changing protein function. If two branches of an evolutionary tree differ in a key phenotype (annual vs. non-annual life history, in this case), the genes under positive selection likely played a role in the evolution of that phenotype. In interspecies comparisons, positive selection on protein-coding sequences results in an increase in the rate of non-synonymous substitutions as compared with random genetic drift. Statistical models based on the ratio of non-synonymous to synonymous substitution rates (d_N/d_S ratio) can identify specific amino acids within a given gene that evolved due to positive selection and are widely used in comparative genomics.

Genome-wide scans for positive selection were performed in several long-living mammals (bats, the naked mole rat, the bowhead whale). However, it is not possible to establish a link between positively selected genes and evolution of longevity because the short-living sister *taxon* (i.e. the most closely related species/clade) may not be available for analysis, making impossible to exclude that of a codon change was selected before longevity evolved (for a discussion see [43]), and it is very often impossible to relate a codon change to one of the several traits that distinguish two taxa (e.g. a positively selected gene in *H. sapiens* may be related to

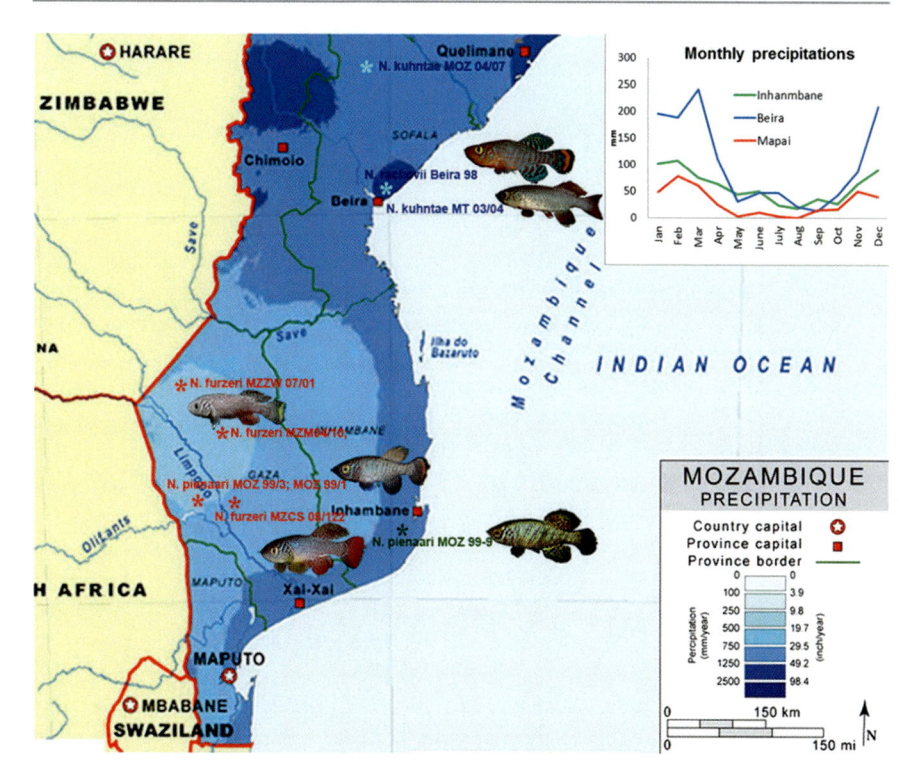

Fig. 15.3 Distribution map for the *Nothobranchius* populations through humid and arid regions of Mozambique. Physical map with annual precipitations was obtained from Stock Map Agency (http://www.stockmapagency.com). Collection points are indicated with asterisks and are color-coded: red indicates semi-arid green intermediate and blue humid habitats. Please note that the collection points of N. rachovii and two of the populations of N. kuhntae are in Beira and are collectively represented by a single asterisk. Inset shows monthly precipitations in Beira, Inhambane and Mapai as examples of humid, intermediate and semi-arid regions (taken from Terzibasi Tozzini, 2013)

longevity, bipedalism, absence of fur, speech, relative brain size or any other trait that distinguishes humans from apes). A direct test would require the manipulation of the selected gene sequence and an analysis of the phenotypic consequences in vivo of this sequence change. This possibility is obviously precluded in long-living mammals.

By performing genome-wide scans for positive selection on six *Nothobranchius* species, *Aphyosemion striatum* and a number related of outgroups [44] positively selected genes have been identified that evolved: i) in the last common ancestor of all *Nothobranchius* (LCA-branch) and therefore in coincidence with evolution of annual life, ii) in coincidence with parallel evolution to an arid habitat or iii) specifically in *N. furzeri* [43, 44]. Here, the most interesting findings of this analysis are listed:

IGF1R. **Insulin/IGF pathway is the prototypical lifespan-regulating pathway** conserved across model organisms. Reducing its activity prolongs lifespan from nematodes to mice [11], hypomorphic *IGF1R* variants were detected in some centenarians and *IGF1R* was previously reported to be under positive selection in the long-lived Brandt's bat. One site under positive selection in the tyrosine kinase domain with a radical substitution (lysine into proline) was detected in the LCA-branch and is therefore possibly linked to evolution of annual life.

XRCC5. *XRCC5* (X-ray repair complementing defective repair in Chinese hamster cells 5; synonyms: Ku80, Ku86) **is a key gene in** the non-homologous end-joining pathway of double-strand **DNA repair**. Variations in *XRCC5* are **associated with human longevity** [45] and loss of a single *XRCC5* allele is sufficient to induce progeria in mice [46]. A comparative study of mammalian transcriptomes revealed that *XRCC5* expression is positively correlated with body mass corrected longevity across a set of 33 mammalian species and one of its interactors, *XRCC6BP1*, is part of a module of co-expressed genes whose expression is correlated with individual lifespan in *N. furzeri* (see [32], Table S14). Previous analyses detected two positively selected sites specific for *N. furzeri* [43], and only one results in a radical substitution (Alanine into Glycine) located in a functionally relevant domain.

ID3. Inhibitor of DNA binding 3 (*ID3*) **is a downstream effector of the TGFβ pathway, a key regulator of** inflammation involved in **a number of age-related diseases** such as tumorigenesis, fibrosis, glaucoma and osteoarthritis. *ID3* shows one positively selected site with a radical substitution of an alanine by glutamine, located in the second exon, which is essential for *ID3* dominant negative function and is nearby the residue 105 whose polymorphism in humans (rs11574) impairs function and is associated with coronary disease, a typical age-dependent disease [47]. Interestingly, the same gene shows in the short-lived species *N. pienaari* a one aa deletion at distance of 10 aa from the positively selected glutamine, suggesting parallel evolution on this gene [42]. To give an idea of the high potential of this model in the context of experimental genetic approaches for the ageing studies, it must be taken in mind that in all three genes described above, a single site is positively selected, and it is technically possible to revert this substitution to the ancestral state by editing the corresponding codon in the *N. furzeri* genome.

Complex I and mitonuclear balance. Respiratory chain complexes are large protein complexes that undergo multistep assembly where nuclearly and mitochondrially encoded components are synthesized, combined and inserted into the mitochondrial inner membrane. The coordination of this process is called **mitonuclear balance** and **is a conserved longevity mechanism** controlled by mitochondrial ribosomal proteins [9].

Genes under positive selection in all steps of mitonuclear balance were identified [44]: transcription from mitochondrial promoters, processing and stabilization of mitochondrial RNAs, mitochondrial translation, assembly of respiratory chain complexes and electron transport chain. Particularly relevant is the pattern of selection on complex I of the respiratory chain and on mitochondrial ribosomal proteins, since manipulation of these genes results in life extension in *C. elegans* [9] and fruit fly. 75 genes under positive selection were detected in the LCA-branch; among these, four code for components of the mitochondrial respiratory chain complex I (GO:0030964, fold enrichment = 14, p = 0.0002, Fisher's exact test). Three further genes of complex I are under positive selection in the *N. pienaari* lineage (fold enrichment = 8.8, p = 0.005, Fisher's exact test). Three mitochondrial ribosome (MRPs) were under positive selection in both the *N. pienaari* and *N. furzeri* lineages (GO:0005761, fold enrichment = 9.1 and 14.7, respectively, p = 0.02 and 0.01, Fisher's exact test, respectively).

Nothobranchius furzeri and Longitudinal Analysis

While genetic heterogeneity clearly influences lifespan, **the relationship between non-genetic variation and individual ageing trajectories is not well understood**. To the best of our knowledge, all available **studies** on individual random variation and ageing have been **performed in the roundworm** *C. elegans*. Pioneering studies showing that the expression level of heat-shock proteins is not heritable but is predictive of individual lifespan [3] demonstrate that phenotypic variability in the absence of genetic heterogeneity can influence ageing. Later studies identified the frequency of "mitoflashes" during an individual's early adult life [4] as predictors of lifespan. Accordingly, knock-down of mitochondrial proteins during development is necessary and sufficient to cause life extension in *C. elegans* [9]). Taken together, these studies **suggest that conditions favouring longevity may be laid out during maturation and early adult life**.

Recently, it has been also realized a longitudinal study of gene expression in individual *N. furzeri* (N = 152) of a strain with median lifespan ~ 8 months by obtaining two fin biopsies at two time points during early adult stage (corresponding to ~ 15% and ~ 30% of maximum life span) to correlate the specific gene expression signatures observed in these individuals during early adult life with their age at death [32]; 45 individuals were selected, based on the age of death, and divided into three groups of 15 individuals each: short-lived (age of death 28–36 weeks), long-lived (age of death 45–50 weeks) and longest-lived (age of death 57–71 weeks).

Then, quantitative variations in gene expression during early adult life were correlated with lifespan. Generally applicable gene enrichment (GAGE) analysis revealed that shorter- and longer-lived individuals differ more in their gene expression at the earliest time point than later (15 differentially expressed pathways vs. 2). This surprising result indicates that conditions for longevity are set relatively **early during lifespan**.

To identify the genes signatures for lifespan, a gene co-expression network was generated using weighted gene co-expression network analysis (WGCNA) . By this approach, large datasets are reduced to modules of co-regulated genes and the correlated expression of the genes within a module can be represented as an *eigengene*. Eigengene expression is correlated with lifespan for each module to identify longevity-related modules and their hub genes. In fact, this approach allowed the **identification of one module of 149 genes whose expression shows the largest absolute correlation with age of death with top enrichments for complex I of the respiratory chain** and mitochondrial small ribosomal subunit. Analysis of the gene network in this module revealed that genes coding for NDUFs and MRPs are tightly co-regulated (Fig. 15.4). This finding is consistent with mouse studies here *Mrps5* expression strongly correlates with expression of oxidative phosphorylation proteins [9] and expression of MRPs is negatively correlated with lifespan across mouse strains [9].

To ask directly whether partial inhibition of the respiratory chain affects fish lifespan, *N. furzeri* MZM-0410 were chronically treated with 15 pM and 150 pM of **a complex I inhibitor** (rotenone, ROT) in the water, corresponding to 0.1% and 1% of the median lethal concentration (LC50) for *N. furzeri*, respectively, starting at mid-age (23 weeks). The lower concentration **induced life span extension** by 15% (log-rank test, p = 0.0181) directly demonstrating the involvement of complex I in regulation of vertebrate lifespan.

Five animals treated with 15 pM ROT were taken after 4 weeks of treatment and RNA-seq was performed from brain, liver and skin samples. These were compared with animals of the same age treated with vehicle and with animals treated with vehicle for 4 weeks starting at age 5 weeks (young controls). Figure 15.5 describes the response to ROT of all genes differentially expressed with age (FDR-corrected p < 0.05, EdgeR and DEseq) detected in comparisons of old vs. young controls. In all three tissues, the vast majority (~ 90%) of genes upregulated during ageing were downregulated by ROT **and** vice versa, i.e. ROT induced **a rejuvenation of the transcriptome** shifting global gene expression towards patterns more typical of young age.

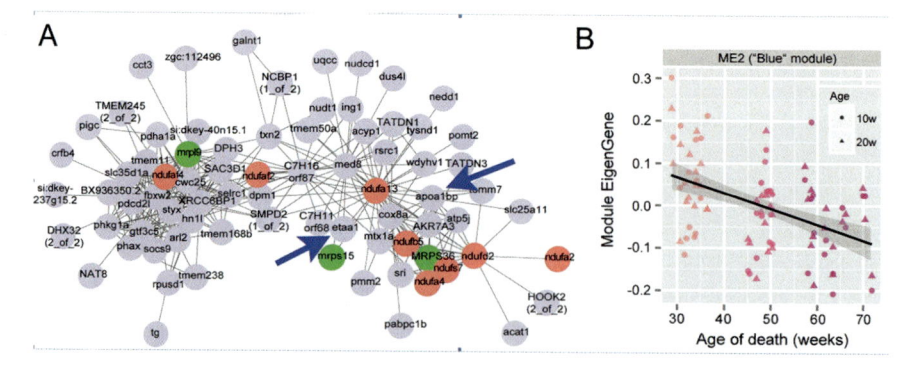

Fig. 15.4 (a) Network of the most connected genes in the longevity module described by [32]. Only edges corresponding to topological overlap >0.1 are shown. Genes of complex I are shown in red and mitochondrial ribosomal genes in green. The arrows points to the position of the positively selected genes ETAA1 and APOA1BP. (b) Expression of the eigengene as a function of age of death (r = −0.46, grey shading indicates the confidence interval of the regression line)

Fig. 15.5 Effects of rotenone (ROT) on age-dependent gene expression. Each dot represents a single differentially expressed gene (DEG). The effect of age is reported on the X axis and the effect of ROT on the Y axis. Negative regression implies that genes upregulated with age are downregulated by ROT and vice versa. From [32]

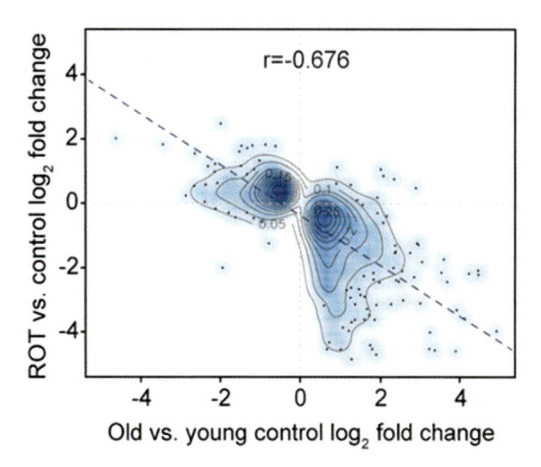

The use of this new model in longitudinal ageing studies, and in particular the existence and comparison of different Nothobranchius species, has brought to light the essential knowledge that genes important for mitonuclear balance seem to act as regulators of longevity from two totally independent approaches: analysis of positive selection across species and longitudinal analysis within a species. This observation highlights the value of the multilevel approach achievable with *Nothobranchius furzeri*.

Nothobranchius furzeri and Quantitative Genetics of Ageing

***N. furzeri* strains show genetic differences in captive lifespan** the GRZ strain, collected in 1969, shows a lifespan of about 3 months [19] and all more recently collected strains of 6–8 months [26]. Genetic mapping (see below) clearly excludes that extremely short lifespan of GRZ is the results of a single recessive mutation, and it must arise from a combination of multiple naturally occurring alleles at different loci.

A quantitative trait loci (QTL) approach was applied to map loci responsible for lifespan differences. By crossing GRZ with a longer-lived strain, a genetic map comprising 355 markers was built and used to identify quantitative trait loci (QTL) for lifespan in a panel of 284 F2 individuals. One significant locus explaining ~ 10% of lifespan variation overlying tens of genes and three further suggestive loci were identified [48]. A second, independent, QTL scan using a much higher density linkage map (8400 markers) and a similar number of individuals also identified only one significant locus (on a different chromosome) that explains a minor proportion of lifespan variation and overlies tens of genes [49]. This approach clearly shows that the basis for interstrain differences in lifespan **is polygenic and** also suggests that **heritability of lifespan is quite low, possibly because age-dependent mortality is determined by the action of several independent biological processes**. It is indeed known that human longevity or lifespan has lower heritability as compared to age-related diseases and organ function [8].

In a second study [7], the **age-related markers lipofuscin deposition (LFD) and apoptotic cell counts (APC)** were quantified in histological sections of F1 and F2 hybrids (287 fish) and a genome-wide scan for QTL with 253 markers was performed. Surprisingly, LFD and APC appeared to be markers of different age-dependent biological processes and to be controlled by different genetic mechanisms. Indeed, inheritance of LFD showed hyperdominance and inheritance of APC showed additivity. More directly, these traits **were uncorrelated in F2 hybrids** [7]. While only two loci, explaining 10% of variance, were identified for LFD, APC was mapped to five loci collectively explaining ~ 30% of the total variance.

The results of APC mapping are encouraging as they suggest that, as in humans, simpler biological processes can be easier to map genetically than longevity. **In how many different biological processes can ageing be decomposed and what are the best biomarkers of these processes?** Using histological techniques to address this question is unpractical and prohibitively time-consuming. Genome-wide expression analysis may provide a suitable approach, and, indeed, *Nothobranchius furzeri* seem to represent the perfect animal model to address this research.

References

1. de Cabo R, Carmona-Gutierrez D, Bernier M, Hall MN, Madeo F. The search for antiaging interventions: from elixirs to fasting regimens. Cell. 2014;157:1515–26.
2. Pincus Z, Smith-Vikos T, Slack FJ. MicroRNA predictors of longevity in Caenorhabditis elegans. PLoS Genet. 2011;7:e1002306.
3. Rea SL, Wu D, Cypser JR, Vaupel JW, Johnson TE. A stress-sensitive reporter predicts longevity in isogenic populations of Caenorhabditis elegans. Nat Genet. 2005;37:894–8.
4. Shen EZ, Song CQ, Lin Y, Zhang WH, Su PF, Liu WY, Zhang P, Xu J, Lin N, Zhan C, Wang X, Shyr Y, Cheng H, Dong MQ. Mitoflash frequency in early adulthood predicts lifespan in Caenorhabditis elegans. Nature. 2014;508:128–32.
5. Harrison DE, Strong R, Sharp ZD, Nelson JF, Astle CM, Flurkey K, Nadon NL, Wilkinson JE, Frenkel K, Carter CS, Pahor M, Javors MA, Fernandez F., Miller RA. Rapamycin fed late in life extends lifespan in genetically heterogeneous mice. Nature. 2009;460:392–5.
6. Valenzano DR, Terzibasi E, Genade T, Cattaneo A, Domenici L, Cellerino A. Resveratrol prolongs lifespan and retards the onset of age-related markers in a short-lived vertebrate. Curr Biol. 2006b;16:296–300.
7. Ng'oma E, Reichwald K, Dorn A, Wittig M, Balschun T, Franke A, Platzer M, Cellerino A. The age related markers lipofuscin and apoptosis show different genetic architecture by QTL mapping in short-lived Nothobranchius fish. Aging (Albany NY). 2014;6:468–80.
8. Steves CJ, Spector TD, Jackson SH. Ageing, genes, environment and epigenetics: what twin studies tell us now, and in the future. Age Ageing. 2012;41:581–6.
9. Houtkooper RH, Mouchiroud L, Ryu D, Moullan N, Katsyuba E, Knott G, Williams RW, Auwerx J. Mitonuclear protein imbalance as a conserved longevity mechanism. Nature. 2013;497:451–7.
10. Cellerino A, Valenzano DR, Reichard M. From the bush to the bench: the annual Nothobranchius fishes as a new model system in biology. Biol Rev Camb Philos Soc. 2016;91:511–33.
11. Fontana L, Partridge L, Longo VD. Extending healthy life span--from yeast to humans. Science. 2010;328:321–6.
12. Johnson SC, Rabinovitch PS, Kaeberlein M. mTOR is a key modulator of ageing and age-related disease. Nature. 2013;493:338–45.

13. Jeck WR, Siebold AP, Sharpless NE. Review: a meta-analysis of GWAS and age-associated diseases. Aging Cell. 2012;11:727–31.
14. Howe K, et al. The zebrafish reference genome sequence and its relationship to the human genome. Nature. 2013;496:498–503.
15. Hwang WY, Fu Y, Reyon D, Maeder ML, Tsai SQ, Sander JD, Peterson RT, Yeh JR, Joung JK. Efficient genome editing in zebrafish using a CRISPR-Cas system. Nat Biotechnol. 2013;31:227–9.
16. Lieschke GJ, Currie PD. Animal models of human disease: zebrafish swim into view. Nat Rev Genet. 2007;8:353–67.
17. Gerhard GS, Kauffman EJ, Wang X, Stewart R, Moore JL, Kasales CJ, Demidenko E, Cheng KC. Life spans and senescent phenotypes in two strains of zebrafish (Danio rerio). Exp Gerontol. 2002;37:1055–68.
18. Hatakeyama H, Nakamura K, Izumiyama-Shimomura N, Ishii A, Tsuchida S, Takubo K, Ishikawa N. The teleost Oryzias latipes shows telomere shortening with age despite considerable telomerase activity throughout life. Mech Ageing Dev. 2008;129:550–7.
19. Valdesalici S, Cellerino A. Extremely short lifespan in the annual fish *Nothobranchius furzeri*. Proc R Soc Lond B Biol Sci. 2003;270(Suppl 2):S189–91.
20. Callaway E. Short-lived fish may hold clues to human ageing. Nature. 2015;528:175.
21. Dance A. Live fast, die young. Nature. 2016;535:453–5.
22. Lakhina V, Murphy CT. Genome sequencing fishes out longevity genes. Cell. 2015;163:1312–3.
23. Lieben L. Genomics: fishing for the ageing secret. Nat Rev Genet. 2016;17:69.
24. Wang AM, Promislow DE, Kaeberlein M. Fertile waters for aging research. Cell. 2015;160:814–5.
25. Di Cicco E, Tozzini ET, Rossi G, Cellerino A. The short-lived annual fish Nothobranchius furzeri shows a typical teleost aging process reinforced by high incidence of age-dependent neoplasias. Exp Gerontol. 2011;46:249–56.
26. Terzibasi E, Valenzano DR, Benedetti M, Roncaglia P, Cattaneo A, Domenici L, Cellerino A. Large differences in aging phenotype between strains of the short-lived annual fish Nothobranchius furzeri. PLoS One. 2008;3:e3866.
27. Hartmann N, Reichwald K, Lechel A, Graf M, Kirschner J, Dorn A, Terzibasi E, Wellner J, Platzer M, Rudolph KL, Cellerino A, Englert C. Telomeres shorten while Tert expression increases during ageing of the short-lived fish Nothobranchius furzeri. Mech Ageing Dev. 2009;130:290–6.
28. Hartmann N, Reichwald K, Wittig I, Drose S, Schmeisser S, Luck C, Hahn C, Graf M, Gausmann U, Terzibasi E, Cellerino A, Ristow M, Brandt U, Platzer M, Englert C. Mitochondrial DNA copy number and function decrease with age in the short-lived fish Nothobranchius furzeri. Aging Cell. 2011;10:824–31.
29. Tozzini ET, Baumgart M, Battistoni G, Cellerino A. Adult neurogenesis in the short-lived teleost Nothobranchius furzeri: localization of neurogenic niches, molecular characterization and effects of aging. Aging Cell. 2012;11:241–51.
30. Baumgart M, Groth M, Priebe S, Appelt J, Guthke R, Platzer M, Cellerino A. Age-dependent regulation of tumor-related microRNAs in the brain of the annual fish Nothobranchius furzeri. Mech Ageing Dev. 2012;133:226–33.
31. Baumgart M, Groth M, Priebe S, Savino A, Testa G, Dix A, Ripa R, Spallotta F, Gaetano C, Ori M, Terzibasi Tozzini E, Guthke R, Platzer M, Cellerino A (2014). RNA-seq of the aging brain in the short-lived fish N. furzeri – conserved pathways and novel genes associated with neurogenesis. Aging Cell.
32. Baumgart M, Priebe S, Groth M, Hartmann N, Menzel U, Pandolfini L, Koch P, Felder M, Ristow M, Englert C, Guthke R, Platzer M, Cellerino A. Longitudinal RNA-Seq analysis of vertebrate aging identifies mitochondrial complex I as a small-molecule-sensitive modifier of lifespan. Cell Systems. 2016;2:122–32.

33. Valenzano DR, Sharp S, Brunet A. Transposon-mediated Transgenesis in the short-lived African killifish Nothobranchius furzeri, a vertebrate model for aging. *G3 (Bethesda)*. 2011;1:531–8.
34. Allard JB, Kamei H, Duan C. Inducible transgenic expression in the short-lived fish Nothobranchius furzeri. J Fish Biol. 2013;82:1733–8.
35. Ripa R, Dolfi L, Terrigno M, Pandolfini L, Arcucci V, Groth M, Terzibasi Tozzini E, Baumgart M, Cellerino A. MicroRNA miR-29 controls a compensatory response to limit neuronal iron accumulation during adult life and aging. BioRXiv. 2016; https://doi.org/10.1101/046516.
36. Harel I, Benayoun BA, Machado BE, Priya Singh P, Hu C-K, Pech MF, Valenzano DR, Zhang E, Sharp SC, Artandi SE, Brunet A. A platform for rapid exploration of aging and diseases in a naturally short-lived vertebrate. Cell,. in press. 2015;160:1013.
37. Terzibasi E, Lefrancois C, Domenici P, Hartmann N, Graf M, Cellerino A. Effects of dietary restriction on mortality and age-related phenotypes in the short-lived fish Nothobranchius furzeri. Aging Cell. 2009;8:88–99.
38. Valenzano DR, Terzibasi E, Cattaneo A, Domenici L, Cellerino A. Temperature affects longevity and age-related locomotor and cognitive decay in the short-lived fish Nothobranchius furzeri. Aging Cell. 2006a;5:275–8.
39. Tozzini ET, Dorn A, Ng'oma E, Polacik M, Blazek R, Reichwald K, Petzold A, Watters B, Reichard M, Cellerino A. Parallel evolution of senescence in annual fishes in response to extrinsic mortality. BMC Evol Biol. 2013;13:77.
40. Hartmann N, Englert C. A microinjection protocol for the generation of transgenic killifish (species: Nothobranchius furzeri). Dev Dyn. 2012a;241:1133–41.
41. Hartmann N, Englert C. A microinjection protocol for the generation of transgenic killifish (species: Nothobranchius furzeri). Dev Dyn. 2012b;241(6):1133–41.
42. Reichwald K, Petzold A, Koch P, Downie BR, Hartmann N, Pietsch S, Baumgart M, Chalopin D, Felder M, Bens M, Sahm A, Szafranski K, Taudien S, Groth M, Arisi I, Weise A, Bhatt SS, Sharma V, Kraus JM, Schmid F, Priebe S, Liehr T, Gorlach M, Than ME, Hiller M, Kestler HA, Volff JN, Schartl M, Cellerino A, Englert C, Platzer M. Insights into sex chromosome evolution and aging from the genome of a short-lived fish. Cell. 2015;163:1527–38.
43. Sahm A, Platzer M, Cellerino A. Outgroups and positive selection: the Nothobranchius furzeri case. Trends Genet. 2016b;32:523–5.
44. Sahm A, Bens M, Platzer M, Cellerino A. Convergent evolution of genes controlling mitonuclear balance in short-lived annual fishes. BioRXiv. 2016a; https://doi.org/10.1101/055780.
45. Soerensen M, Dato S, Tan Q, Thinggaard M, Kleindorp R, Beekman M, Jacobsen R, Suchiman HE, de Craen AJ, Westendorp RG, Schreiber S, Stevnsner T, Bohr VA, Slagboom PE, Nebel A, Vaupel JW, Christensen K, McGue M, Christiansen L. Human longevity and variation in GH/IGF-1/insulin signaling, DNA damage signaling and repair and pro/antioxidant pathway genes: cross sectional and longitudinal studies. *Exp Gerontol*. 2012;47:379–87.
46. Didier N, Hourde C, Amthor H, Marazzi G, Sassoon D. Loss of a single allele for Ku80 leads to progenitor dysfunction and accelerated aging in skeletal muscle. EMBO Mol Med. 2012;4:910–23.
47. Manichaikul A, Rich SS, Perry H, Yeboah J, Law M, Davis M, Parker M, Ragosta M, Connelly JJ, McNamara CA, Taylor AM. A functionally significant polymorphism in ID3 is associated with human coronary pathology. PLoS One. 2014;9:e90222.
48. Kirschner J, Weber D, Neuschl C, Franke A, Bottger M, Zielke L, Powalsky E, Groth M, Shagin D, Petzold A, Hartmann N, Englert C, Brockmann GA, Platzer M, Cellerino A, Reichwald K. Mapping of quantitative trait loci controlling lifespan in the short-lived fish Nothobranchius furzeri- a new vertebrate model for age research. Aging Cell. 2012;11:252–61.
49. Valenzano DR, Benayoun BA, Singh PP, Zhang E, Etter PD, Hu CK, Clement-Ziza M, Willemsen D, Cui R, Harel I, Machado BE, Yee MC, Sharp SC, Bustamante CD, Beyer A, Johnson EA, Brunet A. The African turquoise killifish genome provides insights into evolution and genetic architecture of lifespan. Cell. 2015;163:1539–54.

Part II

Alterations in Metabolism During Aging and Diseases

Aging in Muscle   16

Sunil Pani and Naresh C. Bal

Common Terminology

Atrophy: Decrease in muscle mass and strength. Muscle atrophy can arise due to several pathophysiological conditions such as bedridden state, cancer cachexia, and aging.

Sarcopenia: Involuntary loss in skeletal muscle mass, strength, and function with aging. Generally characterized by muscle atrophy along with a reduction in muscle fiber quality, oxidative stress, reduction in muscle metabolism, and impairment of the neuromuscular junction.

Myasthenia: Increased weakness and fatigue of skeletal muscle. It is often associated with disorder in neuromuscular junction.

Myositis: This refers to conditions causing inflammation in muscles. Infection, autoimmunity, injury, and drug side effects may lead to myositis. General symptoms are swelling, weakness, and pain.

Fatigue: Temporary inability of a skeletal muscle to be recruited optimally. This depends upon an individual's physical fitness, sleep deprivation, and overall health.

Myonecrosis: A myopathy in which a part of skeletal muscle undergoes necrosis. It is mainly caused by infection with bacteria, Clostridium species. This leads to gas accumulation in muscle.

Rhabdomyolysis: A condition where skeletal muscle fibers undergo rapid breakdown, usually due to injury. This leads to release of muscle proteins into the blood and causes kidney failure.

S. Pani · N. C. Bal (✉)
KIIT School of Biotechnology, KIIT University, Bhubaneswar, Odisha, India
e-mail: naresh.bal@kiitbiotech.ac.in

© Springer Nature Singapore Pte Ltd. 2020
P. C. Rath (ed.), *Models, Molecules and Mechanisms in Biogerontology*,
https://doi.org/10.1007/978-981-32-9005-1_16

NEAT: This refers to "Non-Exercise Activity Thermogenesis." NEAT is the sum of energy expended for everything that an individual does, except sleeping, eating, and exercises.

Introduction

By the year 2050, the number of people over 60 years of age will be more than double from 11% currently to ~25% worldwide. Better management of aged population requires systematic understanding of aging and etiology of its associated illnesses. This puts gerontological research back in high priority. Skeletal muscle changes are primary cause of the progressive deterioration in functional capacity associated with the aging process. Several changes are noticed in the skeletal muscle during aging at all levels Several changes, including physiological, structural, molecular, and biochemical, are noticed in the skeletal muscle during aging at all levels. Starting with the fourth decade of life, muscle mass and force-generating capacity decline and by the age of 80, about one-third functional capacity is left. Structurally, muscle fiber type, size, and architecture are altered with advancement of age in both men and women. In addition to skeletal muscle atrophy, progressive accumulation of denervated myofibers is observed with aging. Molecular and biochemical changes in skeletal muscle include impairments in excitation–contraction coupling, actin–myosin cross-bridge interaction, organellar dysfunction, mitochondrial properties (energy production), and regeneration.

Unlike human, most of the rodent models show a distinct path of skeletal muscle aging. The loss of muscle mass among rodents is comparatively slow and starts only after advanced stages of aging. It is also true for the other physiological parameters including grip strength and performance. So, results from studies performed in many other species cannot always be directly extrapolated to human aging. Regardless, the fundamental process of muscle aging is visible in most organisms including the fruit fly *Drosophila melanogaster*. More and more studies are pointing to the fact that extrinsic factors play critical role in muscle aging including stem cell function, inflammatory cells, neuromuscular junctions, and endocrine regulation. With rapid increase in the number of individuals affected with metabolic diseases such as type 2 diabetes, maintenance of skeletal muscle function is the foremost task. This is because muscles (both skeletal and cardiac) are important sites of glucose disposal; their functions are significantly affected in diabetic patients and usually display acceleration in the aging process. Another nascent area of research is "effect of crosstalk" between different organs on muscle aging. Understanding of these complex multifactorial conditions will provide the molecular targets for suitable therapeutic intervention to retard muscle aging and contribute to healthy aging.

In this chapter, we have primarily focused on skeletal muscle, but cardiac muscle has also been taken into account. We have summarized the gender-specific changes in muscle aging. We have elaborately described the current understanding about muscle-insulin resistance and its association with type 2 diabetes. Roles of different processes in muscle aging, such as mitochondrial dysfunction, cytokines, muscle-insulin resistance, fibroblasts, and satellite cells, have been discussed in detail.

Finally, we describe the potential of promotion of physical activity by NEAT and exercise in countering muscle aging.

Changes in Skeletal Muscle during Aging

Morphological Changes

Skeletal muscle undergoes significant alterations both at molecular as well as morphological levels. In general, the skeletal muscle attends maximal growth by the age of 20–25 in humans and starting with the age of 30, a progressive decline is observed. The rate of loss in muscle mass (more than 1% per year) is usually much higher after the age of 60. In addition, cross-sectional area of muscle fibers also undergoes progressive reduction during the aging process; 30% loss in a 70-year-old compared to a 20-year-old. Further, loss of type II (fast-twitch) fibers and increase in type I/type II fiber ratio are reported in several studies on elderly people. As a consequence of all these changes, strength, and performance (i.e., muscle twitch time and twitch force) of the skeletal muscle is impaired. Common cell-level changes during aging are reduction in muscle cell number, sarcoplasmic reticulum (SR) abundance, and calcium-pumping capacity. Interestingly, during aging on one hand muscle mass is reduced; on the other hand, the body tends to accumulate more fat and there is no change in body weight. Hence, reduced muscle mass has to carry similar load, which means they have to work harder and may lead to fatigue. To overcome this situation, there is increased reliance on type I muscle fibers that result in reduction of swiftness of muscle action. Several biochemical and structural changes in aging muscle have also been reported, such as intramuscular fat accumulation, decrease in protein content (whereas in nonmuscle lean tissues, it remains unchanged), decrease in noncontractile area, and decrease in cross-bridging between the fibers in the muscle, and loss of potassium and calcium levels. In conjunction with muscle loss, bone density reduction limits the relative recruitment (force formation) of the muscle. Some of the usual intramyocyte changes during aging are disorganization of sarcomere spacing and change in position of nuclei to become more centralized on the muscle fiber. Further, decrease in the dihydropyridine receptor (DHPR) expression and its coupling with T-tubular ryanodine receptors (RyRs) leading to the impairment of excitability of muscle plasma membrane. As a cumulative effect of the above changes, older people become disabled due to compromised ability to recruit the muscle.

Fiber-Type Distribution

As described earlier, reduction in muscle mass and cross-sectional area of about 30% and 20%, respectively, is witnessed in an 80 year-old man compared to a 20-years old. This observation may be due to the associated changes in fiber type. The skeletal muscles are classified into three different fiber types:

1. **Slow oxidative fibers:** Also called "Type I," these fibers upon activation contract slowly, but can remain contracted for a long time with lesser force. They are rich in mitochondrial content, equipped to meet high-energy demand for a prolonged period of time by using fatty acid as a major substrate. These muscles are enriched with myoglobin and highly vascularized for continued supply of oxygen and substrates, for example, postural muscles. Muscles having higher percentage of slow oxidative fibers are more resistant to fatigue. Sarco/endoplasmic reticulum (SR) Ca2+-signaling in these muscles is quite unique. They express sarco/endoplasmic reticulum calcium ATPase (SERCA)2a, the kinetically slower isoform, along with expression of sarcolipin (SLN). Therefore, such muscles are well suited for being recruited for adaptive thermogenesis.
2. **Fast glycolytic:** Also called "Type IIb," these fibers contract at a very fast rate with high degree of force production. They have very low mitochondrial abundance and use glycolysis (anaerobic) to produce ATP to supply energy for the muscle activity. They tend to exhaust their fuel reserve quickly leading to fatigue; so they can be recruited only for short duration of highly demanding activities. For example, muscles recruited during sprinting. In these muscles, after contraction during the relaxation process the myoplasmic Ca2+ concentration comes back to baseline much faster than other muscle types. This is assisted by the expression of only SERCA1a, which is a kinetically faster isoform, and elimination of SLN expression that would allow SERCA to work unhindered. In many chronic pathophysiological conditions like Duchenne muscular dystrophy, the majority of type IIb fibers display fiber conversion and are replaced by Type I or Type IIa fibers.
3. **Fast oxidative:** Also termed "Type IIa," these fibers have mechanical properties of fast fibers and at the same time have enzymatic/structural machinery capable of carrying out oxidative metabolism. They have intermediate number of mitochondria, greater fatigue resistance than Type IIb fibers, and capacity to generate higher force than the slow oxidative (Type I) fibers. SLN is expressed at low abundance in these muscles. Having the capability to supply energy for muscle activity by both aerobic and anaerobic metabolic processes, type IIa fibers provide flexibility to mammals (and birds) to fine-tune their whole-body energy expenditure to environmental factors such as food availability, temperature, pregnancy, etc.

Usually, fast-twitch (type II) glycolytic has the largest cross-sectional area followed by type II oxidative and slow-twitch (type I) has the smallest area. Initial cross-sectional studies showed alteration in muscle fiber composition resulting from selective loss (and/or atrophy) of type II fibers with increasing age. Further, type I fiber area is preserved because it is used for most daily routine activities and submaximal exercise like walking. In contrast, high-intensity activities (that preferentially recruit type II fibers) become restricted with growing age that result in atrophy of these fibers.

The earliest investigations on muscle fiber composition during aging using muscle biopsies approach suggested increase in percentage of type I fibers with aging.

People in their 20s had 39% type I fibers while 60-year-old participants had 66%. Whereas, Lexall et al. sampled whole muscle tissue using autopsy material and found that the type I fiber percentage was 49%, 52%, and 51% for men in their 20s, 50s, and late 70s, respectively. Several studies have shown that cross-sectional area of type I fibers remains unaltered during aging. Thus, now it is a consensus that number or size (area) of type I fibers remains more or less constant during the process of aging. In contrast, type II fibers exhibit changes in cross-sectional area. While, biopsy studies reported 15% to 25% reductions in the cross-sectional area of type IIA and IIB fibers, autopsy study by Lexall et al. found a 26% reduction in the size of type II fibers from age 20 to 80. Hence, reduction in type II fiber size is a major cause of muscle wasting during the aging process. Lexall et al. using autopsy method on vastus lateralis muscle demonstrated that number of muscle fibers start to decline after the age of 25 and by 80 it is reduced by ~39%. Interestingly, loss of muscle mass among women shows a different trend. A study on pectoralis muscle showed that decrease in fiber number in females starts in their 60s and they rapidly lose up to 25% of fibers by the age of 80.

Muscle fibers are held together by connective tissue. In contrast to muscle fiber reduction, it is reported that connective tissue volume increases between 20% and 40% during aging in animals. Skeletal muscles being purely voluntary are mainly recruited by innervation by motor neurons. Both muscle–nerve synapse and motor neurons undergo some changes. Decrease in the number and activity of motor units are found during aging which impairs the motor control by the central nervous system. In the field of muscle biology, an unresolved but accepted concept is "switching of fiber type" during aging. It is proposed that in the course of aging type II fibers get deinnervated and further reinnervated by type I motor neurons. Due to these changes, there is decrease in most of the physiological properties such as single-fiber intrinsic force generation, twitch contraction time, and maximum shortening speed.

Neuromuscular Junction

Neuromuscular junction (NMJ) is the point of contact between muscle and nerve that integrates its recruitment with the central nervous system. NMJ consists of axon terminal of the motor neuron and motor end plate of the muscle cell which is also called as motor unit. The space between these two is called "synaptic cleft" and is usually 30 nanometers wide. In most part of adult life, the several axon endings remain in constant competition for establishing a NMJ with a particular motor end plate. During the denervation–reinnervation cycle, an individual muscle fiber is transiently disconnected from its motor neuron. Following this, the original motor axon if still intact can get reinnervated or an axon from an adjacent motor neuron can establish a new NMJ by collateral sprouting. Several structural and physiological changes are observed in the NMJ during aging. In most cases, the size of the motor unit increases, whereas the number of units per fiber area decreases. During aging, the denervation–reinnervation cycles lead to disruption of the NMJ components: decreased alignment of the pre- and postsynaptic structures; thinning of terminal axons; altered expression of synaptic membrane proteins; and extensive sprouting of terminal axons. In aging muscle, there is a progressive collapse of the motor units leading to clustering of fibers within the given motor unit. The NMJ instability is also closely associated with aging-mediated muscle atrophy. In addition, the nervous firing rate to muscle and the regenerative abilities of the nervous tissue decrease. Progressive muscle loss during aging and its prevention by submaximal exercise including routine physical activity are still a topic of debate. Two divergent explanations have been put forward. One school of thought holds that there is reduction of regeneration capacity of muscle fibers, which is not supported by ample experimental evidence. The second notion is that regenerative capacity of motor units is significantly impaired during aging, which is supported by several studies.

Several studies using electromyographic (EMG) studies showed that number of active motor units decrease, while the size of low-threshold motor units becomes greater. Also, it was reported that fiber density per motor neuron increases and up to 25% of the motor neurons gradually become nonfunctional during aging. It has also been proposed by some groups that the age-related muscle loss is a result of denervation of muscle fibers, especially for the type II fibers. Denervation leads to lack of recruitment of the fibers that causes progressive muscle atrophy. As an adaptation to minimize atrophy of denervated type II fibers, they tend to gain collateral innervation from the type I motor neurons. Currently, there are several questions on the role of denervation–reinnervation cycle on fiber-type grouping in the course of aging. These include the relative contributions of traumatic injury, axon degeneration, motor neuron death, and NMJ instability.

Adipocyte Accumulation

Skeletal muscle is a major site of increased ectopic adiposity (fat deposition in tissues that normally do not store fat) during aging. While aging-related increased adiposity in liver (a tissue notoriously known for fat deposition) is still under debate that of skeletal muscle is widely accepted and also linked to several other medical complications. Both inter- and intramuscular sites show elevated adipose deposition. Fat infiltration into skeletal muscle has negative impact on muscle strength and performance as well as allied to aging-related skeletal muscle wasting. That is also closely associated with development of metabolic diseases as it is commonly found in people with type 2 diabetes and/or obesity. Interestingly, exercise-mediated metabolic benefits are blunted in type 2 diabetic patients having increased intramuscular adiposity. However, molecular details of induction of insulin resistance by increased intramuscular adiposity remain unclear. The accumulation of fat (especially triglycerides) in myocytes is mostly due to increased fatty acid transporters such as FAT/CD36. Short-term, high-fat diet treatment studies in humans showed higher lipid uptake, but decreased β-oxidation as well as reduction of insulin-mediated downregulation of pyruvate dehydrogenase kinase (PDK) 4 activity favored lipid storage. Apart from insulin resistance, higher lipid accumulation in skeletal muscle compromises mitochondrial functions and suppressed lipid oxidation. Reduction in lipid oxidation causes increased malonyl-coenzyme A (CoA) accumulation which inhibits carnitine palmitoyltransferase 1 (CPT-1) promoting lipid deposition finally insulin resistance.

Although correlations between triglycerides and insulin resistance exist, this association is very complex involving several effector molecules such as uncommon lipids, branched chain amino acids, cytokines, gut microbiome, etc. On the contrary, trained athletes, especially frequent marathon runners, pose a metabolic paradox. They exhibit elevated muscle triglyceride accumulation on one hand, while on the other they exhibit high insulin sensitivity. This is an adaptation that provides the skeletal muscles of the runner with the metabolic flexibility to switch substrate utilization depending on demand and supply. This observation is termed as "athlete's paradox." Recent studies have suggested the appearance of intermuscular adipose tissue as a predictor of cardiovascular disease risk. Additionally, studies on animal models suggest that muscle glucose transporter type 4 (Glut4) mRNA and protein levels are reduced during aging leading to reduced glucose oxidation.

Vascularization

In skeletal muscles, blood is delivered to cells via primary arteries progressively bunching into smaller vessels and capillaries. Oxidative muscles (both slow- and fast-oxidative) have higher capillary density than fast-glycolytic muscles. Increase in plasma insulin after meal leads to a modest increase in resting skeletal muscle blood flow via endothelial nitric oxide (NO)-dependent mechanism. Whereas, during exercise blood flow to the contracting skeletal muscles can increase up to 80-fold compared to rest via multiple mechanisms. The details of these mechanisms by which blood flow to the muscle capillary occurs are not fully understood. Vascular endothelial growth factor A (VEGF-A) is known to play the major role in skeletal muscle vascularization in response to exercise training. VEGF-A is stored in small vesicles present between the myofibrils and inside cells located in the interstitial fluid between capillaries and muscle fibers such as pericytes. Recent studies have suggested the role of other factors in vascularization of skeletal muscle including vascular endothelial growth factor B (VEGF-B), neuropilin 1, transforming growth factor beta-1 (TGF-β1), and hypoxia-inducible factor (HIF) 1.

It is observed that blood flow into skeletal muscle increased via enhanced vascularization and vasodilation during physiological conditions like diet, exercise, and cold. However, in aged people usually these responses are significantly compromised. In young, healthy adults (in their 20s), both mixed and protein-rich meals led to upregulation in microvascular blood flow in muscle within 1 h via insulin-dependent signaling. This response was blunted in individuals in their 40s and was completely absent in people in their 70s. Several studies have suggested that this age-related impairment in skeletal muscle vascularization contributes to the development of insulin resistance. Further, it has been found that a physically active lifestyle (regular exercise) improves muscle vascular response, thereby countering the onset of metabolic disorders. Some investigators have suggested that endothelial dysfunction is the underlying mechanism for the age-related impairment in muscle vascular response in both oxidative and glycolytic skeletal muscles. Also, aging weakens sympathetic α-adrenergic control of skeletal muscle circulation and the same is restored by regular exercise training.

Changes in Cardiac Muscle during Aging

Morphological Changes

The human heart weighing about 300 grams beats 100,000 times a day since its formation around 4th week postfertilization till the end of life. Along with the lungs, heart is protected by the ribs in the thoracic chamber, bottom of which is formed by another muscle called the diaphragm. Heart is made up of a special muscle called cardiac muscle and is autonomic in its uninterrupted contraction-relaxation process. This is due to the presence of specialized cardiac cells called "sino-atrial node" that generate spontaneous rhythmic impulse for the heartbeat. Therefore, any major morphological and/or anatomical changes are detrimental for the individual. However, in aging people several changes, which may primarily arise due to pathological conditions, have been reported. In healthy adults, some commonly observed changes are: increased left ventricular wall thickness; diastolic dysfunction; proliferation of cardiac fibroblasts impairing heart muscle function; large artery as well as myocardial stiffening; progressive loss of myocytes by apoptosis and/or necrosis. Arterial stiffening poses a resistance against the heart and it has to perform at a higher force to eject blood during contraction. This results in remodeling of the ventricle, also known as "ventricular–vascular-coupled stiffening." All these changes increase oxygen consumption of the cardiomyocytes, but the metabolic capacity of the mitochondria decreases leading to energy deficit. With age, oxidative stress increases and capillary density in heart is reduced leading to focal ischemia that causes the death of myocytes. Loss of myocytes from the myocardium is compensated by hypertrophy of the remaining myocytes, which aids in thickening of the left ventricular (LV) wall. In addition to ventricle, the atria also undergo changes with age. Echocardiography measurements in healthy subjects suggest left atrial (LA) size increases, while LA-mediated ventricular filling is decreased. Studies have suggested these changes as risk factors for atrial fibrillation (AF) and heart failure.

Vascularization

Like skeletal muscle, heart also requires vibrant supply of blood via a unique vascular system called "coronary circulation." Healthy aging usually maintains normal vessel-to-tissue ratio and network architecture in the heart. However, the cardiac vascular network gradually becomes nonresponsive to stimuli like norepinephrine, leading to reduced blood flow with aging in both man and animals. Additionally, coping with increased blood flow into the myocardium is compromised in aging people starting with those in around their 50s due to "Atherosclerosis." This is associated with arterial wall thickening due to invasion and deposition of white blood cells (WBC) and enlargement of intimal smooth muscle cell resulting in atherosclerotic plaque, which narrows down the diameter of the vessel. These changes resist the cardiac function and result in focal ischemia that causes irreversible changes in coronary vessels. Ironically, successful neovascularization in senescent heart does not necessarily lead to improved coronary perfusion in many cases. Induction of vascularization to treat ischemic areas therapeutically with angiogenic factors is still being pursued including small molecule, protein, gene, and cell-based approaches. Some glycogen synthase kinase 3β (GSK-3β) inhibitors, which upregulate VEGF expression, have been shown to provide cardioprotection. VEGF-mimetic peptide amphiphilic nanostructures and microRNA-based angiogenic pathway modulation using "miR-30b" are also under trial. Therefore, factors that maintain innate plasticity of the existing vasculature of heart play a more important role in postischemic recovery and preservation of contractility of cardiomyocytes. While diastolic dysfunction is considered to increase, resting systolic dysfunction normally does not increase with age. But, systolic function during physiologically demanding conditions (like intense exercise) is progressively impaired during aging. Major changes in the cardiac vasculature are dilation and thickening of walls of large arteries (mainly intimal media layer) and collagen and calcium deposition, when comparison is made between 20- and 90-year-old people. These modifications in association with ventricular and valvular changes determine the pace of senescence of the heart in an individual-specific manner. Exercise has been proposed to help in maintaining the plasticity of the cardiac vasculature and ameliorate the age-related alteration in heart function. However, several elderly people are exercise intolerant with decreased maximal heart rate (VO2 max). Recent studies are reporting that the number of such aged people is increasing among different ethnic groups around the world. This alarming finding suggests that it is time to revisit some of the heart-related apprehensions in aging population.

Contractile Capacity

Healthy human myocardium is comprised of cellular and extracellular components, which include cardiac myocytes ~75% by volume and ~30–40% by number. Another major component of normal heart is fibroblast cells that are multifunctional in nature contributing to cardiac homeostasis and sustaining the matrix network. Further, fibroblasts play vital role in cardiac repair after myocardial infarction and are also central in cardiac fibrosis development. Aging of myocardium is often associated with fibrosis and left ventricular hypertrophy causing diastolic dysfunction leading to heart failure. Progressive collagen deposition with aging in the vascular wall and interstitial spaces leads to reduction of cardiomyocyte volume causing myocardial stiffness, both in human and animal models. Increased fibrosis and myocardial stiffness along with impaired ventricular relaxation initiate the process for development of diastolic dysfunction. During aging, progressive loss of myocytes and compensatory hypertrophy of remaining cardiomyoctes are observed. Also, the myocardium becomes progressively insensitive to sympathetic stimuli (norepinephrine). All these above-mentioned age-related changes compromise contractile capability of the myocardium in older people. Recent advancements in magnetic resonance imaging employing tissue tagging and other spectroscopic techniques have enabled precise diagnosis of the above age-related changes and have also studied intramyocardial strains and cardiac energetics.

Left ventricular remodeling with increased mass-to-volume ratio and myocardial dysfunction (both systolic and diastolic) are associated with aging. But, if these symptoms are observed in early life the cardiovascular risk of the individual is higher in older age. After the age of 20, left ventricular diastole and early diastolic filling rate slows down up to 50% by the age of 80. In spite of slowing the left ventricular filling early in diastole, more filling occurs in late diastole, by more vigorous atrial contraction leading to atrial hypetrophy, which manifest as a fourth heart sound called the "atrial gallop." A study from Baltimore with healthy participants aged between 20 and 85 years showed that despite significant changes in the diastolic filling pattern in older group, healthy persons in the group exhibited similar left ventricular end–diastolic volume index as their younger counterparts. Several different types of exercise (seated cycling, upright running, and postural changes) studies have been performed by various groups comparing young versus old participants. During vigorous upright exercise, end diastolic filling is greater in older men than younger counterparts, whereas during short-term submaximal seated cycling end diastolic filling is not significantly altered. Interestingly, in women end diastolic volume was similar in response to vigorous exercise in both older and younger groups. Based on exercise studies it has been suggested that cardiorespiratory fitness training reduces age-related myocardial mortality, the risk of heart failure, and cardiac arterial and ventricular stiffening.

Sex-Specific Differences in Muscle Aging

As body metabolism is different between males and females in several ways including the role of sex hormones, muscle aging exhibits distinct phenotypic changes. Hence, gender-specific differences in age-related loss of muscle mass and strength (especially skeletal muscle) are observed. Human females usually experience earlier reduction of muscle strength than males, whereas no significant difference is witnessed in loss of muscle mass during aging. Phillips et al. reported that decline in muscle mass is greater in case of perimenopausal as well as in postmenopausal females, which can be prevented by performing hormonal replacement therapy. Expression of estrogen receptors in muscles drops after menopause. As estrogen receptors can be activated by estrogen and insulin-like growth factor (IGF)-1, drop in its expression leads to loss of muscle mass and strength. In the literature, there are discrepancies regarding gender-related differences in muscle performance or fatigue during aging. Most researchers believe that postmenopausal women have greater fatigue resistance than men during submaximal exercise, whereas during intense exercise this gender-specific difference is not found. It is proposed that older women have smaller muscles and are weaker than older men, their muscle results in less vascular occlusion and local muscle ischemia, thus are more resistant to fatigue. An opposing result was published by Ditor and Hicks who showed that no sex differences were observed in fatigue resistance while using the adductor pollicis muscle. As cardiac muscle performance is higher in aged females, Ca-2+-regulated proteins have been studied in aging animals. Structural proteins, such as myosin heavy chain (MHC) α and β, troponin I, and calsequestrin (CASQ) 1, do not show gender-specific differences. While the expression of important players, such as SERCA isoforms, RYR1, voltage-dependent anion channel (VDAC), and mitochondrial calcium uniporter (MCU), has not been compared in male and females.

Studies have reported that older women, who have not received hormonal replacement therapy, have greater fatigue resistance than age-matched men. It is interpreted that estrogen limits muscle endurance capacity by its influence on substrate utilization of both glycogen and fat. In addition to loss of muscle mass and strength, fiber-type distribution is also altered in older women. Several studies showed increase in type I fiber area in many skeletal muscles during aging in older women. This scenario may originate by two distinct ways: first, by more fiber type switching to type I; second, by better retention of higher type I fibers since young age. However, either of these two routes has not been fully investigated. Another interesting yet unaddressed question regarding fatigue resistance in older women is substrate metabolism, especially substrate preference and role of metabolic enzymes. Estrogen mediates increased storage of fat in the gluteofemoral area, while it inhibits fat storage in the abdominal region in premenopausal women by regulating lipoprotein lipase (LPL) activity that uptakes fatty acid into cells from the blood. This tissue-specific difference in LPL activity is suppressed in postmenopausal women due to cessation of estrogen secretion.

Cardiac aging in females, on the other hand, has been considered to be more prominent than males due to the supposedly cardioprotective properties of estrogen, as premenoposual women are more resistant to cardiovascular diseases (CVD). Based on these apprehensions several hormone (estrogen and estrogen + progestin) replacement therapies (HRT) were conducted in the 1990s. However, HRT resulted in increased incidence of cancer and CVD. This surprising observation led to the finding that most enrolled women were 10 years postmenopause and might have undergone adaptations in most organs to function without estrogen. Application of estrogen after such adaptations might induce abnormal physiological response.

Aging and Insulin Resistance

In the last 50 years, metabolic disorders, including type II diabetes, impaired glucose tolerance (IGT), and obesity, are becoming the major causes of death worldwide surpassing the number of deaths due to pathogenic diseases. Most of the individuals with metabolic disorders display insulin resistance (at least to some extent) and prevalence of such cases is increasing in an exponential rate. Resistance to insulin action generally increases with aging in humans. Insulin receptors are expressed in multiple organs that highlight the pleiotropic effects of insulin signaling. Since the discovery of insulin itself, studies have focused on its signal transduction pathways and its role in nutrient homeostasis in the physiology of insulin resistance and progression of diabetes. The major targets of insulin in mammalian body include skeletal muscle, heart, adipose tissue, liver, and brain. Insulin resistance is either the primary cause or a determinant of the pace of progression of devastating age-related illnesses, including cardiovascular diseases, cancers, and neurological dysfunctions. Although, combined actions of several tissues regulate whole-body insulin sensitivity, muscles (both skeletal and cardiac) are the primary sites for insulin-mediated glucose disposal from the blood.

Insulin Signaling and Glucose Uptake

Insulin consists of two polypeptide chains and is synthesized and secreted by the β-cells of pancreatic islets. It is then carried in the blood and delivered to the target organs by the capillaries by crossing the endothelial barrier. It then binds to the insulin receptors on the target cell surface. As skeletal muscles are an important site of glucose uptake, they abundantly express insulin receptors. However, the delivery of insulin to skeletal muscle cell membrane is a rate-limiting step, which is compromised in obese insulin-resistant people. After insulin binds to its receptor, a series of rapid phosphorylation events occur. First, the insulin receptor is autophosphorylated initiating the signaling cascade leading to tyrosine phosphorylation of insulin receptor substrate (IRS) proteins. Next, the phosphorylated IRS binds to phosphatidylinositol 3-kinase (PI3K) mediating phosphorylation protein kinase B (called Akt). Phosphorylated Akt facilitates phosphorylation of the Akt substrate of 160 kDa (termed AS160). AS160 is also known as Rab GTPase TBC1D4 and contains multiple Akt Ser/Thr phosphorylation motifs. AS160 regulates the GLUT4 translocation rate to the plasma membrane in a phosphorylation-dependent manner mediating insulin-induced glucose uptake.

Any alteration in this signaling pathway results in insulin resistance of the skeletal muscle, which appears with increasing age. Marked decrease in insulin-stimulated Akt phosphorylation and activity in skeletal muscle is witnessed in people around 70 years of age compared to 30 years. Older individuals (both men and women nearing 70 years) showed that lower phosphorylation of AS160 (on sites Ser588 and Thr642) has been also reported when compared to individuals in their 20s. Another mechanism of skeletal muscle insulin resistance has been proposed in which light chain fatty acid (LCFA)-CoA plays the central character. The direct role of LCFA-CoA induced insulin resistance in the skeletal muscle aging, but might play some role indirectly modulating lipid utilization. Several other factors regulate insulin binding to its receptor and thereby are important in the progression of insulin resistance which include exercise, diet, glucocorticoids, thyroid hormones, sex steroids, and cyclic nucleotides. In addition to skeletal muscle, insulin has profound effect on myocardial substrate metabolism including glucose uptake and the response of the heart to workload or ischemia. Although heart generally prefers fatty acid as a substrate, it readily enhances glucose uptake during elevated energy demanding states like exercise. Further, heart is an important site of diet-induced insulin sensitivity. Both these properties are compromised during aging adding to hyperglycemia, hyperlipidemia, insulin resistance, and oxidative stress which predisposes to cardiovascular diseases including congestive heart failure, myocardial infarction, ventricular hypertrophy, and hypercoagulability of blood. Glucose uptake into heart is also promoted by newly identified cytokines such as fibroblast growth factor 21 (FGF21) and adiponectin.

In addition to skeletal muscle and heart, brown adipose tissue (BAT) and white adipose tissue (WAT) are the important sites of glucose uptake and may play crucial role in causation of age-related hyperglycemia. Recent studies have indicated that adipocytes in BAT and WAT are interconvertible depending on energy intake and utilization. The dynamic balance between these two sites and their interplay with the skeletal muscle is an expanding area of research. As brown adipocytes and skeletal myocytes are derived from single progenitor cells and can be under similar metabolic signaling pathway, BAT is an interesting and adaptable site of glucose homeostasis. Insulin and insulin-like growth factor (IGF)-1 are classically known to control adipocyte differentiation by regulating gene expression including that of uncoupling protein 1 (UCP1) in the preadipocytes. In addition, several other factors, including FGF21, sympathomimetics, prostaglandins, peroxisome proliferator-activated receptor c (PPARc) activators, and natriuretic peptides, regulate the interconversion between BAT and WAT properties mediating systemic glucose clearance alleviating obesity-associated hyperglycemia and insulin resistance.

Diabetes and Skeletal Muscle Insulin Resistance

Aging increases the risk of type 2 diabetes in both normal weight and obese individuals. Recent data from the United States showed that fasting glycemia and IGT increased progressively from 20.9%, 46.9%, 67.4% to 75.6% in individuals at 20–39 years, 40–59 years, 60–74 years, and more than 75 years, respectively. Studies show that aging-related loss of skeletal muscle mass and the progression of diabetes can be attenuated by lifestyle intervention strategies such as regular exercise. Specifically, submaximal endurance exercise at a rate of three to five times a week improves insulin sensitivity and also helps in reduction of body fat. Further, short-duration high-intensity resistance exercise performed regularly helps in retention of skeletal muscle mass and strength. Therefore, in the last decades, muscular exercise has become the hallmark of promoting "healthy aging." Insulin resistance in elderly people in part originates from age-related changes in body composition, as accumulation of abdominal and/or ectopic fat is associated with the deterioration in glucose homeostasis. Further, it has been suggested that muscle atrophy and changes in muscle quality, which appear due to aging, also contribute to metabolic dysfunction. Healthy older people (65–70 years) show moderately impaired glucose disposal compared to younger controls in their 30s. However, there is no consensus on the role of aging in induction of skeletal muscle insulin resistance independent of physical activity and body composition. On the other hand, altered insulin action on heart has been proposed to have a close association with impaired glucose homeostasis in diabetic and obese individuals.

Lipid Accumulation and Muscle Insulin Resistance

Ectopic accumulation of fat, especially in the skeletal muscle, is associated with development of insulin resistance. Randle and colleagues for the first time put forward a mechanism for impairment in insulin-stimulated glucose oxidation in muscle. They showed that lipid availability inhibits key glycolytic enzymes, such as pyruvate dehydrogenase, phosphofructokinase, and hexokinase, which results in intracellular accumulation of glucose preventing further glucose uptake. Later studies showed that lipid infusions into muscle result in decline in insulin receptor substrate 1 (IRS-1)-mediated PI3K activity suppressing insulin-stimulated glucose uptake. In humans, metabolic rate declines with age that reduces the rate of fatty acid oxidation resulting in accumulation of LCFA-CoA in active tissues like muscles. LCFA-CoA and its major downstream metabolized lipids such as diacylglycerols (DAG) and ceramides induce insulin resistance in the skeletal muscle. (Fig. 16.1).

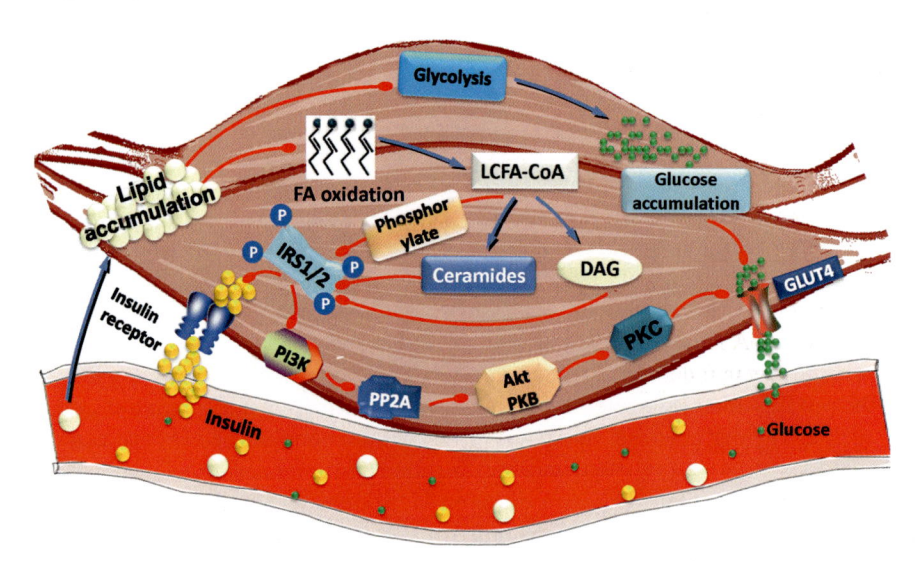

Fig. 16.1 Accumulation of lipid components in muscle during aging. Due to reduction in skeletal muscle activity, energy expenditure usually declines progressively with aging. This leads to accumulation of lipids inside the skeletal muscle, which in turn has an inhibitory effect on insulin-mediated glucose uptake via IRS–Akt pathway. Intramyocellular fatty acids and their metabolic intermediates interfere with different steps of glucose uptake causing insulin resistance leading to metabolic disorders

Role of DAG These are triglyceride metabolic intermediates of LCFA-CoA. A close association between elevated intracellular DAG levels and skeletal muscle insulin resistance has been reported from several studies, which mostly involve protein kinase C (PKC). Skeletal muscles from type 2 diabetic patients show increased protein kinase C-θ concomitant with DAG levels. Aging studies showed higher levels of DAG in blood and muscle, suggesting aging-related progression of diabetes is partly augmented by DAG-dependent pathways in the skeletal muscle.

Role of Ceramides These are sphingolipids, which are synthesized from saturated LCFA-CoA and serine and are generally found in the plasma membranes. Intramyocellular accumulation of ceramides has been reported in obese and diabetic individuals in the last decade igniting investigations into the role of ceramides in metabolic disorders. Exercise training studies showed that overweight participants who responded positively (exercise-induced insulin sensitivity) also exhibited reduction in muscle DAG and ceramide content. Ceramide-induced insulin resistance relies on PKC-α via phosphorylation of IRS-1 complex and protein phosphatase 2A, which enhances dephosphorylation of the Akt/PKB (protein kinase B) complex at an allosteric site, thereby inhibiting glucose uptake into the skeletal muscle.

However, few studies reported only marginal differences in muscle lipid metabolite accumulation between groups with varying insulin sensitivity. On the contrary, evidence is mounting to show that lipid metabolite intermediates regulate insulin-responsiveness in tissues such as skeletal muscle by controlling CPT-1-mediated LCFA-CoA entry into the mitochondria and fat oxidation, which, on the other hand, influence glucose uptake. In this lipid-mediated insulin resistance in skeletal muscle, malonyl-CoA serves as an important fuel sensor. Malonyl-CoA concentration increases with excess carbohydrate supply or muscle inactivity causing inhibition of fatty acid oxidation leading to LCFA-CoA accumulation, which is reduced by starvation and exercise. An antidiabetic drug, pioglitazone, that works as a PPAR-γ agonist is known to reduce skeletal muscle malonyl-CoA level. However, cytosolic level of malonyl-CoA is not always associated with skeletal muscle insulin resistance such as high-fat diet in rodents, type 2 diabetes in man. Understanding how mitochondria/fat metabolism pathway is integrated with carbohydrate metabolism in the skeletal muscle has potential to provide target to induce insulin sensitivity treating diabetes.

Role of Mitochondria in Muscle Aging

Mitochondria, being the energy hub of the cell, takes the central position in muscle metabolism and its biogenesis is very closely integrated with the energy demand of the muscle. Mitochondria working in conjunction with the nucleus plays an important role in aging, disuse-induced muscle atrophy, and cell death signaling. Due to the role of mitochondria as an energy supplier for muscle contraction, they exhibit a very unique pattern of distribution in both skeletal and cardiac myocytes. Two distinct mitochondrial subpopulations are observed in the muscle, based on their bioenergetics state and structure. These are subsarcolemmal mitochondria (SSM), located below the plasma membrane, and intermyofibrillar mitochondria (IFM), arranged between the myofibrils in parallel rows.

Mitochondrial Content, Morphology, and Metabolism

SSM and IFM in the striated muscles form two distinct subpopulations. They possess diverse enzymatic components, display different rates of degradation (via fragmentation and autophagic pathways), and exhibit differential susceptibility toward apoptotic stimuli. It is therefore proposed that they make unique contributions to the pathogenesis of aging-associated sarcopenia and other varieties of muscle atrophies. Number and morphology of mitochondria is dynamically regulated by two interrelated processes called fusion and fission. Mitochondrial fusion is facilitated by Mitofusin (Mfn) 1, Mfn 2, and the Optic Atrophy-1 (OPA1); whereas fission is regulated by fission (Fis) 1, Dynamin-related protein (Drp) 1, and mitochondrial fission factor (MFF). Fusion and fission of mitochondria remains in a dynamic balance in most cells including the myocytes. A change in the mitochondrial dynamics causes alteration in mitochondrial morphology and number. The SSM and IFM in the striated muscles exhibit different metabolic capacity that correlates with the intracellular ATP requirements. The difference between the metabolic capacities of SSM and IFM becomes even clearer after a period of muscular inactivity. SSM produce ROS and decline in number, whereas IFM undergo apoptosis after muscular inactivity due to aging and also denervation. In contrast, muscular activity training and other conditions that promote oxidative metabolism in the skeletal muscle upregulate mitochondrial fusion machinery leading to improved mitochondrial network.

Mitochondrial bioenergetics is significantly altered during the aging process and is associated with age-related muscle atrophy. These changes are considered to originate from muscular inactivity and mitochondrial synthesis and degradation, which might be associated with aging process. Therefore, there is a complex progressive pathophysiological remodeling of the muscle mitochondria during aging. Several molecular events of this process (mitophagic pathway, degradation of cristae structure, etc.) overlap with structural changes to the muscle fibers (change in cross-sectional area, fiber type). The enzymatic machinery that controls glucose and lipid metabolism also undergoes drastic modification during pathophysiological alteration of SSM and IFM. Apart from ATP supply, muscle mitochondria regulate the reduction–oxidation (redox) homeostasis and cellular quality control such as cell death signaling. Researchers have suggested that altered redox balance in muscle causes oxidative damage to the mitochondrial DNA and causes significant alteration in its metabolism. Overall, mitochondrial content, morphology, and metabolism depend on dynamics (rate of fission vs. fusion), mitophagy, and biogenesis (including protein synthesis) of this unique organelle. All these processes play important roles during aging as they determine the plastic nature of mitochondria.

Mitochondrial Plasticity in Aging Skeletal Muscle

Mitochondria is a very dynamic organelle that can swiftly undergo biogenesis in response to intracellular energy demand like exercise training. Several groups have shown that muscle mitochondrial biogenic capacity declines with aging and can be retarded by application of exercise. Interestingly, exercise responsiveness also declines with aging. This happens due to alteration in the signal transducing pathways that regulate mitochondrial biogenesis primarily governed by peroxisome proliferator-activated receptor gamma 1 (PPAR-γ), especially peroxisome proliferator-activated receptor gamma coactivator 1-alpha (PGC-1α). Major energy sensors, such as adenosine monophosphate kinase (AMPK) and sirtuin 1 (Sirt-1), regulate PGC-1α activity linking energy demand and mitochondrial biogenesis. Studies have shown that chronic electrical stimulation, which recruits muscle contractile activity and mimics exercise training, also induces mitochondrial biogenesis. This response is also blunted in animals in very advanced age. It has been shown that a failure to upregulate PGC-1α and other signaling pathways, such as PPARs and ERRs, is responsible for blunted mitochondrial plasticity in the skeletal muscle. Due to the involvement of Sirt-1, some researchers have suggested that gene silencing subsequent to changes in DNA methylation and acetylation status might be involved in the regulation of mitochondrial plasticity in muscle. Another overlapping but distinct concept for blunted mitochondrial plasticity in aging exists in the field called "mitochondrial free radical theory of aging (MFRTA)." According to this theory, oxidative damage to mitochondrial DNA causes mitochondrial dysfunction, which is the primary driving force for the aging process. Damaged mitochondria are energetically inefficient and also generate higher levels of reactive oxygen species (ROS) that dampens the cellular quality control mechanisms enhancing apoptotic pathways. The combined effects of these processes promote the age-related muscle atrophy and sarcopenia (Fig. 16.2). Understanding the molecular details of the links between blunted mitochondrial plasticity and aging process may provide druggable targets to pharmacologically retard aging.

Inflammatory Processes and Cytokines in Muscle Aging

Research from last few decades has highlighted the fact that inflammatory cells and signaling molecules modulate the functioning of the skeletal and cardiac muscles. Monocytes enter into the muscle upon signaling including injury, high-fat diet feeding and transform into macrophages leading to modification of muscle function and metabolism. As discussed above, inflammatory processes undergo significant alterations during aging. Therefore, aging-related changes in inflammatory processes affect muscle function and vice versa. Recent studies have suggested such inter-organ-crosstalk that regulates metabolic status is mediated by cytokines. Aging of heart involves cytokines like TGF-β, endothelin-1, angiotensin II, etc., resulting in increased fibrosis of the myocardium. Skeletal muscle fibrosis is prominent in most

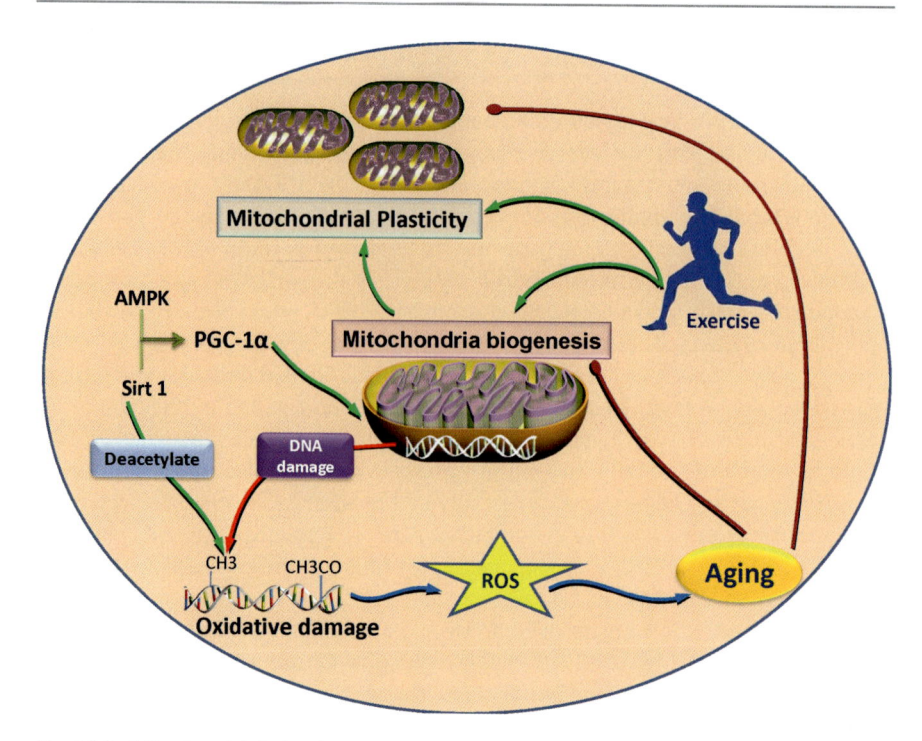

Fig. 16.2 Mitochondrial plasticity in muscle provides protection against aging. Mitochondria being the powerhouse can sense the ambient energy demand. With reduced physical activity during aging, mitochondria gradually lose the ability to promptly respond to meet energy demand. This impairment in mitochondrial plasticity can be worsened by alteration in metabolic sensors like AMPK, Sirt-1, and PGC-1α. Exercise is a booster of energy demand and retards the rate of deterioration in muscle plasticity. It has been noted by several groups that pace of aging is minimized by inclusion of exercise in daily life

of the muscular dystrophies. Increased inflammatory processes in the skeletal muscle have been reported in obese individuals.

In healthy humans, inflammatory cells remove damaged cells after acute injury, thereby facilitating the process of regeneration of damaged part by muscle stem cells known as satellite cells (described in latter sections). Several transcription factors mediating communication between muscle cells and the inflammatory cells have been studied extensively. Nuclear factor kappa beta (NFκB) is expressed in the skeletal muscle and plays a major role in muscle wasting and cachexia. NFκB is normally retained in cytosol by inhibitory IκB. Upon activation of tumor necrosis factor alpha (TNF-α), IκB kinase (IKKβ) mediates the release of NFκB from IκB resulting in nuclear translocation of NFκB and transcription of NFκB-dependent genes. Mice overexpressing IKKβ in muscle-specific manner exhibit muscle wasting. Further, mice with minimized NFκB activity (knockout for p105/p50 subunit) show reduced muscle atrophy. These studies support the role of NFκB in muscle atrophy. In addition to muscle atrophy, NFκB along with TNF-α causes insulin resistance by interfering with IGF1-Akt pathway. Akt function is dependent on the fine balance between two proteins: mTOR (mammalian target of rapamycin) (mediate protein synthesis) and FoxO (forkhead box class O proteins) (mediate protein degradation). The interplay between NFκB and IGF1-Akt pathway is an active area of research. IGF1 overexpression in muscle in mice shows improved muscle growth and regeneration capacity. Interestingly, these mice exhibit angiotensin-induced resistance to muscle atrophy and cardiac cachexia. Intramuscular IGF1 injection counters muscle wasting and aging-related changes in the muscle in animal models including MDX and in humans. Recombinant IGF1 treatment on elderly women has shown improved net muscle protein synthesis and counters muscle loss.

Role of several cytokines has been implicated in the development of sarcopenia and age-related physical decline, which may be worsened by reduction in growth factor levels. Interleukin 6 (IL-6) is one such cytokine whose plasma level increases with aging. It is believed that IL-6 level is associated with decreased muscle performance and increased muscle loss based on studies on geriatric patients. Although the exact mechanism of action of IL-6 in aging-associated muscle atrophy is unclear, elevated plasma IL-6 level has been reported in several disorders such as obesity, smoking, and type 2 diabetes. Another cytokine, TNF-α, is also known to induce aging-related loss of muscle mass. Recent studies show that TNF-α enhances myocyte apoptosis via caspase pathway in rodent models, thereby mediating age-related muscle loss. In contrast, circulating IGF1 level declines progressively during aging starting in early 30s. It has been proposed that decrease in IGF-1 with age can be used as a predictor of age-related muscle loss. Exhaustive exercise, but not during regular physical activity, leads to increased production of another myokine called "IL-8" in skeletal muscle leading to elevated plasma levels. A recent study reported eightfold higher expression of interleukin 8 (IL-8) in skeletal muscle arteries from elderly individuals compared with younger adults. These data indicate the role for IL-8 in vascularization process, which might play important role in the recruitment of skeletal muscle function.

Another cytokine "Leptin" secreted by white fat cells is also known to regulate muscle metabolism and health. Studies using leptin-deficient (ob/ob) mice have demonstrated the key role of leptin in maintenance of muscle mass and performance. Interestingly, expression of leptin receptor in the skeletal muscles is upregulated in disuse-related and aging-related atrophy. Injection of leptin alone and in conjunction with several other biomolecules, such as microRNAs (miRNAs), has been tested to reverse aging-related muscle wasting. Due to its profound effects on food intake and body weight control, leptin was tested as an anti-obesity modality, but became unsuccessful owing to its pleiotropic effects on several organs. Leptin lessens intramuscular fat by augmenting oxidation of fatty acids, thereby enhancing glucose uptake and energy expenditure in skeletal muscle. It is worth noting that recent application of recombinant leptin in lipodystrophy patients resulted in sustained improvements in glycemic status. This is further considered as a pharmacological option in some diabetic obese individuals and has implication for elderly population. Several other cytokines are emerging as mediators of muscle health status and may play important role in aging. These cytokines include FGF21, fibroblast growth factor 2 (FGF2), myonectin (CTRP15), interleukin 10 (IL-10), interleukin 15 (IL-15), growth related oncogene (GRO)-α also called chemokine (C–X–C motif) ligand 1 (CXCL-1), brain-derived neurotrophic factor (BDNF), leukemia inhibitory factor (LIF), and monocyte chemotactic protein (MCP)-1. Role of various cytokines in regulation of skeletal muscle aging is still poorly understood. Defining the functions of cytokines will help in deciphering how different organs communicate with each other, thereby regulating whole-body metabolism during health and disease conditions.

Fibroblast in Muscle Aging

Aging displays progressive fibrosis of many organs including kidney, liver, lungs, and muscles. Myocardia as well as vascular walls of coronary arteries show collagen deposition along with increase in fibrotic tissues both in mice and in human. Collagen content increases significantly in old animals (rodents and rabbits) when compared to young controls. Relative percentage of type III collagen in myocardium of male has been reported to increase with aging. Increased fibrosis, a major basis of myocardial stiffness and along with impaired relaxation, causes diastolic dysfunction. Even healthy individuals display some degree of changes in cardiac anatomy and function progressively with aging. Cardiomyocytes undergo hypertrophy during aging and with loss of aortic elasticity alter hemodynamic function of the heart observed in older people. Age-related impairment in myocardial contractility is due to enhanced collagen deposition in the interstitial and perivascular space. Based on histological analysis it has been suggested that progressive loss of cardiomyocytes due to necrotic and apoptotic cell death is a primary cause for cardiac aging. Collagen accumulation requires complex posttranslational events. This starts with synthesis and secretion of procollagen by the fibroblasts into the pericellular space. These are then stabilized into fibrillar collagen by enzyme lysyl oxidase that increases its tensile strength. Level of collagen accumulation in a tissue is measured by analyzing hydroxylysylpyridinoline (HP) residues. Left ventricular HP increases fivefold in old rats compared to younger groups. Exercise training has been shown to significantly lower collagen cross-linking when compared to nontrained controls in rats. Interestingly, glucose links myocardial collagens nonenzymatically, producing advanced glycation endproducts (AGEs). Currently, AGEs are considered as an important marker of cardiac aging.

Fibroblasts in skeletal muscles, on the other hand, play a very crucial role in regeneration and repair processes. These cells localized in the interstitial areas serve as the major source for extracellular matrix (ECM). Deposition of ECM in the skeletal muscle has been observed in several pathophysiological conditions including muscle tearing, wasting, and regeneration. It is currently suggested that fibroblasts modulate the pace of satellite cell proliferation and muscle regeneration by secretion of ECM. Studies showed that coculture of satellite cells with fibroblasts robustly stimulates myogenesis of satellite cells, thus suggesting existence of cross-talk between them. During recovery from muscle injury, it has been shown that fibroblast number increases by fourfold, preferentially surrounding regenerating muscle fibers. It is suggested that fibroblasts are essential for muscle regeneration; that is, satellite cell differentiation and fusion. Therefore, defining this intercellular cross-talk during physiological and pathological scenarios will provide critical insight on how we can manipulate muscle remodeling to ameliorate aging-related skeletal muscle illnesses.

Muscle Plasticity and Regeneration

Studies in the last few decades have demonstrated that skeletal muscle has remarkable plasticity and can regenerate to recover from small injuries that occur regularly during daily life. Skeletal muscle also can recover from a variety of injuries like acute accidental injuries or strains from sport. Within a few weeks of a major injury, skeletal muscle structure is almost completely restored. This regeneration capacity of the skeletal muscle is due to its resident stem cells termed as satellite cells (SCs), located below the basal lamina of the muscles. Injury activates the SCs to undergo proliferation and fusion to formation of new muscle fibers. The high regeneration capacity of the skeletal muscle (at least in part) is due to metabolic flexibility and mitochondrial function, which are affected by aging and insulin resistance. Although regeneration ability of muscle is retained up to old age, it rapidly deteriorates toward the later part of old age, primarily due to reduced capacity to accelerate the signaling and gene regulatory pathways necessary for the metabolic adaptations. In many cases, regeneration causes remodeling of skeletal muscle structure such as branching in fibers, formation of scar tissue, or development of new myotendinous junctions. Chances of modified structures progressively increase during aging. Several research groups have been trying to boost muscle regeneration by pharmacological, genetic, and cellular approaches to counter age-related muscle loss.

The SCs remain in a dormant state and upon activation, they start to express the paired box protein (Pax7) and enter proliferation state. Next, Pax7 expression declines and the transcription factors MyoD and myogenin upregulate. These transcriptional programs guide SCs to become the muscle-specific cell type and start to express embryonic myosin heavy chain. Skeletal muscle regeneration program is regulated by intrinsic and extrinsic factors. Evidence is growing to show that immune cells including macrophages, eosinophils, and regulatory T-cells also contribute to the muscle regeneration process (Fig. 16.3). Several proteins of skeletal muscle regeneration process have been targeted to develop therapy for muscular dystrophy and aging. One such protein is mitsugumin 53 (MG53), a member of the tripartite motif (TRIM) family, so it is also called TRIM72. Others include mammalian target of rapamycin complex 1 (MTORC1), regulatory associated Protein of mTOR complex 1 (RPTOR), forkhead box class O proteins (FOXO), and the adenosine monophosphate-activated protein kinase (AMPK). These interventions are especially effective when coadministered along with endurance exercise. Skeletal muscle regeneration also depends on Ca2+-dynamics. Hence, few studies have tried

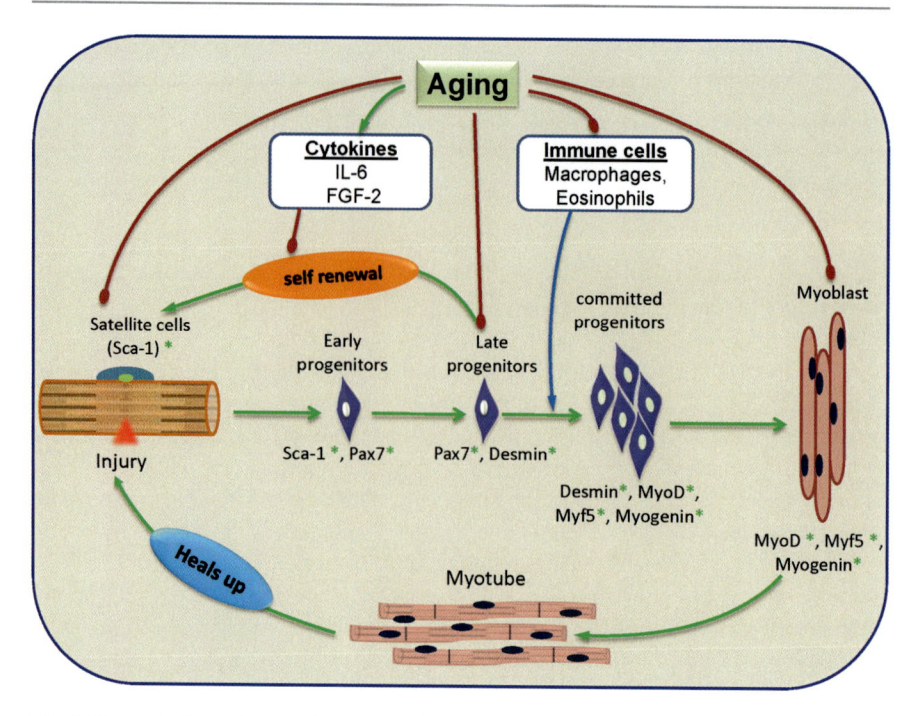

Fig. 16.3 Aging impairs muscle regenerative capacity. Muscles, especially the skeletal muscle, face injury regularly and undergo a regenerative pathway specifically guided by sequential appearance of different factors. For regeneration, muscles have a group of stem-cell population that is activated by injury to undergo proliferation and fusion leading to formation of new myotubes. During the process, several cytokines like IL-6 and FGF2 and immune cells influence the commitment events. Aging has effect on stem-cell and immune cell populations, myoblast, and circulating cytokines. The final outcome is aging impairs muscle regeneration. Also age-related metabolic disorders worsen the muscle regeneration process and vice versa. It is believed that factors that promote muscle plasticity will reduce ill-effects of aging

to target intracellular Ca2+-signaling, either directly via modulating RyR and SERCA or indirectly via anabolic steroids like testosterones. In contrast to skeletal muscle, cardiac muscle does not possess a stem cell pool and has poor regenerative ability. Therefore, any injury to the myocardium results in the formation of fibrotic scar and such scar usually increases with age and affects pump function. Taking insight from skeletal muscle, several studies have tried to improve cardiac muscle regeneration.

Role of Satellite Cells and Aging

Satellite cells (SCs), stem cells of the skeletal muscle, are present in a quiescent state located between the basal lamina and the sarcolemma of muscle fibers. Muscle injury acts as a signal for the SCs to enter into cell cycle and simultaneously proliferate, differentiate, migrate, and fuse leading to formation of new myofibers. Some of the proliferating SCs retain stemness and return to the quiescence state, thereby replenishing the SCs' pool. Apart from intrinsic program, several extrinsic factors, like inflammatory cells, stromal cells, neuronal signals, and ECM components, guide SCs through the regeneration process. Recent studies have identified stem cells other than SCs in the skeletal muscle that include pericytes (NG2-/-, CD146-/-, PDGFR-/-), fibroadipogenic progenitor cells (FAPs, platelet-derived growth factor receptor (PDGFR)), and muscle-derived mesenchymal stem cells (mMSCs, stem cell antigen-1 (Sca-1)) located within the interstitium. Although these cell types contribute to regeneration either directly or indirectly, their relative roles are not currently well understood.

As described earlier, the skeletal muscle regenerative capacity progressively decreases with aging. Primary cause of this is reduction of the pool of SCs and other stem cells. In addition, the myogenic capacity of individual SCs also declines during aging. Studies using in vitro culture have demonstrated that SCs from aged mice proliferate and differentiate at a slower rate than those from younger mice. Other factors that influence the muscle regeneration ability include increased intramuscular fat accumulation and reduction of cross-sectional area of muscle fibers. Age-related shortening of telomere has been proposed to explain reduction of stemness of SCs, however it is still to be fully established. Studies have shown that SCs from aged individuals exhibit upregulation in the janus kinase/signal transducer and activator of transcription (JAK/STAT) signaling pathway which is associated with increased IL-6 level. Elevated stat3 function is believed to cause reduction in regenerative potential of SCs. Another cytokine, FGF2, whose level increases with aging acts via p38α/β signaling and reduces the self-renewal capacity of the SCs. Increased p38α/β level in SCs has been shown to enhance the expression of several senescence markers. Although exact mechanisms still remain debatable, it can be concluded that SCs rely on a fine balance of both intrinsic and extrinsic factors, which is disturbed by aging. So, age-related loss of self-renewal capacity of SCs is irreversible and cannot be fully restored by artificial interventions.

Using Exercise and NEAT to Counter Muscle Aging

Compromised skeletal muscle functions are the major causes of reduced physical activity by the aging population. Various properties of muscle, such as size, aerobic fitness, strength, and function, seem to reduce with increasing age. It is usually reported that while we cannot stop aging, but certainly healthy aging can reduce the sufferings of the elderly people. As energy expenditures, both in active and resting conditions, decrease with aging, several researchers have proposed the elevation of energy expenditure by introduction of higher physical activity in lifestyle. However, there are several challenges that need serious consideration. Toward this effect, two interrelated yet distinct concepts are emerging: one is exercise and other is nonexercise activity thermogenesis (NEAT).

Several weight loss programs have highlighted the application of high-intensity exercise. But, intense exercise only attenuates and does not stop the above age-related changes as these are also observed in aging athletes. Therefore, losing weight with intense aerobic exercise only becomes challenging without any calorie restriction. Although the energy expenditure still remains substantial ($\approx$500–900 kcal/h), it becomes a difficult task to find the time and motivation to perform intensive aerobic trainings needed to reduce body weight significantly. In contrast, resistant exercise shows a very minimal loss in weight as the energy expenditure is very less in both male ($\approx$200–500 kcal/h) and female ($\approx$120–250 kcal/h). While the aerobic exercise shows relatively high-energy expenditure compared to resistance exercise, larger increase in total energy expenditure (TEE) is observed in resistant training. Also resting energy expenditure (REE) increases about 7–8% probably due to augment in muscle hypertrophy, whereas minimal increase is observed in case of intense aerobic exercise. Hence, application of regular exercise might be beneficial only for a part of population who are more active, but may not be equally fruitful for inactive individuals.

Recent studies have indicated that individuals pursuing intensive aerobic training show a decreased NEAT. In contrast, untrained but active individuals exhibit higher NEAT. These data imply that intensive exercise training regimen improves the energy efficiency of the individual, thereby reducing the usefulness of the same in weight loss. It is now suggested that increased recruitment of NEAT rather than intense exercise during early stages of aging (may be in 50s) is more beneficial for achieving healthy aging. Energy cost of all activities by the body when not engaged in sleeping or exercise is loosely termed as NEAT. It includes all trivial physical activities such as walking, typing, driving, working in the backyard, and fidgeting. Currently, it is not clear as to what is the percentage of NEAT in an individual's daily metabolic rate. But, it is increasingly believed that culturally (by change in lifestyle) promoting NEAT is an effective way to manage healthy aging. A major dilemma with NEAT is that it increases with overfeeding and decreases with underfeeding. In spite of this, researchers are trying to identify factors that contribute to NEAT, including hypothalamic, inflammatory (cyto/chemokines), muscular, which will pave the way to modulate weight balance and aging by targeting NEAT. Physical training, which includes all the activities above the level of seating, technically close to NEAT has been suggested as increasing the energy expenditure (Fig. 16.4). These studies postulated systematic manipulation of training frequency, intensity, and duration improves the oxygen consumption (20–30% in age groups within 60–80) significantly. Muscle fibers' nature was shifted more toward aerobic side as switching of type IIB fibers to type IIA fibers, increased capillary density, and improvement in the respiratory capacity of mitochondria were observed in the muscle. Increase in capillary density is another important adaptation because of physical training. This increases the capillary-to-fiber ratio leading to decrease in distance of diffusion for oxygen. Aging-associated decline in VO2 max was reduced in response to physical training as elevated level aerobic enzymes and capillary density was observed in elderly groups.

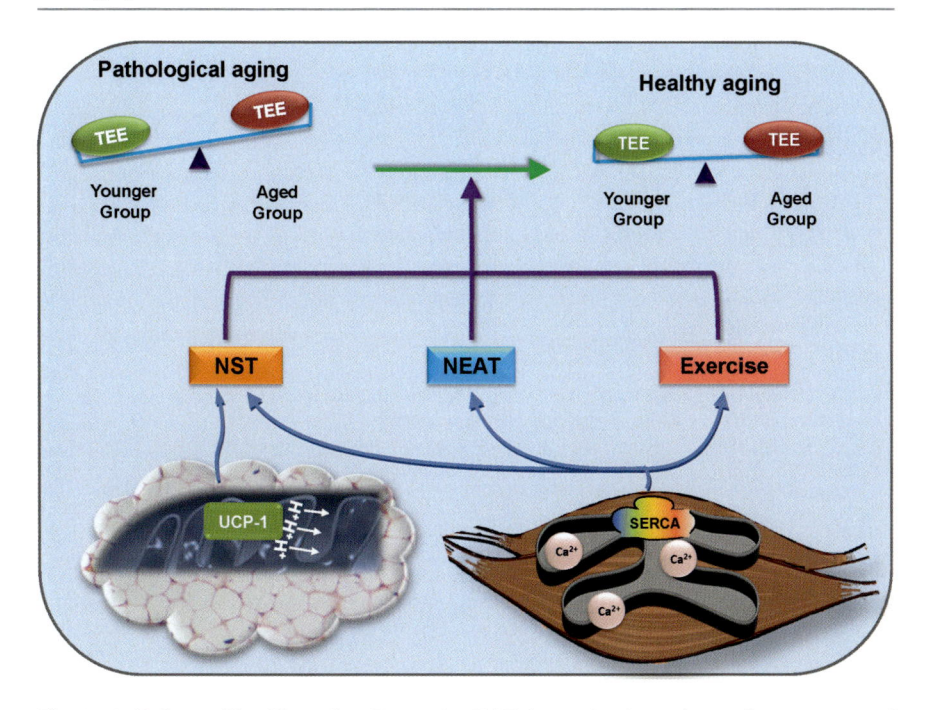

Fig. 16.4 Drivers of healthy aging. Brown fat (BAT) is a major determinant of energy status of mammals, specifically rodents. BAT is the major site of non-shivering thermogenesis (NST) via a mitochondrial protein called uncoupling protein 1 (UCP1). Recent discovery of existence of BAT in adult humans ignited the field of metabolism to utilize this mechanism to counter metabolic disorders including aging. Although role of skeletal muscle recruitment via exercise is established to promote healthy aging, the actual mechanism for this remains ill defined. This is primarily because the skeletal muscle is also recruited during NST and non-exercise activity thermogenesis (NEAT). Drawing the line between NST, NEAT, and exercise has been a difficult task. Understanding molecular details of these processes will provide insight to target these processes to increase metabolism and promote healthy aging. *TEE* total energy expenditure

Targeting Noncontractile Metabolism to Counter Muscle Aging

The concept of targeting skeletal muscle metabolism to counter obesity gathered momentum with the discovery of exercise-mimetics in the last decade. One can expect that a pharmacological agent that can selectively activate NEAT can even be more beneficial to increase energy expenditure. It is getting clearer from recent data that NEAT and non-shivering thermogenesis (NST) in muscle have significant overlap. Dissecting muscle NST from NEAT at molecular level is a very difficult task and several groups including ours have only marginally differentiated between these two processes. It is interesting to note that mitochondrial oxidative metabolism probably plays a bigger role in muscle NST than in NEAT. Further, cytosolic, and mitochondrial Ca^{2+}-dynamics is inherently associated with NST. SERCA isoforms and their regulators such as SLN have been explored as targets to develop pharmacological agents to enhance NST in muscle. Alternatively, if Ca2+-ion channels can be activated to elevate cytosolic Ca2+ level precisely to activate SERCA activity only can be of therapeutic value. Based on our studies, we propose yet another way to selectively enhance muscle NST is by promoting mitochondrial network by specifically targeting fusion machinery. Because exercise and NEAT improve muscle health, selective activation of NST will not only be useful to control obesity, but also be helpful in reducing bad effects of aging on the skeletal muscle, thereby ensuring healthy aging (Fig. 16.4). These research developments provide enthusiasm that soon we can expect medicines targeting NST in muscle to counter ill-effects of aging.

Acknowledgments This work was supported in part by the Ramalingaswamy Re-entry Fellowship from the Department of Biotechnology (DBT), India, and Research Award under File No. ECR/2016/001247 by the Science and Engineering Research Board (SERB), Department of Science and Technology (DST), India to N.C.B.

Suggested Reading

Books and book chapters

1. Askanas V, Engel WK. Muscle Aging, Inclusion-Body Myositis and Myopathies: Blackwell Publishing Ltd; 2011. https://doi.org/10.1002/9781444398311.
2. Lynch GS. Sarcopenia – Age-Related Muscle Wasting and Weakness. Dordrecht: Springer; 2011.
3. Mobbs CV, Hof PR. Body Composition and Aging. Basel: S. Karger AG; 2010. https://doi.org/10.1159/isbn.978-3-8055-9522-3.
4. Zinner C, Sperlich B. Marathon Running: Physiology, Psychology, Nutrition and Training Aspects: Springer International Publishing; 2016. https://doi.org/10.1007/978-3-319-29728-6.

Research Article

1. Kenny GP, Groeller H, McGinn R, Flouris AD. Age, human performance, and physical employment standards. Appl Physiol Nutr Metab. 2016;41(6 Suppl 2):S92–S107.
2. Kirkendall DT, Garrett WE. The effects of aging and training on skeletal muscle. Am J Sports Med. 1998;26(4):598–602.
3. Garg K, Boppart MD. Influence of exercise and aging on extracellular matrix composition in the skeletal muscle stem cell niche. J Appl Physiol. 2016;121(5):1053–8.
4. Hunter SK, Pereira HM, Keenan KG. The aging neuromuscular system and motor performance. J Appl Physiol. 2016;121(4):982–95.
5. López-Lluch G, Navas P. Calorie restriction as an intervention in ageing. J Physiol. 2016;594(8):2043–60.
6. Hariharan N, Sussman MA. Cardiac aging – Getting to the stem of the problem. J Mol Cell Cardiol. 2015;83:32–6.
7. AJ LB, Hoying JB. Adaptation of the coronary microcirculation in aging. Microcirculation. 2016;23(2):157–67.
8. Lee CE, McArdle A, Griffiths RD. The role of hormones, cytokines and heat shock proteins during age-related muscle loss. Clin Nutr. 2007;26(5):524–34.

9. Hamrick MW, ME MG-L, Frechette DM. Fatty infiltration of skeletal muscle: Mechanisms and comparisons with bone marrow adiposity. Front Endocrinol (Lausanne). 2016;7:69.
10. Knowlton AA, Korzick DH. Estrogen and the female heart. Mol Cell Endocrinol. 2014;389(1–2):31–9.
11. Ma Y, Li J. Metabolic shifts during aging and pathology. Compr Physiol. 2015;5(2):667–86.
12. Aon MA, Bhatt N, Cortassa SC. Mitochondrial and cellular mechanisms for managing lipid excess. Front Physiol. 2014;5:282.
13. Demontis F, Piccirillo R, Goldberg AL, Perrimon N. Mechanisms of skeletal muscle aging: insights from Drosophila and mammalian models. Dis Model Mech. 2013;6(6):1339–52.
14. Hood DA, Tryon LD, Carter HN, Kim Y, Chen CC. Unravelling the mechanisms regulating muscle mitochondrial biogenesis. Biochem J. 2016;473(15):2295–314.
15. Carter HN, Chen CC, Hood DA. Mitochondria, muscle health, and exercise with advancing age. Physiology (Bethesda). 2015;30(3):208–23.
16. Hunter GR, Plaisance EP, Carter SJ, Fisher G. Why intensity is not a bad word: Optimizing health status at any age. Clin Nutr. 2017; https://doi.org/10.1016/j.clnu.2017.02.004.
17. Ji LL, Kang C. Role of PGC-1α in sarcopenia: Etiology and potential intervention – A mini-review. Gerontology. 2015;61(2):139–48.
18. Dumont NA, Bentzinger CF, Sincennes MC, Rudnicki MA. Satellite cells and skeletal muscle regeneration. Compr Physiol. 2015;5(3):1027–59.
19. Nguyen N, Sussman MA. Rejuvenating the senescent heart. Curr Opin Cardiol. 2015;30(3):235–9.
20. Cartee GD, Hepple RT, Bamman MM, Zierath JR. Exercise promotes healthy aging of skeletal muscle. Cell Metab. 2016;23(6):1034–47.
21. Crescenzo R, Bianco F, Mazzoli A, Giacco A, Liverini G, Iossa S. Skeletal muscle mitochondrial energetic efficiency and aging. Int J Mol Sci. 2015;16(5):10674–85.
22. Shadrin IY, Khodabukus A, Bursac N. Striated muscle function, regeneration, and repair. Cell Mol Life Sci. 2016;73(22):4175–202.

Metabolic Diseases and Aging

17

Arttatrana Pal and Pramod C. Rath

Introduction

The manifestation of aging is associated with progressive deterioration in organs or tissues both in structure and function of various molecular, cellular, and tissue components that can be influenced by genetic, hormonal, environmental, and metabolic factors. A number of theories have been proposed to explain the aging process in different organisms from lower invertebrates to higher mammals, such as loss of telomere and shortening of chromosomes, biomolecules damage, accumulation of damaged nucleic acids in cells, and dysfunction of cellular organelles, namely, endoplasmic reticulum (ER) and mitochondria. However, it is well established that aging is a major risk factor for the progression of various metabolic diseases including obesity, insulin resistance, and diabetes. More importantly, all these complicated metabolic diseases are linked to molecular and cellular levels in the endocrine, hepatic, renal, respiratory, cardiovascular systems, and central nervous system (CNS) leading to aging. Diabetes, often referred by physicians as diabetes mellitus (DM), describes a group of metabolic diseases characterized by chronic hyperglycemia or high blood sugar due to insulin deficiency or inadequate insulin production due to the loss of pancreatic β cell by viral infection or autoimmune disease in the case of type 1 diabetes mellitus (T1DM). Unlike T1DM, people with type 2 diabetes mellitus (T2DM) produce insulin; however, the insulin their pancreas secretes is either not enough or the body is unable to recognize the insulin or body's cells do

---

A. Pal (✉)
Department of Zoology, School of Life Sciences, Mahatma Gandhi Central University, Motihari, Bihar, India
e-mail: arttatrana@yahoo.com

P. C. Rath (✉)
Molecular Biology Laboratory, School of Life Sciences, Jawaharlal Nehru University, New Delhi, India
e-mail: pcrath@mail.jnu.ac.in

© Springer Nature Singapore Pte Ltd. 2020
P. C. Rath (ed.), *Models, Molecules and Mechanisms in Biogerontology*, https://doi.org/10.1007/978-981-32-9005-1_17

not respond properly. While obesity, considered as a disorder of energy homeostasis, is associated with risk for many diseases alone, such health complications include pathologic changes involving coronary heart disease, renal failure, retinopathy, neuropathy, vascular diseases, myocardial infarction, and stroke.

Recently, a great progress has been made on the association between various metabolic diseases and aging. Pharmacological intervention and genetic manipulations of key signaling pathways involved in the regulation of blood glucose and energy metabolism, such as insulin, oxidative-nitrosative stress, advanced glycation end product (AGE), and mammalian target of rapamycin (mTOR) signaling pathways, have been shown to improve the extension of lifespan in diverse model organisms. Moreover, the maintenance of telomere length, alleviated ER-stress, reduced mitochondrial dysfunctions including increased mitochondrial stress, and/or respiratory chain activity could be primary longevity determinants. Consistent with this, regular aerobic exercise, caloric restriction (CR), reduced somatotropic signaling, adenosine monophosphate-activated protein kinase (AMPK) activation, mitogen-activated protein kinase (MAPK) signaling, mTOR and insulin/IGF-1 signaling, restoration of cellular antioxidants capacity, and restoration of biomolecules from damage are some of the best-known interventions that keep human subjects healthy and may extend their lifespan. In addition, CR facilitates the degradation of damaged organelles, nucleic acids, and protein aggregates in cells by induction of autophagy. Current evidence shows that autophagy is required for ER stress-associated apoptosis and mitochondrial turnover and thus may mediate the integration of the insulin/IGF-1 and Akt/mTOR signaling pathways with other cellular machineries in regulating metabolic disease-related aging.

Metabolic Syndrome and Aging

If we look back in time, we will observe that it was immensely tough for humans to gather food for survival throughout the globe. Fortunately, recent industrialization has succeeded to eradicate the basic food problem by making many choices of calorie-rich food easily available at doorstep any time with little physical effort. However, overnourishment in people having calorie-rich food has led to diseases, hormonal imbalances, and health problems by introducing a worldwide outbreak of obesity and T2DM. More importantly, a cluster of metabolic risk factors and pathophysiological abnormalities including obesity, insulin resistance, impaired glucose tolerance, dyslipidemia, and high blood pressure, which are interconnected and called metabolic syndrome (MS), regulates telomere length and organelles dysfunctions, often precedes health problems, and minimizes lifespan (Fig. 17.1). Metabolic syndrome precedes several disorders in different organs and tissues in human body resulting from metabolic dysregulation. Among all these metabolic abnormalities, insulin resistance alone interferes with human health worldwide. Basically, these metabolic diseases are characterized by a compromised ability of insulin to control glucose and lipid metabolism, which results in decreased glucose disposal to muscle, increased hepatic glucose production, and overt postprandial hyperglycemia. Without proper treatment/management, insulin resistance led to development of T2DM and related complications, finally leading to organ or tissue damage, which

Fig. 17.1 Components of metabolic syndrome and their possible pathophysiological links with aging

Risk Factors
Genetic
Metabolic
Endocrine
Environmental
Lifestyle

→ **Metabolic syndrome**
Obesity, Impaired Glucose tolerance, Insulin resistance, Hypertension, Dislipidimia

↓

Telomere length shortening ⇌ **Diabetes**

Cell organelles damage

↓

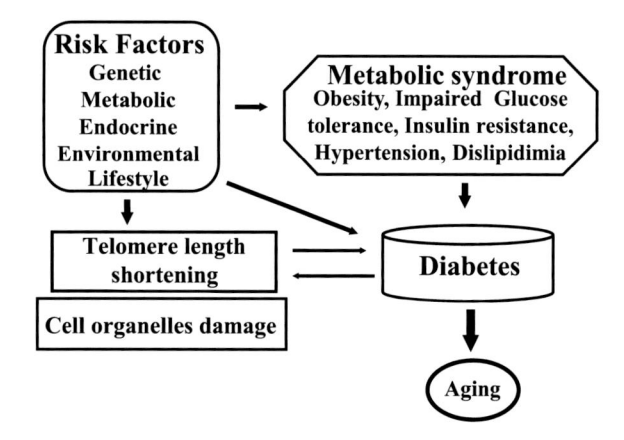

Aging

Table 17.1 Clinical criteria for the diagnosis of metabolic syndrome in human subjects [1]

Metabolic parameters	ATP III	WHO	IDF	Diabetes
Abdominal obesity or large waistline (cm)				
Men: waist circumference	>102	>102	>94	
Women: waist circumference	>88	>88	>80	
High fasting blood sugar (mg/dL)	>110, <126	>110	>100	>130
High blood pressure (mmHg)	>130/85	140/90	>130/85	
High triglyceride level (mg/dL)	150	150	150	
Reduced high-density lipoprotein (HDL) cholesterol (mg/dL)				
Men	<40	<35	<40	
Women	<50	<39	<50	

ATP III Adult Treatment Panel III based on the National Education Program, *WHO* World Health Organization, *IDF* International Diabetes Foundation

are some of the prominent causes of aging. First of all, the unified agreement about the definition of MS was drawn up during a meeting organized by the International Diabetes Federation (IDF) in 2005. As stated in this, MS is the major contributing factor in obesity, measured by waist circumference and body mass index. Subsequently, scientific community defined MS with different names such as syndrome X, cardiometabolic syndrome, IR syndrome, or CHAOS (coronary artery disease, hypertension, atherosclerosis, obesity, and stroke, used in Australia). A patient is diagnosed with MS, if three or more of the major criteria are detected from the categories given in Table 17.1 [1]. Till date it is not clear or difficult for diagnosis of MS by a clinician or physician, and it represents an important pathophysiological combination to study metabolism in human subjects and other mammals. According to clinical diagnosis, the presence of MS in human subjects leads to an increased risk of T2DM and cardiovascular complications, in the form of coronary or peripheral atherosclerosis and heart failure. At the bottom line, MS patients have increased all-cause mortality and a shortened lifespan compared with the general population. Therefore, it is successively recognized that MS is associated with early aging, which is of predominant importance considering the worldwide growing epidemic of MS.

Metabolism and Cell Aging

In the process of aging, organs or tissues are considered to be induced by alterations in metabolism in different biochemical and molecular pathways at the levels of small molecules, proteins and metabolic homeostasis, signaling patterns, organelles dysfunctions, free radical accumulation, and inter-/intra-tissue communications that ultimately decline or shift during cellular aging (Fig. 17.2). More importantly, normal aging partly results from the functional decline of adult stem cells, which are essential for the maintenance and regeneration of organs or tissues that exhibit high rates of cell turnover and regenerative reserve, including skin, gastrointestinal epithelium, and hematopoietic system. Also, there is a link between oxidative-nitrosative stress and metabolic pathways and how these converge on important cell organelle like mitochondria to compromise energy maintenance and drive aging. The regenerative capacity of stem cells, telomere length, and the link between oxidative-nitrosative stress and organelles functions are currently being actively researched to establish theories on metabolism and cellular aging.

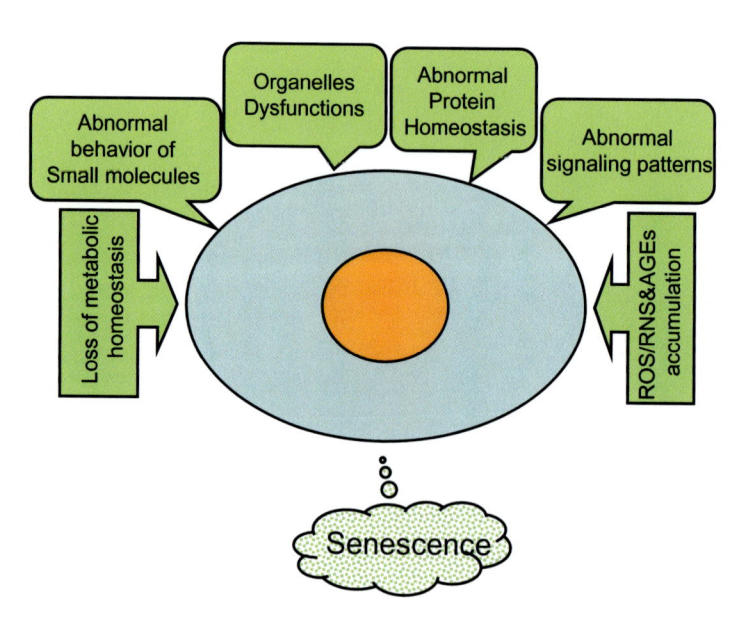

Fig. 17.2 Complex series of biochemical and molecular changes in cellular compartments force cells to early senescence

Tissue-Specific Metabolic Disorder and Aging

CNS/Brain and Aging

Glucose is the major metabolic fuel source for the healthy brain and it is transported to the CNS from the periphery via facilitative glucose transporters. While the brain can neither synthesize nor store glucose for extended periods of time, it is essential that proper glucose regulation to CNS is achieved in the periphery to ensure appropriate glucose transport and these processes may be disrupted in MS. Human subjects with diabetes or high blood sugar in middle age are more likely to have brain cell loss, and loss of cell-cell communications, as well as problems with memory and thinking skills, than people who never have diabetes or high blood sugar or who develop it in old age. Nevertheless, noninvasive brain imaging techniques recently providing more authentic information on brain anatomy, white matter hyperintensities, and function have indicated abnormal brain cell architecture and functional abnormalities associated with metabolic disease like diabetes. In general, patients with diabetes have a strong effect on brain structure and function influencing the whole body system and that might negatively influence longevity and quality of life. On the other hand, CNS plays a crucial role in regulating glucose homeostasis in higher mammals through hepatic gluconeogenesis, glycogenolysis, and pancreatic function. Interestingly, all these remarkable metabolic activities are largely mediated by central regulation of the autonomic nervous system, which acts in concert with the hypothalamic-pituitary-adrenal (HPA) axis to regulate metabolic responses to changes in energy requirements and plasma glucose concentration. In contrast, critical autonomic regulatory neurons in the hypothalamus and brainstem are responsible for the management of energy homeostasis. However, potential mechanistic mediators of MS-induced cognitive impairment and brain aging by complex pathophysiological features including obesity, hyperglycemia, insulin resistance, impaired glucose homeostasis, hyperlipidemia, and hyperleptinemia lead to dysregulation of HPA axis function and accumulation of free radicals, AGEs, and damaged biomolecules. The consequences of these pathophysiologies in the anatomical changes of the brain, including dendritic/cerebral atrophy, small infarction, white matter lesion, changes in synapse formation, electrophysiological deficits leading to memory decline, psychomotor slowing, and increased dementia risk, lead to early senescence in neuronal cells (Fig. 17.3).

Cardiovascular Complications and Aging

Cardiovascular complications are the most important health problems with people suffering from metabolic disease like diabetes. The most common cause of death in adult individuals is heart disease, mainly ischemic, coronary heart disease, peripheral arterial disease, and stroke, which is also the most common cause of severe disability in younger ones who are suffering from complex MS leading to early aging (Fig. 17.4). Recent studies demonstrate that one reason for poor prognosis in

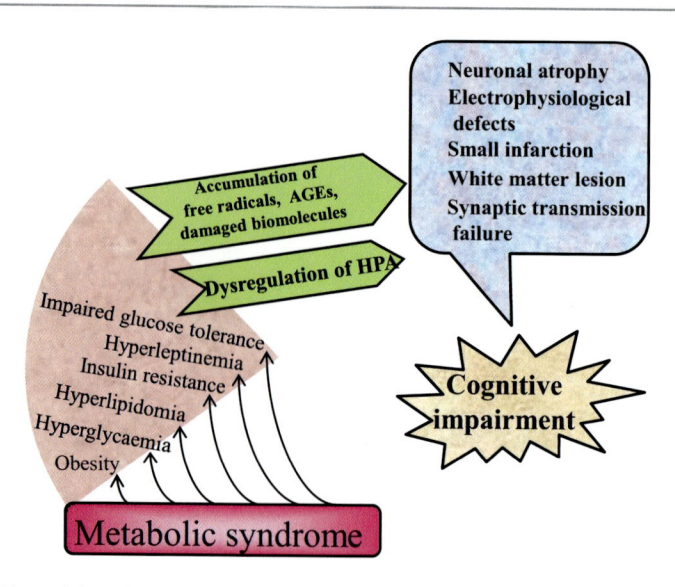

Fig. 17.3 Potential mechanistic mediators of metabolic syndrome-induced cognitive impairment and brain aging

patients with both diabetes and ischemic heart disease seems to be an enhanced myocardial dysfunction leading to accelerated heart failure. This is not only linked with chronic hyperglycemia but also associated with several metabolic factors such as severe coronary atherosclerosis, prolonged hypertension, microvascular disease, glycosylation of myocardial proteins, and autonomic neuropathy. Taking preventive measures may improve diabetic complications, and prevention of atherosclerosis with cholesterol-lowering therapy could prevent or mitigate both structures and functions of diabetic heart and early vascular aging (Table 17.2). Recently, mortality rate increased due to stroke almost three- to fivefold in patients with diabetes, and the most common site of cerebrovascular (CVA) disease in patients with diabetes is the occlusion of small paramedian penetrating arteries. The elderly patients with diabetes are at particularly high risk of morbidity and mortality from CVA and the prevalence of CVA in the older population with diabetes is also higher than those without diabetes. Research on different animal models and human subjects demonstrated that the prevalence of ventricular hypertrophy increases and diastolic function declines with metabolic diseases like diabetes. Similarly, systolic function of the heart is relatively preserved in human subjects at rest, and the maximal exercise capacity decreases with diabetes. The clinical complications related to ventricular and systolic functional changes are likely manifestations of intrinsic cardiac diseases due to diabetes. Both clinicians and researchers focus on inflammatory changes in the blood vessel wall called atherosclerosis in response to many metabolic diseases like obesity and insulin resistance. Atherosclerosis develops early in the course of diabetes and causes endothelial dysfunction and foam cell formation. An increased inflammatory state of immune cells such as monocytes or macrophages is responsible for this accelerated atherosclerosis. Elevated glucose levels in blood have adverse effects on endothelial cell function, leading to atherogenesis by

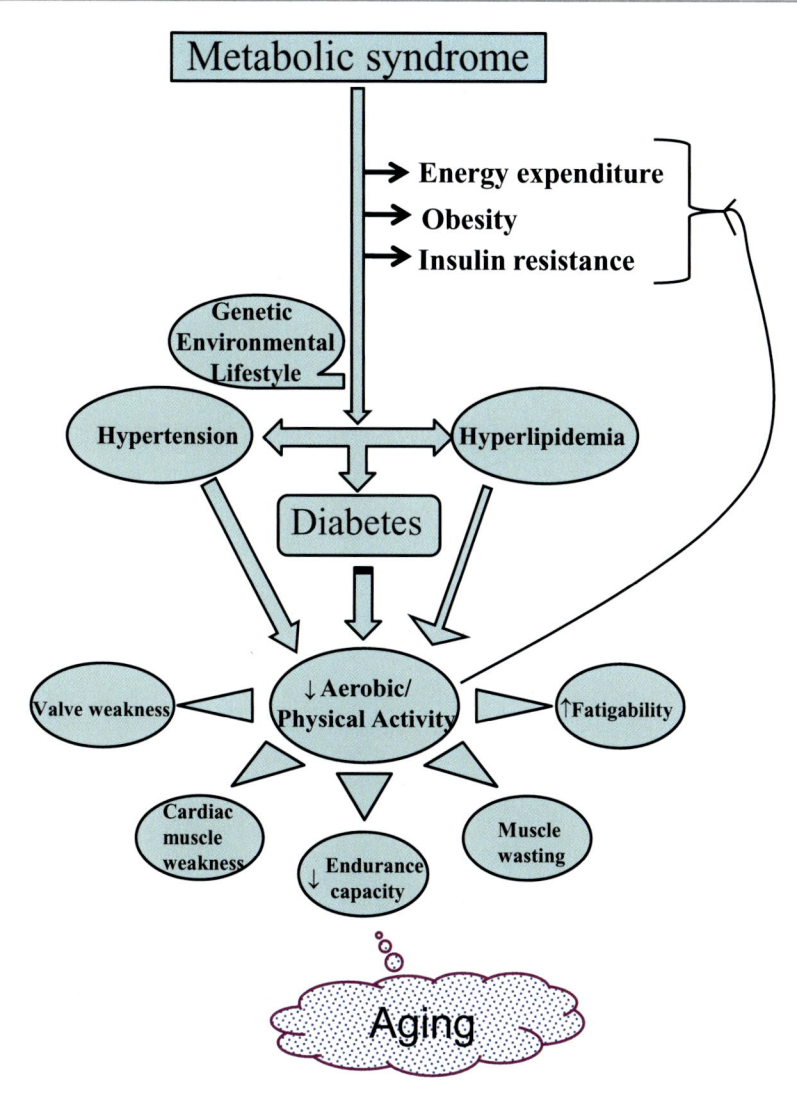

Fig. 17.4 Schematic approach of metabolic syndrome, linking it with cardiovascular complications leading to early aging

glycosylation of lipoproteins and other tissue proteins. More importantly, the development and progression of these lesions in hyperglycemia condition can be prevented by aggressive lipid-lowering therapy and intracellular signal transduction pathways in atherosclerosis associated with diabetes [2]. Some studies have demonstrated that high insulin levels in the circulating system are associated with the development of cardiovascular disease, even though there is evidence of this in non-diabetic subjects and in early stages of diabetes. In that case, insulin may promote atherogenesis by its direct effects on the arterial wall or by effects on activation of macrophages/monocytes, lipid metabolism, blood pressure, or clotting factors.

Table 17.2 Metabolic syndrome-linked cardiovascular changes and early aging in human subjects

Structure	Myocardial	Valvular	Arterial
	Increased myocardial mass	Increased thickness of aortic and mitral leaflets	Increased intimal thickness
	Increased left ventricular wall thickness	Increased circumference of all four valves	Increased collagen content
	Increased deposition of collagen	Calcification of mitral annulus	
Function	**Heart rate**	**Left ventricular**	**Myofibril**
	Decreased heart rate at rest	*Systolic*	Unchanged peak contractile force
	Decreased maximal heart rate during exercise	Unchanged cardiac output	Increased duration of contraction
	Decreased heart rate variability	Increased stroke volume index	Decreased Ca^{2+} uptake by sarcoplasmic reticulum
	Decreased sinus node intrinsic rate	Left ventricular	Decreased r~-adrenergic-mediated contractile augmentation
	Vascular	*Diastolic*	
	Decreased compliance	Decreased left ventricular compliance	
	Increased pulsed-wave velocity	Increased early diastolic	
		Left ventricular filling	

Peripheral Vascular Disease

Peripheral vascular disease (PVD) is a common metabolic disease complication among adults diagnosed with diabetes in older age compared to middle age. Among all, retinopathy is the most common microvascular PVD of diabetes. Studies have demonstrated that in patients with diabetes, men were more likely than women to have diabetic retinopathy. Patients with retinopathy had a longer duration of diabetes, higher HbA1c value, and were more likely to be using insulin [3]. The PVD, like urinary incontinence, can have a negative influence on a diabetic patient's quality of life, physical functioning, mental health, and general health perception, particularly in women with diabetes [4]. On the other hand, in older people, chronic kidney disease (CKD) is a common diabetes-related complication. According to recent analysis of the Kidney Early Evaluation Program (KEEP) database, NHANES data, and billing codes from a sample of the US Medicare population, the prevalence of CKD is consistently higher among patients with diabetes in older age [5]. This is a very common trend in elderly populations throughout the globe with diabetes, including a large proportion of those receiving dialysis for diabetic nephropathy and most of them have multiple comorbid conditions, such as heart disease and peripheral vascular disease. Furthermore, these older populations with diabetes have arteriovenous fistula complications. Elderly people with diabetes have an increased prevalence of geriatric complications such as impairment of multiple physiological systems, leading to functional disability, falls, depression, dementia, loss of vision, and hearing impairment. Different potential factors such as polypharmacy, pain, lower physical activity, functional limitations, and cognitive impairments are related to the increased frequency of falls; even the use of insulin therapy may also be related to the risk of falls in older persons with diabetes. Apart from other complications, loss of vision and hearing impairment simultaneously, both can increase the risk of falls in older people with diabetes, leading to functional disability, feeling isolated, and being more vulnerable to depression [6].

Respiratory Complications and Aging

The effects of aging on the respiratory system due to MS are similar to those that occur in other organs in human subjects suffering from diabetes. Maximum lung function gradually declines with metabolic diseases including decreases in peak airflow and gas exchange, decreases in measures of lung functions such as vital capacity, activity of the respiratory muscles, and effectiveness of lung defense mechanisms. In younger persons, these MS-related changes seldom lead to complicated respiratory symptoms. These changes contribute somewhat to an older person's reduced ability to do exercise. MS-related diseases decreases heart functions, hence it may be a more important cause for such limitation. Additionally, along with MS, older people are at higher risk of developing pneumonia after bacterial or viral infections.

Metabolic Process Linked to Diseases and Aging

Aging process is universally enhanced by alterations in cellular metabolism and related diseases. The development of MS shows a very complex pathophysiology in human subjects and T2DM represents a major component of metabolic alterations of hepatic gluconeogenesis, adipose lipogenesis, (defective) glycogen synthesis, and glucose uptake, they are very commonly observed in aged people with T2DM [7]. More importantly, abdominal obesity is frequently observed with aging and people suffering from diabetes. Many factors are linked to pathologic obesity found in a persons suffering from metabolic disease along with aging, a common cause for this may be expression of inflammatory mediators, which are known to critically influence diabetes.

Alteration of Body Composition and Aging

A very common phenomenon in human populations, irrespective of sex, is the increased adiposity when people are in their 40s or beyond and it may increase, or decrease, or remain unchanged thereafter. A recent medical technique such as computational tomography scan reveals that with increasing age, parallel subcutaneous fat (SF) decreases and visceral fat (VF) increases in both men and women. On the other hand, the VF accumulation in both these men and women is associated with insulin resistance and diabetes, which is a major risk factor for cardiac complications, coronary artery disease, stroke, and finally leading to death [8]. Similar to human aging, different models of aged laboratory rodents also develop increased fat mass, with a disproportionate increase in VF compared to SF mass [9]. If we compare human subjects with rodents, VF and SF are biologically distinct in terms of cascades of gene expression and downstream targets of secretory profiles of adipokines and pro-inflammatory mediators, all of which can contribute to the pathogenesis of diabetes and its age-associated chronic complications [10]. Augmented release of free fatty acids is another risk factor of VF accumulation both in humans and in rodents due to MS, which can reach the liver via hepatic portal circulation and promptly interfere with hepatic insulin action. Muzumdar et al. [11] demonstrated in rodents that surgical removal of VF has shown to restore insulin sensitivity, improve lipid profiles, decrease hepatic triglycerides, and prolong lifespan. Adiponectin, another metabolic regulator derived from adipose tissue, is linked to aging in persons with MS including diabetes. More importantly, adiponectin is an insulin sensitizer with anti-inflammatory properties and a potent activator of AMP-activated protein kinase (AMPK) as opposed to other fat-derived cytokines [12]. Sarcopenia or loss of skeletal muscle mass represents another unfavorable phenotypic change observed more frequently in elderly people and it is a risk factor in those who are suffering from MS like obesity and diabetes [13]. Lee et al. [14] demonstrated that insulin resistance is another contributor to the decline in muscle quantity and quality in human subjects, as it has been associated with reduced skeletal muscle strength, reduced protein synthesis rates, and accelerated skeletal muscle loss.

Abnormal Endocrine Function

Declines in different hormones have been linked to the MS-related aging process, exclusively in people suffering from diabetes. Both human subjects and rodents studies revealed the potential for hormone replacement strategies to modulate features of metabolic syndrome-regulated aging. Rossouw et al. [15] have reported that the estrogen and progesterone replacement in older women would modulate several key age-related complications. However, this study was stopped early in women, because it increased the risk of many physiological complications like cardiovascular problems, cognitive decline, breast cancer, and many other health problems. This technique did not work for common people and mostly it failed. The interactions among hormones in young females are important for reproductive behavior and then it is different in postmenopausal women. More importantly, among all the sex hormones, estrogen may interact in a harmful way with MS environment, and these women are highly at risk with increase in inflammation, senescent cells, and breakdown in repair mechanism of cells. On the other hand, alteration of hormonal deficiencies in women with MS was also limited to ovarian steroids. Also, there was no replacement of growth hormone, IGF-1, or important circulating peptides to more youthful levels. Growth hormone secretion markedly decreases with age and diabetes, resulting in a concomitant decline in IGF-1 concentrations in humans [16]. In addition, reversal of insulin/IGF-1 signaling (IIS) in rodents has led to extended lifespan [17], and functional mutations of human IGF-1R have been linked to better longevity [18]. In human subjects, thyroid hormones are major regulators of energy expenditure and hypothyroidism is the most common thyroid disorder in older people. Studies have reported that people with diabetes have a higher prevalence of thyroid disorders compared to the normal population and the rate of postpartum thyroiditis in diabetic patients is three times more than that of normal women [19]. In laboratory conditions, experimentally induced hypothyroidism in young rodents has resulted in extended lifespan, whereas inducing hyperthyroidism reduces longevity [20]. Moreover, hypothyroidism may modulate lifespan by lowering metabolic rate, core body temperature, and oxygen consumption, thereby reducing generation of oxygen and nitrogen free radicals and associated oxidative-nitrosative stress. Collectively, longevity is negatively related to MS, and also linked to hormones regulating metabolism with physiological manifestations.

Decline in Cellular Antioxidant Capacity and Aging

Oxidative-nitrosative stress is a significant cause of glucose intolerance and insulin resistance, and there is substantial evidence supporting the idea that antioxidant treatments in laboratory animal models prevent age-associated metabolic dysfunction. Hoehn et al. [21] reported that treatment of mice with mitochondria-targeted superoxide dismutase MnTBAP prevented the development of insulin resistance and glucose intolerance caused by high-fat diet feeding. Another study by Ilkun et al. [22] in genetically obese mice, revealed that MnTBAP treatment also improved glucose homeostasis. Moreover, mitochondria-targeted antioxidant SS31 showed a similar protection from high-fat diet in rats [23]. Several genetic models of increased antioxidant enzyme expression have also shown to protect from obesity-induced insulin resistance. Overexpression of MnSOD, catalase, and peroxiredoxin 3 has been shown to preserve glucose homeostasis during high-fat diet feeding. Lee et al. showed that specific mitochondrial CAT mice are protected from age-associated declines in insulin due to reduced mitochondrial H_2O_2 production and reduced accumulation of oxidative damage [24]. Long-term treatment of antioxidants vitamin C or butylated hydroxytoluene improved insulin response of adipocytes isolated from old rats [25]. However, antioxidants may be beneficial for glucose metabolism in different animal models, but most studies on antioxidant therapy in human subjects have been largely inconclusive.

Decline in Mitochondrial Function and Aging

Aging is associated with progressive loss in mitochondrial number and function in people suffering from MS and it is more critical in diabetic patients. Mitochondria are the major source for the overproduction of reactive oxygen species (ROS) in diabetes, leading to oxidative damage of macromolecules, and one of the fundamental causes of cellular aging and senescence. However, the relationships among mitochondrial dysfunction, glucose homeostasis, and disease have been another area of intense investigation in aging process. Intriguingly, how impaired mitochondrial function in metabolically active tissues, including liver and skeletal muscle, may impinge on insulin signaling is complex, but it is thought to involve reduced or incomplete ß-oxidation of fatty acid substrates. Montgomery and Turner [26] demonstrated an association between IR and impaired glucose tolerance with decreased mitochondrial oxidative activity and ATP synthesis in elderly and diabetes patients. There is a separate chapter in this book about the roles of mitochondria in the aging process.

Increase in Advanced Glycation End Products (AGEs) and Aging

Basically, AGEs are formed in normal metabolic process in human tissues, which increase with different pathophysiological conditions and more frequently in elderly people and diabetes. After their formation in different organs/tissues, these free or unbound or protein-bound AGEs are freely available in blood circulation and in due course of time, they are removed from the body through enzymatic clearance and renal excretion. Different research reports on different animal models demonstrated that MS enhanced the imbalance between the formation and natural clearance of AGEs in the human body, leading to the accumulation of AGEs in different organs/tissues, and it accelerated the aging process [27]. In many age-related metabolic diseases, the AGE accumulation increased many folds and it was a significant contributing factor in degenerative processes, especially in renal failure, blindness, and cardiovascular diseases [28, 29]. Clinical studies also reported that elevated AGE levels in patients with diabetes are most likely due to an excessive elevation of glucose concentration in the blood, which consequently accelerates the glycation of proteins and caused multiorgan complications [28, 29]. Basically, there are two primary sources for AGEs. The first source is the food we eat, the browning of food is a cooking technique and achieved by heating or cooking sugars with proteins in the absence of water. Since grains, vegetables, fruits, and meat all have proteins, this browning effect is an indication of AGEs. The second source for AGE products happens inside the human body through normal metabolism and aging.

AGEs and Organ/Tissue Complications

In MS like diabetes, AGE formations are accompanied by increased free radical activity that contributes to the biomolecular damage, apoptosis, and finally organ failure sometimes associated with aging. Exactly, what AGEs are doing in cellular compartments so that cells are entering into the process of senescence is a question for both the clinicians and researchers. Simply, inside the tissues or cells, AGEs act as mediators and they can initiate a wide range of abnormal responses such as the inappropriate expression of growth factors, alterations in growth dynamics, accumulation of extracellular matrix that regulates apoptosis through decreased solubility, elasticity, and enzymatic activity in long-lived proteins such as collagen [30]. A number of these factors changes in different tissue collagen such as skin, in which it appears to be accelerated in metabolic diseases in humans and more complications occur in old age. More importantly, AGE cross-linking reactions occurring in collagen, contribute to diabetic cardiomyopathy such as vascular stiffening and myocardial dysfunction [31]. In cell membrane region, non-cross-linking effects are exerted by binding of AGEs to the receptor for AGEs (RAGEs). Different cells express RAGE, and the interaction with AGEs elicits activation of intracellular signaling cascades, gene expression, and production of pro-inflammatory mediators such as interleukin (IL)-6 and tumor necrosis factor alpha (TNF-α) (Fig. 17.5). Moreover, pro-inflammatory mediators exhibit powerful proteolytic activity whereby collagen becomes more vulnerable and tissue elasticity decreases at the peripheral level [32]. Similarly, at the central level, interaction between AGEs, amyloid beta and hyperphosphorylated tau protein, induces cells such as microglia and astrocytes to upregulate the production of oxygen- and nitrogen free radicals and pro-inflammatory mediators, which affects neuronal cell functions [32].

AGEs and Skeletal Tissues

Major chemically characterized AGEs are pentosidine for fluorescent cross-linking and carboxymethyl-lysine (CML) for non-cross-linking. The cross-linking AGEs are considered as being involved in the pathophysiology of metabolic diseases and responsible for an increasing proportion of insoluble extracellular matrix and thickening of tissue as well as increasing mechanical stiffness and loss of elasticity and finally it accelerates arthritis. On the other hand, in metabolic diseases, higher levels of AGEs are reported in patients with osteoporosis, thereby increasing the risk of bone fractures.

AGEs and Lens

Glycation product and AGEs of various derivations are generated significantly in metabolic disease this causes alterations in lens fiber membrane integrity and tertiary structure of lens crystalline proteins, which is fully responsible for cataract formation. Many studies have reported that dicarbonyl compounds such as glyoxal and methylglyoxal are enhanced in diabetes and aging process, leading to AGE cross-links on α-crystallins with resultant loss of chaperone activity, increased αβ-crystallin content, and dense aggregate formation [33]. Recent evidence suggests that a close association exists between advanced glycation, many metal ions, and generation of oxygen/nitrogen free radicals during metabolic disease-related cataract formation, where AGE formation on lens protein crystallins leads to binding of redox-active copper, which in turn catalyzes ascorbate oxidation [34].

Keratinocytes
Proliferation ↓
Apoptosis ↑
ROS/RNS ↑
MMPs ↑
TIPM ↓
NFκB ↑
proinflammatory mediators ↑
α2β1-integrin ↓
Cell renewal ↓
Epidermal homeostasis ↓

Vascular endothelial cells
VCAM, ICAM, E-selectin ↑
Permeability ↑
TNFα, IL-6 ↑
MCP-1 ↑
Induction of proinflammatory mediators and recruitment of immune cells

Fibroblasts
Proliferation ↓
Apoptosis ↑
ECM synthesis ↓
MMPs ↑
NFκB ↑
ROS/RNS ↑
Contractile properties ↓
NOX ↑
Cell renewal ↓
Dermal homeostasis ↓
Skin contractile function ↓

Extracellular matrix proteins
(collagen, fibronectin, elastin)
Crosslinking
Resistance to MMP degradation
Impaired assembly of macromolecules to normal 3D structures
Defect cross-talking to cells
Elasticity ↓
Stiffness ↑
Resistance to repair mechanisms
Tissue permeability ↓

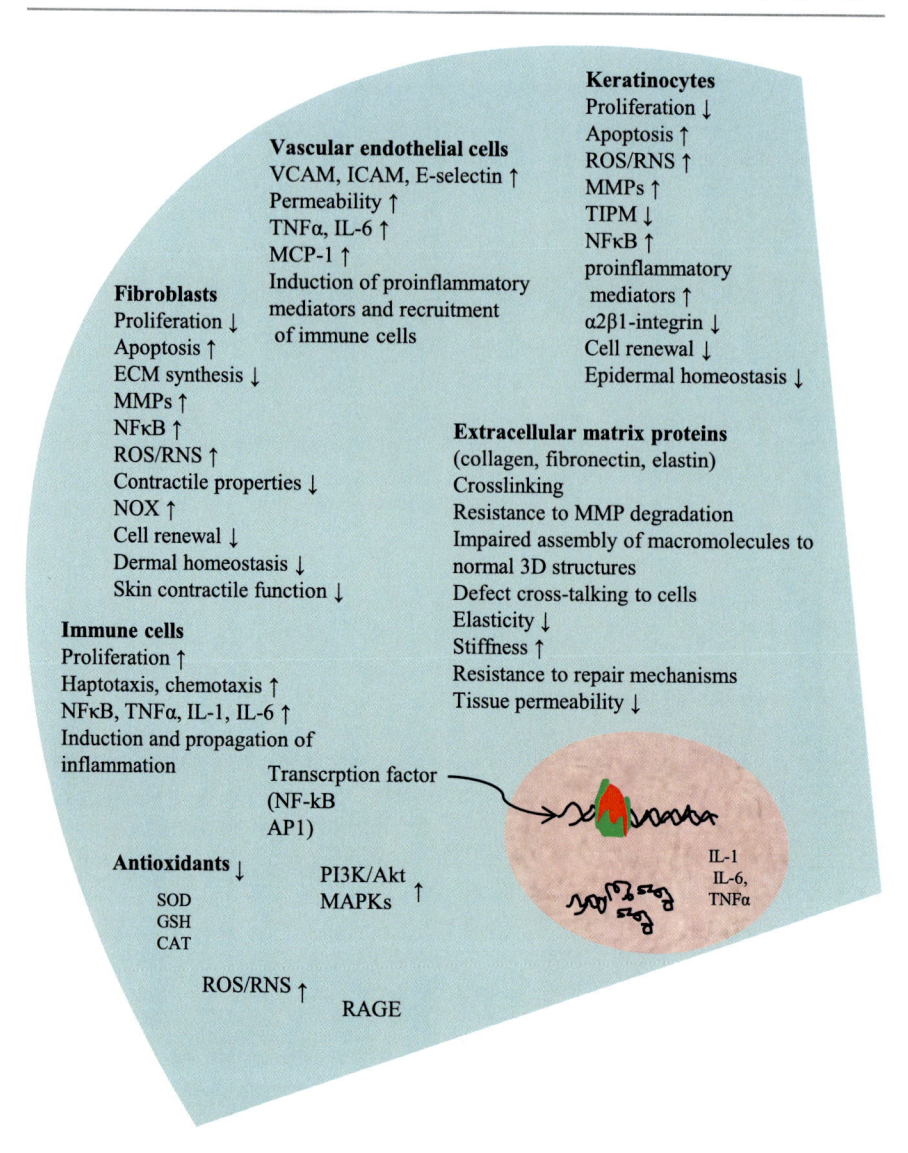

Immune cells
Proliferation ↑
Haptotaxis, chemotaxis ↑
NFκB, TNFα, IL-1, IL-6 ↑
Induction and propagation of inflammation

Transcription factor (NF-kB AP1)

Antioxidants ↓
SOD
GSH
CAT

PI3K/Akt
MAPKs ↑

ROS/RNS ↑
RAGE

IL-1
IL-6,
TNFα

Fig. 17.5 Metabolic syndrome-induced AGE-formation occurs in intra- and extra-cellular environment. They can react with biomolecules in cellular compartments as well as extracellularly, through alteration of physicochemical properties of cellular proteins, decreased cell proliferation, increased apoptosis and senescence, induction of oxidative-nitrosative stress and pro-inflammatory mediators as well as other pathways

AGEs and Cornea/Vitreous

Basically, the Maillard reaction has a major role in altering corneal architecture and biochemistry during MS and aging. Diabetic keratopathy is very common in patients suffering from diabetes, where AGEs play a major role for complications on thickening of the stroma and basement membranes, recurrent erosions, corneal edema, and morphological alterations in the epithelial and endothelial layers. These types of alterations in human diabetic cornea are accompanied by decreased protein stability in the stroma and basal lamina and increased immunoreactive AGEs, which have been partially characterized as pentosidine and CML [35]. In patients who are suffering from diabetes, their Bowman's membrane is heavily glycated. At the same time, in vitro AGE-modified substrates can significantly reduce corneal epithelial cell adhesion and spreading, possibly by disruption of integrin and non-integrin receptor-matrix interactions, which has obvious pathogenic implications for recurrent erosions [36]. In human diabetic patients, AGEs have been accumulating in higher level in vitreous region of eye, where the structural changes to the vitreous such as liquefaction and posterior vitreous detachment are associated with aging. In terms of AGE-mediated vitreous pathology, studies have demonstrated that glycation can induce abnormal cross-links between vitreal collagen fibrils leading to dissociation from hyaluronan and resultant destabilization of the gel structure.

Retina

In diabetic patients, the retinal microvasculature becomes progressively dysfunctional in different architectural regions of the retina such as retinal ganglion cells, retinal pericytes, endothelial cells, and the optic nerve leading to capillary closure and retinal ischemia in response to accumulation of AGEs. Research reports demonstrate that AGEs are localized in vascular basement membranes and retinal pericytes after 8 months of experimental diabetes in rats [37]. Moreover, studies have pointed out that when nondiabetic animals are infused with preformed AGE-albumin, these adducts accumulate around and, within the pericytes, colocalize with AGE receptors, induce basement membrane thickening, and cause breakdown of the inner blood-retinal barrier. In human subjects with different pathophysiological conditions, it has been reported that the levels of serum AGEs, and also the glycoxidation product CML, correlate with the degree of diabetic retinopathy. Many investigations on retinal vascular cells have provided important insights into the action of AGEs and its adducts in both in vitro and in vivo models, especially on their receptors, receptor interactions, and how they contribute to retinal dysfunction in diabetic complications.

Proteasome and Aging

MS and its associated metabolic dysregulations cause organ- or tissue damages by ubiquitin-proteasomal system (UPS), which is the main regulator of both the functional and dysfunctional protein pool of mammalian cells. There are many consequences of metabolic dysregulation in patients suffering from metabolic diseases like diabetes such as chronic redox shift to more oxidative-nitrosative state, i.e., a low-grade systemic inflammation that increases ROS (reactive oxygen species)/

RNS (reactive nitrogen species) formation, lipid peroxidation, protein oxidation, formation of AGEs, glycosylation, S-glutathionylation, mitochondrial stress, endoplasmic reticulum (ER) stress, unfolded protein response (UPR), expression of transcription factors, and release of cytokines and chemokines- these are already known to affect the highly redox-regulated UPS [38, 39]. Many research reports demonstrated that maintaining a highly functional UPS positively correlates with increased health and lifespan, hence modulating the UPS function may be an effective approach to preventing MS-related detrimental consequences.

Metabolic Signaling Networks and Aging

Metabolic-regulated aging is considered to be solely the result of wear and tear, in fact it is governed by specific signaling pathways in the cellular compartments (Fig. 17.6). Simple manipulation of genetic and environmental factors can drastically extend lifespan, suggesting that several of these signaling pathways control longevity in response to changes in the surroundings. Here we summarize the key signaling pathways identified in MS that regulate aging and longevity.

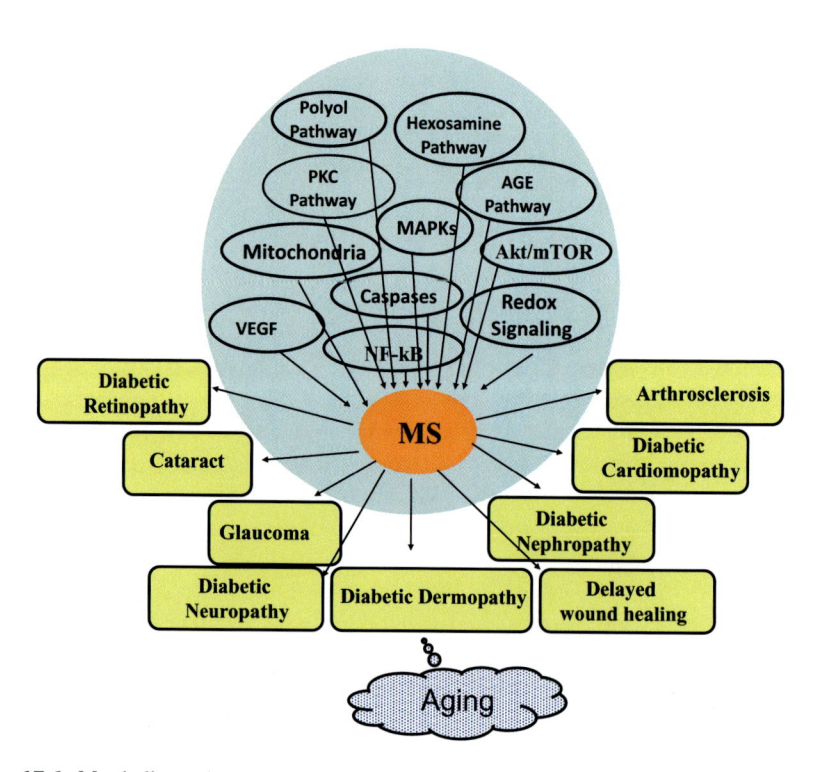

Fig. 17.6 Metabolic syndrome-regulated signaling networks and aging

Hormonal Signaling

Insulin/Insulin-Like Signaling and Metabolic Diseases

Patients with T1DM suffer from insulin deficiency, and effective insulin therapy can lower hyperglycemia and reduce many pathophysiological complications. However, patients with T2DM are noninsulin-dependent, in which intensive insulin therapy lowers blood glucose, but increases body weight, cardiovascular risk, and other medical complications also [40]. In these patients, intensive insulin therapy does not provide much protective action to vital organs in adults, and majority of patients with T2DM have minimized lifespan and finally die of multiorgan failure. More importantly, understanding the mechanisms responsible for insulin action and resistance and finding an effective management of metabolic syndrome, like diabetes and its associated diseases leading to organ/tissue dysfunction have important clinical implications.

Insulin Signaling and Action

Over the last decades, studies on insulin and its action have made a breakthrough in the area of diabetes and biomedical research for clinical applications. Innovative attempts through different advanced molecular techniques on recombinant insulin production and purification, DNA- and protein sequencing, crystallography, and radioimmunoassay allowed cloning of the genes encoding insulin receptor and insulin receptor substrate proteins as well as their molecular actions were identified.

Insulin Receptor

Insulin receptor belongs to the receptor tyrosine kinase (RTK) protein-family. It is made up of glycoproteins consisting of an extracellular α-subunit and a transmembrane β-subunit. Both α- and β-subunits act as an allosteric enzyme; however, α-subunit inhibits tyrosine kinase activity of the β-subunit. Once insulin comes in contact with the cell membrane and binds to the α-subunit, it results in receptor dimerization to form the $\alpha2\beta2$ receptor complex. Soon after the receptor dimerization, the receptor tyrosine kinase (RTK) activation occurs at different tyrosine sites (Tyr1158, Try1162, and Tyr1163) of the β-subunit. The tyrosine kinase activation recruits and phosphorylates several substrates including the insulin receptor substrate (IRS) proteins: IRS1-4, Shc, Grb-2-associated protein (Gab1), Dock1, Cbl, and APS adaptor proteins in the inner side of the cell membrane. All these molecules act like adaptor proteins and provide specific docking sites for recruitment of many downstream signaling proteins, leading to activation of both Ras/MAP kinases and PI3K/Akt signaling cascades [41]. On the other hand, both the subunits of insulin receptor and insulin-like growth factor-1 receptor (IGF-1R) form heterodimers (IR/IGF1R), activating downstream signaling molecules in the cellular compartments.

On the one hand, insulin receptor forms a hybrid complex with Met, a transmembrane tyrosine kinase cell surface receptor for hepatocyte growth factor (HGF) and it is structurally related to the insulin receptor [42]. On the other hand, the IR/Met hybrid complex results in strong signal output by activating insulin receptor-regulated downstream signaling cascades and mediating the metabolic effects of insulin [42]. Studies have demonstrated in animal models that IRS protein and the docking proteins for insulin receptor provide interfaces that insulin, IGF-1, or HGF signaling propagate and engage similar intracellular signaling components. Moreover, insulin activates the Ras/MAP kinases pathways through insulin receptor, which mediates the effect on mitogenesis and cellular growth/proliferation. Similarly, insulin activates PI3Kinase and generates phosphatidylinositol (3, 4, 5)-triphosphate (PIP3), a second messenger activating 3-phosphoinositide-dependent protein kinase-1 and 3-phosphoinositide-dependent protein kinase-2 (PDK1 and PDK2), which in turn phosphorylate the protein kinase Akt (T308 and S473)/mTOR and ribosomal S6Kinase that mediate the effect of IRS1/2 on aging (Fig. 17.7).

Paradox of Insulin Resistance and Aging

In humans, insulin resistance (IR) accompanied by diabetes has clearly been implicated as a major risk factor for multiple age-related diseases. In rodents, improved longevity and multiple features of delayed aging have been described, where insulin sensitivity is increased by genetic mutation [43]. Genetic manipulations of laboratory animals for the expression of key IIS pathway are associated with extension of lifespan. Interestingly, there is a substantial difference between aging of lower invertebrate groups and mammalian aging. However, there are enough evidences in different model organisms to understand the role of attenuated IIS signaling in the mechanisms of aging and longevity that is not limited to the lower taxa, but is evolutionarily conserved up to the mammals. In humans, decreasing IR with pharmacological drugs/inhibitors is a major strategy to relieve demands on pancreatic β-cells; however, it may lead to damage of other tissues with many side effects. Furthermore, different research reports demonstrated that IR has great potential for antioxidant defense mechanism and it reduces the enhanced stress and may contribute to increased longevity of invertebrates with reduced IIS and may help explain the extension of longevity observed in some genetic mutant mice with IR. Consequently, pharmacological application regulates the IR action in humans that may help to minimize the hazards or complications associated with diabetes.

Insulin Resistance and Metabolic Diseases

After taking a meal, insulin secretion starts from the pancreatic β-cells to control the systemic nutrient homeostasis by promoting anabolic processes in a variety of tissues. Insulin has diverse functions in different tissues and cell types including stimulating glucose influx into muscle and fat, protein and glycogen synthesis in the muscle and liver, and lipid synthesis and storage in the liver and adipose tissue, while

Fig. 17.7 Insulin action via insulin receptor substrate 1/2 (IRS1/2), which activates the PI3K/Akt/PDKs/mTOR/S6K kinase pathways via a feedback loop in the fat, muscle, and liver cells. The mTOR/S6K kinase pathway in turn inactivates IRS1/2, thus causing IR and leading to aging

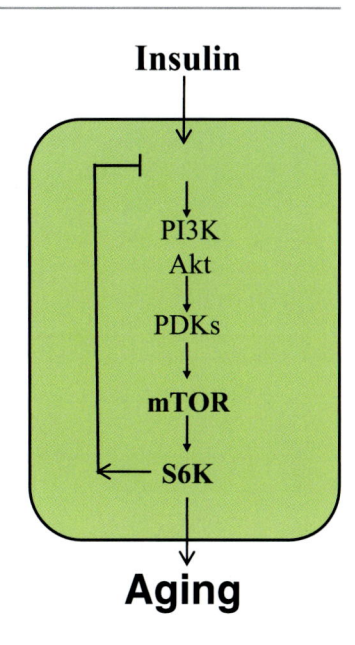

it inhibits fatty acid oxidation, glycogenolysis, and gluconeogenesis, as well as apoptosis and autophagy in insulin-responsive tissues. On the other hand, during the fasting state, insulin secretion decreases, and tissues coordinate with counter-regulatory hormones, glucagon in the liver and fat, in favor of using fatty acids largely derived from adipocyte lipolysis for energy generation and maintenance of glucose homeostasis throughout the body. It is the tough time for the cells for the metabolic adaptations, during the transit from fasting to the postprandial state, they are tightly controlled by insulin under physiological conditions. This adaptive transition reflects the action of insulin and insulin-responsive organs, while it is largely blunted in organs with insulin resistance preceding the development of T2DM [44].

IRS1 and IRS2 in Metabolic Diseases

Different genetic manipulation experiments were carried out in different animal models to better understand the role of insulin receptors, IRS1 and IRS2, in controlling growth and nutrient homeostasis [45]. Earlier studies reported that mice lacking the insulin receptor genes were born with slight growth retardation, but rapidly developed MS-like hyperglycemia and hyperinsulinemia, followed by diabetic ketoacidosis and early postnatal death [46]. Experiments showed that both IRS1- and IRS2-null mice were embryonic lethal indicating that they are essential for life in these animals [47]. Interestingly, a previous study by Araki et al. [48] showed that systemic IRS1-null mice displayed growth retardation and peripheral resistance to insulin and IGF-1, mainly in skeletal muscle, but avoided developing diabetes because of IRS2-dependent pancreatic β-cell growth and compensatory insulin secretion in these mice. Similarly, a study by Withers et al. [49] demonstrated that systemic IRS2-null mice displayed metabolic defects in the liver, muscle, and adipose tissues, but developed diabetes secondary to pancreatic β-cell failure. Further, many recent studies highlighted tissue-specific gene manipulation in mice, providing new insights into the action of the insulin receptor and control of glucose homeostasis and body weight [50]. Experiments showed that mice lacking insulin receptor in the liver, pancreatic β-cells, adipose tissue, or brain developed MS and its associated diseases like hyperglycemia, hyperlipidemia, hyperinsulinemia, and obesity [51]. Moreover, reconstitution of insulin receptor in the liver, β-cells, and brain rescued from diabetes in the mice lacking the insulin receptor and prevented the premature postnatal death [52], suggesting that the liver, pancreatic β-cells, and brain are crucial for the maintenance of glucose homeostasis by insulin. Many studies also demonstrated in rodent models that deletion of IRS1 and IRS2 genes in the liver and cardiac muscle prevented activation of hepatic Akt/Foxo1 phosphorylation and resulted in the development of hyperglycemia, hyperinsulinemia, insulin resistance, and hypolipidemia [53, 54] (Table 17.2).

IRS1/IRS2 in PI3K/Akt/MAP Kinase-Regulated Metabolic Diseases

The IRS1/IRS2 are strongly associated with PI3K/Akt activation and moderately associated with MAP kinase activity; however, deletion of IRS1 and IRS2 causes subjective PI3K inactivation and sustained MAP kinase activation in different specific organs such as the liver and heart of the mice [53, 54]. In different animal models, the differential PI3K inactivation and MAPK activation by the loss of IRS1/IRS2 may provide a fundamental mechanism that illuminates the prevalence of insulin resistance and its association with metabolic diseases like T2DM, obesity, and cardiovascular dysfunction.

PI3K/Akt/Foxo1 in Metabolic Diseases

The studies in animal models and humans also showed that activation of PI3K/Akt pathway plays a central role in metabolic regulation. Hepatic inactivation of very specific genes like *PI3K*, *PDK1*, *Akt*, and *mTORC2* is sufficient for the induction of metabolic diseases like hyperglycemia, hyperinsulinemia, and hypolipidemia [55]. Specifically, mice lacking Akt2 developed T2DM [56], and patients with mutation of Akt2 have T2DM [57]. Many studies have revealed that expression of constitutively active Foxo1 and Akt with specific sites mutated to alanine, blocking phosphorylation either in the liver causing insulin resistance or in the heart resulted in embryonic lethality in mice [58]. However, Foxo1 inactivation either in the liver of mice with T2DM resulted in reversal of hyperglycemia [55], or in the heart of these animals with T2DM prevented heart failure [59]. This suggests that Foxo1 activation was both sufficient and necessary for the induction of hyperglycemia and organ failure following insulin resistance or T2DM in these experimental mice.

Insulin Resistance by Hyperinsulinemia

Several lines of evidences from recent research demonstrated that hyperinsulinemia has profound effects in inducing insulin resistance as given below:

1. *In myocardium and adipocytes:* prolonged insulin treatment was sufficient for preventing acute insulin action on Foxo1 phosphorylation or Glut4 cellular membrane trafficking [54].
2. *In liver:* insulin inhibited IRS2 gene transcription and promoted IRS2 ubiquitination/degradation in murine embryonic fibroblasts [60, 61].
3. *In liver:* insulin stimulated mTORC1 leading to IRS2 ubiquitination and the mTORC1 inhibitor rapamycin completely prevented insulin or IGF-1-induced IRS2 degradation [60, 61].
4. *In liver:* genetic knockout of hepatic S6K, a downstream target of mTORC1, reduced insulin resistance, enhancing IRS1/IRS2 gene expression and preventing diabetes in mice [62].
5. *In liver:* genetic knockout of mTORC2 in mice resulted in a diabetic phenotype and long-term treatment with rapamycin blocked mTORC2-mediated Akt phosphorylation/activation [63].
6. *In skeletal muscle and liver:* hyperinsulinemic treatment induced insulin resistance and was associated with oxidative stress and mitochondrial dysfunction in T1DM mice [64].
7. *IRS1/IRS2 expression:* their decreased expression was observed in tissues of animals and patients with hyperinsulinemia or T2DM [54].
8. *MAP kinases:* it induced IRS serine/threonine phosphorylation and degradation, particularly when animals were fed with a high-fat diet.
9. *p38α MAP kinase activation:* prolonged insulin treatment in cardiomyocytes mediated insulin resistance by increasing IRS1/IRS2 phosphorylation and degradation [54].
10. *p38 MAP kinase:* it mediated induction of inflammatory cytokines that promoted insulin resistance [65, 66].
11. *PKC isoforms:* PKCδ/PCKθ has an important role in inducing IRS serine/threonine phosphorylation, resulting in insulin resistance in tissues following administration of high-fat diet [67].
12. *Protein kinases:* many protein kinases were found to be associated, it will be important to identify in mouse- or human genome sequences, and activation mechanisms under different metabolic disease conditions for induction of IRS serine/threonine phosphorylation and inactivation of insulin signaling.

Insulin Resistance and Foxo1 Activation

In normal physiological conditions, Akt phosphorylates Foxo1, and promotes Foxo1 cytoplasmic retention by ubiquitination process. However, in metabolic disease like hyperglycemia and insulin resistance, Akt dephosphorylated the nuclear

Foxo1-S253, which was detected in liver and heart of animals with T2DM [68]. Genetic knockout of Foxo1 in the liver (L-DKO mice and db/db mice) reduced hepatic glucose production and ameliorated diabetes [53] and in the heart (High Fat Diet-fed mice) prevented heart failure [59, 68]. These animal experiments suggested that IRS/Akt/Foxo1 signaling cascades are critical in the nutrient homeostasis and organ survival in rodents. Furthermore, aberrant Foxo1 activation in laboratory animals disrupts metabolic homeostasis and promotes organ failure, by regulating expression of a number of target genes. There is a separate chapter on nutrient sensing and aging.

Insulin Resistance and Metabolic Diseases

Insulin Resistance in CNS and Obesity

The hypothalamus of the mammalian forebrain regulates circadian rhythm of the normal health. A decade of research has provided enough insights into diabetes-related cognitive dysfunction in CNS/brain. However, the biochemical and molecular mechanisms operating under such conditions are unclear about how the metabolic perturbations of diabetes correlate to the hypothesized anatomic substrates in different major areas of the brain, associated with diabetic neuropathy and the imbalance in energy homeostasis (Fig. 17.8). In turn, the contribution to cognitive changes from the pathological lesions in the brain is not fully clear. Basically, appetite is tightly controlled by insulin action in the CNS. Studies have demonstrated that genetic manipulations of neuron-specific insulin receptor resulted in overweight and insulin resistance in the mice. A low dose of insulin delivery in these mice by the intracerebroventricular infusion decreased both food intake and hepatic glucose production, which are blocked by PI3Kinase inhibitors. To understand the functional significance of the brain insulin signaling, studies on laboratory animals have reported that knockout of IRS2 in the hypothalamus, resulted in hyperglycemia and obesity [69]; however, deletion of IRS1 in the hypothalamus did not respond in

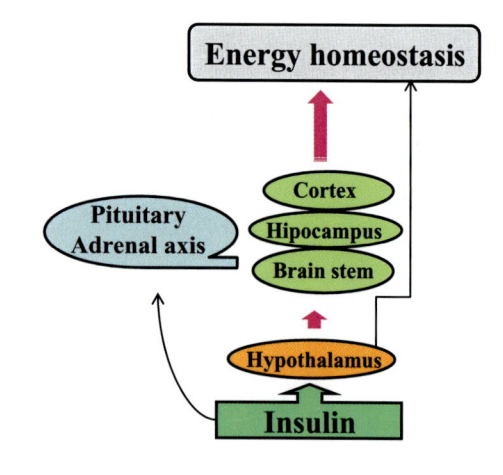

Fig 17.8 Schematic representation of insulin-dependent regulation of energy homeostasis by interaction of different sensitive areas of the brain and pituitary-adrenal axis

the young mice. Apart from insulin, leptin, an adipocyte-derived hormone, also inhibits food intake through CNS. Leptin receptors in neurons are activating many signaling cascades. In the brain, the leptin and insulin, both promote IRS2 tyrosine phosphorylation and PI3K activation. The deletion of IRS2 in leptin receptor-expressing neurons caused metabolic disease like diabetes and obesity, in which Foxo1 inactivation completely reversed the metabolic dysfunction [70].

Insulin Resistance in Adipose Tissue and Inflammation

Like other metabolic diseases, hyperlipidemia is a key feature of the MS, that results from insulin resistance in adipose tissue. In mammals, insulin promotes fat cell differentiation, enhances adipocyte glucose uptake, and inhibits adipocyte lipolysis. Studies have demonstrated that deletion of *TORC2* in mice exhibited hyperglycemia, hyperinsulinemia, failure to suppress lipolysis in response to insulin, elevated circulating fatty acids and glycerol, and insulin resistance in skeletal muscle and liver. Recent studies by Boucher and Kahn [71] in mice lacking insulin receptor in adipose tissue, created by introducing the adiponectin promoter-driven Cre/loxP system, developed severe MS, including lipoatrophic diabetes, hyperglycemia, hyperinsulinemia, hyperlipidemia, and liver steatosis. Therefore, when insulin action fails in fat cells of mammals, adipocyte development is retarded and lipids are unable to get converted from carbohydrates, for storage. Consequently, glucose and lipids are redistributed to different organs of the body through blood circulation, resulting in hyperlipidemia and fatty organs. Apart from normal function, adipose tissue is also playing a key role to secrete hormones and cytokines, including TNF-α, IL-6, leptin, adiponectin, and many others, influencing food intake, systemic insulin sensitivity, and nutrient homeostasis. On the other hand, obesity from fat-expansion disrupts a proper balance of cytokine and hormone generation, promoting insulin resistance. In obesity, cytokines like TNF-α, IL-6, and leptin are markedly increased, whereas adiponectin is significantly reduced, that has anti-inflammatory effect on enhancing insulin sensitivity [72]. Overexpression of IKK-beta in the liver tissue of mice is sufficient for inducing insulin resistance and diabetes. TNF-α reduces IRS1 protein by activation of JNK or S6K, resulting in insulin resistance. Therefore, suppression of inflammatory mediators, increases the insulin sensitivity and improves MS in T2DM. However, the outcome of anti-inflammatory therapy in treating insulin resistance deserves a cautionary note for several reasons as follows:

1. Inflammation is involved in deploying and mobilizing immune cells, leukocytes to defend against infections or toxins.
2. Many inflammatory mediators reduce body weight and increase energy expenditure. For example, overexpression of IL-6 in the liver of mice increased energy expenditure and insulin sensitivity [73].
3. During physical exercise, inflammatory mediators are secreted resulting in inhibition of anabolic metabolism, such as insulin action, and promoting catabolic metabolism, such as fat lipolysis, to meet fuel requirements for muscle.

4. NF-κB is essential for hepatocyte proliferation and survival, and gene-knock out mice lacking the p65 subunit of NF-κB died of liver failure during embryonic development [74].
5. Inflammation not only triggers pro-inflammatory responses but also activates anti-inflammatory processes.

Together, the above findings suggest that a balance between inflammation and anti-inflammation is required for proper insulin action and nutrient homeostasis under normal physiological conditions. Therefore, correcting the imbalance of hormones, nutrients, and inflammation may provide opportunities and challenges for the prevention and treatment of MS-like diabetes.

Insulin Resistance in Liver and Hyperglycemia

In the human body, insulin secretion from pancreas regulates macromolecules' synthesis in the liver and suppresses hepatic glucose production by inhibiting gluconeogenesis; however, insulin resistance severely affects the liver by causing hyperglycemia. Animal studies revealed that genetic manipulation of either *IRS1* or *IRS2* specifically in liver tissues, somehow maintained glucose homeostasis. On the other hand, deletion of both *IRS1* and *IRS2* in L-DKO (liver double knock-out) mice blocked insulin or feeding-induction upon Akt and Foxo1 phosphorylation. Finally, the signaling cascades regulation in this mice resulted in unrestrained gluconeogenesis for hepatic glucose production, and diabetes, with a reduction in hepatic lipogenesis and blood lipids [75]. Other studies highlighted that a high-fat diet ingestion severely impaired IRS2 expression and tyrosine phosphorylation in hepatocytes of liver-specific IRS1 null-mice and the mice developed severe diabetes [75]. At the same time, overnutrition or a high-fat diet ingestion in mice can modify intracellular signaling cascades, affecting IRS2 expression and its functionality, thus altering metabolic gene expression, and impairing glucose homeostasis. Hepatic insulin resistance in the L-DKO mice also affects, directly the insulin resistance in other delicate organs. Guo et al. [75] showed that the L-DKO mice not only demonstrated inhibition of the hepatic Akt signaling cascade but also blunted brain intracerebroventricular (ICV) insulin action on reducing the hepatic glucose production in the ICV-clamp experiments. More studies revealed insulin resistance to be involved in cardiac complications as follows:

(a) Mice (L-DKO) exhibited features of heart-failure, likely to be secondary to hyperinsulinemia, resulting due to cardiac IRS1 and IRS2 suppression [54].
(b) Mice lacking hepatic insulin receptor displayed pro-atherogenic lipoprotein profiles with reduced high-density lipoprotein cholesterol and very low-density lipoprotein particles and, within 12 weeks of being placed on an atherogenic diet, they developed severe hypercholesterolemia.

From the above observations, it is very clear that hepatic insulin resistance is sufficient to produce dyslipidemia and increased risk of atherosclerosis and cardiac dysfunction.

Insulin Resistance in Cardiac Tissues and Heart Failure

Loss of IRS1 and IRS2 acts differently depending upon the organs or tissues, for example, in the liver and brain, it resulted in hyperglycemia, whereas in the heart and pancreas, it resulted in organ failure. Like brain, the heart is an insulin-responsive and energy consuming organ and it requires a constant metabolic-fuel supply to maintain intracellular energy for proper myocardial function. However, gene manipulation of cardiac IRS1 and IRS2 in mice diminished cardiac Akt and Foxo1 phosphorylation and resulted in heart failure and death of these animals [54]. Similarly, deletion of both IRS1 and IRS2 in skeletal and cardiac muscle of mice caused heart failure. Also diminished Akt and Foxo1 phosphorylation in the skeletal muscle of the mice had normal blood glucose and insulin sensitivity, suggesting that insulin resistance in skeletal muscle is not necessary for a disruption of glucose homeostasis in the mice. To promote proper cardiac function and longevity, cardiac muscle requires either IRS1 or IRS2 for the maintenance of endogenous Akt activity and Foxo1 inactivation. In mice, cardiac overexpression of Foxo1 resulted in heart failure [58], and also clinical observation showed this in heart failure in human subjects. Moreover, deletion of both *IRS1* and *IRS2*, following chronic insulin stimulation and p38 MAK activation, contributes to insulin resistance in the heart [54]. Based on the recent studies, IRS1 and IRS2 have major roles in control of cardiac homeostasis, metabolism, and function. This concept was based on the following three-phase observations:

1. *Phase I:* Insulin stimulates glucose transport and oxidation, ensuring effective cardiac utilization of glucose as a substrate for the supply of energy, reduction of IRS2 protein was observed in the mouse liver and heart, compared to those mice in the fasting state [75].
2. *Phase II:* When insulin resistance occurs, the heart undergoes adaptive responses to limit glucose utilization by insulin-dependent pathway and respond to lipid oxidation by less insulin-dependent pathway. At that situation, the heart is capable of generating energy for myocardial contraction and changes in gene expression patterns, with unaltered cardiac morphology. During this crucial period, the metabolic adaptation and remodeling compensate for cardiac energy demand, even without unconcealed indications of heart failure.
3. *Phase III:* When maladaptive metabolic remodeling occurs in H-DKO mice, there is a lack of compensation for cardiac energy demand, secondary to loss of IRS1 and IRS2, with Akt inactivation, utilization of both glucose and fatty acids being restrained, resulting in hyperlipidemia and cardiac energy deficiency and sudden death [54]. Here, the affected heart may exhibit a loss of mitochondrial biogenesis and functions, a process required for fatty acid and glucose utilization via mitochondrial oxidative phosphorylation.

Insulin Resistance in Pancreas and β-Cell Regeneration

Pancreatic β-cell is an important lifeline for human health, and failure of its true function make life difficult with the development of diabetes. Pancreas has different types of cells and has different physiological roles for these cells individually, such as β-cells secrete insulin reducing blood glucose and α-cells secrete glucagon increasing the blood glucose level to meet the metabolic requirements of the body. In metabolic process, insulin secretion is regulated by glucose availability after a meal that enhances glucose-stimulated insulin secretion in healthy humans [76], and at the same time, studies on mice show that insulin receptor-lacking β-cells have impaired insulin secretion. Many signal transduction studies reported that genetic manipulation of *IRS2* in mice resulted in pancreatic β-cell failure and diabetes, whereas Foxo1 inactivation in *IRS2*-null mice prevented loss of β-cells and diabetes, also [77]. Therefore, these studies concluded that IRS2/Foxo1 signaling or Foxo1-inactivation is mainly required for the β-cells' survival. Additionally, *IRS2* deletion in β-cells triggered β-cell repopulation or regeneration in aged mice, leading to a restoration of insulin secretion and resolution of diabetes. As a result, Foxo1-activation following IRS2-inactivation in β-cells promoted β-cell regeneration or differentiation. On the other hand, many studies demonstrated that Foxo1-inactivation in β-cells resulted in reduced β-cell mass, hyperglycemia, and hyperglucagonemia, owing to de-differentiation of β-cells into progenitor-like cells or pancreatic α-cells [78]. Thus, to prevent the development of diabetes, we have to be very careful to keep our pancreatic cells safe and any abnormality at the cellular level, may result in the imbalance of the hormones, insulin and glucagon, and this may provide a potential strategy to prevent diabetes.

Insulin Resistance in Skeletal Muscle Tissue and Shortened Lifespan

The skeletal muscle takes an advantage for its remarkable metabolic flexibility to consume and store glucose and converts glycogen and triglycerides, a process stimulated by insulin. In animal studies, it revealed that muscular insulin receptor-lacking led to the display of elevated fat mass, serum triglycerides, and free fatty acids. However, at the same time, blood glucose, serum insulin, and glucose tolerance had no effect. Therefore, people suffering from diabetes have insulin resistance in muscle tissues, that contributes to the altered fat metabolism, while tissues other than muscle in the same individual appear to be more involved in insulin-regulated glucose disposal. Studies on laboratory animals showed that mice lacking mTORC2 exhibited decreased insulin-stimulated Akt phosphorylation and increased glucose uptake and mild glucose intolerance, whereas mice lacking mTORC1 developed many physiological complications including muscular dystrophy, mild glucose intolerance, and shortened lifespan. Insulin action in skeletal muscle has a key role in the control of lifespan, studies have reported that *IRS1* and *IRS2* both knockout mice in skeletal and cardiac muscles, died at 3 weeks of age, with a much shorter lifespan than mice lacking both IRS1 and IRS2 in only cardiac muscle, the latter died at 7 weeks of age [54]. At the same time, in the same animals, absence of both IRS1 and IRS2 in the skeletal and cardiac muscles did not develop hyperglycemia or hyperinsulinemia, although the insulin-induced glucose uptake was diminished. Another molecule, AMP takes major role in skeletal muscle, its activation causes activation of AMP-dependent protein kinase (AMPK) and stimulates glucose uptake in an insulin-independent manner, it occurs by phosphorylating and activating the Rab-GAP family member AS160, which promotes Glut4 translocation [79]. Also in cellular compartments, AMPK induces acetyl-CoA carboxylase (ACC) phosphorylation and inhibits ACC activity, thus preventing the conversion of acetyl-CoA to malonyl-CoA, and disrupting lipid synthesis and enhancing fatty acid oxidation [80]. Collectively, skeletal muscle takes the plasticity of controlling glucose homeostasis and longevity by regulating different signaling molecules. Moreover, skeletal muscle actively secretes many chemical messengers, specifically irisin, a hormone that systemically regulates glucose homeostasis and MS-like obesity; it would be of interest to determine if a skeletal muscle-derived hormone affects longevity in animals. There is a separate chapter on the muscle and aging.

Insulin Resistance in Vascular Endothelium and Glucose Homeostasis

Insulin resistance also had great effects on vascular endothelium (VE) in people who are suffering from diabetes. Basically, MS like insulin resistance stimulates many severe complications in VE such as vasoconstriction that promotes hypertension, atherosclerosis, impairs systemic insulin sensitivity, and glucose homeostasis. Studies in animal models revealed that inactivation of insulin receptor in VE diminished insulin-induced eNOS expression and blunted aortic vasorelaxant responses to acetylcholine and calcium ionophore in normal mice, and accelerated atherosclerosis in apolipoprotein E-null mice [81]. Insulin acts as a vasodilator that mediates PI3K-dependent signaling pathways and stimulates nitric oxide production in VE. Specific knockout of *IRS2* or both *IRS1* and *IRS2* in mice VE led to reduced endothelial Akt and eNOS activation and impaired skeletal muscle glucose uptake and systemic insulin resistance [82]. Also, Foxo activation plays a key role in stimulating endothelial cell dysfunction in genetic manipulation of *IRS2* or both *IRS1* and *IRS2* in VE. However, deletion of *Foxo1*, *Foxo3*, and *Foxo4* in the endothelium of mice showed different pathophysiological changes including enhanced eNOS activation, reduced inflammation, and oxidative stress in the endothelial cells and prevented atherosclerosis in high-fat diet administered or low-density lipoprotein (LDL) receptor-null mice [83]. In contrast, endothelium-targeted deletion of insulin receptor or *Foxo* genes in rodents barely disrupted glucose homeostasis. However, studies also showed that endothelium-targeted deletion of the transcription factor related transcriptional enhancer factor-1 (*RTEF-1*) increased the blood glucose levels and insulin resistance and RTEF-1 has the potential for interaction with the insulin response element and Foxo1 in these cells [84]. Collectively, from the above studies, it shows that VE serves as an organ that potentially regulates glucose homeostasis.

Insulin Resistance in Bone and Glucose Homeostasis

Insulin plays a major role in the formation of bone and differentiation of osteoblasts that synthesize osteocalcin, which regulates pancreatic insulin secretion and systemically regulates glucose homeostasis. However, mice lacking insulin receptor in the osteoblasts exhibited diverse abnormal pathophysiological functions including reduced bone formation, increased peripheral adiposity, and insulin resistance, primarily due to reduced gene expression and activity of osteocalcin [85]. Moreover, insulin may stimulate osteocalcin in osteoblasts by suppressing Foxo1, which affects bone remodeling and control of glucose homeostasis. In rodents, Foxo1 inhibits osteocalcin expression and activity by increasing expression of Esp, a tyrosine phosphatase protein. On the other hand, osteoblast-specific *Foxo1*-null mice have increased levels of osteocalcin expression and insulin production and reduced levels of blood glucose [86]. Collectively, these studies suggest that bone serves as an endocrine organ in the control of glucose homeostasis, through bone-pancreas crosstalk, in which Foxo1 plays a key role in insulin action, regulating osteocalcin expression and activity in the osteoblasts.

Neurohormonal Regulation

Renin-Angiotensin-Aldosterone System (RAAS)

Metabolic diseases regulate RAAS activation in different organs or tissues specifically critical for cardiovascular diseases, including hypertension, coronary heart disease, and congestive heart failure, as well as artrial fibrillation. In fact, angiotensin II (Ang II) plays a critical role in cardiomyocyte hypertrophy and apoptosis, it increases cardiac fibrosis, and impairs cardiomyocyte relaxation [87]. Studies have shown that cardiac Ang II concentrations increase significantly in the aged rodent hearts, probably it is related to the increased tissue levels of Ang II-converting enzyme (ACE). Although the mechanism of increased ACE in MS and its impact on heart are not well understood, long-term inhibition with angiotensin receptor blockers and disruption of angiotensin receptor type I have been shown to reduce age-dependent cardiac pathology and prolong the animals' survival.

Adrenergic Signaling

Chronic activation of adrenergic signaling is very critical for different organs or tissues in persons, who are struggling in life with metabolic diseases. Basically, activation of adrenergic signaling enhances metabolic demand, specifically to increase the heart rate, contractility, blood pressure, and wall stress, and inhibition of β-adrenergic signaling by β-blockers provides survival benefit in patients with heart failure. Moreover, metabolic diseases stimulate β-adrenergic receptors by adenylate cyclase type 5 (AC5), which is the major form in the heart. Studies have shown that disruption of *AC5* was shown to have prolonged lifespan in the mice, and that was mediated through Raf-1/pMEK/pERK pathway, which confers protection against cellular stress, including oxidative stress, decreased levels of circulating growth hormone (GH), and cardiac aging, including age-dependent cardiac hypertrophy, systolic dysfunction, apoptosis, and fibrosis [88].

Growth Hormone Signaling

Hormonal signaling pathways are extremely potent regulators of human health and lifespan, but metabolic diseases play different roles in human body systems, because they coordinate the longevity of several delicate organs by acting in a systematic manner. Metabolic actions of GH, secreted from the pituitary gland, are diverse and tissue-specific, that regulate carbohydrate and lipid metabolism via complex interactions with insulin and IGF-1. Experimental mouse models showed that the extension of longevity was likely to be due to defects in the ability of the pituitary gland to secrete GH in these animals. A null mutation in the GH receptor (*GHR–/–*) displayed increased in lifespan of the mice [89]; however, GH-overexpressing transgenic mice showed a longevity shorter than wild-type mice. The *Snell* and *Ames dwarf* (genes encoding the pituitary transcription factors PIT1 and PROP1) mutations in the mice extended their lifespan. Few studies have demonstrated that the *Ames dwarf* mice and the *GHR–/–* mice have reduced levels of circulating IGF-1, fasting insulin, and glucose, raising the possibility that the increased longevity of these mice was mediated by the insulin/IGF1 signaling.

Klotho Signaling

Like other hormones and their actions, klotho, is a cell surface protein, whose extracellular domain can act as a circulating hormone [90]. Interestingly, disruption of the expression of *klotho* accelerated aging in mice. However, overexpression of klotho showed extension of lifespan in the mice, but klotho has been found to repress insulin/IGF1 signaling and to regulate phosphate and calcium homeostasis by affecting fibroblast growth factor 23 (FGF23) and the Na+/K+-ATPase.

Steroid Signaling

In flies, the steroid hormone plays interesting role in lifespan, this was known by mutation of the insulin-like receptor. However, the effects of steroids or steroid receptors on overall lifespan in mammals are not clear, but the steroid dehydroepiandrosterone sulfate has been found to be associated with increased lifespan in many primates and humans [91].

Akt/mTOR Signaling and Metabolic Diseases

The mammalian target of rapamycin (mTOR) is a highly conserved cytoplasmic kinase that regulates cell growth and metabolism in response to mitogens, nutrients, growth factors, and hormones including insulin and cytokines. Basically, mTOR pathway is essential for physiological functions of the young individuals; however, in older age, mTOR plays differently, driving cellular quiescence into senescence. In particular, senescent cells show abnormal behaviors, such as hyperfunctional, hypersecretory, pro-inflammatory, and insulin resistant properties leading to age-related diseases. More importantly, both glucose and insulin activate mTOR. There are two distinct complexes called complex 1 (mTORC1) and complex 2 (mTORC2) of the evolutionarily conserved mTOR protein. In metabolic disease like diabetes, mTORC gets phosphorylated and activated through the MAP kinases, Akt and PKC, controlling cell survival and energy homeostasis [92]. Of the two complexes, specifically mTORC1 interacts with the adaptor protein and is activated by RhebGTPase, via suppression of TSC2, following Akt-activation. Recent research publications show that mTORC1 promotes lipogenesis via phosphorylating a phosphatidic acid phosphatase, Lipin 1 and nuclear translocation of Lipin 1 stimulates Srebp1c and lipogenesis. On the other hand, Akt activates mTORC2, which promotes expression and activation of the Srebp1 transcription factor, leading to lipid and cholesterol synthesis [93]. Hagiwara et al. [92] demonstrated that mTORC2 and PDK1 suppress the Foxo1, that promotes gluconeogenesis, mediating the effect of insulin on suppression of the hepatic glucose production. Most importantly, Akt regulates metabolism and survival by controlling expression of a number of genes through transcription factors, of which Foxo1 is the most critical one, and Akt/Foxo1 phosphorylation serves as powerful indicators for insulin sensitivity on metabolic regulation in a variety of cells and tissues. Abnormal growth of small blood capillaries in the retinal part of the eye contributes to retinopathy. For this, in retinal pigment epithelial cells, vascular endothelial growth factor (VEGF) takes the priority and stimulates angiogenesis via the insulin/mTOR pathway leading to blood-retinal barrier breakdown. Similarly, studies on laboratory animals showed that rapamycin decreased renal hypertrophy in diabetic mice and slowed down the progression of diabetic kidney disease in the rats. Activation of mTOR regulates pancreatic β-cell mass. However, hyperfunction of mTOR due to diabetes, eventually causes β-cell failure. In early diabetic state, mTOR stimulates β-cell functions, causing hyperfunction and slowly the chronic hyperstimulation of mTOR renders the β-cells resistant to IGF-1 and insulin, fostering cell death.

Mitochondria and ROS Signaling

Among all the cell organelles, mitochondria act as master organelle in the regulation of cellular energy levels, ROS production, and cell survival. However, they are equally important in overproduction of mitochondrial ROS, by metabolic risk factors determining aging and lifespan of cells in metabolic disease conditions (Fig. 17.9). All the mitochondria-related mechanisms are deregulated by MS, so that the cells forcefully enter into senescence. Genetic studies in worms revealed

Metabolic Risk Factors
↓
Mitochondrial ROS↑

| Insulin resistance | Visceral Obesity | Hypertension | Hyperlipidomia | Hyperglycaemia |

Metabolic syndrome

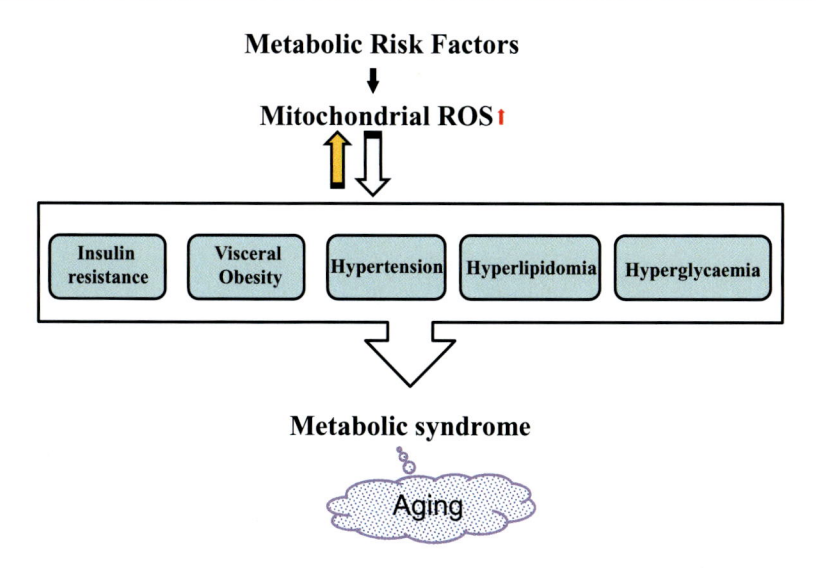

Aging

Fig. 17.9 Schematic relationship between the metabolic risk factors and the mitochondrial ROS-production, which accelerates metabolic syndrome and aging

that few specific genes of the electron transport chain (ETC) in mitochondria directly control lifespan of organisms, and specifically mutations in Clk-1, which encodes a protein required for the biosynthesis of ubiquinone, an essential cofactor in the ETC, extended the worm's lifespan and that was independent of insulin signaling. On the other hand, lack of one allele of the *Clk-1* gene improved the lifespan in mice, which suggests that ubiquinone plays a conserved role in lifespan regulation. Feng et al. [94] demonstrated that mutation of the iron-sulfur protein (ISP-1) of the mitochondrial complex III increased the worm's lifespan. Moreover, many recent studies in RNAi-directed experiments against several other components of the ETC, including the *nuo-2*, NADH/ubiquinone oxidoreductase; *cyc-1*, cytochrome c reductase; and *cco-1*, cytochrome c oxidase, as well as against the mitochondrial ATP synthase (*atp-3*), increased the worm's lifespan [95]. Another molecule, p66shc is present within the mitochondrial intermembrane space and oxidizes cytochrome *c*, thereby generating ROS, and deletion of the *p66shc* gene caused increase in the mouse lifespan compared with its wild-type littermates. In the aging process related to MS, mitochondrial DNA damage, mitochondrial DNA point mutations, and decrease in the mitochondrial DNA-copy number increased manyfolds the metabolic disease in animals. In addition, protein-carbonyls and lipid-peroxidation in the mitochondria, indicative of the mitochondrial protein- and lipid-oxidative damages, significantly increased in the metabolic disease in such individuals. Collectively, these studies suggest that reducing the energy production and increasing the production of ROS associated with the electron transfer and other mitochondrial complications are crucial for longevity, which are also critical for patients who are suffering from MS. There is a separate chapter on mitochondria and aging.

Genome Surveillance Pathways

DNA Repair and Telomere Pathways

Healthy life is maintained when repair of nuclear and mitochondrial DNA lesions are properly carried out by efficient ways in the normal lifespan of an organism. Unfortunately, mutations in a number of DNA repair genes due to MS cause early aging. For human subjects, these mutations in the DNA repair genes and in a gene encoding a protein involved in the nuclear architecture are responsible for the physiological deficiencies in the early age known as the progeroid syndromes [96]. Apart from the typical sequences of genes in DNA, telomere maintenance mechanisms in the chromosomes are equally important for a normal lifespan of cells and tissues in all eukaryotic organisms. Basically, telomeres are tandem repeats of a DNA sequence rich in G bases, bound by a six-protein complex known as shelterin present at the ends of chromosomes (Fig. 17.10). The length of the telomeric repeats in the DNA sequence and the integrity of the telomere-binding protein complex are both essential for the chromosome end-protection and genomic stability in any type of cells under normal and metabolic disease conditions. In addition, telomere length and integrity are regulated by a number of epigenetic modifications, leading to the higher order control of the telomere functions. Genetic manipulation of the telomerase enzyme gene, i.e., in the *Terc* knockout mice, demonstrated that the telomerase was required for the telomere maintenance in mammals, as well as its importance for metabolic disease, like diabetes. In a normal healthy chromosome, telomere synthesis requires two important molecules including telomerase reverse transcriptase (TERT) and telomerase RNA component (TERC). However, research

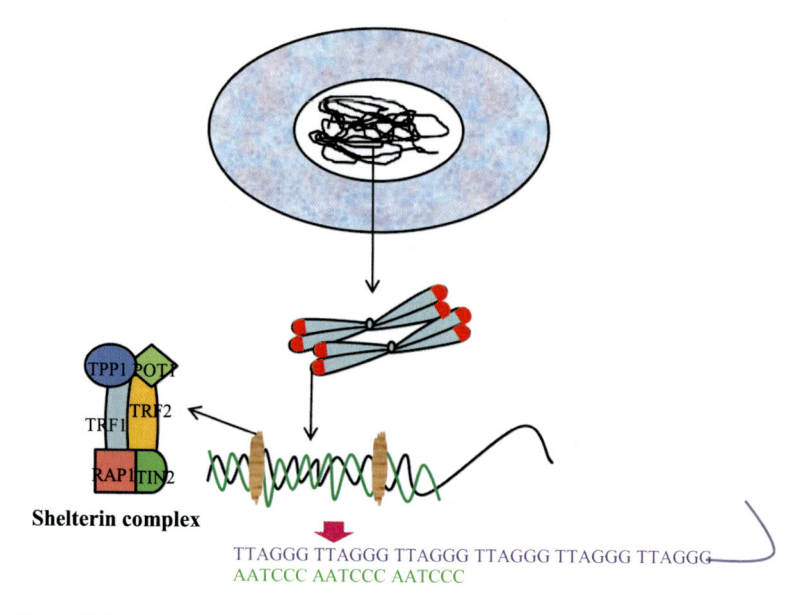

Fig. 17.10 Schematic presentation of the mammalian telomere

evidences suggested that in mice lacking the *TERC* gene (*mTR−/−*), it displayed signs of accelerated aging at the sixth generation of the mice and the fact that rapid aging was only observed in the later generations of the telomerase-deficient mice, suggested that it is the telomere length or accumulation of the DNA damage, rather than the absence of the enzyme, that caused the effects of early aging. It is clear from the above studies that DNA repair and telomere maintenance in the chromosomes are essential components for normal health and lifespan. On the other hand, telomerase-deficient cells exhibited an accelerated telomere shortening that eventually led to loss of telomere protection in the chromosomes and end-to-end chromosome fusions, and this was limiting for the mouse's longevity. There is a separate chapter on the telomere biology and aging.

Telomere Length in Metabolic Disease

MS like obesity and insulin resistance influence telomere length. Recently, Peter and Nilsson [97] demonstrated in a large population-based study that short telomeres were associated with an increased risk of metabolic abnormalities. In their study, they showed that shortening of leukocyte telomeres was considered to be a molecular marker for aging and it was suggested to be linked with increased risk of cardiovascular and metabolic disease. Similarly, Zhao et al. [98] investigated the association of leukocyte telomere length at baseline with future risk of diabetes over an average follow-up period of over 5 years and found that individuals in the lowest quartile of leukocyte telomere length were at almost twice the risk of developing diabetes compared to those with longer telomeres. Their study highlighted a nonlinear association between telomere length and diabetes-risk in which the increased risk was largely confined to those with the shortest telomere length. Studies also suggest that in atherosclerosis, telomere length probably contributes as a primary abnormality. Collectively, this apparent threshold effect is consistent with the hypothesis that there is a critical limit of telomere length, beyond that it induces cellular aging.

Telomere Length in Diabetes

There is apparent evidence concerning to the short telomeres and T2DM. Many studies have been cross-sectional in nature, and so many ambiguities come to mind: Do the metabolic perturbances of T2DM cause telomere attrition, or do shorter telomeres lead to T2DM? There is answer for all these plausible biological hypotheses supporting both scenarios [99]. Shorter telomeres are observed in diabetic patients, where in both T1DM and T2DM cases, β-cell failure is the final trigger. Shorter telomeres may lead to premature β-cell senescence, resulting in reduced β-cell mass and subsequent impaired insulin secretion and glucose tolerance [99]. Therefore, one could hypothesize that critically short telomeres contribute to the onset of diabetes by eliciting senescent phenotypes in β-cells. In fact, experimental evidence in animal study suggests that telomerase is important in maintaining glucose homeostasis. On the contrary, elevated blood glucose levels increase oxidative-nitrosative stress, potentially interfering with telomerase function and resulting in shortened telomeres [100]. Moreover, telomeres were shorter in patients with only impaired glucose tolerance compared to the controls and even shorter in T2DM patients [101]. In addition, telomere shortening has been linked to diabetic complications and associated diseases including diabetic nephropathy, microalbuminuria, and epithelial cancers, while telomere shortening seems to be attenuated in patients with well-controlled diabetes.

Telomere Length in Cardiovascular Disease and Diabetes

Sometimes, diabetes patients suffering with coronary heart disease, and stroke, show manifestations of shorter telomeres compared to patients with diabetes or other metabolic disease. A study done by Adaikalakoteswari et al. demonstrated that among T2DM patients, those with atherosclerotic plaques had the shorter telomeres [101]. Similarly, Olivieri et al. [102] showed that T2DM patients with myocardial infarction had shorter telomeres than T2DM subjects free of myocardial infarction. Collectively, these annotations suggest that decreased telomere length, either caused by the common risk factors between diabetes-related cardiovascular disease and inherited short telomeres, possibly reflects greater tissue aging and greater prevalence of senescent phenotypes in various tissues.

Metabolic Strategies to Delay Aging

Regular Exercise

Low aerobic or physical exercise and mechanical lifestyle are correlated with increased MS leading to aging exclusively in younger generation. Along with impaired organelle functions, specifically mitochondrial task may be an important mechanism for low aerobic capacity and metabolic risk factors that accompany the MS. Regular physical endurance exercise improves the insulin action and glucose tolerance in insulin-resistant patients, it increases mitochondrial size, number, and stimulates mitochondrial biogenesis by increasing expression of genes such as *PGC-1*, *NOS*, *NRF-1*, and *TFAM*. Thus, regular exercise can improve the age-associated reduction in expression of mitochondrial genes, and mitochondrial biogenesis is restored with aerobic exercise and it may improve the human lifespan, specially of those who are suffering from metabolic diseases.

Caloric Restriction

One of the strongest observations by different researchers and physicians in the biology of aging is the ability of caloric restriction (CR) in daily food habits along with aerobic

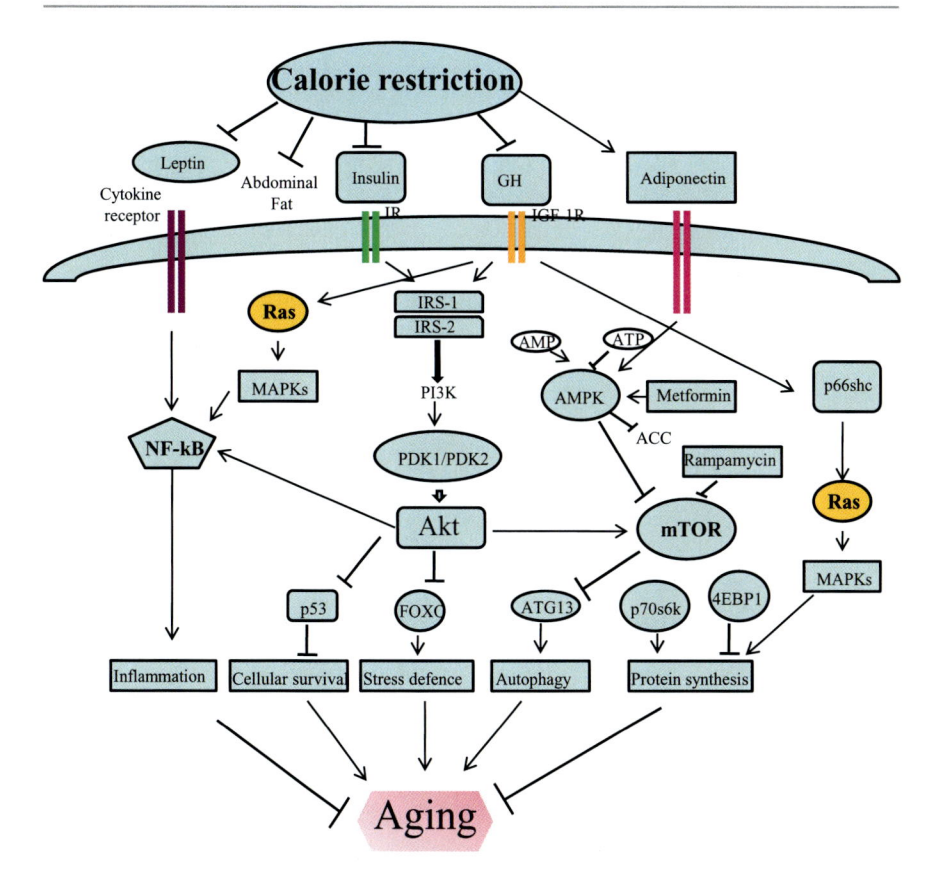

Fig. 17.11 Signaling network regulating through caloric restriction (CR) represents the most robust intervention to extend lifespan

exercise to delay or prevent a range of age-related processes by minimizing metabolic complications and it can significantly extend human lifespan. Subsequently, the CR paradigm has served as a critical research model in the laboratory for uncovering possible modulators of aging in both simple and complex organisms. More importantly, CR regulates insulin secretion and glucose action, secretion of thyroid hormones, reproductive hormones, and GH/IGF-1 levels. In response to these, numerous downstream cellular signaling cascades are engaged. The collective response of these signaling pathways to CR is believed to promote cellular fitness and ultimately longevity via activation of autophagy, stress defense mechanisms, and survival pathways while attenuating pro-inflammatory mediators and cellular growth (Fig. 17.11). Furthermore, there is evidence supporting that lifespan extension can be achieved with pharmacologic approaches, such as rapamycin, via mTOR signaling blockade, and metformin, which seems to be a robust stimulator of AMPK activity. In addition, in mammals, the CR phenotype includes prevention of some of the potentially harmful changes that are typically related to aging, such as increased adiposity and VF or impaired hepatic and peripheral insulin action. Understanding the molecular characteristics of various experimental animal models exposed to CR include numerous other changes in the transcriptome, metabolome, and proteome, as well as increases in stress hormones such as corticosterone or cortisol depending on the animal species.

Restoration of Cellular Antioxidants and Attenuated Oxidative Stress

One of the defects in metabolic syndrome and its associated diseases is excess cellular ROS and RNS accumulation. This is due to loss of cellular antioxidant capacity that damages biomolecules and cell organelles, more specifically mitochondrial components, resulting in reduced efficiency of the electron transport chain. Recent research evidence indicates that reduced mitochondrial function caused by lowering antioxidant capacity and increased free radicals accumulation are related to fatigue, a common complaint of metabolic syndrome patients [103]. This can be accomplished, in part, by neutralizing excess free radicals with various types of antioxidants or increasing free radical scavenging systems. Many recent studies in laboratory animals demonstrated that the dietary use of antioxidants, trace metal ions, vitamins, and herbal extracts separately or together, alter the course of MS progression and inhibit the progression of MS-associated diseases and prolong lifespan. Moreover, lipid replacement therapy (LRT) administered as a nutritional supplement with antioxidants can prevent excess oxidative-nitrosative damage, restore mitochondrial electron transport function and other cellular membrane functions, and reduce metabolic disease like fatigue. Therefore, LRT plus antioxidant supplements may be considered for MS patients who suffer from various degrees of fatigue [104]. However, under the conditions of overnutrition and lack of physical activity, typical of those fostering the MS, chronic overactivation of the redox signaling pathways may contribute to aging.

Pharmacological Intervention

Mitochondrial function, mitochondrial biogenesis, fission and fusion, and stimulation of mitochondrial events may have beneficial effects in metabolic diseases. The beneficial effects of thiazolidinediones and synthetic PPAR ligands have been reported to reduce insulin resistance in the liver, adipocytes, and heart, as well as pancreatic cell function and endothelial dysfunction. The molecular mechanism of insulin-sensitizing activity for thiazolidinediones and metformin may, in part, improve mitochondrial function by reducing oxidative-nitrosative stress and stimulating mitochondrial biogenesis through activation of AMPK/PGC-1/NRF signaling pathway. Furthermore, these drugs may reduce actions of angiotensin and angiotensin receptor, increase insulin sensitivity, block angiotensin-induced oxidative-nitrosative stress, and improve mitochondrial functions. Recent research publications demonstrated that superoxide scavenger is able to ameliorate neuronal, cardiac, and vascular dysfunction, normalize angiotensin-induced insulin resistance, and improve mitochondrial functions. Collectively, pharmacological intervention can stimulate mitochondrial biogenesis by reducing oxidative-nitrosative stress, such as metformin, angiotensin receptor blockers, and antioxidants and they may have beneficial effects on MS-associated diseases and aging.

Reduced Somatotropic Signaling

Human lifespan and longevity have been closely associated with MS-associated diseases and somatotropic signaling. Many studies on laboratory animals showed that with reduced function of the somatotropic axis, including *Ames* and *Snell dwarf* mice and mice lacking the GH receptor, all of which have decreased plasma IGF-1 concentrations. Different animal models and human studies have also showed that reduced IGF-1 signaling may cause reduced risk of many metabolic diseases as well as improved longevity. In addition, functional mutations have been identified in the human *IGF-1R* gene that results in altered IGF-1 signaling and it is more common in centenarians than in younger control subjects [105]. Studies have demonstrated that low level of IGF-1 in humans have been associated with increased risk for CVD, stroke, T2DM, and osteoporosis. However, mutations in the GH receptor (*GHR*) gene with Ecuadorian individuals led to severe GHR, and circulating IGF-1 deficiencies called the syndrome of Laron dwarfism, it had reduced risk for T2DM, presumably due to the absence of the anti-insulinemic action of GH. Collectively, these studies suggest that optimizing the IGF-1 to promote healthy aging in humans is more complex than it was originally appreciated and it will require a greater understanding of its array of interactions and tissue specificity in order to strike the right balance throughout the lifespan.

Attenuation of mTOR Signaling

The mTOR signaling pathway is closely linked to components of IIS pathway and it plays a pivotal role in the energy metabolism and glucose homeostasis and aging. Aberrant activation of mTOR signaling has been linked to several age-related diseases including T2DM, leading to studies on the role of this pathway in metabolism, aging, and lifespan. Inhibition of the mTOR signaling pathway by genetic or pharmacological intervention protects against diet-induced obesity and enhances insulin sensitivity and extends lifespan. In fact, disruption or systemic knockdown of mTOR signaling has yielded diverse, tissue-specific effects in many rodents, resulting in longevity [106]. For instance, specific gene manipulation of raptor, a component of mTOR complex in adipose tissue, protects against diet-induced obesity, whereas deletion of raptor in skeletal muscle is deleterious, resulting in a muscular dystrophy phenotype. In the liver, overexpression of dominant negative raptor improves insulin sensitivity, whereas inhibition of mTOR in the pancreas decreases insulin production by islets, leading to hypoinsulinemia and glucose intolerance [107]. Increased mTOR activity in the vascular endothelial cells, brain, and retina

Fig. 17.12 Insulin-regulated mTOR signaling and retinopathy. Pretreatment with rapamycin may prevent negative effects of insulin therapy against diabetic retinopathy

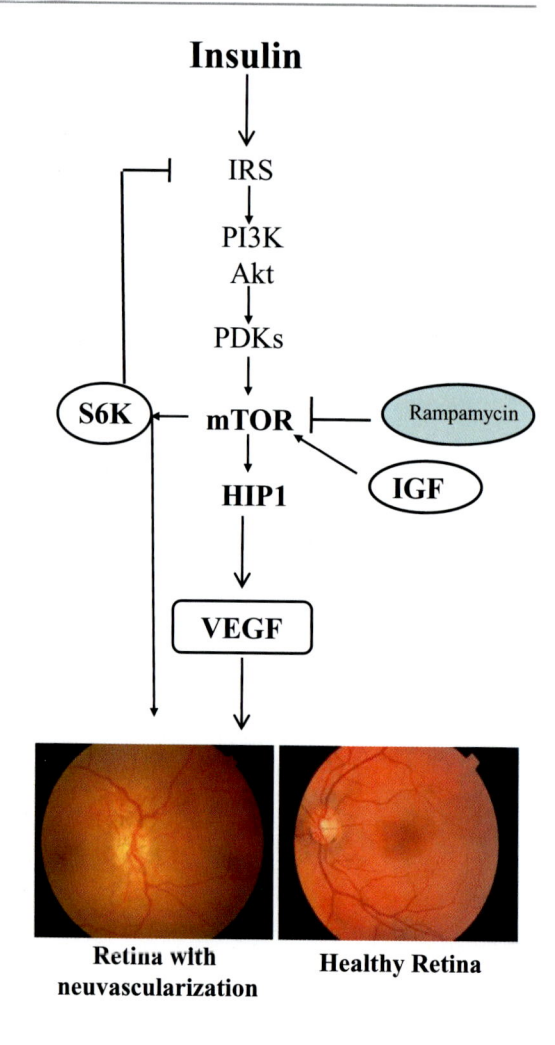

activates S6K via IRS/Akt pathway in the hypothalamus leading to decreased appetite via modulation of insulin, leptin, and ciliary neurotrophic factor and specifically developing neovascularization in the retina (Fig. 17.12). Rapamycin can inhibit the action in rodents and use in humans is also limited by its numerous side effects, including hyperglycemia, dyslipidemia, immunosuppression, vasospasm, and renal failure. Furthermore, human skeletal muscle from older adults has been shown to impair mTOR1 activation and protein synthesis in response to an acute resistance exercise for short period [108]. Thus, systemic blockade of the mTOR pathway could elicit an aging phenotype in MS and associated diseases.

Anti-AGEs Therapeutic Strategies

Prevention of AGE-mediated cell toxicity has been a key strategy in the prevention of MS and its associated diseases leading to some age-related pathology. Deterrence of AGEs interacting with their receptors or other proteins in cellular compartments is a valid therapeutic approach. There have been ranges of approaches, which seek to prevent AGE formation in cellular compartments, reduce AGE effects on cells, and even break established AGE cross-links with biomolecules. An important pharmacological AGE inhibitor, aminoguanidine, can prevent diabetic vascular complications in experimental laboratory animals, while clinical trials of aminoguanidine are shown to effectively reduce AGE-Hb while leaving HbA1c unaffected. On the other hand, aminoguanidine and/or other related AGE inhibitors may eventually find a place in the management of diabetics or in individuals at risk of age-related sequelae. Vascular basement membrane (BM) thickening is a fundamental structural alteration of small blood vessels in diabetes and by specific antibodies against glycated albumin, it has been shown to prevent vascular BM thickening in diabetic mice. Also, the use of the AGE binding properties of lysozyme has succeeded in reducing AGE levels in diabetic patients with kidney disease. Additionally, elucidation of AGE receptor signal transduction pathways may also offer intracellular strategies to control receptor-mediated protection from toxic tissue contents of AGEs and reverse hyperglycemia-related complications.

Stem Cells and the Promise of Pancreas Regeneration

Diabetes is strongly associated with VF, which is metabolically more active than SF, which is actively interfere to release non-esterified fatty acids and cytokines that weaken the insulin actions on its target tissues. Corresponding to this action, typically pancreatic β-cells adapt to a reduction in insulin sensitivity by increasing insulin secretion through increased insulin synthesis and increased β-cell mass. However, diabetes ruin the β-cell action leading to hyperglycemia that breaks the closed feedback loop between the β-cell and its target tissue. In this regard, it is important to understand the pancreatic β-cell contribution in the maintenance of the pancreatic β-cell pool against the increased metabolic demands associated with diabetes that may provide new therapeutic targets for treating diabetes. Embryonic stem (ES) cell research has developed novel therapies to restore or replace pancreas that has been damaged through diabetic injury.

Future Direction and Conclusion

The incidence of the metabolic syndrome, which is a cluster of metabolic risk factors associated with diabetes, obesity, and insulin resistance, is dramatically growing day by day throughout the globe in younger generation rather than only in the old people. This disorder consists of a bunch of metabolic complications, such as hypertriglyceridemia, hyper-low-density lipoproteins, hypo-high-density lipoproteins, insulin resistance, abnormal glucose tolerance, and hypertension, that – in combination with genetic, hormonal, environmental susceptibility, and abdominal obesity – are risk factors for diabetes, vascular inflammation, atherosclerosis, and renal, liver, lung, CNS/brain and heart diseases. For all these metabolic complications and its associated diseases, the causes are excess of ROS/RNS, AGE formation within or outside the cells, cellular antioxidant depletion, activation of metabolic signaling network etc. They cause damage to cell organelles like mitochondrial components and initiate degradative processes throughout the body and these toxic reactions contribute significantly to the aging process. To minimize all these toxic substances generated due to MS, it requires management at the patient level starting from detection to therapy. In spite of the availability of efficient sensing technologies, there is a need for better and improved technologies for keeping pace with the demand to treat the MS. Although, recently, nanotechnology is in its infancy for MS management, as an example it has a proven track record at various levels in diverse organs/tissues for efficiently dealing with the above health complications. The surface plasmon resonance and photothermal and fluorescence properties of nanoparticles have an extremely important role in various techniques for sensing, imaging, delivery, and therapies in managing MS and aging [109].

References

1. Guo S. Insulin signaling, resistance, and the metabolic syndrome: insights from mouse models into disease mechanisms. J Endocrinol. 2014;220:T1–T23.
2. Kumar P, Swain MM, Pal A. Hyperglycemia-induced inflammation caused down-regulation of 8-oxoG-DNA glycosylase levels in macrophages is mediated by oxidative-nitrosative stress-dependent pathways. Int J Biochem Cell Biol. 2016;73(2016):82–98.
3. Zhang X, Saaddine JB, Chou CF, Cotch MF, Cheng YJ, Geiss LS, et al. Prevalence of diabetic retinopathy in the United States, 2005–2008. JAMA. 2010;304:649–56.
4. Corriere M, Rooparinesingh N, Rita Rastogi Kalyani RR. Epidemiology of diabetes and diabetes complications in the elderly: an emerging public health burden. Curr Diab Rep. 2013;2013(6):10.
5. Stevens LA, Li S, Wang C, Huang C, Becker BN, Bomback AS, et al. Prevalence of CKD and comorbid illness in elderly patients in the United States: results from the Kidney Early Evaluation Program (KEEP). Am J Kidney Dis. 2010;55(3 Suppl 2):S23–33.
6. Kalyani RR, Saudek CD, Brancati FL, Selvin E. Association of diabetes, comorbidities, and A1C with functional disability in older adults: results from the National Health and Nutrition Examination Survey (NHANES), 1999–2006. Diabetes Care. 1999;33(5):1055–60.
7. Höhn A, König J, Jung T. Metabolic syndrome, redox state, and the proteasomal system. Antioxid Redox Signal. 2016;25:902–17.
8. Barzilai N, Huffman DM, Muzumdar RH, Bartke A. The critical role of metabolic pathways in aging. Diabetes. 2012;61:1315–22.
9. Huffman DM, Barzilai N. Role of visceral adipose tissue in aging. Biochim Biophys Acta. 2009;1790:1117–23.
10. Atzmon G, Yang XM, Muzumdar R, Ma XH, Gabriely I, Barzilai N. Differential gene expression between visceral and subcutaneous fat depots. Horm Metab Res. 2002;34:622–8.
11. Muzumdar R, Allison DB, Huffman DM, et al. Visceral adipose tissue modulates mammalian longevity. Aging Cell. 2008;7:438–40.

12. Berg AH, Scherer PE. Adipose tissue, inflammation, and cardiovascular disease. Circ Res. 2005;96:939–49.
13. Evans WJ, Paolisso G, Abbatecola AM, et al. Frailty and muscle metabolism dysregulation in the elderly. Biogerontology. 2010;11:527–36.
14. Lee CG, Boyko EJ, Strotmeyer ES, et al. Osteoporotic Fractures in Men Study Research Group. Association between insulin resistance and lean mass loss and fat mass gain in older men without diabetes mellitus. J Am Geriatr Soc. 2011;59:1217–24.
15. Rossouw JE, Anderson GL, Prentice RL, et al. Writing Group for the Women's Health Initiative Investigators. Risks and benefits of estrogen plus progestin in healthy postmenopausal women: principal results from the Women's Health Initiative randomized controlled trial. JAMA. 2002;288:321–33.
16. Rincon M, Muzumdar R, Atzmon G, Barzilai N. The paradox of the insulin/IGF-1 signaling pathway in longevity. Mech Ageing Dev. 2004;125:397–403.
17. Russell SJ, Kahn CR. Endocrine regulation of ageing. Nat Rev Mol Cell Biol. 2007;8:681–91.
18. Suh Y, Atzmon G, Cho MO, et al. Functionally significant insulin-like growth factor I receptor mutations in centenarians. Proc Natl Acad Sci U S A. 2008;105:3438–344257.
19. Patricia W. Thyroid disease and diabetes. Clin Diab Winter 2000;18(1).
20. Ooka H, Shinkai T. Effects of chronic hyperthyroidism on the lifespan of the rat. Mech Ageing Dev. 1986;33:275–82.
21. Hoehn KL, Salmon AB, Hohnen-Behrens C, Turner N, Hoy AJ, Maghzal GJ, Stocker R, Van Remmen H, Kraegen EW, Cooney GJ, et al. Insulin resistance is a cellular antioxidant defense mechanism. Proc Natl Acad Sci. 2009;106:17787–92.
22. Ilkun O, Wilde N, Tuinei J, Pires KM, Zhu Y, Bugger H, Soto J, Wayment B, Olsen C, Litwin SE, Abel ED. Antioxidant treatment normalizes mitochondrial energetics and myocardial insulin sensitivity independently of changes in systemic metabolic homeostasis in a mouse model of the metabolic syndrome. J Mol Cell Cardiol. 2015;285:104–16.
23. Anderson EJ, Lustig ME, Boyle KE, Woodlief TL, Kane DA, Lin C, Kang L, Rabinovitch PS, Szeto HH, Houmard JA, Cortright RN, Wasserman DH, Neufer PD. Mitochondrial H2O2 emission and cellular redox state link excess fat intake to insulin resistance in both rodents and humans. J Clin Invest. 2009;119(3):573–81.
24. Lee HY, Choi CS, Birkenfeld AL, Alves TC, Jornayvaz FR, Jurczak MJ, Zhang D, Woo DK, Shadel GS, Ladiges W, et al. Targeted expression of catalase to mitochondria prevents age-associated reductions in mitochondrial function and insulin resistance. Cell Metab. 2010;12:668–74.
25. Moustafa SA, Webster JE, Mattar FE. Effects of aging and antioxidants on glucose transport in rat adipocytes. Gerontology. 1995;41:301–7.
26. Montgomery MK, Nigel TN. Mitochondrial dysfunction and insulin resistance: an update. Endocr Connect. 2015;4(1):R1–R15.
27. Luevano-Contreras C, Chapman-Novakofski K. Dietary advanced glycation end products and aging. Nutrients. 2010;2:1247–65.
28. Rahmadi A, Steiner N, Munch G. Advanced glycation endproducts as gerontotoxins and biomarkers for carbonyl-based degenerative processes in Alzheimer's disease. Clin Chem Lab Med. 2011;49:385–91.
29. Gasser A, Forbes JM. Advanced glycation: implications in tissue damage and disease. Protein Pept Lett. 2008;15:385–91.
30. Cox TR, Erler JT. Remodeling and homeostasis of the extracellular matrix: implications for fibrotic diseases and cancer. Dis Model Mech. 2011;4(2):165–78.
31. Singh VP, Bali A, Singh N, Jaggi AS. Advanced glycation end products and diabetic complications. Korean J Physiol Pharmacol. 2014;18(1):1–14.
32. Wang W, Tan M, Yu J, Tan L. Role of pro-inflammatory cytokines released from microglia in Alzheimer's disease. Ann Transl Med. 2015;3(10):136.
33. Farrukh AS, Sharkey E, Creighton D, et al. Maillard reactions in lens proteins: methylglyoxal-mediated modifications in the rat lens. Exp Eye Res. 2000;70:369–80.

34. Saxena P, Saxena AK, Cui XL, et al. Transition metal-catalyzed oxidation of ascorbate in human cataract extracts: possible role of advanced glycation end products. Invest Ophthalmol Vis Sci. 2000;41:1473–81.
35. Kaji Y, Usui T, Oshika T, et al. Advanced glycation end products in diabetic corneas. Invest Ophthalmol Vis Sci. 2000;41:362–8.
36. Stitt AW. Advanced glycation: an important pathological event in diabetic and age related ocular disease. Br J Ophthalmol. 2001;85:746–53.
37. Stitt AW, Li YM, Gardiner TA, et al. Advanced glycation endproducts (AGEs) co-localise with AGE-receptors in the retinal vasculature of diabetic and AGE-infused rats. Am J Pathol. 1997;150:523–32.
38. Bonet-Costa V, Pomatto LC, Davies KJA. The proteasome and oxidative stress in Alzheimer's disease. Antioxid Redox Signal. 2016;25:886–901.
39. Kumar P, Raman T, Swain MM, Mishra R, Pal A. Hyperglycemia-induced oxidative-nitrosative stress induces inflammation and neurodegeneration via augmented tuberous sclerosis complex-2 (TSC-2) activation in neuronal cells. Mol Neurobiol. 2017;54:238–54.
40. Greer EL, Brunet A. Signaling network in aging. J Cell Sci. 2008;121(4):407–12.
41. White MF. Insulin signaling in health and disease. Science. 2003;302:1710–1.
42. Fafalios A, Ma J, Tan X, Stoops J, Luo J, Defrances MC, Zarnegar R. A hepatocyte growth factor receptor (Met)-insulin receptor hybrid governs hepatic glucose metabolism. Nat Med. 2011;17:1577–84.
43. Ford ES, Giles WH, Dietz WH. Prevalence of the metabolic syndrome among US adults: findings from the third National Health and Nutrition Examination Survey. JAMA. 2002;287:356359.
44. Johnson AM, Olefsky JM. The origins and drivers of insulin resistance. Cell. 2013;152:673–84.
45. Guo S. Molecular basis of insulin resistance: the role of IRS and Foxo1 in the control of diabetes mellitus and its complications. Drug Discov Today Dis Mech. 2013;10:e27–33.
46. Oshi RL, Lamothe B, Cordonnier N, Mesbah K, Monthioux E, Jami J, Bucchini D. Targeted disruption of the insulin receptor gene in the mouse results in neonatal lethality. EMBO J. 1996;15:1542–7.
47. Withers DJ, Burks DJ, Towery HH, Altamuro SL, Flint CL, White MF. Irs-2 coordinates Igf-1 receptor-mediated beta-cell development and peripheral insulin signalling. Nat Genet. 1999;23:32–40.
48. Araki E, Lipes MA, Patti ME, Bruning JC, Haag B 3rd, Johnson RS, Kahn CR. Alternative pathway of insulin signalling in mice with targeted disruption of the IRS 1 gene. Nature. 1994;372:186–90.
49. Withers DJ, Gutierrez JS, Towery H, Burks DJ, Ren JM, Previs S, Zhang Y, Bernal D, Pons S, Shulman GI, et al. Disruption of IRS-2 causes type 2 diabetes in mice. Nature. 1998;391:900–4.
50. Rask-Madsen C, Kahn CR. Tissue-specific insulin signaling, metabolic syndrome, and cardiovascular disease. Arterioscler Thromb Vasc Biol. 2012;32:2052–9.
51. Boucher J, Kahn CR. Differential role of insulin and IGF-1 receptors in brown and white adipose tissue and development of lipoatrophic diabetes. Diabetes. 2013;62:A37.
52. Lin HV, Accili D. Reconstitution of insulin action in muscle, white adipose tissue, and brain of insulin receptor knockout mice fails to rescue diabetes. J Biol Chem. 2011;286:9797–804.
53. Dong XC, Copps KD, Guo S, Li Y, Kollipara R, DePinho RA, White MF. Inactivation of hepatic Foxo1 by insulin signaling is required for adaptive nutrient homeostasis and endocrine growth regulation. Cell Metab. 2008;8:65–76.
54. Qi Y, Xu Z, Zhu Q, Thomas C, Kumar R, Feng H, Dostal DE, White MF, Baker KM, Guo S. Myocardial loss of IRS1 and IRS2 causes heart failure and is controlled by p38alpha MAPK during insulin resistance. Diabetes. 2013;62:3887–900.
55. Lu M, Wan M, Leavens KF, Chu Q, Monks BR, Fernandez S, Ahima RS, Ueki K, Kahn CR, Birnbaum MJ. Insulin regulates liver metabolism in vivo in the absence of hepatic Akt and Foxo1. Nat Med. 2012;18:388–95.

56. Cho H, Mu J, Kim JK, Thorvaldsen JL, Chu Q, Crenshaw EB 3rd, Kaestner KH, Bartolomei MS, Shulman GI, Birnbaum MJ. Insulin resistance and a diabetes mellitus-like syndrome in mice lacking the protein kinase Akt2 (PKB beta). Science. 2001;292:1728–31.
57. George S, Rochford JJ, Wolfrum C, Gray SL, Schinner S, Wilson JC, Soos MA, Murgatroyd PR, Williams RM, Acerini CL, et al. A family with severe insulin resistance and diabetes due to a mutation in AKT2. Science. 2004;304:1325–8.
58. Evans-Anderson HJ, Alfieri CM, Yutzey KE. Regulation of cardiomyocyte proliferation and myocardial growth during development by FOXO transcription factors. Circ Res. 2008;102:686–94.
59. Battiprolu PK, Hojayev B, Jiang N, Wang ZV, Luo X, Iglewski M, Shelton JM, Gerard RD, Rothermel BA, Gillette TG, et al. Metabolic stress-induced activation of FoxO1 triggers diabetic cardiomyopathy in mice. J Clin Invest. 2012;122:1109–18.
60. Guo S, Dunn SL, White MF. The reciprocal stability of FOXO1 and IRS2 creates a regulatory circuit that controls insulin signaling. Mol Endocrinol. 2006;20:3389–99.
61. Rui L, Fisher TL, Thomas J, White MF. Regulation of insulin/insulin-like growth factor-1 signaling by proteasome-mediated degradation of insulin receptor substrate-2. J Biol Chem. 2001;276:40362–7.
62. Bae EJ, Xu J, Oh DY, Bandyopadhyay G, Lagakos WS, Keshwani M, Olefsky JM. Liver-specific p70 S6 kinase depletion protects against hepatic steatosis and systemic insulin resistance. J Biol Chem. 2012;287:18769–80.
63. Hagiwara A, Cornu M, Cybulski N, Polak P, Betz C, Trapani F, Terracciano L, Heim MH, Ruegg MA, Hall MN. Hepatic mTORC2 activates glycolysis and lipogenesis through Akt, glucokinase, and SREBP1c. Cell Metab. 2012;15:725–38.
64. Liu HY, Cao SY, Hong T, Han J, Liu Z, Cao W. Insulin is a stronger inducer of insulin resistance than hyperglycemia in mice with type 1 diabetes mellitus (T1DM). J Biol Chem. 2009;284:27090–100.
65. Li M, Georgakopoulos D, Lu G, Hester L, Kass DA, Hasday J, Wang Y. p38 MAP kinase mediates inflammatory cytokine induction in cardiomyocytes and extracellular matrix remodeling in heart. Circulation. 2005;111:2494–502.
66. Shoelson SE, Lee J, Goldfine AB. Inflammation and insulin resistance. J Clin Invest. 2006;116:1793–801.
67. Bezy O, Tran TT, Pihlajamaki J, Suzuki R, Emanuelli B, Winnay J, Mori MA, Haas J, Biddinger SB, Leitges M, et al. PKCdelta regulates hepatic insulin sensitivity and hepatosteatosis in mice and humans. J Clin Invest. 2011;121:2504–17.
68. Altomonte J, Richter A, Harbaran S, Suriawinata J, Nakae J, Thung SN, Meseck M, Accili D, Dong H. Inhibition of Foxo1 function is associated with improved fasting glycemia in diabetic mice. Am J Physiol Endocrinol Metab. 2003;285:E718–28.
69. Taguchi A, Wartschow LM, White MF. Brain IRS2 signaling coordinates life span and nutrient homeostasis. Science. 2007;317:369–72.
70. Sadagurski M, Leshan RL, Patterson C, Rozzo A, Kuznetsova A, Skorupski J, Jones JC, Depinho RA, Myers MG Jr, White MF. IRS2 signaling in LepR-b neurons suppresses FoxO1 to control energy balance independently of leptin action. Cell Metab. 2012;15:703–12.
71. Boucher J, Kahn CR. Differential role of insulin and IGF-1 receptors in brown and white adipose tissue and development of lipoatrophic diabetes. Diabetes. 2013;62:A37.
72. Romeo GR, Lee J, Shoelson SE. Metabolic syndrome, insulin resistance, and roles of inflammation–mechanisms and therapeutic targets. Arterioscler Thromb Vasc Biol. 2012;32:1771–6.
73. Ye J, McGuinness OP. Inflammation during obesity is not all bad: evidence from animal and human studies. Am J Physiol Endocrinol Metab. 2013;304:E466–77.
74. Malato Y, Ehedego H, Al-Masaoudi M, Cubero FJ, Bornemann J, Gassler N, Liedtke C, Beraza N, Trautwein C. NF-kappaB essential modifier is required for hepatocyte proliferation and the oval cell reaction after partial hepatectomy in mice. Gastroenterology. 2012;143(6):1597—1608. e1511.

75. Guo S, Copps KD, Dong X, Park S, Cheng Z, Pocai A, Rossetti L, Sajan M, Farese RV, White MF. The Irs1 branch of the insulin signaling cascade plays a dominant role in hepatic nutrient homeostasis. Mol Cell Biol. 2009;29:5070–83.
76. Rhodes CJ, White MF, Leahy JL, Kahn SE. Direct autocrine action of insulin on beta-cells: does it make physiological sense? Diabetes. 2013;62:2157–63.
77. Nakae J, Park BC, Accili D. Insulin stimulates phosphorylation of the forkhead transcription factor FKHR on serine 253 through a Wortmannin-sensitive pathway. J Biol Chem. 1999;274:15982–5.
78. Kitamura T. The role of FOXO1 in beta-cell failure and type 2 diabetes mellitus. Nat Rev Endocrinol. 2013;9:615–23.
79. Pehmoller C, Treebak JT, Birk JB, Chen S, Mackintosh C, Hardie DG, Richter EA, Wojtaszewski JF. Genetic disruption of AMPK signaling abolishes both contraction- and insulin-stimulated TBC1D1 phosphorylation and 14-3-3 binding in mouse skeletal muscle. Am J Physiol Endocrinol Metab. 2009;297:E665–75.
80. Hoehn KL, Turner N, Swarbrick MM, Wilks D, Preston E, Phua Y, Joshi H, Furler SM, Larance M, Hegarty BD, et al. Acute or chronic upregulation of mitochondrial fatty acid oxidation has no net effect on whole-body energy expenditure or adiposity. Cell Metab. 2010;11:70–6.
81. Rask-Madsen C, Li Q, Freund B, Feather D, Abramov R, Wu IH, Chen K, Yamamoto-Hiraoka J, Goldenbogen J, Sotiropoulos KB, et al. Loss of insulin signaling in vascular endothelial cells accelerates atherosclerosis in apolipoprotein E null mice. Cell Metab. 2010;11:379–89.
82. Kubota T, Kubota N, Kumagai H, Yamaguchi S, Kozono H, Takahashi T, Inoue M, Itoh S, Takamoto I, Sasako T, et al. Impaired insulin signaling in endothelial cells reduces insulin-induced glucose uptake by skeletal muscle. Cell Metab. 2011;13:294–307.
83. Tsuchiya K, Tanaka J, Shuiqing Y, Welch CL, DePinho RA, Tabas I, Tall AR, Goldberg IJ, Accili D. FoxOs integrate pleiotropic actions of insulin in vascular endothelium to protect mice from atherosclerosis. Cell Metab. 2012;15:372–81.
84. Messmer-Blust AF, Philbrick MJ, Guo S, Wu J, He P, Li J. RTEF-1 attenuates blood glucose levels by regulating insulin-like growth factor binding protein-1 in the endothelium. Circ Res. 2012;111:991–1001.
85. Ferron M, Wei J, Yoshizawa T, Del Fattore A, DePinho RA, Teti A, Ducy P, Karsenty G. Insulin signaling in osteoblasts integrates bone remodeling and energy metabolism. Cell. 2010;142:296–308.
86. Rached MT, Kode A, Silva BC, Jung DY, Gray S, Ong H, Paik JH, DePinho RA, Kim JK, Karsenty G, et al. FoxO1 expression in osteoblasts regulates glucose homeostasis through regulation of osteocalcin in mice. J Clin Invest. 2010;120:357–68.
87. Domenighetti AA, Wang Q, Egger M, et al. Angiotensin II-mediated phenotypic cardiomyocyte remodeling leads to age-dependent cardiac dysfunction and failure. Hypertension. 2005;46:426–32.
88. Yan L, Vatner DE, O'Connor JP, et al. Type 5 adenylyl cyclase disruption increases longevity and protects against stress. Cell. 2007;130:247–58.
89. Coschigano KT, Holland AN, Riders ME, List EO, Flyvbjerg A, Kopchick JJ. Deletion, but not antagonism, of the mouse growth hormone receptor results in severely decreased body weights, insulin and IGF-1 levels and increased lifespan. Endocrinology. 2003;144:3799–810.
90. Shiraki-Iida T, Aizawa H, Matsumura Y, Sekine S, Iida A, Anazawa H, Nagai R, Kuro-o M, Nabeshima Y. Structure of the mouse klotho gene and its two transcripts encoding membrane and secreted protein. FEBS Lett. 1998;424:6–10.
91. Roth GS, Lane MA, Ingram DK, Mattison JA, Elahi D, Tobin JD, Muller D, Metter EJ. Biomarkers of caloric restriction may predict longevity in humans. Science. 2002;297:811.
92. Hagiwara A, Cornu M, Cybulski N, Polak P, Betz C, Trapani F, Terracciano L, Heim MH, Ruegg MA, Hall MN. Hepatic mTORC2 activates glycolysis and lipogenesis through Akt, glucokinase, and SREBP1c. Cell Metab. 2012;15:725–38.

93. Yecies JL, Zhang HH, Menon S, Liu S, Yecies D, Lipovsky AI, Gorgun C, Kwiatkowski DJ, Hotamisligil GS, Lee CH, et al. Akt stimulates hepatic SREBP1c and lipogenesis through parallel mTORC1-dependent and independent pathways. Cell Metab. 2011;14:21–32.
94. Feng J, Bussière F, Siegfried HS. Mitochondrial electron transport is a key determinant of life span in *Caenorhabditis elegans*. Dev Cell. 2001;1:633–44.
95. Lee YH, Giraud J, Davis RJ, White MF. c-Jun N-terminal kinase (JNK) mediates feedback inhibition of the insulin signaling cascade. J Biol Chem. 2003;278:2896–902.
96. Kudlow BA, Kennedy BK, Monnat RJ, RJ. Werner and Hutchinson–Gilford progeria syndromes: mechanistic basis of human progeroid diseases. Nat Rev Mol Cell Biol. 2007;8:394–404.
97. Peter M, Nilsson PM. Genetics: telomere length and the metabolic syndrome—a causal link? Nat Rev Endocrinol. 2014;10:706–7.
98. Zhao J, Zhu Y, Lin J, et al. Short leukocyte telomere length predicts risk of diabetes in American Indians: the strong heart family study. Diabetes. 2013;63:354–62.
99. Elks CE, Scott RA. The long and short of telomere length and diabetes. Diabetes. 2014;63(1):65–7.
100. Serra V, Grune T, Sitte N, Saretzki G, von Zglinicki T. Telomere length as a marker of oxidative stress in primary human fibroblast cultures. Ann N Y Acad Sci. 2000;908:327–30.
101. Adaikalakoteswari A, Balasubramanyam M, Ravikumar R, Deepa R, Mohan V. Association of telomere shortening with impaired glucose tolerance and diabetic macroangiopathy. Atherosclerosis. 2007;195:83–9.
102. Olivieri F, Lorenzi M, Antonicelli R, et al. Leukocyte telomere shortening in elderly Type2DM patients with previous myocardial infarction. Atherosclerosis. 2009;206:588–93.
103. Garth L. and Nicolson GL. (2007). Metabolic syndrome and mitochondrial function: molecular replacement and antioxidant supplements to prevent membrane peroxidation and restore mitochondrial function J Cell Biochem. 100:1352–1369
104. Nicolson GL, Ellithorpe R. Lipid replacement and antioxidant nutritional therapy for restoring mitochondrial function and reducing fatigue in chronic fatigue syndrome and other fatiguing illnesses. J Chronic Fatigue Syndr. 2006;13:57–68.
105. Suh Y, Atzmon G, Cho MO, et al. Functionally significant insulin-like growth factor I receptor mutations in centenarians. Proc Natl Acad Sci. 2008;105:3438–42.
106. Harrison DE, Strong R, Sharp ZD, et al. Rapamycin fed late in life extends lifespan in genetically heterogeneous mice. Nature. 2009;460:392–5.
107. Polak P, Hall MN. mTOR and the control of whole body metabolism. Curr Opin Cell Biol. 2009;21:209–18.
108. Fry CS, Drummond MJ, Glynn EL, et al. Aging impairs contraction-induced human skeletal muscle mTORC1 signaling and protein synthesis. Skelet Muscle. 2011;1:11.
109. Si S, Pal A, Jagdeep M, Satapathy S. Gold nanostructure materials in diabetes management. J Phys D Appl Phys. 2017;50(13):134003.

Interplay Between Nutrient-Sensing Molecules During Aging and Longevity

Ibanylla Kynjai Hynniewta Hadem, Teikur Majaw, and Ramesh Sharma

Introduction

Several types of nutrient-sensing molecules have been recognized, and in mammals, the major nutrient-sensing molecules include mTOR, GCN2, AMPK, and sirtuins. Most of these pathways are conserved in eukaryotes. In addition to nutrient sensing, these pathways are regulated by varied hormones [1]. Interestingly, different mechanisms are employed by the mammalian cells to sense the various types of macronutrients (glucose, amino acids, and lipids). When the availability of food is high, the nutrient-sensing pathway triggers anabolism and storage, whereas, during nutrient scarcity, it engages in energy-producing catabolic pathways to ensure diverse homeostatic mechanisms. The different sensors that detect macronutrient can be either direct or indirect [2]. Deregulation of nutrient-sensing pathways represents one of the hallmarks of aging [3].

Amino Acid Sensing

The most studied nutrient-sensing strategy involves the sensing of amino acids. mTORC1 (mammalian/mechanistic target of rapamycin complex 1) plays a key role in detecting abundance of amino acids, although the mechanism is indirect and indistinct. Scarcity of amino acids is directly detected by GCN2, a kinase that binds to the uncharged tRNA molecule [1]. The synthesis of protein is energetically demanding; hence, the cells are equipped to sense the changes in amino acid availability precisely. During extreme nutrient deprivation, proteins are degraded through

---

Ibanylla Kynjai Hynniewta Hadem and Teikur Majaw contributed equally with all other contributors.

I. K. H. Hadem · T. Majaw · R. Sharma (✉)
Department of Biochemistry, North-Eastern Hill University, Shillong, Meghalaya, India
e-mail: sharamesh@gmail.com

proteasome-mediated process or through autophagy to generate amino acid stores. These amino acids are further recycled to generate glucose or ketone bodies to meet the different energy requirements of the body, particularly the brain [2]. Besides its importance for protein synthesis, amino acids are essential for other processes such as tryptophan (precursor for NAD^+ and serotonin biosynthesis), methionine (plays a role in one-carbon transfer reactions), glutamate (neurotransmitter), and many others that are intermediate metabolites in several cellular processes. Lower organisms can synthesize all the amino acids from carbon skeleton; however, higher organisms require the essential amino acids in their diets [4].

GCN2 Senses Amino Acid Starvation

The deficiency of any of the 20 amino acids cannot be compensated by the others; thereby the cells are capable of detecting the lack of any of the amino acids. This is mediated through the presence of the enzyme GCN2 (kinase general control nonderepressible 2) [5]. During protein synthesis, the amino acids are added to the growing peptide chain by the ribosome through the binding of specific transfer RNAs (tRNAs) that is covalently linked to its respective amino acid [6]. Uncharged tRNA acts as a surrogate marker for low levels of amino acids. Under nutrient scarcity, the low levels of amino acids increase the accumulation of uncharged tRNAs in the cell. Thus, this leads to inefficient translation, and the cells prevent this process by triggering adaptive responses that lead to inhibition of translation initiation. In bacteria, there is a direct control on transcription, whereas in eukaryotes uncharged tRNAs activate a signal transduction pathway through the direct interaction with GCN2 [4, 7]. This protein has a HisRS (histidyl-tRNA synthetase) domain that binds to all uncharged tRNAs irrespective of its amino acid specificity. GCN2 on binding to the uncharged tRNA undergoes a conformational change that leads to its kinase activation through homodimerization and autophosphorylation [8, 9]. Besides its autoactivation, the only other protein that GCN2 target through phosphorylation is the eukaryotic initiation factor 2 alpha (eIF2α), a key early activator of translation initiation. The phosphorylation of eIF2α inhibits the efficient initiation of translation at the methionine start codon, thereby preventing global protein synthesis [7, 10]. However, there is a selective translation of certain mRNA having specific regulatory elements in their 5′-UTRs (untranslated regions), a phenomenon known as translational derepression. An example of a derepressed mRNA is the one that codes for GCN4 transcription factor. The GAAC (general amino acid control) pathway acts through GCN4 to regulate the activation of transcription of amino acid biosynthetic and transport genes [11, 12].

In mammals, depletion of essential amino acids including leucine, histidine, tryptophan, or lysine in the diet leads to a rapid activation of GCN2 [13]. Additionally, GCN2 can be activated due to depletion of amino acids caused by other non-dietary-related mechanisms such as during acute stress caused by trauma or sepsis, which can lead to increased production of nitric oxide catalyzed by nitric oxide synthase leading to decreased level of arginine [14, 15]. GCN2 can be activated by the depletion of amino acids caused via dietary limitation and enzymatic action or through pharmacological means which results in the accumulation of uncharged tRNAs. One of the effectors of GCN2 is ATF4 (an orthologue of the bacterial GCN4). Unless there is dietary deprivation of essential amino acids, mice, where GCN2 protein is absent, are normal [16]. However, cells lacking ATF4 need excess of non-essential amino acids, and ATF4 knockout mice exhibit several developmental abnormalities and are generally smaller than their wild-type counterparts [17]. Besides being activated by low amino acid levels, glucose deprivation and UV irradiation can also activate GCN2 [18].

mTOR Senses Amino Acid Abundance

The ability of cells to sense nutrient sufficiency involves the TOR signaling cascade. mTOR is a serine/threonine kinase involved in a number of cellular processes including cell growth. In mammals, mTOR is present in two structurally and functionally distinct complexes, namely, mTORC1 and mTORC2. mTORC1 is inhibited by rapamycin and sensitive to changes in nutrient levels. However, mTORC2 is neither nutrient sensitive nor inhibited by rapamycin (unless it is under prolonged treatment of rapamycin, which then prevents mTORC2 assembly by sequestering the mTOR subunit in some cell type) [19, 20]. mTOR is the catalytic subunit of mTORC1 which has other subunits including a negative regulator, proline-rich AKT/PKB substrate 40 kDa (PRAS40), and a regulatory-associated protein of mTOR (RAPTOR). These are involved in substrates and regulatory protein binding. In addition, mTORC1 has two other complexes which are common with mTORC2, Dep domain mTOR-interacting protein (Deptor), and the positive mTORC1 regulator mammalian lethal with sec-13 protein 8 (mLST8, also known as GβL). mTOR is also regulated by growth factors where it coordinates various anabolic processes like protein and lipid synthesis [21, 22]. Although the direct sensor for the abundance of amino acids is unknown, however, for mTORC1 signaling, the presence of amino acids is an absolute necessity [23]. Growth factors and other triggers cannot effectively activate mTOR when the levels of amino acids are limiting. The low levels of leucine and arginine are particularly effective in downregulating mTOR pathway [23–25]. Additionally, glutamate has also been implicated in the regulation of mTOR activation [26]. The molecular mechanisms of mTORC1 activation have been extensively studied although the exact mechanism has not been established. Besides being regulated by the presence of amino acids and growth factors, mTORC1 is also regulated by glucose levels and stress [27]. During nutrient sufficiency, mTORC1 is activated, whereas nutrient deprivation leads to the inhibition of the mTORC1 signaling pathway resulting in a well-coordinated regulation of cellular processes involved in cell growth and metabolism [28]. mTORC1 signals through ribosomal S6 kinase (S6K) and eIF4E binding protein (4EBP) to promote protein synthesis. Recent studies indicate that mTORC1 activates the translation of specific mRNAs that encode for proteins involved in translation, cell proliferation, and metabolism [15, 29]. mTORC1 is also implicated in mitochondrial function and is important for the maintenance of mitochondrial oxidative phosphorylation (OXPHOS) [30].

The best-described pathway through which amino acids signal and activate mTORC1 is via the Rag GTPases [23, 31]. In mammals, there are four Rag proteins, Rag A or Rag B form heterodimers with Rag C or Rag D, and these are localized on the surface of the lysosomes [27]. When the levels of amino acids are high, this promotes the loading of GTP to the Rag A/B proteins and the loading of GDP to the Rag C/D proteins; the heterodimer Rag A/B^{GTP}-Rag C/D^{GDP} then recruits mTOR to the lysosomes by interacting through the RAPTOR subunit. On the surface of the lysosome, mTOR co-localized with LAMP2 (lysosomal-associated membrane protein) [28, 32]. The recruitment of mTOR to the lysosome brings it into close proximity with the small GTPases Rheb protein, an activator of mTOR [33]. The guanine status of the Rag GTPases is vital for the recruitment and activation of mTORC1 [32]. Additionally, Rheb is crucial for the activation of mTORC1 by amino acids [34] and Rheb is in turn activated by growth factors via the PI3K-AKT-TSC pathway [33]; hence activation of mTORC1 is dependent on both signals from growth factors and amino acids [32]. The Rag GTPases do not have the membrane-targeting complex, and hence, it is anchored to the lysosome surface via a complex named Ragulator [35]. This complex promotes guanine nucleotide exchange of the Rag GTPases, thereby activating them [32], and depletion of the Ragulator complex interrupts the interaction of the Rag GTPases with lysosome and prevents mTORC1 activation [28]. Besides the Ragulator complex, other proteins have been shown to regulate the Rag GTPases. These include the leucine sensor leucyl-tRNA synthetase [36, 37], the adaptor protein p62 that plays a role in the formation of active Rag heterodimers [38], and during nutrient scarcity a negative regulator of mTORC1 activation; SH3 domain-binding protein 4 (SH3BP4) binds to the inactive Rag GTPase complex to inhibit the formation of active Rag complex [39].

Although the mechanism by which mTORC1 responds to amino acid signals is not precisely understood, however, it has been suggested that lysosomes may play a key role in an "inside-out" type of signaling mechanism [40]. The involvement of lysosomes in the sensing of amino acids was first suggested from the observation that yeast vacuoles accumulate nutrients, such as amino acids [41], that leads to the recruitment of mTORC1, a mechanism which is conserved in yeast [42]. Lysosomal degradation and phagocytosis cause an increase in intralysosomal nutrient levels and thus provide energy to protists such as *D. discoideum* [43], which further suggest the involvement of lysosomes in nutrient sensing. Finally, during nutrient depletion, mTORC1 is inactivated which leads to initiation of autophagy, a catabolic process where cellular components are degraded in the lysosomes and vacuoles to generate amino acids and other nutrients. However, after prolonged starvation, mTORC1 is reactivated possibly due to generation of amino acids by autophagy, and such activation is important for autophagy termination. Thus, the sufficient level of amino acids in the lysosome may be an indirect sensor resulting in mTORC1 recruitment and activation that in turn regulate various cellular processes [2, 27, 44]. There is a coordinated action between GCN2 and mTOR in nutrient sensing; when amino acid levels are low, GCN2 is activated and mTOR is inhibited to ensure that only those mRNAs that are important for cellular adaptation to nutrient starvation are translated.

Lipid Sensing

Sensing mechanisms for lipid biomolecules are primarily mediated by membrane-bound proteins particularly those associated with the endoplasmic reticulum. Among those, the primary lipid sensors are GPR120, GPR40, GPR41, GPR43, GPR84, CD36, and SCAP-SREBP complex and the HMG-CoA reductase. GPR120, GPR 40, GPR41, GPR43, and GPR84 are G-protein-coupled receptors. Interestingly, these GPR sensors have different functions in the context of types of fatty acid sensing capability. For example, GPR120 and GPR40 mediate the signal transduction for the presence of long-chain unsaturated fatty acids [45, 46], GPR41 and GPR43 sense the presence of short-chain fatty acids [47], while GPR84 mediates signal for the presence of medium-chain fatty acids [48]. CD36, GPR120, and GPR40 are expressed within the oral epithelium that is associated with gustatory perception [49]. CD36 (cluster of differentiation 36) is expressed mostly in the taste bud cells and GPR120 is expressed in the white adipose tissue. It is interesting to note that sensing of lipids by these sensors triggers a cascade of signal transductions that integrate a mechanism associated with maintenance of systemic homeostasis. For instance, CD36 activation due to presence of unsaturated fatty acids initiates the pancreatic-biliary excretion to prepare for digestion of dietary fats [49]. Similarly, interactions of GPR120 with unsaturated fatty acids trigger the activation of PI3K (phosphatidylinositide 3 kinase) and Akt (protein kinase B), which in turn mediate the glucose uptake [50]. These observations indicate that the interactions between various signaling pathways precisely maintain cellular homeostasis and disturbance in any one, create an imbalance that leads to a disease state. For example, disruption of GPR120 is associated with obesity in human [51], which indicates that signal transduction pathway is crucial in systemic regulation of homeostasis.

The presence of cholesterol is detected by binding of SCAP (SREBP1 cleavage activating protein), which is a membrane-bound protein in the endoplasmic reticulum in complex with SREBPs (sterol regulatory element-binding protein) [52, 53]. In the presence of high cholesterol, SCAP binds to cholesterol and undergoes a conformational change that triggers a dissociation of SREBPs. The cholesterol-bound SCAP in turn bind to another protein, INSIG (insulin-induced gene), which is an anchor for SCAP and SREBP within ER membranes [54]. The INSIG protein interacts with HMG-CoA (3-hydroxy-3-methyl-glutaryl-coenzyme A). This interaction triggers degradation of HMG-CoA reductase, a rate-limiting enzyme in the synthesis of cholesterol [55]. When the level of cholesterol is scarce, the SCAP-SREBP complex remains bound and gets translocated into the Golgi apparatus wherein the SREBP is cleaved by a specific protease. The cleaved fragment of SREBP is translocated into the nucleus and regulates the expression of a gene associated with cholesterol anabolic pathway such as synthesis of HMG-CoA reductase [56]. Thus, the lipid-sensing mechanism is a synergistic response in close proximity with cholesterol synthesis pathway.

Glucose Sensing

Like other sensing mechanisms for amino acids and lipids, sensing mechanisms for the carbohydrates particularly glucose do exist. Glucose intake, mobilization, and utilization are efficiently controlled by the interplay of several mechanisms that are involved at both intracellular and extracellular levels. This involves a network of hormones like insulin and glucagon, enzymes, and membrane receptors [2]. Extracellular sensing mechanism for glucose is mediated by glucose transporter presence in many cell membranes.

These glucose transporters, called GLUT, which comprise of 14 isoforms in the human genome, consist of 12 membrane-spanning domains [57]. These transporters are expressed differentially in various cell types and are classified based on their affinity and rate of transport for glucose [58, 59]. Among these, GLUT2 has higher Km for glucose, which makes GLUT2 an ideal glucose sensor since it is operational during a period of hyperglycemic conditions. For instance, the Km value of GLUT1 and GLUT4 for glucose is 1 mM and 5 mM, respectively, while that of GLUT2 is 20 mM [60]. While other glucose transporters are saturated even during fasting glycemia, GLUT2 can still transport glucose across cell membrane efficiently. In addition, GLUT2 transports glucose, bidirectionally, into the cell during transient hyperglycemic state and outside the cell into circulation when intrahepatic glucose is high; thus it serves as a key regulator of glucose homeostasis [2]. Therefore, it is obvious that a deregulated expression and function of this transporter may lead to various disease states. The transport of glucose into the cells is tightly linked with that of insulin. Insulin regulates transport of glucose into the insulin-sensitive cells via enriching concentration of the glucose transporters rather than increasing their activity [61, 62].

Once glucose is inside the cells, the next level of sensing mechanism involves the protein glucokinase, which constitutes the intracellular glucose sensor. Glucokinase is an enzyme that catalyzed the rate-limiting step in glycolysis, hence, regulating glucose consumption or storage. Glucokinase phosphorylates glucose into glucose-6-phosphate (G6P) to direct it for utilization. Among all enzymes that are involved in glucose metabolism, glucokinase is the only enzyme that acts as a glucose sensor during glucose abundance because of its low affinity for glucose. Glucokinase is activated only when the level of glucose is relatively high (around 120 mg dl^{-1}, or 7 mM, or greater). Glucokinase is principally expressed in liver cells [63], which make it a principal regulator of glucose homeostasis. During glucose limitation, glucokinase becomes inactive and permits export of glucose from the liver for utilization to support the energetic demand of the brain and muscles.

It is interesting to note that AMPK and mTOR may also play a role in glucose sensing. Since glucose metabolism determines the cellular AMP/ATP ratio and AMPK is activated under high AMP/ATP ratio, therefore, it is possible that glucose availability may be sensed by AMPK enzyme [64]. Low glucose levels are mainly detected by sensory neurons that are present in both the brain [65, 66] and periphery [67, 68]. Although the mechanism is not well understood, however, neurons, present in the ventromedial hypothalamus (VMH), are activated when circulating glucose level is increased (glucose-excited, GE) and become inactivated when glucose level is decreased (glucose-inhibited, GI) [69]. AMPK has been shown to play a key role as glucose sensor in these neurons. In them, AMPK detects the hypoglycemic condition and triggers the glucose counter-regulatory response. This is achieved via promoting the release of counter-regulatory hormones (adrenaline and glucagon) that stimulate glucose production by the liver [70]. Recent evidence also has implicated mTOR protein to behave as a glucose sensor through the Rag GTPases [71]; however, the mechanism of action remained to be elucidated.

Energy Sensing

In addition to sensing the availability of macronutrients, the cells are also capable of sensing the alteration of nutrient levels through the changes in the level of certain metabolites such as $NAD^+/NADH$ or the AMP/ATP.

AMPK Senses AMP/ATP

Almost every physiological process requires energy. This energy is provided by hydrolysis of ATP, and the free energy release during this process is used to drive many biochemical reactions crucial for cell survival and sustenance. Increased ATP consumption will increase the level of AMP concentration in the cell. The AMP/ATP ratio therefore signals the energy status of the cell. Thus, it is logical that there should be a balance between ATP production and utilization so that cell functionality is sustained. Although a small number of metabolic enzymes exist that sense directly the AMP/ATP ratio, namely, glycogen phosphorylase and phosphofructokinase that are activated by increased AMP/ATP ratio and fructose-1,6-bisphosphatase that is inhibited by increased AMP/ATP ratio [64], however, the principal sensor that plays a key role in this context is an enzyme called AMP-activated protein kinase (AMPK). It is a heterotrimeric protein consisting of three subunits, namely, a catalytic alpha (α), scaffolding beta (β), and a regulatory gamma (γ) subunit. AMPK is known as the fuel sensor of the cell, working to ensure that energy homeostasis of the cell is maintained under conditions of cellular stresses such as exercise, ischemia, starvation, hypoxia, and glucose starvation [72, 73]. The mechanism by which AMPK senses the low energy status is through binding of the AMP or ADP to the (γ) subunit at the nucleotide-binding site present at the C-terminal. This binding causes the gamma (γ) subunit to interact with the linker region of the alpha (α) subunit. This pulls the autoinhibitory domain on the alpha (α) subunit away from the kinase domain. This induces conformational changes that protect AMPK from being deactivated (dephosphorylation) by the phosphatases [73–75]. The downstream effects of AMPK activation involve a cascade of metabolic processes that promote the production of ATP while at the same time inhibiting those that consume ATP. The output is to preserve energy homeostasis. The catabolic events toward ATP generation include increased fatty acid oxidation, increased glucose transport, and enhanced autophagy but inhibit, simultaneously, anabolic processes such as fatty acid and protein synthesis that require ATP consumption [73]. The increased glucose uptake in the tissues is through increased membrane expression of glucose transporters such as GLUT4 (muscle cells and liver) and GLUT1 (in other cells) [76]. Similarly, AMPK enhances fatty acid uptake via increased expression of fatty acid transporters such as CD36 at the plasma membrane [77]. AMPK promotes catabolism of glucose by activation of glycolysis via phosphorylation of 6-phosphofructo-2-kinase/fructose-2,6-biphosphatase (PFKFB). The phosphorylated PFKFB catalyzes the generation of fructose-2,6-bisphosphate, which is a key allosteric activator of the glycolytic enzyme 6-phosphofructo-1-kinase (PFK1) [78]. AMPK promotes fatty acid oxidation in the mitochondria by phosphorylation and inactivation of the acetyl-CoA carboxylase (ACC2 also called ACCβ), triggering a drop in concentration of ACC product, malonyl CoA, which is an inhibitor of fatty acid transport into mitochondria [79]. On the other side, AMPK prevents anabolic ATP-consuming biosynthetic processes to conserve ATP. This is achieved by phosphorylation and inactivation of regulatory process associated with protein, carbohydrate, and fatty acid biosynthesis. For example, it phosphorylates ACC1 to inhibit fatty acid synthesis [80]. Interestingly, AMPK inhibits the activity of mTOR and, thus, coordinately maintains cellular homeostasis through enhanced autophagy and inhibiting protein synthesis. This inhibition of mTOR is mediated through phosphorylation of tuberous sclerosis complex 2 (TSC2) and regulatory-associated protein of mTOR (RAPTOR) [81, 82]. One of the most important signaling cascades mediated by the activated AMPK is the direct phosphorylation and activation of peroxisome proliferator-activated receptor-gamma coactivator 1 alpha (PGC-1α), a transactivator and metabolic regulator of energy metabolism. PGC-1α is a "master regulator" of mitochondria biogenesis. The impact of PGC-1α on energy homeostasis is through coordinating expression of genes that stimulate mitochondrial biogenesis [83, 84].

Given that AMPK forms the central regulator of energy homeostasis crucial for cell survival, deregulation of its activity may be associated with many pathophysiological states including age-related diseases such as atherosclerosis, neurodegenerative disease, diabetes, and cancer. For instance, AMPK inhibits cancer and tumorigenesis [85]. The anticancer property of AMPK is elicited through inhibition of mTOR and PTEN activity as well as blocking cyclin-dependent kinase activity. Additionally, AMPK phosphorylates eEF-2 kinase to induce autophagy. The overall outcome is the cell cycle arrest and protein synthesis inhibition that prevents cell growth and proliferation [85]. Therefore, growing interest is focused on preventive capacity of many interventions that maintain AMPK activity, and one of those is dietary restriction (DR).

Sirtuins Sense NAD⁺/NADH Level

Nicotinamide adenine dinucleotide (NAD⁺) and its reduced form NADH are main electron carriers that play a key role in many metabolic pathways such as glycolysis and TCA cycle and are involved in the electron transport chain. They also have a role as cofactors in many other processes such as protein deacetylation executed by sirtuins [86]. Because of its role in various metabolic pathways, the ratio of NAD^+ to NADH reflects the metabolic status of the cell and changes in this ratio can regulate many proteins that are sensitive to it. An example of such changes is during muscle differentiation where the NAD^+ to NADH ratio decreases that leads to a concomitant decrease in Sir2 [87]. Alteration in the level of NAD(H) has several consequences including cell metabolism, cell survival, and aging [88]. NAD^+ is generated by the action of the enzyme NAMPT (nicotinamide phosphoribosyltransferase); NAMPT converts nicotinamide and 5′-phosphoribosyl-pyrophosphate to nicotinamide mononucleotide (NMN), a key intermediate of NAD^+. The NMN generated is then adenylated by NMN adenylyltransferases (NMNAT 1–3) to produce NAD^+ [89]. In the nucleus and mitochondria, the NAD^+ biosynthetic pathway via NAMPT is a well-known process that controls sirtuin activity [90].

Sirtuins are mostly lysine-specific NAD^+-dependent protein deacetylases that were first recognized to catalyze the removal of acetyl group from histone proteins. There are seven types of sirtuins in mammals that are localized in different compartments of the cell. SIRT1, SIRT2, SIRT6, and SIRT7 are present in the nucleus; SIRT1 and SIRT2 are found in the cytoplasm as well. SIRT3, SIRT4, and SIRT5 are localized in the mitochondria [91, 92]. The varied localization of sirtuins suggests that they have other non-histone substrates. During nutrient scarcity, the NAD^+/NADH ratio increases leading to the activation of sirtuins. Sirtuins in the presence of NAD^+ catalyzed the transfer of an acetyl group that is supposed to be removed from the target protein to the NAD^+ cofactor, and the products are a deacetylated protein, an OAADPR (O-acetyl-ADP-ribose), and an NAD-derived nicotinamide [93]. Sirtuins are involved in many cellular functions including energy metabolism, DNA repair, cellular senescence, stress, inflammation, neurodegeneration, and cancer.

The presence of sirtuins in different cellular compartments allows them to sense the changes in energy level in the nucleus, the cytoplasm, and the mitochondria contributing to their versatile functions [94]. Availability of nutrients such as glucose increases the glycolytic rate and hence reduces the NAD^+/NADH ratio. Because all sirtuins are NAD^+ dependent, they have low activity when the supply of nutrients is high. However, during nutrient depletion and when the glucose level is low, cellular energy decreases leading to an increase in NAD^+/NADH ratio; this causes an increase in the activity of sirtuins [95]. It has been shown that in human liver, almost all the enzymes involved in metabolic pathways such as glycolysis, gluconeogenesis, tricarboxylic acid, and urea cycles are acetylated. For instance, enzymes involved in fatty acid and glycogen metabolic pathways have also been found to be acetylated. Acetylation of the enzymes in the glycolytic, tricarboxylic acid, and fatty acid oxidation pathways seems to have a positive effect, while acetylation of the enzymes involved in gluconeogenic pathway is negative [96]. It has been hypothesized that since sirtuins are activated when nutrients are limited, they may be involved in regulating the energy status of the cell by deacetylating these acetylated metabolic enzymes; however, it has not been proven until date [97].

Nutrient-Sensing Molecules, Aging, and Dietary Restriction

Aging is defined as the progressive loss of biological function with the passage of time. It is an irreversible alteration in the physiology of the organism leading to failure of tissues, organs, and the organism as a whole. Aging is associated with altered nutrient-sensing mechanism manifested by the changes in the expression level of many nutrient-associated sensors. These changes may result in altered metabolism [98] and rising incidence of certain age-related diseases such as decreased bone density, insulin resistance, and increased inflammation. Osteoarthritis and osteoporosis are age-associated diseases [99, 100]. Growth and aging may seem to be opposite in nature; however, it may be noted that when growth ceases, aging follows. The same factors that regulate growth are implicated to drive aging, and manipulations of these factors that decrease growth result in decreased aging and prolonged lifespan [101]. The process of aging is primarily dependent on the genetic and epigenetic factors. Besides these, the environment to which the organism is exposed also affects the process of aging and longevity. One of the environmental factors that has been extensively studied and shown to influence aging and longevity is the limitation in intake of nutrients (dietary restriction) [102, 103]. Several longevity genes and pathways have been identified that are involved in lifespan extensions. The nutrient-sensing molecules and pathways are those that are implicated to affect the aging process and can modulate longevity. One of the pathways that have been shown to consistently influence aging in a wide spectrum of animals is the insulin-like growth factor (IGF)-1 signaling cascades [104, 105]. Alteration of this pathway by dietary restriction (DR), a process where the intake of food is reduced without causing malnutrition, has beneficial lifespan effect on animals ranging from yeast to humans. There is a positive association between mutations of genes of IGF-1 signaling cascade with extension of lifespan in humans [106, 107]. The effect of dietary restriction on IGF-1 signaling pathway is mediated through the FoxO transcription factors [105, 108]. In addition, one of the effectors of the growth factor is TOR. TOR activity is sensitive to many environmental cues including nutrition availability and hormone actions [22]. It integrates the signal from these factors to elicit several cellular processes. mTOR is also considered to be one of the key modulators in aging and age-related diseases. Its deregulation can cause cancer and lead to many age-associated diseases. Inhibition of this pathway imparts protection to several age-associated pathologies and extends lifespan in several organisms [109]. Through its downstream substrates, S6K and 4E-BP, mTOR promotes protein synthesis and is a master regulator of growth and metabolism. It also negatively regulates autophagy [110]. mTOR is implicated in its role on aging and longevity through the regulation of protein synthesis and autophagy [111]. The extension of lifespan observed as a result of DR has been shown to be mediated through the mTOR signaling cascades in various organisms [112]. In *C. elegans*, lowering the expression of proteins involved in mRNA translation such as eIF4e and S6K extends the lifespan of the organism. As protein synthesis is an expensive mechanism, it has been suggested that the reduction in protein synthesis may cause a decrease in energy consumption hence decreasing respiration and ROS production [111,

113–115]. Indeed mTORC1 signaling has been shown to enhance mitochondrial function by increasing the transcription of genes responsible for it [30]. This may lead to increase the production of energy required for anabolic processes. During nutrient abundance, mTOR signaling promotes ribosome biogenesis, and the number of ribosomes indicates the capacity of cell to undergo protein synthesis and enhanced cell growth and division, but during nutrient depletion, mTOR signaling is inhibited and there is increase in lifespan, which may be due to a decreased ribosome production [116]. The role of mTOR on autophagy is evident from studies which showed that inactivation of autophagy gene reverses the lifespan-extending mechanism of mTOR inhibition [117]. When there is loss of function of autophagy gene in *C. elegans*, there are increased tissue aging and decreased lifespan [118]. Similarly, flies and yeast have a shorten lifespan when the autophagy gene is mutated suggesting that loss of autophagy increases aging [119, 120]. DR in worms, flies, and rodents induces autophagy since mTOR signaling is impaired as amino acid levels are low [121–123]. The decrease in autophagy with age can be rescued by dietary restriction in mice [124].

Another protein that senses amino acid deprivation, GCN2, does not affect aging under normal conditions. Studies reveal that when GCN2 gene is deleted, there is no effect on growth, but under amino acid limitation, there is reduction in the lifespan of these wild-type animals [16]. The disruption of *gcn2* gene in a DR model of worm (eat-2 mutants) reverts the enhanced lifespan of the organism. Additionally, the extended lifespan achieved by inactivation of the TOR substrate was also affected by deletion of GCN2. In *C. elegans*, TOR negatively regulates a transcription factor PHA-4/FoxA, which is responsible for DR-mediated longevity in adults [125]. During DR and amino acid deprivation when TOR is inactivated, GCN2 is required for the activation of *pha-4*. PHA-4 target genes include stress survival and autophagy genes, and these are activated in a GCN2-mediated manner upon amino acid starvation causal to stress survival and longevity of worms [16]. Therefore, the lifespan extension due to mTOR inhibition may in part be dependent on GCN2 activation.

Certain evidence has implicated the role of lipid sensors in age-related diseases. For example, studies in model organisms indicate that mice lacking GPR120 developed insulin resistance, glucose intolerance, and enhanced hepatic lipogenesis with decreased adipocyte differentiation [51]. The role of GPR120 has been shown to mediate anti-inflammatory reactions and enhance insulin-sensitizing and antidiabetic effect via suppressing macrophage-induced tissue inflammation [50]. Similarly, studies have also indicated that mice deficient of GPR40 exhibit osteoporotic features. GPR40 is expressed in bone cells and is, therefore, implicated to play a role in mediating the effects of fatty acids on bone remodeling and bone density [126]. In another study, GPR40 is implicated to mediate anti-inflammation reaction in osteoarthritis. Chondrocytes from GPR40-deficient mice exhibit increased secretion of inflammatory mediators. Therefore, there are good evidences that targeting these lipid receptors may improve the quality of life in the elders and decrease mortality. Apart from these, there is an age-dependent change in the level of cholesterol sensors in both rodents and human. The changes in the levels of these sensors are responsible for the increased circulating level of cholesterol during aging, a condition called hypercholesterolemia that is well associated with cardiac pathology. During aging, nuclear SREBPs rise, indicating that cholesterol synthesis increases during aging. INSIG proteins which inhibit SREBPS and HMG-CoA reductase (HMG-CoAR) activity are also reduced [127]. In aged rats, the hepatic HMG-CoAR, the rate-limiting enzyme of cholesterol biosynthesis, is fully active, and its degradation rate is slow [128, 129]. Several evidences indicate the beneficial effect of DR to modulate the age-related changes relevant to lipid metabolism. Studies have implicated that DR reduces the circulating level of cholesterol and its precursors such as lanosterol and lathosterol in the brain of old Wistar rats [130]. This beneficial effect of DR in maintenance of cholesterol homeostasis is brought about by the reducing activity of HMG-CoA reductase and other related proteins associated with regulation of its activity and expression such as transcription factor SREBPs and INSIG proteins [127]. All these changes of the lipid sensor that regulate cholesterol synthesis may play a role in triggering and aggravating age-related disorders, and DR may alleviate such disorders by maintaining cholesterol homeostasis.

Energy and metabolic homeostasis determine the longevity and cellular senescence. As aging is associated with reduced energy production, therefore, factors that regulate processes associated with efficient energy production may play a vital role in determining health span and lifespan of an individual. Among those, alteration in the glucose transporters (GLUT) may have an impact in determining the accelerating rate of aging process because these regulate the efficient transport of glucose in and out of the cells. Thus, directly determine the availability of glucose for energy production. For instance, GLUT8, glucose transporters expressed highly in testis but also in the brain, heart, skeletal muscle, spleen, prostate, and intestine cells, decrease significantly during aging in mouse's testis [131–133]. This indicates that reduced glucose availability may be responsible for decreased energy production to drive testosterone synthesis during aging, which is well connected with the age-related decrease of spermatogenesis in male reproductive system. Age-dependent decrease of GLUT does not necessarily indicate that they directly regulate the aging process. However, increased levels of these transporters are also associated with increasing incidence of many age-related diseases. For example, in aged lungs, GLUT1 play a key role in glycolysis induction, which is associated with increasing incidence of age-related lung fibrosis in mouse and human [134]. As the animal ages, one of the main characteristics is the increasing incidence of insulin secretion complications. Alteration in the glucose sensing in the β-cells both in human and animal models has become a prime focus. Reports have indicated glucose-induced insulin secretion decrease drastically as the animal ages. These complications have been characterized and associated with the age-dependent decline expression of GLUT2 in the pancreatic β-cells [135]. A reduced expression of GLUT2 in these insulin-secreting cells may be the etiological factor in diabetes and many age-related metabolic pathophysiological conditions. Alteration of hepatic glucokinase activity also has a profound effect on circulating glucose and hepatic glucose flux. Age-related change in the enzyme will determine the efficient utilization of glucose. In diabetic conditions, the levels of this enzyme reduced significantly which is responsible for the increased glucose circulating level, and normalizing expression of this enzyme increases glucose sensitivity and improves hepatic glucose metabolisms [136]. Thus, the interplay of various glucose sensing molecules both at the extracellular and intracellular levels will determine the physiological state of an organism. Disturbing the link that connects these sensing molecules will lead to disease conditions as witnessed in diabetic conditions.

Maintenance of energy homeostasis forms an integral part of driving many physiological processes, and imbalance in energy status may provoke onset of diseases and jeopardize healthy aging. AMPK is a metabolic energy sensor that coordinates and integrates energy balancing, and as such, correct regulation of this sensor is associated with health and survival. Under energy stress, AMPK triggers a cascade of biochemical signals that stimulate ATP production; thus it effectively regulates energy metabolism and cellular homeostasis [137]. Hence, AMPK acts as a cellular energy housekeeper. Therefore, it is plausible that age-associated diseases are linked to perturbed energy homeostasis and deregulated AMPK. For instance, declined AMPK activity during aging provokes diseases such as cardiovascular diseases and metabolic syndrome [138]. Since AMPK triggers fatty acid oxidation via enhanced mitochondrial functions, reduced mitochondrial oxidative phosphorylation activity has been implicated with increased incidence of insulin resistance and type 2 diabetes during aging [139, 140]. AMPK is also indicated to control lifespan of an organism as evident by the fact that stimulation of AMPK activity is associated with extension of the lifespan of various model organisms [141, 142]. On the other hand, inhibition of AMPK is linked to disturbed energy homeostasis that jeopardized both health span and lifespan [143]. From these findings, it is evident that AMPK acts as a mediator of lifespan extension, but whether it acts alone or in concert with other AMPK-linked signaling networks needs further studies. Nevertheless, it is apparent by a plethora of studies that dietary restriction delays aging process and extends lifespan in a broad spectrum of model organisms by pathways that are under control of AMPK activity. These signaling networks include inhibition of CRTC-1/CRE, NF-kB, and mTOR, while it activates SIRT1, ULK1, p53, and FoxO. Various studies have indicated that responsiveness of AMPK activation declines with age [144]. This entailed that reduced activity of AMPK in one hand would attenuate autophagic clearance through enhanced activity of mTOR, while in the other, it leads to reduced activity of sirtuins, FoxO transcription factor, and p53 that are well associated with maintenance of homeostasis during stress responses [145]. DR elicits lifespan extension through its physiological low energy state; therefore, AMPK makes an appealing mechanistic link between energy status, DR, and lifespan by its ability to integrate and crosstalk with other signaling molecules to promote longevity.

Sirtuins are one of the main regulators of aging and the antiaging roles of sirtuins are evolutionarily conserved. The availability of NAD^+ decreases with aging, so does the activity of sirtuins. This reduction in sirtuins activity affects both the cellular and systemic communications in an organism and may play a contributory role in the development of age-associated pathologies [146, 147]. Sirtuins are involved in the regulation of several biological processes in the body. Besides its deacetylase enzyme action, it has many other activities such as ADP-ribosyltransferase, demaloynylase, desuccinylase, deglutarylase, long-chain deacylase, and lipoamidase. All these enzyme activities require the cofactor NAD^+. SIRT1 is the most studied sirtuins and is the closest homolog to the yeast Sir2. The overexpression of SIRT1 in the brain delays aging and extends the lifespan of both male and female mice [148]. SIRT1 expression in the brain, particularly the hypothalamus, is important for longevity in mammals [149], whereas whole-body expression of SIRT6 in transgenic mice exhibits lifespan extension although this was observed only in males [150]. The relation between NAD^+ and sirtuin is vital during aging and longevity. The interaction between NAD^+ and sirtuin is dependent on the biosynthesis of NAD^+, the regulation of sirtuin activity by NAD^+ derivatives/substrate, and the degree of usage by other NAD^+-dependent mechanisms. The alteration in the biosynthesis of NAD^+ affects the activity of sirtuin. With age, there is a decline in the level of NAMPT and the synthesis of NAD^+ in various organs of mammals [151, 152]; moreover, during diseased state the NAMPT-mediated NAD^+ biosynthesis declines [153, 154]. This reduction in NAD^+ results in the decrease of sirtuin activity with age, which may be responsible for the development of many age-related diseases. NAD^+ substrates and derivatives such as nicotinamide and NADH can also modulate sirtuin activity. The NAD^+/NADH ratio has been shown to be a regulator of sirtuin activity as NADH is a competitive inhibitor for sirtuin [155]. Other enzymes also utilize NAD^+ for catalyzing a reaction, and one of the enzymes that compete with SIRT1 for NAD^+ is PARP (poly-ADP-ribose polymerases) [156, 157]. It has been shown that in aging worms and mice, PARP is continually activated, and this leads to increase in poly-ADP-ribosylation of cellular proteins. This activation of PARP may result in depletion of NAD^+, thereby decreasing sirtuin activity [146, 158]. When the availability of NAD^+ is low and sirtuin activity decreases with age, there is a deregulation of mitochondrial metabolic function through the stability and increased level of HIF-1α, which in turn leads to defective TFAM expression, a gene responsible for coding mitochondrial transcription factor [151]. SIRT1 low activity also decreases PGC-1α and FoxO1 function, and this, in turn, causes decrease in oxidative metabolism, mitochondrial biogenesis, and antioxidant defense pathways [84, 159]. In mammals, SIRT1 can activate mitochondrial unfolded protein response such as HSP60, which are implicated in aging and longevity [160]. The increased in SIRT1 activity has been observed in both mice and humans under dietary restriction [94]. Increased SIRT1 activity can lead to increase in deacetylation of several protein pathways such as AMPK signaling, PGC-1α, and autophagy [161]. Enhanced SIRT1 activity also increases OXPHOS activity which otherwise declines during aging [151]. In addition, SIRT3 is also involved in the beneficial DR-mediated effect on mitochondrial function [162]. SIRT3 can regulate ROS levels through the activation of FoxO transcription factor resulting in enhanced expression of antioxidant proteins such as SOD (superoxide dismutase) [163]. During different types of stress including DR, the biosynthetic pathway of NAD^+ is activated by the induction of NAMPT, which causes an increase in sirtuin activity [164]. Hence, the failure in the connection between NAD^+ and sirtuins leads to mitochondrial dysfunction, one of the hallmarks of aging.

Conclusion

Nutrient sensing is central to survival and sustenance of an organism. The cells are highly capable of sensing the diverse range of nutrients available in nature to ensure proper functionality of the organism. The changes in the availability of nutrients detected by cells with extreme precision ascertain that there is a coordinated integration of the nutrient-sensing pathway to maintain cellular homeostasis (Fig. 18.1). Disturbance of any of these nutrient-sensing molecules and pathways creates an

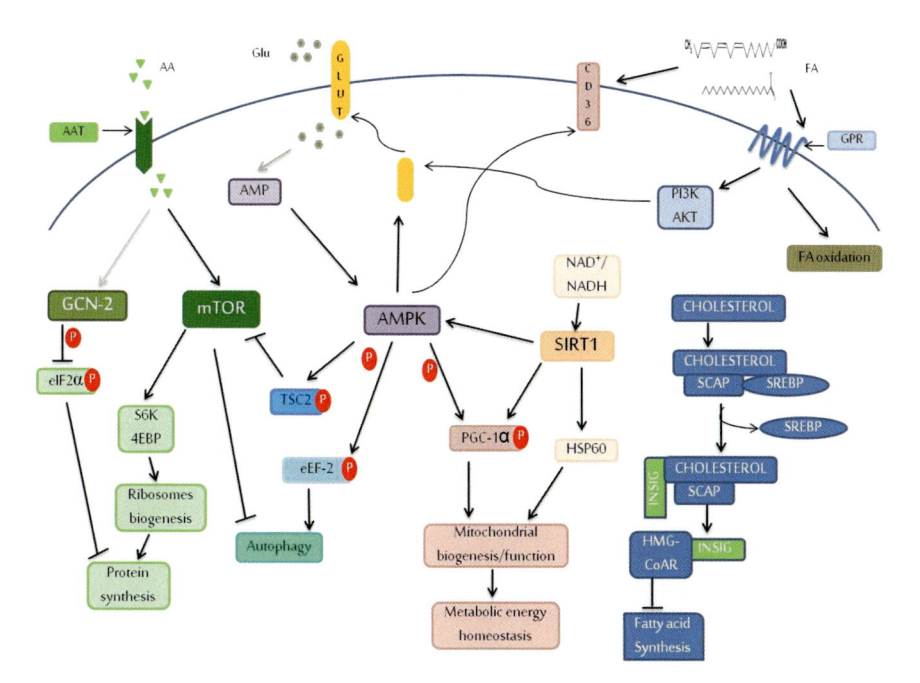

Fig. 18.1 Interplay of nutrient-sensing pathway. Nutrient-sensing molecules can be receptors (GPR, CD36) or transporters (GLUT) present at the plasma membrane or intracellular proteins (mTOR, GCN2, AMPK, SIRT, HMG-CoAR). The nutrients detected by these molecules can communicate with each other to ensure an overall homeostasis of the cell. The signals detected trigger a cascade of intracellular signaling, which integrates to regulate various cellular processes. Deregulation of the nutrient-sensing pathway enhances the aging process and intervention by DR promotes healthy aging and longevity. [Gray bold arrow depicts during nutrient deprivation and black bold arrow indicates during nutrient abundance.] Abbreviation: *AA* amino acids, *AAT* amino acid transporter, *GCN2* kinase general control nonderepressible 2, *mTOR* mammalian/mechanistic target of rapamycin, *eIF2α* eukaryotic initiation factor 2 alpha, *S6K* ribosomal protein S6 kinase, *4EBP* eIF4E-binding protein, *Glu* glucose, *GLUT* glucose transporter, *AMP* adenosine monophosphate, *AMPK* AMP kinase, *TSC2* tuberous sclerosis complex 2, *eEF-2* eukaryotic elongation factor 2, *PGC-1α* peroxisome proliferator-activated receptor gamma coactivator 1-alpha, NAD^+ nicotinamide adenine dinucleotide (oxidized), *NADH* nicotinamide adenine dinucleotide (reduced), *SIRT1* sirtuin 1, *HSP60* heat shock protein 60, *FA* fatty acids, *PI3K* phosphatidylinositide 3-kinase, *Akt* protein kinase B, *GPR* G-protein-coupled receptors, *CD36* cluster of differentiation 36, *SCAP* SREBP cleavage-activating protein, *SREBP* sterol regulatory element-binding protein, *INSIG* insulin-induced gene, *HMG-CoAR* 3-hydroxy-3-methyl-glutaryl-coenzyme A reductase

Table 18.1 List of nutrient-sensing molecules and the changes during aging and DR

Nutrient sensors	Nutrients sense for	Tissues present in	Aging/age-related diseases	DR	References
mTOR	Amino acids, glucose	Most tissues	↑	↓	[1, 111, 112]
GCN2	Amino acids	Most tissues	–	↑	[5, 16]
GLUT	Circulating glucose	Heart, liver, hypothalamus and many other tissues	↓	↑	[58, 59, 140, 165]
IGF-1	Glucose, amino acids	Most tissues	↓	↓↓	[104, 106, 107]
Hypoxia-inducible factor 1α (HIF-1α)	Oxygen sensing, glucose sensing	Hypothalamus	↑	–	[166, 167]
CD36	Unsaturated fatty acids	Taste buds cell, adipose tissues	↑	–	[49, 168, 169]
GPR	Fatty acids	Adipose tissue, large intestine, lungs	↓	–	[51, 170]
HMG-CoA reductase	Cholesterol		↑	↓	[127–129]
AMPK	Energy sensing, glucose sensing	Most tissues	↓	↑	[64, 73, 138]
Sirtuins	NAD+/NADH	Most tissues SIRT1: Cytosol/nucleus SIRT2: Cytosol SIRT3: Mitochondria SIRT4: Mitochondria SIRT5: Mitochondria SIRT6: Nucleus SIRT7: Nucleus	↓	↑	[94, 146, 147]

imbalance leading to a pathological state (Table 18.1). Nutrient availability promotes biosynthesis and storage, while starvation supports breakdown and salvage of stored nutrients to provide the necessary energy for the body. During aging, there is a deregulation of the nutrient-sensing cascade, and this is associated with the development of several age-related diseases. In fact, deregulated nutrient sensing is one of the characteristics of aging. These nutrient-sensing molecules mediate the process of growth and development and maintenance of systemic homeostasis. These same molecules are responsible for accelerating the rate of aging when the growth rate has reached its plateau. Interventions either pharmaceutical or nutraceuticals that can modulate and maintain the healthy interplay of these pathways can be a novel target to ameliorate the deteriorating process of aging and promote healthy lifespan. Among the interventions, dietary restriction offers a promising tool to healthy aging and longevity. Some of the presently available pharmaceuticals that have been shown to mimic DR and improve health span include resveratrol and

STACs that enhance sirtuin activity, rapamycin and its related drugs that inhibit mTOR, and metformin that activates AMPK.

Acknowledgments Authors thank the Department of Biochemistry, North-Eastern Hill University, Shillong, for the research facilities. We thank the UGC-SAP program in Biochemistry, the DBT Infrastructure, and Biotech Hub to NEHU for financial support.

References

1. Chantranupong L, Wolfson RL, Sabatini DM. Nutrient-sensing mechanisms across evolution. Cell. 2015;161:67–83.
2. Efeyan A, Comb WC, Sabatini DM. Nutrient sensing mechanisms and pathways. Nature. 2015;517:302–10.
3. López-Otín C, Blasco MA, Partridge L, Serrano M, Kroemer G. The hallmarks of aging. Cell. 2013;153:1194–217.
4. Gallinetti J, Harputlugil E, Mitchell JR. Amino acid sensing in dietary-restriction-mediated longevity: roles of signal-transducing kinases GCN2 and TOR. Biochem J. 2013;449:1–10.
5. Dong J, Qiu H, Garcia-Barrio M, Anderson J, Hinnebusch AG. Uncharged tRNA activates GCN2 by displacing the protein kinase moiety from a bipartite tRNA-binding domain. Mol Cell. 2000;6:269–79.
6. Ibba M, Soll D. Aminoacyl-tRNA synthesis. Annu Rev Biochem. 2000;69:617–50.
7. Diallinas G, Thireos G. Genetic and biochemical evidence for yeast GCN2 protein kinase polymerization. Gene. 1994;143:21–7.
8. Berlanga JJ, Santoyo J, De Haro C. Characterization of a mammalian homolog of the GCN2 eukaryotic initiation factor 2alpha kinase. Eur J Biochem. 1999;265:754–62.
9. Scheuner D, Song B, McEwen E, Liu C, Laybutt R, Gillespie P, Saunders T, Bonner-Weir S, Kaufman RJ. Translational control is required for the unfolded protein response and in vivo glucose homeostasis. Mol Cell. 2001;7:1165–76.
10. Narasimhan J, Staschke KA, Wek RC. Dimerization is required for activation of eIF2 kinase Gcn2 in response to diverse environmental stress conditions. J Biol Chem. 2004;279:22820–32.
11. Hinnebusch AG. Evidence for translational regulation of the activator of general amino acid control in yeast. Proc Natl Acad Sci USA. 1984;81:6442–6.
12. Bushman JL, Asuru AI, Matts RL, Hinnebusch AG. Evidence that GCD6 and GCD7, trans-lational regulators of GCN4, are subunits of the guanine nucleotide exchange factor for eIF-2 in Saccharomyces cerevisiae. Mol Cell Biol. 1993;13:1920–32.
13. Wek SA, Zhu S, Wek RC. The histidyl-tRNA synthetase-related sequence in the eIF-2a protein kinase GCN2 interacts with tRNA and is required for activation in response to starvation for different amino acid. Mol Cell Biol. 1995;15:4497–506.
14. Munn DH, Sharma MD, Baban B, Harding HP, Zhang Y, Ron D, Mellor AL. GCN2 kinase in T cells mediates proliferative arrest and anergy induction in response to indoleamine 2,3-dioxygenase. Immunity. 2005;22:633–42.
15. Thoreen CC, Chantranupong L, Keys HR, Wang T, Gray NS, Sabatini DM. A unifying model for mTORC1-mediated regulation of mRNA translation. Nature. 2012;485:109–13.
16. Vlanti A, Rousakis A, Syntichaki P. GCN2 and TOR converge on aging. Aging. 2013;5:584–5.
17. Yang X, Matsuda K, Bialek P, Jacquot S, Masuoka HC, Schinke T, Li L, Brancorsini S, Sassone-Corsi P, Townes TM, Hanauer A, Karsenty G. ATF4 is a substrate of RSK2 and an essential regulator of osteoblast biology; implication for Coffin-Lowry Syndrome. Cell. 2004;117:387–98.
18. Deng J, Harding HP, Raught B, Gingras AC, Berlanga JJ, Scheuner D, Kaufman RJ, Ron D, Sonenberg N. Activation of GCN2 in UV-irradiated cells inhibits translation. Curr Biol. 2002;12:1279–86.

19. Phung TL, Ziv K, Dabydeen D, Eyiah-Mensah G, Riveros M, Perruzzi C, Sun J, Monahan-Earley RA, Shiojima I, Nagy JA, Lin MI, Walsh K, Dvorak AM, Briscoe DM, Neeman M, Sessa WC, Dvorak HF, Benjamin LE. Pathological angiogenesis is induced by sustained Akt signaling and inhibited by rapamycin. Cancer Cell. 2006;10:159–70.
20. Sarbassov DD, Ali SM, Sengupta S, Sheen JH, Hsu PP, Bagley AF, Markhard AL, Sabatini DM. Prolonged rapamycin treatment inhibits mTORC2 assembly and Akt/PKB. Mol Cell. 2006;22:159–68.
21. Garami A, Zwartkruis FJ, Nobukuni T, Joaquin M, Roccio M, Stocker H, Kozma SC, Hafen E, Bos JL, Thomas G. Insulin activation of Rheb, a mediator of mTOR/S6K/4E-BP signaling, is inhibited by TSC1 and 2. Mol Cell. 2003;11:1457–66.
22. Laplante M, Sabatini DM. mTOR signaling in growth control and disease. Cell. 2012;149:274–93.
23. Sancak Y, Peterson TR, Shaul YD, Lindquist RA, Thoreen CC, Bar-Peled L, Sabatini DM. The Rag GTPases bind raptor and mediate amino acid signaling to mTORC1. Science. 2008;320:1496–501.
24. Hara K, Yonezawa K, Weng QP, Kozlowski MT, Belham C, Avruch J. Amino acid sufficiency and mTOR regulate p70 S6 kinase and eIF-4E BP1 through a common effector mechanism. J Biol Chem. 1998;273:14484–94.
25. Avruch J, Long X, Ortiz-Vega S, Rapley J, Papageorgiou A, Dai N. Amino acid regulation of TOR complex 1. Am J Physiol Endocrinol Metab. 2009;296:592–602.
26. Duran RV, Oppliger W, Robitaille AM, Heiserich L, Skendaj R, Gottlieb E, Hall MN. Glutaminolysis activates Rag-mTORC1 signaling. Mol Cell. 2012;47:349–58.
27. Kim J, Guan KL. Amino acid signaling in TOR activation. Annu Rev Biochem. 2011;80:1001–32.
28. Jewell JL, Guan KL. Nutrient signaling to mTOR and cell growth. Trends Biochem Sci. 2013;38:233–42.
29. Hsieh AC, Liu Y, Edlind MP, Ingolia NT, Janes MR, Sher A, Shi EY, Stumpf CR, Christensen C, Bonham MJ, Wang S, Ren P, Martin M, Jessen K, Feldman ME, Weissman JS, Shokat KM, Rommel C, Ruggero D. The translational landscape of mTOR signalling steers cancer initiation and metastasis. Nature. 2012;485:55–61.
30. Cunningham JT, Rodgers JT, Arlow DH, Vazquez F, Mootha VK, Puigserver P. mTOR controls mitochondrial oxidative function through a YY1-PGC-1alpha transcriptional complex. Nature. 2007;450:736–40.
31. Kim E, Goraksha-Hicks P, Li L, Neufeld TP, Guan KL. Regulation of TORC1 by Rag GTPases in nutrient response. Nat Cell Biol. 2008;10:935–45.
32. Sancak Y, Bar-Peled L, Zoncu R, Markhard AL, Nada S, Sabatini DM. Ragulator-Rag complex targets mTORC1 to the lysosomal surface and is necessary for its activation by amino acids. Cell. 2010;141:290–303.
33. Saucedo LJ, Gao X, Chiarelli DA, Li L, Pan D, Edgar BA. Rheb promotes cell growth as a component of the insulin/TOR signalling network. Nat Cell Biol. 2003;5:566–71.
34. Long X, Ortiz-Vega S, Lin Y, Avruch J. Rheb binding to mammalian target of rapamycin (mTOR) is regulated by amino acid sufficiency. J Biol Chem. 2005;280:23433–6.
35. Bar-Peled L, Schweitzer LD, Zoncu R, Sabatini DM. Ragulator is a GEF for the Rag GTPases that signal amino acid levels to mTORC1. Cell. 2012;150:1196–208.
36. Bonfils G, Jaquenoud M, Bontron S, Ostrowicz C, Ungermann C, De Virgilio C. Leucyl-tRNA synthetase controls TORC1 via the EGO complex. Mol Cell. 2012;46:105–10.
37. Han JM, Jeong SJ, Park MC, Kim G, Kwon NH, Kim HK, Ha SH, Ryu SH, Kim S. Leucyl-tRNA synthetase is an intracellular leucine sensor for the mTORC1-signaling pathway. Cell. 2012;149:410–24.
38. Duran A, Amanchy R, Linares JF, Joshi J, Abu-Baker S, Porollo A, Hansen M, Moscat J, Diaz-Meco MT. p62 is a key regulator of nutrient sensing in the mTORC1 pathway. Mol Cell. 2011;44:134–46.
39. Kim YM, Stone M, Hwang TH, Kim YG, Dunlevy JR, Griffin TJ, Kim DH. SH3BP4 is a negative regulator of amino acid-Rag GTPase-mTORC1 signaling. Mol Cell. 2012;46:833–46.

40. Zoncu R, Bar-Peled L, Efeyan A, Wang S, Sancak Y, Sabatini DM. mTORC1 senses lysosomal amino acids through an inside-out mechanism that requires the vacuolar H(+)-ATPase. Science. 2011;334:678–83.
41. Kitamoto K, Yoshizawa K, Ohsumi Y, Anraku Y. Dynamic aspects of vacuolar and cytosolic amino acid pools of Saccharomyces cerevisiae. J Bacteriol. 1988;170:2683–6.
42. Binda M, Péli-Gulli MP, Bonfils G, Panchaud N, Urban J, Sturgill TW, Loewith R, De Virgilio C. The Vam6 GEF controls TORC1 by activating the EGO complex. Mol Cell. 2009;35:563–73.
43. Neuhaus EM, Almers W, Soldati T. Morphology and dynamics of the endocytic pathway in Dictyostelium discoideum. Mol Biol Cell. 2002;13:1390–407.
44. Yu L, McPhee CK, Zheng L, Mardones GA, Rong Y, Peng J, Mi N, Zhao Y, Liu Z, Wan F, Hailey DW, Oorschot V, Klumperman J, Baehrecke EH, Lenardo MJ. Termination of autophagy and reformation of lysosomes regulated by mTOR. Nature. 2010;465:942–6.
45. Itoh Y, Kawamata Y, Harada M, Kobayashi M, Fujii R, Fukusumi S, Ogi K, Hosoya M, Tanaka Y, Uejima H, Tanaka H, Maruyama M, Satoh R, Okubo S, Kizawa H, Komatsu H, Matsumura F, Noguchi Y, Shinohara T, Hinuma S, Fujisawa Y, Fujino M. Free fatty acids regulate insulin secretion from pancreatic beta cells through GPR40. Nature. 2003;422:173–6.
46. Hirasawa A, Tsumaya K, Awaji T, Katsuma S, Adachi T, Yamada M, Sugimoto Y, Miyazaki S, Tsujimoto G. Free fatty acids regulate gut incretin glucagon-like peptide-1 secretion through GPR120. Nat Med. 2005;11:90–49.
47. Tazoe H, Otomo Y, Kaji I, Tanaka R, Karaki SI, Kuwahara A. Roles of short-chain fatty acids receptors, GPR41 and GPR43 on colonic functions. J Physiol Pharmacol. 2008;2:251–62.
48. Wang J, Wu X, Simonavicius N, Tian H, Ling L. Medium-chain fatty acids as ligands for orphan G protein-coupled receptor GPR84. J Biol Chem. 2006;281:34457–64.
49. Laugerette F, Passilly-Degrace P, Patris B, Niot I, Febbraio M, Montmayeur JP, Besnard P. CD36 involvement in orosensory detection of dietary lipids, spontaneous fat preference, and digestive secretions. J Clin Invest. 2005;115:3177–84.
50. Oh DY, Talukdar S, Bae EJ, Imamura T, Morinaga H, Fan W, Li P, Lu WJ, Watkins SM, Olefsky JM. GPR120 is an omega-3 fatty acid receptor mediating potent anti-inflammatory and insulin-sensitizing effects. Cell. 2010;142:687–98.
51. Ichimura A, Hirasawa A, Poulain-Godefroy O, Bonnefond A, Hara T, Yengo L, Kimura I, Leloire A, Liu N, Iida K, Choquet H, Besnard P, Lecoeur C, Vivequin S, Ayukawa K, Takeuchi M, Ozawa K, Tauber M, Maffeis C, Morandi A, Buzzetti R, Elliott P, Pouta A, Jarvelin MR, Körner A, Kiess W, Pigeyre M, Caiazzo R, Van Hul W, Van Gaal L, Horber F, Balkau B, Lévy-Marchal C, Rouskas K, Kouvatsi A, Hebebrand J, Hinney A, Scherag A, Pattou F, Meyre D, Koshimizu TA, Wolowczuk I, Tsujimoto G, Froguel P. Dysfunction of lipid sensor GPR120 leads to obesity in both mouse and human. Nat Med. 2012;483:350–4.
52. Brown AJ, Sun L, Feramisco JD, Brown MS, Goldstein JL. Cholesterol addition to ER membranes alters conformation of SCAP, the SREBP escort protein that regulates cholesterol metabolism. Mol Cell. 2002;10:237–45.
53. Radhakrishnan A, Sun L-P, Kwon HJ, Brown MS, Goldstein JL. Direct binding of cholesterol to the purified membrane region of SCAP: mechanism for a sterol-sensing domain. Mol Cell. 2004;15:259–68.
54. Yang T, Espenshade PJ, Wright ME, Yabe D, Gong Y, Aebersold R, Goldstein JL, Brown MS. Crucial step in cholesterol homeostasis: sterols promote binding of SCAP to INSIG-1, a membrane protein that facilitates retention of SREBPs in ER. Cell. 2002;110:489–500.
55. Song BL, Sever N, DeBose-Boyd RA. Gp78, a membrane-anchored ubiquitin ligase, associates with Insig-1 and couples sterol-regulated ubiquitination to degradation of HMG CoA reductase. Mol Cell. 2005;19:829–40.
56. Ye J, DeBose-Boyd RA. Regulation of cholesterol and fatty acid synthesis. Cold Spring Harb Perspect Biol. 2011;3:pii: a004754.
57. Mueckler M, Caruso C, Baldwin SA, Panico M, Blench I, Morris HR, Allard WJ, Lienhard GE, Lodish HF. Sequence and structure of a human glucose transporter. Science. 1985;229:941–5.

58. Mueckler M. Facilitative glucose transporters. Eur J Biochem. 1994;219:713–25.
59. Walmsley AR, Barrett MP, Bringaud F, Gould GW. Sugar transporters from bacteria, parasites and mammals: structure-activity relationships. Trends Biochem Sci. 1998;23:476–81.
60. Thorens B, Mueckler M. Glucose transporters in the 21st century. Am J Physiol Endocrinol Metab. 2010;298:141–5.
61. Furtado LM, Somwar R, Sweeney G, Niu W, Klip A. Activation of the glucose transporter GLUT4 by insulin. Biochem Cell Biol. 2002;80:569–78.
62. Watson RT, Kanzaki M, Pessin JE. Regulated membrane trafficking of the insulin-responsive glucose transporter 4 in adipocytes. Endocr Rev. 2004;25:177–204.
63. Iynedjian PB, Pilot PR, Nouspikel T, Milburn JL, Quaade C, Hughes S, Ucla C, Newgard CB. Differential expression and regulation of the glucokinase gene in liver and islets of Langerhans. Proc Natl Acad Sci USA. 1989;86:7838–42.
64. Hardie DG, Ross FA, Hawley SA. AMPK: a nutrient and energy sensor that maintains energy homeostasis. Nat Rev Mol Cell Biol. 2012;13:251–62.
65. Ritter S, Dinh TT, Zhang Y. Localization of hindbrain glucoreceptive sites controlling food intake and blood glucose. Brain Res. 2000;856:37–47.
66. Frizzell RT, Jones EM, Davis SN, Biggers DW, Myers SR, Connolly CC, Neal DW, Jaspan JB, Cherrington AD. Counterregulation during hypoglycemia is directed by widespread brain regions. Diabetes. 1993;42:1253–61.
67. Donovan CM. Portal vein glucose sensing. Diabetes Nutr Metab. 2002;15:308–12.
68. Hevener AL, Bergman RN, Donovan CM. Novel glucosensor for hypoglycemic detection localized to the portal vein. Diabetes. 1997;46:1521–5.
69. Routh VH. Glucose-sensing neurons: are they physiologically relevant? Physiol Behav. 2002;76:403–13.
70. McCrimmon RJ, Shaw M, Fan X, Cheng H, Ding Y, Vella MC, Zhou L, McNay EC, Sherwin RS. Key role for AMP-activated protein kinase in the ventromedial hypothalamus in regulating counterregulatory hormone responses to acute hypoglycemia. Diabetes. 2008;57:444–50.
71. Efeyan A, Zoncu R, Chang S, Gumper I, Snitkin H, Wolfson RL, Kirak O, Sabatini DD, Sabatini DM. Regulation of mTORC1 by the Rag GTPases is necessary for neonatal autophagy and survival. Nature. 2013;493:679–83.
72. Kahn BB, Alquier T, Carling D, Hardie DG. AMP-activated protein kinase: ancient energy gauge provides clues to modern understanding of metabolism. Cell Metab. 2005;1:15–25.
73. Cameron KO, Kurumbail RG. Recent progress in the identification of adenosine monophosphate-activated protein kinase (AMPK) activators. Bioorg Med Chem Lett. 2016;26:5139–48.
74. Chen L, Wang J, Zhang YY, Yan SF, Newmann D, Schattner U, Wang ZX, Wu JW. AMP-activated protein kinase undergoes nucleotide-dependent conformational changes. Nat Struct Mol Biol. 2012;19:716–8.
75. Hardie DG, Schaffer BE, Brunet A. AMPK: An energy-sensing pathway with multiple inputs and outputs. Trends Cell Biol. 2016;26:190–201.
76. Sakamoto K, Holman GD. Emerging role for AS160/TBC1D4 and TBC1D1 in the regulation of GLUT4 traffic. Am J Physiol Endocrinol Metab. 2008;295:29–37.
77. Habets DDJ, Coumans WA, Hasnaoui ME, Zarrinpashneh E, Bertrand L, Viollet B, Kiens B, Jensen TE, Richter EA, Bonen A, Glatz JFC, Luiken JJFP. Crucial role for LKB1 to AMPKα2 axis in the regulation of CD36-mediated long-chain fatty acid uptake into cardiomyocytes. Biochim Biophys Acta. 2009;1791:212–9.
78. Marsin AS, Bertrand L, Rider MH, Deprez J, Beauloye C, Vincent MF, Van den Berghe G, Carling D, Hue L. Phosphorylation and activation of heart PFK-2 by AMPK has a role in the stimulation of glycolysis during ischaemia. Curr Biol. 2000;10:1247–55.
79. Merrill GF, Kurth EJ, Hardie DG, Winder WW. AICA riboside increases AMP-activated protein kinase, fatty acid oxidation, and glucose uptake in rat muscle. Am J Phys. 1997;273:1107–12.

80. Muoio DM, Seefeld K, Witters LA, Coleman RA. AMP-activated kinase reciprocally regulates triacylglycerol synthesis and fatty acid oxidation in liver and muscle: evidence that sn-glycerol-3-phosphate acyltransferase is a novel target. Biochem J. 1999;338:783–91.
81. Alers S, Löffler AS, Wesselborg S, Stork B. Role of AMPK-mTOR-Ulk1/2 in the regulation of autophagy: cross talk, shortcuts, and feedbacks. Mol Cell Biol. 2012;32:2–11.
82. Gwinn DM, Shackelford DB, Egan DF, Mihaylova MM, Mery A, Vasquez DS, Turk BE, Shaw RJ. AMPK phosphorylation of raptor mediates a metabolic checkpoint. Mol Cell. 2008;30:214–26.
83. Puigserver P, Spiegelman BM. Peroxisome proliferator-activated receptor-gamma coactivator 1 alpha (PGC-1 alpha): transcriptional coactivator and metabolic regulator. Endocr Rev. 2003;24:78–90.
84. Rodgers JT, Lerin C, Haas W, Gygi SP, Spiegelman BM, Puigserver P. Nutrient control of glucose homeostasis through a complex of PGC-1alpha and SIRT1. Nature. 2005;434:113–8.
85. Rehman G, Shehzad A, Khan AL, Hamayun M. Role of AMP-activated protein kinase in cancer therapy. Arch Pharm (Weinheim). 2014;347:457–68.
86. Koch-Nolte F, Fischer S, Haag F, Ziegler M. Compartmentation of NAD^+-dependent signalling. FEBS Lett. 2011;585:1651–6.
87. Fulco M, Schiltz RL, Iezzi S, King MT, Zhao P, Kashiwaya Y, Hoffman E, Veech RL, Sartorelli V. Sir2 regulates skeletal muscle differentiation as a potential sensor of the redox state. Mol Cell. 2003;12:51–62.
88. Houtkooper RH, Canto C, Wanders RJ, Auwerx J. The secret life of NAD^+: an old metabolite controlling new metabolic signaling pathways. Endocr Rev. 2010;31:194–223.
89. Imai S. Nicotinamide phosphoribosyltransferase (Nampt): a link between NAD biology, metabolism, and diseases. Curr Pharm Des. 2009;15:20–8.
90. Revollo JR, Grimm AA, Imai S. The NAD biosynthesis pathway mediated by nicotinamide phosphoribosyltransferase regulates Sir2 activity in mammalian cells. J Biol Chem. 2004;279:50754–63.
91. Frye RA. Phylogenetic classification of prokaryotic and eukaryotic Sir2-like proteins. Biochem Biophys Res Commun. 2000;273:793–8.
92. Haigis MC, Guarente LP. Mammalian sirtuins-emerging roles in physiology, aging, and calorie restriction. Genes Dev. 2006;20:2913–21.
93. Krejčı́ A. Metabolic sensors and their interplay with cell signalling and transcription. Biochem Soc Trans. 2012;40:311–23.
94. Chang HC, Guarente L. SIRT1 and other sirtuins in metabolism. Trends Endocrinol Metab. 2014;25:138–45.
95. Haigis MC, Sinclair DA. Mammalian sirtuins: biological insights and disease relevance. Annu Rev Pathol. 2010;5:253–95.
96. Zhao S, Xu W, Jiang W, Yu W, Lin Y, Zhang T, Yao J, Zhou L, Zeng Y, Li H, Li Y, Shi J, An W, Hancock SM, He F, Qin L, Chin J, Yang P, Chen X, Lei Q, Xiong Y, Guan KL. Regulation of cellular metabolism by protein lysine acetylation. Science. 2010;327:1000–4.
97. Guarente L. The logic linking protein acetylation and metabolism. Cell Metab. 2011;14:151–3.
98. Majaw T, Sharma R. Arginase I expression is upregulated by dietary restriction in the liver of mice as a function of age. Mol Cell Biochem. 2015;407:1–7.
99. Jakob F, Seefried L, Schwab M. Age and osteoporosis. Effects of aging on osteoporosis, the diagnostics and therapy. Internist (Berl). 2014;55:755–61.
100. Monfoulet LE, Philippe C, Mercier S, Coxam V, Wittrant Y. Deficiency of G-protein coupled receptor 40, a lipid-activated receptor, heightens in vitro- and in vivo-induced murine osteoarthritis. Exp Biol Med (Maywood). 2015;240:854–66.
101. Blagosklonny MV, Hall MN. Growth and aging: a common molecular mechanism. Aging. 2009;1:357–62.
102. Sharma R. Dietary restriction and its multifaceted effects. Curr Sci. 2004;87:1203–10.
103. Sharma R, Dkhar P. Interventions for healthy aging. In: Sanchetee P, editor. Textbook on geriatric medicine. Hyderabad: Indian Academy of Geriatrics, Paras Medical Publisher; 2014. p. 812–5.

104. Hadem IKH, Sharma R. Age- and tissue-dependent modulation of IGF-1/PI3K/Akt protein expression by dietary restriction in mice. Horm Metab Res. 2016;48:201–6.
105. Hadem IKH, Sharma R. Differential regulation of hippocampal IGF-1-associated signaling proteins by dietary restriction in aging mouse. Cell Mol Neurobiol. 2017; https://doi.org/10.1007/s10571-016-0431-7.
106. Kojima T, Kamei H, Aizu T, Arai Y, Takayama M, Nakazawa S, Ebihara Y, Inagaki H, Masui Y, Gondo Y, Sakaki Y, Hirose N. Association analysis between longevity in the Japanese population and polymorphic variants of genes involved in insulin and insulin-like growth factor 1 signaling pathways. Exp Gerontol. 2004;39:1595–8.
107. Suh Y, Atzmon G, Cho MO, Hwang D, Liu B, Leahy DJ, Barzilai N, Cohen P. Functionally significant insulin-like growth factor I receptor mutations in centenarians. Proc Natl Acad Sci USA. 2008;105:3438–42.
108. Anselmi CV, Malovini A, Roncarati R, Novelli V, Villa F, Condorelli G, Bellazzi R, Puca AA. Association of the FOXO3A locus with extreme longevity in a southern Italian centenarian study. Rejuvenation Res. 2009;12:95–104.
109. Johnson SC, Rabinovitch PS, Kaeberlein M. mTOR is a key modulator of ageing and age-related disease. Nature. 2013;493:338–45.
110. Mizushima N, Levine B, Cuervo AM, Klionsky DJ. Autophagy fights disease through cellular self-digestion. Nature. 2008;451:1069–75.
111. Hands SL, Proud CG, Wyttenbach A. mTOR's role in ageing: protein synthesis or autophagy? Aging. 2009;1:586–97.
112. Ruetenik A, Barrientos A. Dietary restriction, mitochondrial function and aging: from yeast to humans. Biochim Biophys Acta. 2015;1847:1434–47.
113. Hansen M, Taubert S, Crawford D, Libina N, Lee SJ, Kenyon C. Lifespan extension by conditions that inhibit translation in Caenorhabditis elegans. Aging Cell. 2007;6:95–110.
114. Tohyama D, Yamaguchi A, Yamashita T. Inhibition of a eukaryotic initiation factor (eIF2B-delta/F11A3.2) during adulthood extends lifespan in Caenorhabditis elegans. FASEB J. 2008;22:4327–37.
115. Dkhar P, Sharma R. Late-onset dietary restriction modulates protein carbonylation and catalase in cerebral hemispheres of aged mice. Cell Mol Neurobiol. 2014;34:307–13.
116. Mayer C, Grummt I. Ribosome biogenesis and cell growth: mTOR coordinates transcription by all three classes of nuclear RNA polymerases. Oncogene. 2006;25:6384–91.
117. Hansen M, Chandra A, Mitic LL, Onken B, Driscoll M, Kenyon C. A role for autophagy in the extension of lifespan by dietary restriction in C elegans. PLoS Genet. 2008;4:e24.
118. Tóth ML, Sigmond T, Borsos É, Barna J, Erdélyi P, Takács Vellai K, Orosz L, Kovács AL, Csikós G, Sass M, Vellai T. Longevity pathways converge on autophagy genes to regulate life span in Caenorhabditis elegans. Autophagy. 2008;4:330–8.
119. Simonsen A, Cumming RC, Brech A, Isakson P, Schubert DR, Finley KD. Promoting basal levels of autophagy in the nervous system enhances longevity and oxidant resistance in adult Drosophila. Autophagy. 2008;4:176–84.
120. Alvers AL, Fishwick LK, Wood MS, Hu D, Chung HS, Dunn WAJ, Aris JP. Autophagy and amino acid homeostasis are required for chronological longevity in Saccharomyces cerevisiae. Aging Cell. 2009;8:353–69.
121. Scott RC, Schuldiner O, Neufeld TP. Role and regulation of starvation-induced autophagy in the Drosophila fat body. Dev Cell. 2004;7:167–78.
122. Morck C, Pilon M. C. elegans feeding defective mutants have shorter body lengths and increased autophagy. BMC Dev Biol. 2006;6:39.
123. Wohlgemuth SE, Julian D, Akin DE, Fried J, Toscano K, Leeuwenburgh C, Dunn JWA. Autophagy in the heart and liver during normal aging and calorie restriction. Rejuvenation Res. 2007;10:281–92.
124. Cavallini G, Donati A, Gori Z, Pollera M, Bergamini E. The protection of rat liver autophagic proteolysis from the age related decline co-varies with the duration of anti-ageing food restriction. Exp Gerontol. 2001;36:497–506.

125. Panowski SH, Wolff S, Aguilaniu H, Durieux J, Dillin A. PHA-4/Foxa mediates diet-restriction-induced longevity of C. elegans. Nature. 2007;447:550–5.
126. Wauquier F, Philippe C, Léotoing L, Mercier S, Davicco MJ, Lebecque P, Guicheux J, Pilet P, Miot-Noirault E, Poitout V, Alquier T, Coxam V, Wittrant Y. The free fatty acid receptor G protein-coupled receptor 40 (GPR40) protects from bone loss through inhibition of osteoclast differentiation. J Biol Chem. 2013;288:6542–51.
127. Martini C, Pallottini V, Cavallini G, Donati A, Bergamini E, Trentalance A. Caloric restrictions affect some factors involved in age-related hypercholesterolemia. J Cell Biochem. 2007;101:235–43.
128. Marino M, Pallottini V, D'Eramo C, Cavallini G, Bergamini E, Trentalance A. Age-related changes of cholesterol and dolichol biosynthesis in rat liver. Mech Ageing Dev. 2002;123:1183–9.
129. Pallottini V, Montanari L, Cavallini G, Bergamini E, Gori Z, Trentalance A. Mechanisms underlying the impaired regulation of 3-hydroxy-3-methylglutaryl coenzyme A reductase in aged rat liver. Mech Ageing Dev. 2004;125:633–9.
130. Smiljanic K, Vanmierlo T, Djordjevic AM, Perovic M, Ivkovic S, Lütjohann D, Kanazir S. Cholesterol metabolism changes under long-term dietary restrictions while the cholesterol homeostasis remains unaffected in the cortex and hippocampus of aging rats. Age (Dordr). 2014;36:9654.
131. Carayannopoulos MO, Chi MM, Cui Y. GLUT8 is a glucose transporter responsible for insulin-stimulated glucose uptake in the blastocyst. Proc Natl Acad Sci USA. 2000;97:7313–8.
132. Doege H, Schürmann A, Bahrenberg G, Brauers A, Joost HG. GLUT8, a novel member of the sugar transport facilitator family with glucose transport activity. J Biol Educ. 2000;275:16275–80.
133. Banerjee A, Anuradha, Mukherjee K, Krishna A. Testicular glucose and its transporter GLUT 8 as a marker of age-dependent variation and its role in steroidogenesis in mice. J Exp Zool A Ecol Genet Physiol. 2014;321:490–502.
134. Cho SJ, Moon JS, Lee CM, Choi AM, Stout-Delgado HW. GLUT-1-dependent glycolysis is increased during aging-related lung fibrosis and phloretin inhibits lung fibrosis. Am J Respir Cell Mol Biol. 2016; https://doi.org/10.1165/rcmb.2016-0225OC.
135. Novelli M, De Tata V, Bombara M, Bergamini E, Masiello P. Age-dependent reduction in GLUT-2 levels is correlated with the impairment of the insulin secretory response in isolated islets of Sprague-Dawley rats. Exp Gerontol. 2000;35:641–51.
136. Torres TP, Catlin RL, Chan R, Fujimoto Y, Sasaki N, Printz RL, Newgard CB, Shiota M. Restoration of hepatic glucokinase expression corrects hepatic glucose flux and normalizes plasma glucose in zucker diabetic fatty rats. Diabetes. 2009;58:78–86.
137. Kharbhih WJ, Sharma R. Inorganic pyrophosphatase of cardiac and skeletal muscle is enhanced by dietary restriction in mice during aging. In: Rath PC, Sharma R, Prasad S, editors. Topics in biomedical gerontology. Singapore: Springer Nature; 2017. p. 117–22.
138. Steinberg GR, Kemp BE. AMPK in health and disease. Physiol Rev. 2009;89:1025–78.
139. Petersen KF, Befroy D, Dufour S, Dziura J, Ariyan C, Rothman DL, DiPietro L, Cline GW, Shulman GI. Mitochondrial dysfunction in the elderly: possible role in insulin resistance. Science. 2003;300:1140–2.
140. Ma Y, Li J. Metabolic shifts during aging and pathology. Compr Physiol. 2015;5:667–86.
141. Onken B, Driscoll M. Metformin induces a dietary restriction-like state and oxidative stress response to extend C. elegans healthspan via AMPK, LKB1, and SKN-1. PLoS ONE. 2010; https://doi.org/10.1371/journal.pone.0008758.
142. Funakoshi M, Tsuda M, Muramatsu H, Hatsuda H, Morishita S, Aigaki T. A gain-of-function screen identifies wdb and lkb1 as lifespan-extending genes in Drosophila. Biochem Biophys Res Commun. 2011;405:667–72.
143. Viollet B, Andreelli F, Jørgensen SB, Perrin C, Flamez D, Mu J, Wojtaszewski JF, Schuit FC, Birnbaum M, Richter E, Burcelin R, Vaulont S. Physiological role of AMP-activated protein kinase (AMPK): insights from knockout mouse models. Biochem Soc Trans. 2003;31:216–9.

144. Reznick RM, Zong H, Li J, Morino K, Moore IK, Yu HJ, Liu ZX, Dong J, Mustard KJ, Hawley SA, Befroy D, Pypaert M, Hardie DG, Young LH, Shulman GI. Aging-associated reductions in AMP-activated protein kinase activity and mitochondrial biogenesis. Cell Metab. 2007;5:151–6.
145. Salminen A, Kaarniranta K. AMP-activated protein kinase (AMPK) controls the aging process via an integrated signaling network. Ageing Res Rev. 2012;11:230–41.
146. Imai S, Guarente L. It takes two to tango: NAD^+ and sirtuins in aging/longevity control. Aging Mech Dis. 2016;2:16017. https://doi.org/10.1038/npjamd.2016.17.
147. Hadem IKH. Dietary and age-dependent regulation of insulin-like growth factor-1 and its related signaling in mice. PhD dissertation, North-Eastern Hill University, Shillong, India. 2016.
148. Satoh A, Brace CS, Rensing N, Cliften P, Wozniak DF, Herzog ED, Yamada KA, Imai S. Sirt1 extends life span and delays aging in mice through the regulation of Nk2 homeobox 1 in the DMH and LH. Cell Metab. 2013;18:416–30.
149. Giblin W, Skinner ME, Lombard DB. Sirtuins: guardians of mammalian healthspan. Trends Genet. 2014;30:271–86.
150. Kanfi Y, Naiman S, Amir G, Peshti V, Zinman G, Nahum L, Bar-Joseph Z, Cohen HY. The sirtuin SIRT6 regulates lifespan in male mice. Nature. 2012;483:218–21.
151. Gomes AP, Price NL, Ling AJ, Moslehi JJ, Montgomery MK, Rajman L, White JP, Teodoro JS, Wrann CD, Hubbard BP, Mercken EM, Palmeira CM, De Cabo R, Rolo AP, Turner N, Bell EL, Sinclair DA. Declining NAD(+) induces a pseudohypoxic state disrupting nuclear-mitochondrial communication during aging. Cell. 2013;155:1624–38.
152. Stein LR, Imai S. Specific ablation of Nampt in adult neural stem cells recapitulates their functional defects during aging. EMBO J. 2014;33:1321–40.
153. Ghosh D, Levault KR, Brewer GJ. Relative importance of redox buffers GSH and NAD(P) H in age-related neurodegeneration and Alzheimer disease-like mouse neurons. Aging Cell. 2014;13:631–40.
154. Jukarainen S, Heinonen S, Rämö JT, Rinnankoski-Tuikka R, Rappou E, Tummers M, Muniandy M, Hakkarainen A, Lundbom J, Lundbom N, Kaprio J, Rissanen A, Pirinen E, Pietiläinen KH. Obesity is associated with low NAD/SIRT pathway expression in adipose tissue of BMI-discordant monozygotic twins. J Clin Endocrinol Metab. 2015;101:275–83.
155. Lin SJ, Ford E, Haigis M, Liszt G, Guarente L. Calorie restriction extends yeast life span by lowering the level of NADH. Genes Dev. 2004;18:12–6.
156. Bai P, Cantó C, Oudart H, Brunyánszki A, Cen Y, Thomas C, Yamamoto H, Huber A, Kiss B, Houtkooper RH, Schoonjans K, Schreiber V, Sauve AA, Menissier-de Murcia J, Auwerx J. PARP-1 inhibition increases mitochondrial metabolism through SIRT1 activation. Cell Metab. 2011;13:461–8.
157. Imai S, Guarente L. NAD^+ and sirtuins in aging and disease. Trends Cell Biol. 2014;24:464–71.
158. Mouchiroud L, Houtkooper RH, Moullan N, Katsyuba E, Ryu D, Canto C, Mottis A, Jo YS, Viswanathan M, Schoonjans K, Gaurente L, Auwerx J. The NAD^+/Sirtuin pathway modulates longevity through activation of mitochondrial UPR and FOXO signalling. Cell. 2013;154:430–41.
159. Brunet A, Sweeney LB, Sturgill JF, Chua KF, Greer PL, Lin Y, Tran H, Ross SE, Mostoslavsky R, Cohen HY, Hu LS, Cheng HL, Jedrychowski MP, Gygi SP, Sinclair DA, Alt FW, Greenberg ME. Stress-dependent regulation of FOXO transcription factors by the SIRT1 deacetylase. Science. 2004;303:2011–5.
160. Papa L, Germain D. SirT3 regulates the mitochondrial unfolded protein response. Mol Cell Biol. 2014;34:699–710.
161. Baur JA, Ungvari Z, Minor RK, Le Couteur DG, de Cabo R. Are sirtuins viable targets for improving healthspan and lifespan? Nat Rev Drug Discov. 2012;11:443–61.
162. Hebert AS, Dittenhafer-Reed KE, Yu W, Bailey DJ, Selen ES, Boersma MD, Carson JJ, Tonelli M, Balloon AJ, Higbee AJ, Westphall MS, Pagliarini DJ, Prolla TA, Assadi-Porter F, Roy S, Denu JM, Coon JJ. Calorie restriction and SIRT3 trigger global reprogramming of the mitochondrial protein acetylome. Mol Cell. 2013;49:186–99.

163. Sundaresan NR, Gupta M, Kim G, Rajamohan SB, Isbatan A, Gupta MP. Sirt3 blocks the cardiac hypertrophic response by augmenting Foxo3a-dependent antioxidant defense mechanisms in mice. J Clin Invest. 2009;119:2758–71.
164. Song J, Ke SF, Zhou CC, Zhang SL, Guan YF, Xu TY, Sheng CQ, Wang P, Miao CY. Nicotinamide phosphoribosyltransferase is required for the calorie restriction-mediated improvements in oxidative stress, mitochondrial biogenesis, and metabolic adaptation. J Gerontol A Biol Sci Med Sci. 2014;69:44–57.
165. Cartee GD, Dean DJ. Glucose transport with brief dietary restriction: heterogenous responses in muscles. Am J Phys. 1994;266:946–52.
166. Wang H, Wu H, Guo H, Zhang G, Zhang R, Zhan S. Increased hypoxia-inducible factor 1alpha expression in rat brain tissues in response to aging. Neural Regen Res. 2012;7:778–82.
167. Chen TT, Maevsky EI, Uchitel ML. Maintenance of homeostasis in the aging hypothalamus: the central and peripheral roles of succinate. Front Endocrinol (Lausanne). 2015;6:7.
168. Sheedfar F, Sung MM, Aparicio-Vergara M, Kloosterhuis NJ, Miquilena-Colina ME, Vargas-Castrillón J, Febbraio M, Jacobs RL, de Bruin A, Vinciguerra M, García-Monzón C, Hofker MH, Dyck JR, Koonen DP. Increased hepatic CD36 expression with age is associated with enhanced susceptibility to nonalcoholic fatty liver disease. Aging (Albany NY). 2014;6:281–95.
169. Abumrad NA, el Maghrabi MR, Amri EZ, Lopez E, Grimaldi PA. Cloning of a rat adipocyte membrane protein implicated in binding or transport of longchain fatty acids that is induced during preadipocyte differentiation. Homology with human CD36. J Biol Chem. 1993;268:17665–8.
170. Miyauchi S, Hirasawa A, Iga T, Liu N, Itsubo C, Sadakane K, Hara T, Tsujimoto G. Distribution and regulation of protein expression of the free fatty acid receptor GPR120. Naunyn Schmiedeberg's Arch Pharmacol. 2009;379:427–34.

Protein Aggregation, Related Pathologies, and Aging

19

Karunakar Kar, Bibin G. Anand, Kriti Dubey, and Dolat Singh Shekhawat

Protein Aggregation and Related Diseases

Every protein, irrespective of its aggregation propensity, is synthesized in the cellular protein synthesis machinery. Once the nascent polypeptide chain is synthesized in the ribosome, the unstructured chain undergoes a protein folding process by which it gains the necessary secondary and tertiary interactions leading to formation of a soluble and functionally active native structure (Fig. 19.1). The properly folded native structure of a protein is vital to carry out its biological functions which are critical for normal cellular metabolic processes. Proteins play several important roles in life processes, such as cellular and structural repair, defense mechanism, metabolism of hormones and enzymes, molecular transportation, conducting stimuli, building nutrients, and energy production [1, 2]. However, sometimes these protein molecules can be trapped into an aggregation pathway due to many factors such as mutation, deletion, degradation, and protein misfolding, as illustrated in Fig. 19.1 [3]. Biological fate of many functional protein species that are trapped in aggregation pathway may have severe direct and indirect consequences on the normal cell physiology. For instance, due to loss of native conformation during protein aggregation, the physiological concentrations of the proteins are predicted to deplete, which may trigger protein-deficient diseases (Tables 19.1 and 19.2). Coincidently, many protein deficiency diseases are observed in the old age. The other direct effect of protein aggregation is the accumulation of cytotoxic amyloid fibrils. Recent reports have confirmed that smaller molecular weight on-pathway oligomers, produced during aggregation, are more toxic than the mature amyloid fibrils [5–7]. Typical

---

K. Kar (✉)
School of Life Sciences, Jawaharlal Nehru University, New Delhi, India
e-mail: kkar@mail.jnu.ac.in

B. G. Anand · K. Dubey · D. S. Shekhawat
Department of Bioscience and Bioengineering, Indian Institute of Technology, Jodhpur, India

© Springer Nature Singapore Pte Ltd. 2020
P. C. Rath (ed.), *Models, Molecules and Mechanisms in Biogerontology*,
https://doi.org/10.1007/978-981-32-9005-1_19

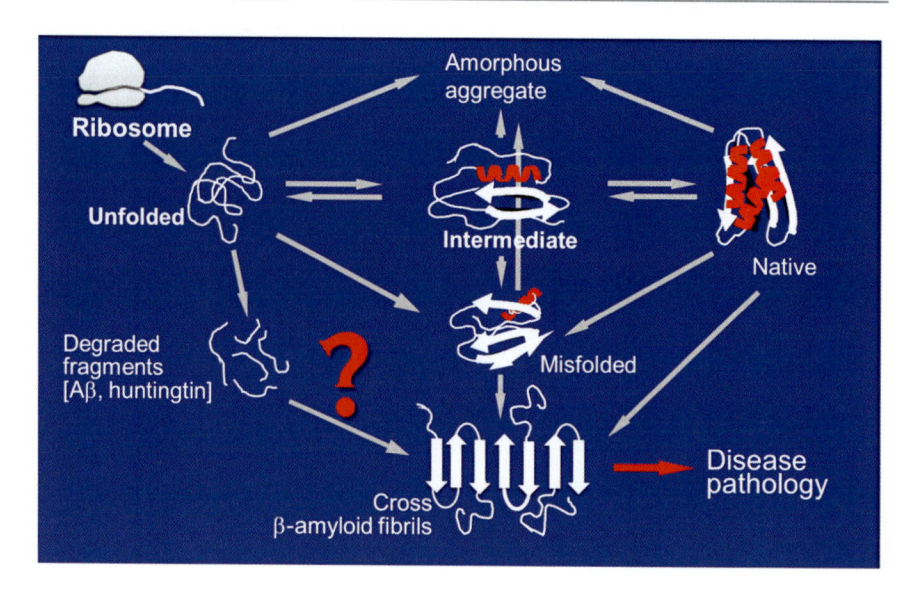

Fig. 19.1 Schematic representation of protein aggregation mechanism, showing possible pathways that trap proteins into an aggregation process leading to the formation of cross-β-structured amyloid fibrils or irregular amorphous aggregates [3, 56, 57]

oligomers and higher order mature fibrils generated due to aggregation of insulin molecules under in vitro conditions are shown in Fig. 19.2.

Cells possess certain molecular mechanisms that are engaged in either repairing or degrading misfolded proteins to prevent the process of protein aggregation. These molecular machineries are known as molecular chaperons. For example, Hsp70, Hsp40, and clusterins [8–11] are the cellular chaperones which help in proper folding of protein polypeptide chains so that they can arrive at the desired biologically active native structures. In addition to guiding proper folding of nascent polypeptide chains, these chaperones also help in the transport of proteins across membranes [12, 13]. The level of chaperones, particularly from Hsp family, is regulated in the cell to match the level of protein synthesis in the cytosol [14]. However, under stress conditions (such as pH stress, temperature stress, and salt stress), the expression of Hsp molecules is believed to rise beyond the normal level [13]. In addition to the prevention of protein misfolding and protein aggregation, there exists another kind of chaperones that are known as disaggregases, which can effectively solubilize the preformed formed aggregated structures of proteins. It has been reported that Hsp104 disaggregase, with the help of other Hsp molecules, can promote the disassembly of protein aggregates [15, 16]. In mammals, Hsp 110 has been identified as one type of such chaperones which can facilitate disaggregation process, leading to fragmentation of α-synuclein-generated toxic amyloid fibrils into nontoxic soluble species [17].

Table 19.1 List of selected amyloidogenic proteins, the respective diseases, and the sequences of the aggregation-prone regions of the proteins[a]

Name of parent peptide/protein	Pathophysiological conditions	Short active aggregation-prone sequence
Islet amyloid polypeptide	Diabetes mellitus (type II)	FGAIL
β-amyloid peptide	Alzheimer's disease	QKLVFF,LPFFD,LVFFA.
Lactadherin	Aortic medial amyloid	NFGSVQFV
Gelsolin	Finnish hereditary amyloidosis	SFNNGDCCFILD
Serum amyloid A	Chronic inflammation amyloidosis	SFFSFLGEAFD
Thyroid carcinoma peptide	Thyroid carcinoma	DFNKF, DFNK
Human muscle acylphosphatase amyloid	Human muscle acylphosphatase	RVQGVCFRMTEDEARSKLEYSNFSIRY
Calcitonin	Medullary thyroid carcinoma	YTQDFNKFFHTFPPQTAIGV
BRI	Neuronal dysfunction and dementia	FENKFAVFAIRHF
β2-microglobulin	Dialysis-associated renal amyloidosis	DWSFYLLYTEFT
PrP	Creutzfeldt–Jakob disease	PHGGGWGQ

[a]Adapted from Anand et al. *Scientific Reports* 2017

In addition to chaperone machineries which protect the proteins from going into an aggregation pathway, cells also possess other mechanisms of protein degradation that can prevent protein misfolding and protein aggregation. One of such protective systems is ubiquitin–proteasome system in the cell [18, 19]. Soluble misfolded proteins are first conjugated to the ubiquitin with the help of different ligases (E1, E2, and E3). Such ubiquitinated protein, with the help of other chaperones, is delivered to the proteolytic chamber of the cell, where the misfolded corrupt proteins are degraded into nontoxic soluble components or fragments [18, 19].

It is possible that despite the presence of cellular protective mechanism, some aggregation-prone peptides and proteins can still form toxic amyloid structures leading to the onset of several amyloid-related diseases. It is also possible that as the age increases, the efficacies of these cellular protective machineries may gradually deplete, and such condition may enhance the aggregation process. How does the process of protein aggregation begin? What are the changes in their conformational properties that trigger aggregation? Is there any effective strategy to prevent aggregation process? Clear answers to these fundamental questions are required not only to understand the mechanism of protein aggregation but also to find effective strategies for prevention of aggregation-linked diseases.

Table 19.2 Physiological concentration of the selected blood proteins, their functions, and related pathophysiology

Protein	About protein	Function	Protein pathophysiology
BSA	Most abundant serum protein, MW 66.5 kDa, synthesis site liver (12–25 g/day), half-life – 16 h	Blood coagulation, transports hormones, fatty acids, drugs, etc. Maintains oncotic pressure, prevents photodegradation of folic acid	Peripheral and pulmonary edema, delayed wound healing, nephrotic syndrome, hepatic cirrhosis, heart failure, and malnutrition
Insulin	Synthesized in the pancreas within the β cells of the islets of Langerhans. MW 5.8 kDa, half-life 4–5 h	Carbohydrate and fat metabolism, facilitates the packing of glucose into fat cells as triglycerides	Localized amyloid deposition, poor glycemic control, acanthosis nigricans, hyperinsulinemia is directly related to cancer, obesity, type-II diabetes, hypertension, arthrosclerosis, chronic inflammation, cardiovascular disease, prostate enlargement
Cyt c	Located in the mitochondrial intermembrane. MW 12 kDa, half-life 5–8 min	Electron transport, intrinsic type II apoptosis, scavenges reactive oxygen species, oxidizes cardiolipin during apoptosis	Lewy bodies and other neurodegenerative disorders [38]
Lysozyme	Synthesis in osteoclasts MW 14.3 kDa, half-life 4 h	Hydrolyzing the glycosidic bond, immune functions	Prominent amyloid nephropathy, nephrotic syndrome and sicca syndrome, spontaneous splenic rupture, cholestasis, liver failure, massive hepatic hemorrhage
Myoglobin	Synthesis in cardiac myocytes and oxidative skeletal muscle fibers. MW 16.7 kDa	Oxygen storage, serve as a buffer of intracellular PO_2, facilitated O_2 diffusion, scavenger of NO in heart, normal muscle development and function	Muscular dystrophy, rhabdomyolysis

Adapted from Dubey et al. (2017) *Scientific Report*

Neurodegenerative Diseases

Alzheimer's Disease

Over the past hundred years, life-threatening diseases like dementia and Alzheimer's disease are known to us, which mostly appear in the old age. Every year, large number of clinical reports related to dementia is recorded, and this number is rising exponentially, which is a great concern [20]. According to World Alzheimer Report

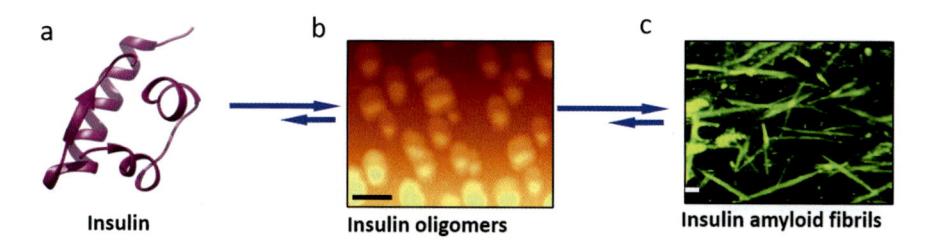

Fig. 19.2 Aggregation of soluble insulin into insoluble amyloid fibrils. (**a**) Native structure of insulin monomer (PDB ID: 1TRZ). (**b**) Atomic force microscopy images showing morphology of on-pathway spheroidal oligomers, scale bar 50 nm. (**c**) Morphology of Thioflavin T-stained mature amyloid fibrils of insulin, scale bar 10 μm

[20], by the year 2015, almost 46.8 million patients were suffering from dementia and Alzheimer's diseases, and by the year 2050, this count will rise to 131.5 million, which seems an alarming situation. Until now, no effective cure is available for such neurodegenerative diseases, neither to treat nor to suppress the progression of the disease. Alzheimer's disease pathology involves formation of two types of amyloid aggregates: (1) formation of extracellular plaques in the brain tissues that originates from the β-sheet-rich Aβ peptides [21] and (2) formation of intracellular neurofibrillary tangles [22] which are mostly originated from amyloid aggregation of tau protein. The mechanistic understanding of the process of amyloid formation of Aβ peptide is largely unknown. The Aβ peptide, which is amyloid prone, is a byproduct resulting from the cleavage of amyloid precursor protein (APP), which is a single pass transmembrane protein mostly present in the brain tissues [23, 24]. APP protein has been directly linked to several metabolic events through a combinatorial action of specific proteases and intramembranous secretases [23]. APP is known to have multiple cleavage sites for different cleaving enzymes whose action results in the accumulation of different peptide segments including Aβ peptide. One of such processing pathways of APP involves its cleavage by α- and γ-secretase and that action yields the p3 peptide, a nonamyloidogenic segment. Another processing of APP involves its cleavage by β- and γ-secretase that yields amyloidogenic Aβ peptide [25, 26]. Thus, a heterogeneous mixture of Aβ peptides, whose length varies from 27 to 49 residues [27, 28], is predicted to accumulate in the body system. Reports have proposed that the Aβ homeostasis is deregulated in the pathophysiology of early onset of the Alzheimer's disease [29]. Aggregation propensity of the Aβ peptides is also affected by mutations in the peptide sequence [30, 31]. Hence, targeting the aggregation of Aβ peptides has become one of the effective strategies to prevent the onset of Alzheimer's disease.

Prion Diseases

Prion diseases are often described as infectious amyloid pathologies [32]. The normal prion protein (PrP^C), in its disease form (PrP^{SC}), attains an abnormal conformation which is known to trigger the amyloid formation, leading to formation of toxic prion plaques in neuronal tissues [32, 33]. The prion diseases are infectious because it is transmissible across species and the infectious entity is in fact the aggregates of abnormal PrP^{SC} species. It has been proposed that the PrP^{SC} aggregates, after infection, readily interact with the normal PrP^C of the individual, and such interaction alters the native conformation of the PrP^C protein into an aggregation-prone conformation, triggering an aggressive aggregation process and subsequent amyloid plaque formation [34]. Besides formation of plaque deposits in the brain tissues, another hall mark of prion disease is the formation of large vacuoles, a process identified as spongiosis. Hence, prion diseases are also mentioned as transmissible spongiform encephalopathies [3].

Polyglutamine Diseases

Polyglutamine diseases are amyloid-linked neurodegenerative pathologies caused by autosomal dominant mutation that leads to abnormal expansion of genetic codon CAG [35–37]. Since the codon CAG represents the amino acid glutamine (Q), these mutated gene products result in proteins with expanded glutamine repeats (polyglutamine or polyQ). There are about nine neurodegenerative diseases which have been identified as polyglutamine diseases (Table 19.3). As shown in the Table 19.3, all the gene products would result in the respective proteins with extended polyQ-tract in their sequence, and when the number of glutamine residues surpasses the

Table 19.3 List of neurodegenerative diseases related to polyglutamine (polyQ) amyloid formation [4]

Polyglutamine disorder	Gene	Symptoms	Threshold polyQ length
Haw River syndrome	Atrophin 1	Dementia, ataxia	35–49
Huntington's disease	Huntingtin	Chorea, dementia	35–37
Spinobulbar muscular atrophy	Androgen receptor	Weakness	36–38
Spinocerebellar ataxia type 1	ATXN1	Ataxia	35–49
Spinocerebellar ataxia type 2	ATXN2	Ataxia	32–77
Spinocerebellar ataxia type 3	ATXN3	Ataxia	40–55
Spinocerebellar ataxia type 6	CACNA1A	Ataxia	18–21
Spinocerebellar ataxia type 7	ATXN7	Ataxia	17–38
Spinocerebellar ataxia type 17	TATA Box Binding protein	Ataxia	42–47

Reference: Gusella et al. Nature Rev. Neurosci. 2000

threshold number, there is a high risk for the onset of polyglutamine diseases. The polyglutamine amyloid pathologies include several neurodegenerative diseases such as spinal and bulbar muscular atrophy, Huntington's disease, and spinocerebellar ataxias [39, 40]. The length of the polyQ stretch is inversely related to the age of onset of the disease [41–43]. In case of Huntington's disease, the threshold length of polyQ tract is considered to be ~35–37 glutamine repeats that corresponds to the age of ~50–55 years for the disease onset [42, 43]. For example, if the number of glutamine residues increases to 100, there would be no survival at the beginning of life. The thumb rule in polyQ diseases is that more the number of glutamine residues in the polyQ tract of the protein sequence, lower the age of onset of the disease [4, 42].

The culprit protein for Huntington's disease is believed to be the exon1 protein, which comprises a 17 residue N-terminal region, a polyQ repeating region, and a longer C-terminal region.

It is considered that the polyQ sequence is inherently amyloidogenic and in the mature amyloid structures of exon1, the polyQ regions show interdigitated cross-β conformation [43–45].

Parkinson's Disease

Parkinson disease is considered as one of the most severe neurodegenerative pathology that results in lethal neuronal defects, including tremors, bradykinesia (slower movement), and impaired balance and coordination in the affected body parts [3, 46]. As the age of the patient progresses, the complications related to the Parkinson's disease (PD) worsen. Parkinson's disease causes degeneration of nerve cells in brain, particularly affecting the substantia nigra region of the brain [46, 47]. Due to loss of neuronal cells, the production of dopamine, a vital neurotransmitter [48], decreases in the specific segments of the brain, which controls mostly movement and coordination-related functions of important body parts. One of the foundational reasons for the neuronal cell death linked to Parkinson's disease is the formation of Lewy bodies [49–51]. Lewy bodies are defined as intracellular inclusions of amyloid nature generated as a result of the aggregation of a protein α-synuclein [51]. Recent observations have also confirmed that the presence of Lewy bodies is not only restricted to mid brain region but also in other areas of brain including the stem region and olfactory bulb region [38]. These areas of the brain are associated with nonmotor functions such as sleep regulation [52, 53]. Further, it has been reported that there is a decrease in the intestinal dopamine cells, which can predispose gastrointestinal symptoms in PD patients [54]. The pathophysiology of Parkinson's disease involves several metabolic defects that include α-synuclein proteostasis, calcium signaling, oxidative stress, and neuroinflammation [46, 47]. Due to this reason, using various biomedical detection tools (such as PET), it has become possible for early diagnosis of Parkinson's disease. Since dopamine-regulated metabolic pathways have a direct relevance to PD, potential drugs that substitute dopamine are preferred in the treatment of Parkinson's disease. Targeting the cellular transport and amyloid aggregation of α-synuclein has been recently shown as a potential approach for the treatment. Though the above strategies have shown enormous potential to prevent Parkinson's disease, however, early detection of PD seems central to the treatment of this lethal disease. Moreover, α-synuclein aggregation leading to the formation of Lewy bodies in PD has also attracted many researchers to find potential strategies to either reduce gene expression of α-synuclein or prevent the onset of α-synuclein aggregation. Recent research reports from Austrian biotech AFFiRiS have proposed a potential vaccine against PD, which is in the first stages of clinical trials, that is designed to bind to α-synuclein and to eliminate it from the affected brain tissues [55].

Nonneurodegenerative Amyloid Diseases

Protein aggregation is also involved in nonneurodegenerative diseases with severe lethality by affecting the tissues other than brain tissues [56, 57]. Similar to neurodegenerative diseases, conformational changes in any specific proteins toward an amyloid-prone structure are believed to be one of the foundational events for the onset of such nonneurodegenerative diseases. Examples of such nonneurodegenerative disease are IAPP-linked type II diabetes [58], crystallins-linked cataracts [59], and nonneuropathic systemic amyloidosis [60].

Type II Diabetes

Type II diabetes is an age-linked disease that is primarily caused by abnormal regulation glucose metabolism in the body system. This metabolic defect has been ascribed to the aggregation of the islet amyloid polypeptide (IAPP) that leads to formation of toxic amyloid species capable of killing pancreatic islet β cells. In normal physiology, insulin is secreted from pancreatic β cells for the regulation of glucose metabolism. The protein IAPP exists in the insulin secretory granules of the pancreatic β cells, and it is always co-secreted with insulin. The actual role of IAPP is largely unknown; however, it is believed that IAPP plays a major role in maintaining glucose homeostasis [61–64]. IAPP is inherently amyloidogenic because it shows a high propensity to undergo an aggregation process under in vitro systems [65]. Furthermore, accumulation of IAPP-generated amyloid plaques has been observed in most of the patients suffering from type II diabetes. Not only IAPP, insulin protein is also known to undergo an amyloid aggregation that often yields toxic amyloid species both in vitro and in vivo conditions [66–70]. The storage of insulin in the vials which is used by the diabetic patients is also a big concern as the insulin has been reported to form toxic amyloid aggregates, possibly due to prolonged storage and exposure to varied external factors such as temperature [70].

Cataract

Cataract is a common nonneurodegenerative disease in which individuals suffer from blindness. This disease affects almost more than 50% of individuals over the age of ~70, and it has been reported that about 100 million people have been suffering from age-related cataract [71, 72]. Normally, the lens can stay transparent throughout life, as there is no protein turnover or synthesis. In cataracts, soluble proteins of the lens tend to aggregate and form insoluble amyloid structures. Hence, formation of such aggregated structures adversely affects the lens transparency leading to blindness. Though cataracts are not considered as a life-threatening disease, this problem is directly or indirectly linked to several medical and social consequences [59]. More than one-third of the lens components are made up of the molecular chaperones αA-crystallin and αB-crystallin. These crystallin proteins are identified as molecular chaperones whose presence in the lens is necessary for maintaining the solubility of other lens proteins. As the age of an individual increases, crystallins undergo extensive posttranslational modifications, and with such modification, these proteins become aggregation prone, triggering an aggressive aggregation process [73, 74]. Sometimes, mutations are also known to convert native conformation of soluble crystallins into abnormal aggregation-prone conformation, triggering the formation of opaque aggregates [59]. For example, the Arg120Gly mutation in αB-crystallin is known to trigger early onset of cataract [75, 76]. The ability of the crystallins to convert into fibrils under destabilizing conditions suggests that this process could contribute to the development of cataract with aging.

Nonneuropathic Systemic Amyloidosis

Amyloid aggregation in multiple tissue components sometimes leads to lethal severities, which are identified as nonneuropathic systemic amyloidosis [77], such as amyloid light-chain (AL) amyloidosis and serum amyloid A (SAA) amyloidosis. AL amyloidosis is the most common form of systemic amyloidosis disorder, which is caused by the amyloid aggregation of the peptides derived from light chains of immunoglobulin. Generation of such amyloid-prone fragments from light chains of immunoglobulin proteins has been reported to occur in multiple organs [78]. These aggregation-prone light-chain peptides are produced in the bone marrow, and upon release, they are internalized by cells of diverse tissue components in the body system, and eventually, such cells are adversely affected as the peptide forms toxic amyloid aggregates [79]. Various vital organs including liver, heart, kidney, and pancreas have been reported to be affected by the occurrence of AL amyloidosis [78, 79]. Another unique nonneuropathic systemic amyloidosis is caused by serum amyloid A protein (SAA). SAA is usually considered as an amyloid-prone protein which can undergo an aggregation process to yield toxic amyloid species [80]. The kidney tissues are mainly affected by SAA amyloidosis, which very often leads to renal failure [56]. Furthermore, transthyretin-related (TTR) amyloidosis is also considered as a big treat to normal physiology. Mutations in the wild-type TTR have been considered as one of the factors to trigger familial amyloid cardiomyopathy [81]. The wild-type TTR can sometimes, due to misfolding, undergo aggregation that can cause nongenetic amyloid diseases, particularly triggering the onset of senile systemic amyloidosis (SSA) [82, 83].

Does Coaggregation of Proteins Exist in Biology?

In nature, most proteins coexist in different microenvironments or subcellular compartments and intermolecular interactions between them such as protein–protein interactions and formation of protein complexes are crucial for normal cellular metabolism. Hence, it is important to understand what effect the aggregation process of a particular protein would have on the biophysical properties of other neighboring proteins. The question is whether amyloid formation of one type of protein can trigger amyloid aggregation process in another type of protein. Though amyloid formation of a specific protein or peptide is usually known to cause the related disease, it is largely unknown whether such aggregation-prone peptides/proteins can induce an aggregation process in other proteins. Recent investigations have reported the coexistence of two different amyloid-linked diseases in individual patients [84–86]. For example, occurrence of both Alzheimer's disease (AD) and Huntington's disease (HD) has been observed in the same patient [84]. Coexistence of Huntington's disease and amyotrophic lateral sclerosis (ALS) has also been reported in individuals in a recent study [85]. Lauren et al. have shown that prion proteins and amyloid-β oligomers can interact with each other [87], and in yet another study, Ohara et al. reported the coaggregation of α-synuclein and Cu/Zn superoxide dismutase protein [86]. In a recent in vitro study, it was demonstrated that amyloid aggregates of one type of protein can effectively trigger amyloid aggregation process in other proteins [69, 88]. The same study has revealed in vitro coaggregation and cross-seeding between lysozyme, bovine serum albumin, insulin, and cytochrome c [69]. In another research work involving in vitro experiments, it has been shown that phenylalanine amyloid fibrils can effectively trigger aggregation of other globular proteins leading to the formation of cytotoxic fibrils of amyloid nature [88]. Since amyloid-linked inclusions, plaques, and tangles are usually considered as heterocomponent entities, despite their origin from one type of proteins, the study of cross-seeding and coaggregation between proteins/peptides seems important to improve our mechanistic understanding of amyloid-related pathologies.

Mechanism of Protein Aggregation

The process of amyloid formation of most proteins is believed to follow a nucleation-dependent propagation [89, 90] in which protein monomers self-assemble to form stable oligomers, capable of recruiting and driving aggregation of other monomers, sequestering an aggregation process. If the population of such initial nucleation

19 Protein Aggregation, Related Pathologies, and Aging

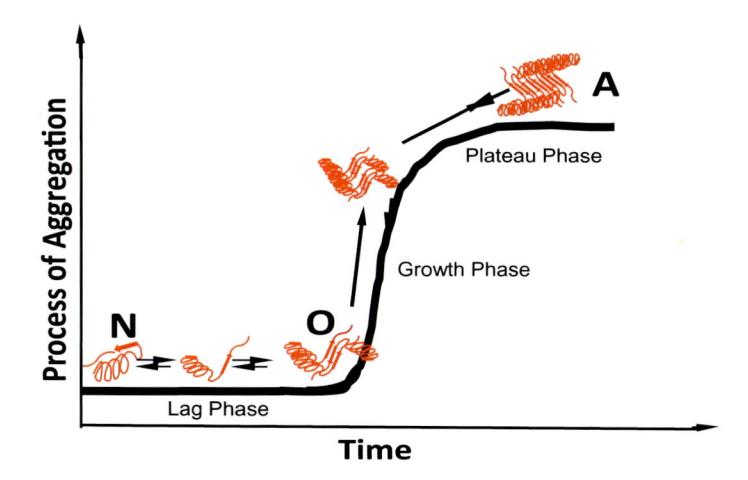

Fig. 19.3 Schematic representation of the nucleation and propagation process during protein aggregation as a function of time. N, is the native conformation of a protein; O, is the intermediate oligomeric state; and A, is the aggregated mature higher order structures

capable oligomers rises above the critical concentration, the aggregation reaction is favored, leading to conversion of soluble protein monomers into insoluble amyloid fibrils [90] However, in some cases, the protein monomer itself can act as a nucleus such as in the case of the amyloid fibril formation of polyQ peptides [43, 45, 91]. Simple polyQ peptides also undergo spontaneous aggregation process under physiological conditions and the kinetics of such aggregation process increases with the increase in the number of glutamine residues. Dr. Wetzel's group has proposed a nucleation growth mechanism for aggregation of polyQ peptides in which the structure of the nucleus possibly has a duplex β-sheet architecture [45, 92]. For longer polyQ peptides, with more than ~25 Qs, the nucleus structure can be attained by the single peptide, possibly by taking one or two reverse turns to arrive at the required β-sheet conformation. However, for short polyQ peptides, with less than ~25 Qs, at least four peptides would have to come together for the formation the required nucleus structure to begin aggregation, thus reducing the aggregation propensity of the short polyQ peptides [45]. The requirement of a β-hairpin structure to begin aggregation of polyQ peptides has been further validated by recent reports on rapid aggregation of short polyQ peptides with inbuilt β-hairpin motifs in the peptide sequence [93, 94]. Interestingly, strategically designed β-hairpin-based inhibitors, particularly aimed to interfere with the intermolecular H-bonding between polyQ peptides, have shown strong inhibition effect against polyQ aggregation under in vitro conditions [93, 94].

In general, the nucleation process during protein aggregation mainly comprises three distinct phases: lag phase, time taken by proteins monomers to form oligomers; growth phase, exponential growth of the formation of aggregates; and plateau phase, indicating completion of aggregation (Fig. 19.3). Recent studies have proposed that the on-pathway oligomers are more toxic than the mature amyloid fibers

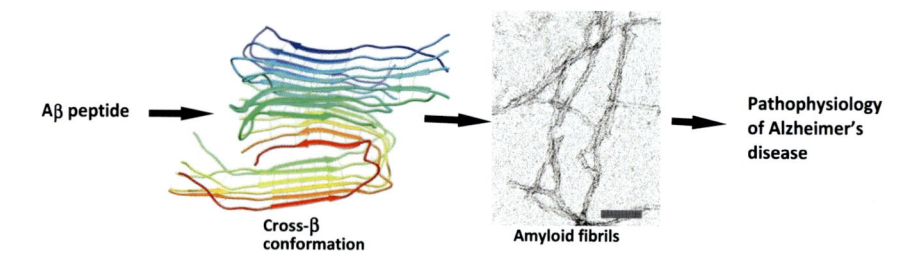

Fig. 19.4 Schematic representation of the proposed mechanism for the aggregation of Aβ (1–42) peptide related to Alzheimer's disease. The peptide Aβ is believed to form cross-β architecture consisting of an ordered arrangement of β sheets (PDB ID: 2LMN) that becomes the core structureof the amyloid fibrils [101]. (Reference: Luhrs et al. 2005, PNAS)

[5, 6, 95]. Therefore, researchers are now more interested in understanding the process of oligomerization and the mechanistic understanding of their cytotoxic effects. The structural conformation of most protein oligomers is typical cross-β type [57, 96–98]. When proteins undergo an aggregation process, they generally form stable higher order structures with diverse morphologies, such as disordered amorphous aggregates and highly ordered cross-β-structured amyloid fibrils (as schematically shown in Fig. 19.1) [99]. Formation of the cross-β conformation is fundamentally driven by viable noncovalent interactions between two partially unfolded amyloidogenic chains. The partially unfolded protein species are known to trigger self-association of proteins. Such interactions are believed to be stabilized by hydrophobic–hydrophobic interactions mediated through exposed hydrophobic groups [100]. The cross-β structure is usually stabilized by stable H-bonds mediated through the functional groups present in the backbone of the polypeptide chain.

Further, these cross-β structures are also stabilized by interactions between the functional groups of the side chains, which include π–π stacking, hydrophobic interactions, and salt bridges [101, 102]. Figure 19.4 shows the cross-β structure (PDB ID: 2LMN) of the Aβ peptide [103], that is believed to be the core structure of the amyloid entities observed in Alzheimer's disease. All these noncovalent interactions are vital to the formation of supramolecular protein assemblies.

Strategies to Target Protein Aggregation Process

Prevention of amyloid-linked diseases, including both neuropathic and nonneuropathic amyloidosis, largely depends on how well the mechanism of such aggregation process is understood. Notably, as indicated in Fig. 19.5, there can be three possible strategies one could propose: (1) to target the gene product and block the synthesis of the amyloidogenic proteins, (2) to inhibit the initiation of the aggregation of the proteins, and (3) to promote disassembly of the higher order amyloid structures such as plaques and inclusion bodies.

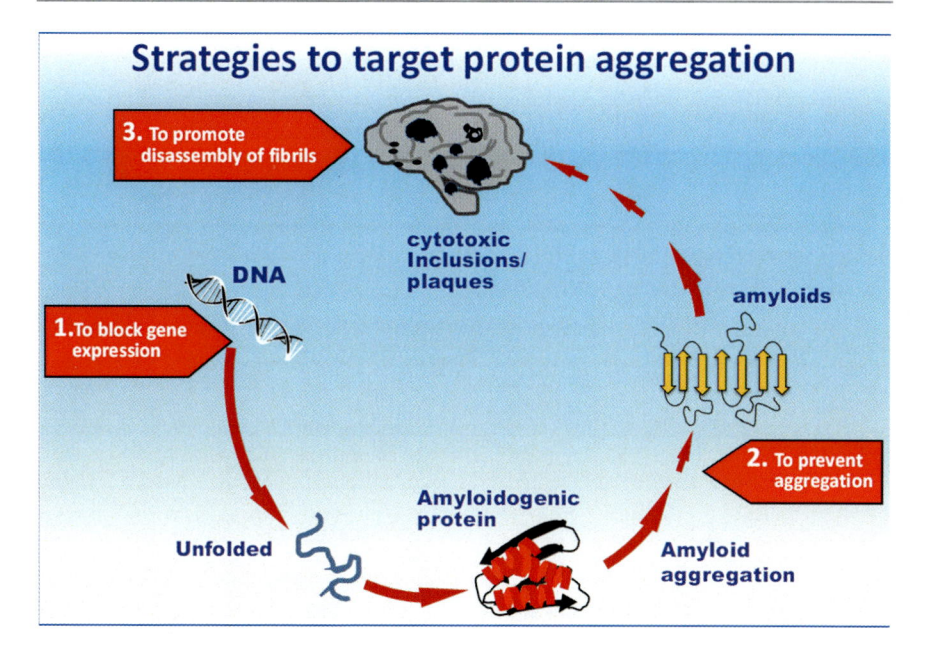

Fig. 19.5 Schematic representation of different possible targets to prevent protein aggregation linked diseases: (1) to target the gene product and block the synthesis of the amyloidogenic proteins; (2) to inhibit the initiation of the aggregation process; (3) to promote the disassembly of higher order amyloid structures such as plaques and inclusions

Though targeting gene expression of the amyloidogenic proteins sounds beneficial, with this approach the essential roles of the protein, if any, may be lost, which could result in various side effects. For treatment of Alzheimer's disease, one of the straightforward strategies is to reduce the production of amyloidogenic Aβ (1–42) peptide. Though the exact connection between Aβ peptide and pathophysiology of AD remains largely unknown, recent reports on the successful trials of potential drugs designed to target the Aβ amyloids have shown promising results in reducing the AD disease progression. One of such candidates to target AD pathology is Gantenerumab that has been reported to result in the reduction of the plaque mass in a dose-dependent manner [104]. This drug molecule is an anti-Aβ antibody which has inherent property to specifically bind to Aβ plaques [104]. Gantenerumab is currently undergoing Phase II and III clinical trials, and it seems to be a potential drug in treatment of Alzheimer's disease. Another approach to target AD is inhibition of amyloid formation through potential candidates that can block the activity of both γ-secretase and β-secretase enzymes. Blocking of these secretase enzymes would reduce the cleavage of APP, preventing the formation amyloid-prone Aβ peptides [105].

Since normal globular proteins can also undergo amyloid formation upon exposure to aggregating conditions, sometimes such aggregation process may directly or indirectly lead to severe medical complications (Table 19.2). Normal globular proteins such as lysozyme, serum albumin, insulin, and myoglobin are usually considered as convenient model systems to study amyloid formation of proteins, mostly under in vitro conditions. Much information on both structural and aggregation properties of these globular proteins is available in literature [106–108]. Aggregation of lysozyme is known to induce systemic amyloidosis [109], whereas insulin fibril assembly aggregation is known to trigger site-specific localized amyloidosis [110]. Since insulin is a vital supplement for the diabetic patients, amyloid formation of insulin is also a great concern during its storage as therapeutics agents, in the syringe vials [111]. Furthermore, due to increase in the local concentration of insulin at the site of injection, formation of toxic insulin amyloid aggregated has been reported in patients, particularly at the site of arterial walls [112]. Due to all these reasons, it is important to have a clear and solid strategy to prevent the aggregation of these globular proteins. These approaches may possibly involve (i) inhibition of oligomerization, (ii) stabilization of protein monomers, and (iii) depolymerizing the already existing oligomers/mature amyloid structures [113]. These globular proteins have been used as convenient model systems to study amyloid formation, and finding suitable inhibitors against such process is important for amyloid research. Several potential compounds including single metabolites [114–116], natural products [68, 114], proteins, and small peptides [93, 117–119] have been found to protect proteins against amyloid fibril formation. Recent studies have revealed that strategically designed nanoparticles can also be used to target amyloid formation of proteins [66, 67, 120–122]. Citrate-capped gold nanoparticles of size greater than ~20 nm are known to delay the growth rate of α-synuclein aggregation [123]. Metallic nanoparticles, surface functionalized with natural compounds such as capsaicin, show inhibition effect against amyloid formation of globular protein. Interestingly, the anti-amyloid efficacy of some natural compounds such as capsaicin and piperine was found to be enhanced when they are coated with the nanoparticles [66, 120]. In a recent study, albumin-modified magnetic nanoparticles were found to promote the depolymerization of insulin amyloid fibrils [121]. Successful synthesis of biologically compatible gold nanoparticles using various amino acids has been reported recently [66, 67]. Gold nanoparticles coated with different amino acids have also showed substantial effect on serum albumin absorption and cytotoxicity [19], and in some studies, these amino acid-coated gold nanoparticles were found to be capable of both the prevention of amyloid aggregation and the dissociation of already formed amyloid aggregates [67, 121]. For instance, tyrosine- and tryptophan-coated gold nanoparticles have been shown to be effective in both inhibiting insulin aggregation and triggering disassembly of insulin amyloids [67]. Sometimes, the inhibitors not only prevent the onset of the aggregation process but also protect the cells against amyloid-induced toxicity [66, 68]. A recent biophysical study (Fig. 19.6) has revealed the anti-amyloidogenic activity of eugenol (a potential natural plant product), which can prevent the process of amyloid aggregation of proteins and protect the human erythrocytes against amyloid-induced toxicity [68].

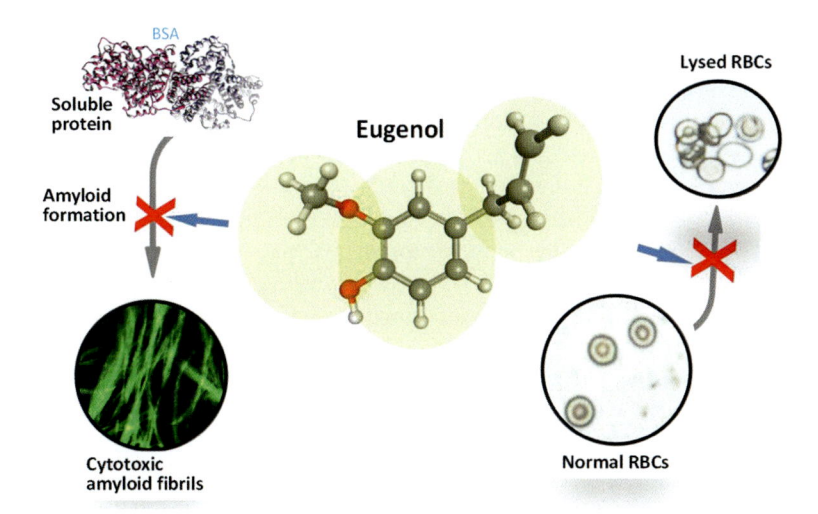

Fig. 19.6 Schematic representation showing the anti-amyloidogenic property of the natural compound eugenol, which can effectively prevent in vitro amyloid formation of serum albumin and can also protect erythrocytes from amyloid-induced hemolysis [68]. (Adapted from Dubey et al. (2017), Scientifc Reports)

The Relationship Between Aging and Protein Aggregation

Increase in the age of an organism is known to affect both the protein quality control systems [14, 124–126] and the cellular defense machinery against reactive oxygen species [127]. As discussed in the beginning of this chapter, a series of neurodegenerative diseases appear during aging, such as Alzheimer's, Huntington's, and Parkinson's diseases [60]. It is believed that the critical cellular processes linked to protein folding and protein homeostasis are adversely affected due to the accumulation of pathogenic amyloid entities, such as plaques, tangles, inclusions, and Lewy bodies [128–130]. Since amyloid formation in pathologies such as Huntington's disease is believed to be toxic gain of function, it is predicted that disease severities during old age would rise due to both the loss of protein quality control mechanism and the accumulation of toxic amyloids. The question of whether loss of protein function, toxic gain of cytoplasmic function, or a combination of both can cause neurodegeneration is largely unknown [131]. Dr. Dupuis group, based on a recent study on mouse models, has proposed that mislocalization of cytoplasmic FUS (a RNA-binding protein whose aggregation is linked to amyotrophic lateral sclerosis and frontotemporal dementia) can lead to nuclear loss of function and can also trigger motor neuron death due to a toxic gain of functions [131]. Aging is fundamentally associated with the progressive decline of protein homeostasis, and the mechanism behind the age-linked alterations in the proteome composition in the body system remains largely unknown. However, using convenient model systems such as *C. elegans* and yeast cells, researcher have been able to reveal key insights into the relationship between the aging process and protein aggregation [132–134]. Walther et al. in an extensive study on nematodes have revealed that one-third of proteins change in abundance at least two fold during aging and one of the factors behind such proteome imbalance is the process of protein aggregation [135]. The same study has suggested that sequestering proteins in chaperone-enriched aggregates can be a protective strategy to reduce the proteostasis decline during nematode aging [135]. In yet another study involving yeast cells, it has been proposed that as the yeast grows older, its sensitivity to reproductive pheromones reduces drastically. It is believed that the activity of histone deacetylate Sir2 is reduced during old age, which affects the metabolism at the chromatic level, particularly affecting the mating loci. Recent investigation by Schlissel et al. has proposed another mechanism that involves aggregation of Whi3, a RNA-binding protein [136]. Whi3 possesses a glutamine-rich tract in its sequence that is predicted to promote amyloid aggregation as the age progresses. Interestingly, deletion of such aggregation-prone glutamine-rich region of Whi3 in yeast cells resulted in increased replicative life span [136]. Another experimental evidence to highlight the direct link between aging and amyloid aggregation of proteins was revealed by Lithgow et al., in which compounds that have been known to specifically bind the amyloid fibrils can substantially increase the life span of the *C. elegans* [132]. Dr. Lithgow's group has revealed that Thioflavin T (a dye that specifically binds to amyloids fibrils) has the potential to increase the life span of *C. elegans*. Their study also reveals that Thioflavin T is also capable of suppressing amyloid-linked paralysis and reducing the levels of soluble aggregation-prone oligomeric proteins [132]. One of the suggested mechanisms behind the increased life span of the nematodes is the direct influence of the amyloid-binding dye on the regulation of crucial age-related transcription factors HSF-1 and SKN-1 [137, 138].

Conclusions

We age because the metabolic efficacies of cells decline as a function of time; however, the interrelationship between protein aggregation and aging-linked defects still remains as a mystery. If the protein quality control machinery in the cell is adversely affected, concentration of misfolded and aggregation-prone peptides and proteins would rise, leading to the onset of amyloid-linked diseases. However, aggregation of proteins may not be always interpreted as a lethality inducing factor to cellular processes. For example, in a recent investigation on yeast cells, a new type of protein aggregate was discovered, particularly in older cells. Interestingly, the presence of these aggregates has been reported to enhance the protein quality control systems in cell. The researchers propose that these age-related aggregates may contain multiple proteins including prion proteins. If the aging and the onset of neurodegenerative diseases are so directly related, probably every individual would suffer from Alzheimer's disease in his/her old age. However, the existence of healthy individuals in their old age certainly demonstrates the complexities related to aging and protein aggregation-linked diseases. It is possible that aggregation of proteins and peptides is a normal and fundamental process, and it may predispose lethality only when the aggregation occurs in the wrong place in the cell or when the amount of aggregate accumulation rises above the threshold limit, sufficient enough to cause cell damage. It is also possible that aggregation-induced metabolic defects could be manifested at the transcriptional and/or at the translational level, involving a complex interplay between RNA and DNA metabolism, much before the synthesis of amyloidogenic proteins begins. Though we have started gaining more knowledge on the aging and the protein homeostasis, extensive future research works would probably unfold the underlying principles behind the mysterious link between aging and protein aggregation.

References

1. Ouzounis CA, Coulson RM, Enright AJ, Kunin V, Pereira-Leal JB. Classification schemes for protein structure and function. Nat Rev Genet. 2003;4:508–19.
2. Berg JM, Tymoczko JL, Stryer L. Protein Structure and Function, Biochemistry. New York: W.H. Freeman and Company; 2002.
3. Aguzzi A, O'Connor T. Protein aggregation diseases: Pathogenicity and therapeutic perspectives, Nature Reviews. Drug Discov. 2010;9:237–48.
4. Gusella JF, Mac Donald ME. Molecular genetics: Unmasking polyglutamine triggers in neurodegenerative disease. Nat Rev. Neurosci. 2000;1:109–15.
5. Haass C, Selkoe DJ. Soluble protein oligomers in neurodegeneration: Lessons from the Alzheimer's amyloid beta-peptide, Nature Reviews. Mol Cell Biol. 2007;8:101–12.
6. Bemporad F, Chiti F. Protein misfolded oligomers: Experimental approaches, mechanism of formation, and structure-toxicity relationships. Chem Biol. 2012;19:315–27.
7. Kayed R, Head E, Thompson JL, McIntire TM, Milton SC, Cotman CW, Glabe CG. Common structure of soluble amyloid oligomers implies common mechanism of pathogenesis. Science. 2003;300:486–9.
8. Humphreys DT, Carver JA, Easterbrook-Smith SB, Wilson MR. Clusterin has chaperone-like activity similar to that of small heat shock proteins. J Biol Chem. 1999;274:6875–81.
9. Sakahira H, Breuer P, Hayer-Hartl MK, Hartl FU. Molecular chaperones as modulators of polyglutamine protein aggregation and toxicity. Proc Natl Acad Sci U S A. 2002;99(Suppl 4):16412–8.
10. Hartl FU, Hayer-Hartl M. Molecular chaperones in the cytosol: From nascent chain to folded protein. Science. 2002;295:1852–8.
11. Frydman J. Folding of newly translated proteins in vivo: The role of molecular chaperones. Annu Rev Biochem. 2001;70:603–47.
12. Kim YE, Hipp MS, Bracher A, Hayer-Hartl M, Hartl FU. Molecular chaperone functions in protein folding and proteostasis. Annu Rev Biochem. 2013;82:323–55.
13. Kakkar V, Meister-Broekema M, Minoia M, Carra S, Kampinga HH. Barcoding heat shock proteins to human diseases: Looking beyond the heat shock response. Dis Model Mech. 2014;7:421–34.
14. Morimoto RI. Proteotoxic stress and inducible chaperone networks in neurodegenerative disease and aging. Genes Dev. 2008;22:1427–38.
15. Glover JR, Lindquist S. Hsp104, Hsp70, and Hsp40: A novel chaperone system that rescues previously aggregated proteins. Cell. 1998;94:73–82.
16. Shorter J. The mammalian disaggregase machinery: Hsp110 synergizes with Hsp70 and Hsp40 to catalyze protein disaggregation and reactivation in a cell-free system. PloS One. 2011;6:e26319.

17. Gao X, Carroni M, Nussbaum-Krammer C, Mogk A, Nillegoda NB, Szlachcic A, Guilbride DL, Saibil HR, Mayer MP, Bukau B. Human Hsp70 disaggregase reverses Parkinson's-linked alpha-synuclein amyloid fibrils. Mol Cell. 2015;59:781–93.
18. Ciechanover A. Intracellular protein degradation: from a vague idea thru the lysosome and the ubiquitin-proteasome system and onto human diseases and drug targeting, Hematology. American Society of Hematology. Educ Program. 2006;1–12:505–6.
19. Reinstein E, Ciechanover A. Narrative review: Protein degradation and human diseases: the ubiquitin connection. Ann Intern Med. 2006;145:676–84.
20. Alzheimer's Association. 2015 Alzheimer's disease facts and figures. Alzheimer's Dement. 2015;11:332–84.
21. Jeong S. Molecular and cellular basis of neurodegeneration in Alzheimer's disease. Mol Cell. 2017;40(9):613–20.
22. Metaxas A, Kempf SJ. Neurofibrillary tangles in Alzheimer's disease: Elucidation of the molecular mechanism by immunohistochemistry and tau protein phosphoproteomics. Neural Regen Res. 2016;11:1579–81.
23. O'Brien RJ, Wong PC. Amyloid precursor protein processing and Alzheimer's disease. Annu Rev Neurosci. 2011;34:185–204.
24. Bekris LM, Galloway NM, Millard S, Lockhart D, Li G, Galasko DR, Farlow MR, Clark CM, Quinn JF, Kaye JA, Schellenberg GD, Leverenz JB, Seubert P, Tsuang DW, Peskind ER, Yu CE. Amyloid precursor protein (APP) processing genes and cerebrospinal fluid APP cleavage product levels in Alzheimer's disease. Neurobiol Aging. 2011;32(556):e513–23.
25. Vandersteen A, Hubin E, Sarroukh R, De Baets G, Schymkowitz J, Rousseau F, Subramaniam V, Raussens V, Wenschuh H, Wildemann D, Broersen K. A comparative analysis of the aggregation behavior of amyloid-beta peptide variants. FEBS Lett. 2012;586:4088–93.
26. Weidemann A, Eggert S, Reinhard FB, Vogel M, Paliga K, Baier G, Masters CL, Beyreuther K, Evin G. A novel epsilon-cleavage within the transmembrane domain of the Alzheimer amyloid precursor protein demonstrates homology with Notch processing. Biochemistry. 2002;41:2825–35.
27. Vigo-Pelfrey C, Lee D, Keim P, Lieberburg I, Schenk DB. Characterization of beta-amyloid peptide from human cerebrospinal fluid. J Neurochem. 1993;61:1965–8.
28. Takami M, Nagashima Y, Sano Y, Ishihara S, Morishima-Kawashima M, Funamoto S, Ihara Y. Gamma-Secretase: Successive tripeptide and tetrapeptide release from the transmembrane domain of beta-carboxyl terminal fragment. J Neurosci Off J Soc Neurosci. 2009;29:13042–52.
29. Masters CL, Simms G, Weinman NA, Multhaup G, McDonald BL, Beyreuther K. Amyloid plaque core protein in Alzheimer disease and Down syndrome. Proc Natl Acad Sci U S A. 1985;82:4245–9.
30. Chen WT, Hong CJ, Lin YT, Chang WH, Huang HT, Liao JY, Chang YJ, Hsieh YF, Cheng CY, Liu HC, Chen YR, Cheng IH. Amyloid-beta (Abeta) D7H mutation increases oligomeric Abeta42 and alters properties of Abeta-zinc/copper assemblies. PloS One. 2012;7:e35807.
31. Hatami A, Monjazeb S, Milton S, Glabe CG. Familial Alzheimer's disease mutations within the amyloid precursor protein alter the aggregation and conformation of the amyloid-beta peptide. J Biol Chem. 2017;292:3172–85.
32. Kovacs GG, Budka H. Prion diseases: From protein to cell pathology. Am J Pathol. 2008;172:555–65.
33. Prusiner SB. Prions. Proc Natl Acad Sci U S A. 1998;95:13363–83.
34. Halliday M, Mallucci GR. Review: Modulating the unfolded protein response to prevent neurodegeneration and enhance memory. Neuropathol Appl Neurobiol. 2015;41:414–27.
35. Zoghbi HY, Orr HT. Glutamine repeats and neurodegeneration. Annu Rev Neurosci. 2000;23:217–47.
36. La Spada AR, Wilson EM, Lubahn DB, Harding AE, Fischbeck KH. Androgen receptor gene mutations in X-linked spinal and bulbar muscular atrophy. Nature. 1991;352:77–9.
37. Cummings CJ, Zoghbi HY. Trinucleotide repeats: Mechanisms and pathophysiology. Annu Rev Genomics Hum Genet. 2000;1:281–328.

38. Reichmann H, Brandt MD, Klingelhoefer L. The nonmotor features of Parkinson's disease: Pathophysiology and management advances. Curr Opin Neurol. 2016;29:467–73.
39. Klement IA, Skinner PJ, Kaytor MD, Yi H, Hersch SM, Clark HB, Zoghbi HY, Orr HT. Ataxin-1 nuclear localization and aggregation: Role in polyglutamine-induced disease in SCA1 transgenic mice. Cell. 1998;95:41–53.
40. Orr HT, Chung MY, Banfi S, Kwiatkowski TJ Jr, Servadio A, Beaudet AL, McCall AE, Duvick LA, Ranum LP, Zoghbi HY. Expansion of an unstable trinucleotide CAG repeat in spinocerebellar ataxia type 1. Nat Genet. 1993;4:221–6.
41. Sawa A, Wiegand GW, Cooper J, Margolis RL, Sharp AH, Lawler JF Jr, Greenamyre JT, Snyder SH, Ross CA. Increased apoptosis of Huntington disease lymphoblasts associated with repeat length-dependent Mitochondrial depolarization. Nat Med. 1999;5:1194–8.
42. Chen S, Ferrone FA, Wetzel R. Huntington's disease age-of-onset linked to polyglutamine aggregation nucleation. Proc Natl Acad Sci U S A. 2002;99:11884–9.
43. Wetzel R. Physical chemistry of polyglutamine: Intriguing tales of a monotonous sequence. J Mol Biol. 2012;421:466–90.
44. Hoop CL, Lin HK, Kar K, Magyarfalvi G, Lamley JM, Boatz JC, Mandal A, Lewandowski JR, Wetzel R, van der Wel PC. Huntingtin exon 1 fibrils feature an interdigitated beta-hairpin-based polyglutamine core. Proc Natl Acad Sci U S A. 2016;113:1546–51.
45. Kar K, Jayaraman M, Sahoo B, Kodali R, Wetzel R. Critical nucleus size for disease-related polyglutamine aggregation is repeat-length dependent. Nat Struct Mol Biol. 2011;18:328–36.
46. Lotharius J, Brundin P. Pathogenesis of Parkinson's disease: Dopamine, vesicles and alpha-synuclein. Nat Rev. Neurosci. 2002;3:932–42.
47. Dauer W, Przedborski S. Parkinson's disease: Mechanisms and models. Neuron. 2003;39:889–909.
48. Girault JA, Greengard P. The neurobiology of dopamine signaling. Arch Neurol. 2004;61:641–4.
49. Nagahama Y, Okina T, Suzuki N. Neuropsychological differences related to age in dementia with Lewy bodies. Dement Geriatr Cogn Dis Extra. 2017;7:188–94.
50. Tercjak A, Bergareche A, Caballero C, Tunon T, Linazasoro G. Lewy bodies under atomic force microscope. Ultrastruct Pathol. 2014;38:1–5.
51. Spillantini MG, Schmidt ML, Lee VM, Trojanowski JQ, Jakes R, Goedert M. Alpha-synuclein in Lewy bodies. Nature. 1997;388:839–40.
52. Stacy M. Nonmotor symptoms in Parkinson's disease. Int J Neurosci. 2011;121(Suppl 2):9–17.
53. Park A, Stacy M. Dopamine-induced nonmotor symptoms of Parkinson's disease. Parkinsons Dis. 2011;2011:485063.
54. Klingelhoefer L, Reichmann H. The gut and nonmotor symptoms in Parkinson's disease. Int Rev Neurobiol. 2017;134:787–809.
55. Galabova G, Brunner S, Winsauer G, Juno C, Wanko B, Mairhofer A, Luhrs P, Schneeberger A, von Bonin A, Mattner F, Schmidt W, Staffler G. Peptide-based anti-PCSK9 vaccines – An approach for long-term LDLc management. PloS One. 2014;9:e114469.
56. Chiti F, Dobson CM. Protein misfolding, functional amyloid, and human disease. Annu Rev Biochem. 2006;75:333–66.
57. Chiti F, Dobson CM. Amyloid formation by globular proteins under native conditions. Nat Chem Biol. 2009;5:15–22.
58. Jaikaran ET, Clark A. Islet amyloid and type 2 diabetes: From molecular misfolding to islet pathophysiology. Biochim Biophys Acta. 2001;1537:179–203.
59. Moreau KL, King JA. Protein misfolding and aggregation in cataract disease and prospects for prevention. Trends Mol Med. 2012;18:273–82.
60. Knowles TP, Vendruscolo M, Dobson CM. The amyloid state and its association with protein misfolding diseases. Nat Rev. Mol Cell Biol. 2014;15:384–96.
61. Cao P, Marek P, Noor H, Patsalo V, Tu LH, Wang H, Abedini A, Raleigh DP. Islet amyloid: From fundamental biophysics to mechanisms of cytotoxicity. FEBS Lett. 2013;587:1106–18.

62. Abedini A, Schmidt AM. Mechanisms of islet amyloidosis toxicity in type 2 diabetes. FEBS Lett. 2013;587:1119–27.
63. Westermark P, Engstrom U, Johnson KH, Westermark GT, Betsholtz C. Islet amyloid polypeptide: Pinpointing amino acid residues linked to amyloid fibril formation. Proc Natl Acad Sci U S A. 1990;87:5036–40.
64. Marzban L, Soukhatcheva G, Verchere CB. Role of carboxypeptidase E in processing of pro-islet amyloid polypeptide in {beta}-cells. Endocrinology. 2005;146:1808–17.
65. Park K, Verchere CB. Identification of a heparin binding domain in the N-terminal cleavage site of pro-islet amyloid polypeptide. Implications for islet amyloid formation. J Biol Chem. 2001;276:16611–6.
66. Anand BG, Shekhawat DS, Dubey K, Kar K. Uniform, Polycrystalline, and thermostable piperine-coated gold nanoparticles to target insulin fibril assembly. ACS Biomater Sci Eng. 2017;3:1136–45.
67. Dubey K, Anand BG, Badhwar R, Bagler G, Navya PN, Daima HK, Kar K. Tyrosine- and tryptophan-coated gold nanoparticles inhibit amyloid aggregation of insulin. Amino Acids. 2015;47:2551–60.
68. Dubey K, Anand BG, Sekhawat DS, Kar K. Eugenol prevents amyloid fibril formation of proteins and inhibits amyloid induced hemolysis. Sci Rep. 2017;7:40744.
69. Dubey K, Anand BG, Temgire MK, Kar K. Evidence of rapid coaggregation of globular proteins during amyloid formation. Biochemistry. 2014;53:8001–4.
70. Cohen E, Dillin A. The insulin paradox: Aging, proteotoxicity and neurodegeneration. Nat Rev Neurosci. 2008;9:759–67.
71. Wu X, Long E, Lin H, Liu Y. Prevalence and epidemiological characteristics of congenital cataract: A systematic review and meta-analysis. Sci Rep. 2016;6:28564.
72. Foster A, Gilbert C, Rahi J. Epidemiology of cataract in childhood: A global perspective. J cataract Refract Surg. 1997;23(Suppl 1):601–4.
73. Hains PG, Truscott RJ. Post-translational modifications in the nuclear region of young, aged, and cataract human lenses. J Proteome Res. 2007;6:3935–43.
74. Bloemendal H, de Jong W, Jaenicke R, Lubsen NH, Slingsby C, Tardieu A. Ageing and vision: Structure, stability and function of lens crystallins. Progr Biophys Mol Biol. 2004;86:407–85.
75. Perng MD, Muchowski PJ, van Den IP, Wu GJ, Hutcheson AM, Clark JI, Quinlan RA. The cardiomyopathy and lens cataract mutation in alpha B-crystallin alters its protein structure, chaperone activity, and interaction with intermediate filaments in vitro. J Biol Chem. 1999;274:33235–43.
76. Perng MD, Cairns L, van den IP, Prescott A, Hutcheson AM, Quinlan RA. Intermediate filament interactions can be altered by HSP27 and alphaB-crystallin. J Cell Sci. 1999;112(Pt 13):2099–112.
77. Falk RH, Comenzo RL, Skinner M. The systemic amyloidoses. N Engl J Med. 1997;337:898–909.
78. Comenzo RL. Systemic immunoglobulin light-chain amyloidosis. Clin Lymphoma Myeloma. 2006;7:182–5.
79. Marin-Argany M, Lin Y, Misra P, Williams A, Wall JS, Howell KG, Elsbernd LR, McClure M, Ramirez-Alvarado M. Cell damage in light chain amyloidosis: FIBRIL internalization, toxicity and cell-mediated seeding. J Biol Chem. 2016;291:19813–25.
80. Westermark GT, Fandrich M, Westermark P. AA amyloidosis: Pathogenesis and targeted therapy. Annu Rev Pathol. 2015;10:321–44.
81. Gertz MA, Benson MD, Dyck PJ, Grogan M, Coelho T, Cruz M, Berk JL, Plante-Bordeneuve V, Schmidt HHJ, Merlini G. Diagnosis, Prognosis, and therapy of transthyretin amyloidosis. J Am Coll Cardiol. 2015;66:2451–66.
82. Pitkanen P, Westermark P, Cornwell GG 3rd. Senile systemic amyloidosis. Am J Pathol. 1984;117:391–9.
83. Ueda M, Horibata Y, Shono M, Misumi Y, Oshima T, Su Y, Tasaki M, Shinriki S, Kawahara S, Jono H, Obayashi K, Ogawa H, Ando Y. Clinicopathological features of senile systemic

amyloidosis: An ante- and post-mortem study. Mod Pathol Off J U S Canad Acad Pathol, Inc. 2011;24:1533–44.
84. Moss RJ, Mastri AR, Schut LJ. The coexistence and differentiation of late onset Huntington's disease and Alzheimer's disease. A case report and review of the literature. J Am Geriatr Soc. 1988;36:237–41.
85. Tada M, Coon EA, Osmand AP, Kirby PA, Martin W, Wieler M, Shiga A, Shirasaki H, Makifuchi T, Yamada M, Kakita A, Nishizawa M, Takahashi H, Paulson HL. Coexistence of Huntington's disease and amyotrophic lateral sclerosis: A clinicopathologic study. Acta Neuropathol. 2012;124:749–60.
86. Takei Y, Oguchi K, Koshihara H, Hineno A, Nakamura A, Ohara S. alpha-Synuclein coaggregation in familial amyotrophic lateral sclerosis with SOD1 gene mutation. Hum Pathol. 2013;44:1171–6.
87. Lauren J, Gimbel DA, Nygaard HB, Gilbert JW, Strittmatter SM. Cellular prion protein mediates impairment of synaptic plasticity by amyloid-beta oligomers. Nature. 2009;457:1128–32.
88. Anand BG, Dubey K, Shekhawat DS, Kar K. Intrinsic property of phenylalanine to trigger protein aggregation and hemolysis has a direct relevance to phenylketonuria. Sci Rep. 2017;7:11146.
89. Lomakin A, Chung DS, Benedek GB, Kirschner DA, Teplow DB. On the nucleation and growth of amyloid beta-protein fibrils: detection of nuclei and quantitation of rate constants. Proc Natl Acad Sci U S A. 1996;93:1125–9.
90. Xue WF, Homans SW, Radford SE. Systematic analysis of nucleation-dependent polymerization reveals new insights into the mechanism of amyloid self-assembly. Proc Natl Acad Sci U S A. 2008;105:8926–31.
91. Wetzel R. Kinetics and thermodynamics of amyloid fibril assembly. Acc Chem Res. 2006;39:671–9.
92. Misra P, Kodali R, Chemuru S, Kar K, Wetzel R. Rapid alpha-oligomer formation mediated by the Abeta C terminus initiates an amyloid assembly pathway. Nat Commun. 2016;7:12419.
93. Kar K, Baker MA, Lengyel GA, Hoop CL, Kodali R, Byeon IJ, Horne WS, van der Wel PC, Wetzel R. Backbone engineering within a latent beta-hairpin structure to design inhibitors of polyglutamine amyloid formation. J Mol Biol. 2017;429:308–23.
94. Kar K, Hoop CL, Drombosky KW, Baker MA, Kodali R, Arduini I, van der Wel PC, Horne WS, Wetzel R. Beta-hairpin-mediated nucleation of polyglutamine amyloid formation. J Mol Biol. 2013;425:1183–97.
95. Bemporad F, Taddei N, Stefani M, Chiti F. Assessing the role of aromatic residues in the amyloid aggregation of human muscle acylphosphatase. Protein Sci. 2006;15:862–70.
96. Apetri MM, Maiti NC, Zagorski MG, Carey PR, Anderson VE. Secondary structure of alpha-synuclein oligomers: Characterization by Raman and atomic force microscopy. J Mol Biol. 2006;355:63–71.
97. Maiti NC, Apetri MM, Zagorski MG, Carey PR, Anderson VE. Raman spectroscopic characterization of secondary structure in natively unfolded proteins: Alpha-synuclein. J Am Chem Soc. 2004;126:2399–408.
98. Holm NK, Jespersen SK, Thomassen LV, Wolff TY, Sehgal P, Thomsen LA, Christiansen G, Andersen CB, Knudsen AD, Otzen DE. Aggregation and fibrillation of bovine serum albumin. Biochim Biophys Acta. 2007;1774:1128–38.
99. Tyedmers J, Mogk A, Bukau B. Cellular strategies for controlling protein aggregation. Nat Rev. Mol Cell Biol. 2010;11:777–88.
100. Schmittschmitt JP, Scholtz JM. The role of protein stability, solubility, and net charge in amyloid fibril formation. Protein Sci. 2003;12:2374–8.
101. Gazit E. A possible role for pi-stacking in the self-assembly of amyloid fibrils. FASEB J: Off Publ Fed Am Soc Exp Biol. 2002;16:77–83.
102. Tartaglia GG, Cavalli A, Pellarin R, Caflisch A. The role of aromaticity, exposed surface, and dipole moment in determining protein aggregation rates. Protein Sci. 2004;13:1939–41.

103. Luhrs T, Ritter C, Adrian M, Riek-Loher D, Bohrmann B, Dobeli H, Schubert D, Riek R. 3D structure of Alzheimer's amyloid-beta(1-42) fibrils. Proc Natl Acad Sci U S A. 2005;102:17342–7.
104. Ostrowitzki S, Deptula D, Thurfjell L, Barkhof F, Bohrmann B, Brooks DJ, Klunk WE, Ashford E, Yoo K, Xu ZX, Loetscher H, Santarelli L. Mechanism of amyloid removal in patients with Alzheimer disease treated with gantenerumab. Arch Neurol. 2012;69:198–207.
105. Wilcock GK, Black SE, Hendrix SB, Zavitz KH, Swabb EA, Laughlin MA. Efficacy and safety of tarenflurbil in mild to moderate Alzheimer's disease: A randomised phase II trial. Lancet. Neurol. 2008;7:483–93.
106. Swaminathan R, Ravi VK, Kumar S, Kumar MV, Chandra N. Lysozyme: A model protein for amyloid research. Adv Protein Chem Struct Biol. 2011;84:63–111.
107. Gong H, He Z, Peng A, Zhang X, Cheng B, Sun Y, Zheng L, Huang K. Effects of several quinones on insulin aggregation. Sci Rep. 2014;4:5648.
108. Wu G, Robertson DH, Brooks CL 3rd, Vieth M. Detailed analysis of grid-based molecular docking: A case study of CDOCKER-A CHARMm-based MD docking algorithm. J Comput Chem. 2003;24:1549–62.
109. Pepys MB, Hawkins PN, Booth DR, Vigushin DM, Tennent GA, Soutar AK, Totty N, Nguyen O, Blake CC, Terry CJ, et al. Human lysozyme gene mutations cause hereditary systemic amyloidosis. Nature. 1993;362:553–7.
110. Dische FE, Wernstedt C, Westermark GT, Westermark P, Pepys MB, Rennie JA, Gilbey SG, Watkins PJ. Insulin as an amyloid-fibril protein at sites of repeated insulin injections in a diabetic patient. Diabetologia. 1988;31:158–61.
111. Hjorth CF, Norrman M, Wahlund PO, Benie AJ, Petersen BO, Jessen CM, Pedersen TA, Vestergaard K, Steensgaard DB, Pedersen JS, Naver H, Hubalek F, Poulsen C, Otzen D. Structure, aggregation, and activity of a covalent insulin dimer formed during storage of neutral formulation of human insulin. J Pharm Sci. 2016;105:1376–86.
112. Wang F, Hull RL, Vidal J, Cnop M, Kahn SE. Islet amyloid develops diffusely throughout the pancreas before becoming severe and replacing endocrine cells. Diabetes. 2001;50:2514–20.
113. Skovronsky DM, Lee VM, Trojanowski JQ. Neurodegenerative diseases: New concepts of pathogenesis and their therapeutic implications. Annu Rev Pathol. 2006;1:151–70.
114. Kar K, Kishore N. Enhancement of thermal stability and inhibition of protein aggregation by osmolytic effect of hydroxyproline. Biopolymers. 2007;87:339–51.
115. Shiraki K, Kudou M, Fujiwara S, Imanaka T, Takagi M. Biophysical effect of amino acids on the prevention of protein aggregation. J Biochem. 2002;132:591–5.
116. Ghosh R, Sharma S, Chattopadhyay K. Effect of arginine on protein aggregation studied by fluorescence correlation spectroscopy and other biophysical methods. Biochemistry. 2009;48:1135–43.
117. Etienne MA, Aucoin JP, Fu Y, McCarley RL, Hammer RP. Stoichiometric inhibition of amyloid beta-protein aggregation with peptides containing alternating alpha, alpha-disubstituted amino acids. J Am Chem Soc. 2006;128:3522–3.
118. Rajasekhar K, Suresh SN, Manjithaya R, Govindaraju T. Rationally designed peptidomimetic modulators of abeta toxicity in Alzheimer's disease. Sci Rep. 2015;5:8139.
119. Viet MH, Ngo ST, Lam NS, Li MS. Inhibition of aggregation of amyloid peptides by beta-sheet breaker peptides and their binding affinity. J Phys Chem. 2011;B115:7433–46.
120. Anand BG, Dubey K, Shekhawat DS, Kar K. Capsaicin-coated silver nanoparticles inhibit amyloid fibril formation of serum albumin. Biochemistry. 2016;55:3345–8.
121. Siposova K, Kubovcikova M, Bednarikova Z, Koneracka M, Zavisova V, Antosova A, Kopcansky P, Daxnerova Z, Gazova Z. Depolymerization of insulin amyloid fibrils by albumin-modified magnetic fluid. Nanotechnology. 2012;23:055101.
122. Skaat H, Chen R, Grinberg I, Margel S. Engineered polymer nanoparticles containing hydrophobic dipeptide for inhibition of amyloid-beta fibrillation. Biomacromolecules. 2012;13:2662–70.

123. Alvarez YD, Fauerbach JA, Pellegrotti JV, Jovin TM, Jares-Erijman EA, Stefani FD. Influence of gold nanoparticles on the kinetics of alpha-synuclein aggregation. Nano Lett. 2013;13:6156–63.
124. Ben-Zvi A, Miller EA, Morimoto RI. Collapse of proteostasis represents an early molecular event in Caenorhabditis elegans aging. Proc Natl Acad Sci U S A. 2009;106:14914–9.
125. Douglas PM, Dillin A. Protein homeostasis and aging in neurodegeneration. J Cell Biol. 2010;190:719–29.
126. Lopez-Otin C, Blasco MA, Partridge L, Serrano M, Kroemer G. The hallmarks of aging. Cell. 2013;153:1194–217.
127. Finkel T, Holbrook NJ. Oxidants, oxidative stress and the biology of ageing. Nature. 2000;408:239–47.
128. Balch WE, Morimoto RI, Dillin A, Kelly JW. Adapting proteostasis for disease intervention. Science. 2008;319:916–9.
129. Gidalevitz T, Ben-Zvi A, Ho KH, Brignull HR, Morimoto RI. Progressive disruption of cellular protein folding in models of polyglutamine diseases. Science. 2006;311:1471–4.
130. Olzscha H, Schermann SM, Woerner AC, Pinkert S, Hecht MH, Tartaglia GG, Vendruscolo M, Hayer-Hartl M, Hartl FU, Vabulas RM. Amyloid-like aggregates sequester numerous metastable proteins with essential cellular functions. Cell. 2011;144:67–78.
131. Scekic-Zahirovic J, Sendscheid O, El Oussini H, Jambeau M, Sun Y, Mersmann S, Wagner M, Dieterle S, Sinniger J, Dirrig-Grosch S, Drenner K, Birling MC, Qiu J, Zhou Y, Li H, Fu XD, Rouaux C, Shelkovnikova T, Witting A, Ludolph AC, Kiefer F, Storkebaum E, Lagier-Tourenne C, Dupuis L. Toxic gain of function from mutant FUS protein is crucial to trigger cell autonomous motor neuron loss. EMBO J. 2016;35:1077–97.
132. Alavez S, Vantipalli MC, Zucker DJ, Klang IM, Lithgow GJ. Amyloid-binding compounds maintain protein homeostasis during ageing and extend lifespan. Nature. 2011;472:226–9.
133. Kaeberlein M, Burtner CR, Kennedy BK. Recent developments in yeast aging. PLoS Genet. 2007;3:e84.
134. Ray LB. Protein aggregation-mediated aging in yeast. Science. 2017;355:1169–71.
135. Walther DM, Kasturi P, Zheng M, Pinkert S, Vecchi G, Ciryam P, Morimoto RI, Dobson CM, Vendruscolo M, Mann M, Hartl FU. Widespread proteome remodeling and aggregation in aging C. elegans. Cell. 2015;161:919–32.
136. Schlissel G, Krzyzanowski MK, Caudron F, Barral Y, Rine J. Aggregation of the Whi3 protein, not loss of heterochromatin, causes sterility in old yeast cells. Science. 2017;355:1184–7.
137. Tullet JM, Hertweck M, An JH, Baker J, Hwang JY, Liu S, Oliveira RP, Baumeister R, Blackwell TK. Direct inhibition of the longevity-promoting factor SKN-1 by insulin-like signaling in C. elegans. Cell. 2008;132:1025–38.
138. Hsu AL, Murphy CT, Kenyon C. Regulation of aging and age-related disease by DAF-16 and heat-shock factor. Science. 2003;300:1142–5.

Biological Rhythms and Aging

20

Anita Jagota, Kowshik Kukkemane,
and Neelesh Babu Thummadi

Biological Rhythms

A regular and predictable pattern of any biological phenomenon which reoccurs with certain periodicity is considered as biological rhythm [106]. A biological rhythm can be endogenous, where it is controlled by the internal biological clock or exogenous where it is controlled by synchronizing internal cycles with external stimuli. Such stimuli are mostly with respect to the position of Earth to the Sun and to the Moon as well as on the immediate effects of such variations, for example, day alternating with night, high tide alternating with low tide, etc. [33]. Biological rhythms are genetically regulated, temperature independent, and resistant to pharmacological and chemical disruption. Based on the set of cues, the organism entrains and generates rhythms (Fig. 20.1) with varied periodicity such as, circannual rhythms – occurring in cycles of approximately 1 year; circalunar rhythms – occurring in cycles of approximately one lunar cycle; circatidal rhythms – occurring in cycles of approximately one ocean tide; infradian rhythms – occurring in cycles of frequency more than a day (>24 hours (h)); ultradian rhythms – occurring in cycles of frequency less than a day (<24 h) and circadian rhythms – occurring in cycles of approximately 24 h. In anticipation of these day–night phases, living organisms have evolved cellular and physiological rhythms having a periodicity of approximately 24 h known as circadian rhythms [124].

Authors Kowshik Kukkemane and Neelesh Babu Thummadi have equally contributed to this chapter

A. Jagota (✉) · K. Kukkemane · N. B. Thummadi
Neurobiology and Molecular Chronobiology Laboratory, Department of Animal Biology, School of Life Sciences, University of Hyderabad, Hyderabad, India
e-mail: ajsl@uohyd.ernet.in

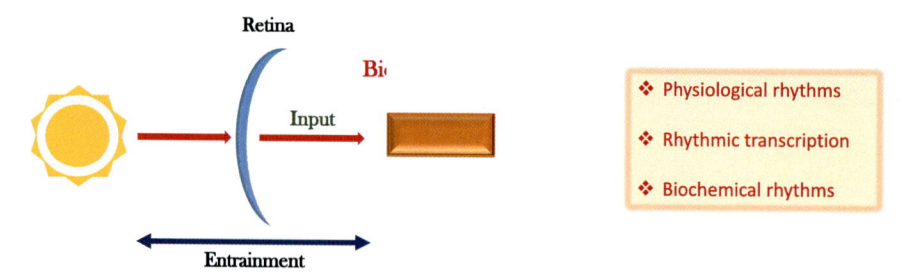

Fig. 20.1 Diagrammatic representation of mammalian circadian timing system. The biological localized in suprachiasmatic nucleus (SCN) gets entrained through photic input and then time information from SCN regulates physiological, biochemical, and transcriptional rhythms and hence synchronizing the entire body processes

Circadian Rhythms

Circadian rhythms regulate physiology and behavior with a period near 24 h. Circadian is derived from a Latin phrase meaning about (*circa*) a day (*dia*). There are several physiological, biochemical and behavioral aspects which follow circadian rhythms that include: sleep/wake cycle, body temperature, hormone secretion, blood pressure, digestive secretions, levels of alertness, etc. [69]. Exogenously they are set and entrained by zeitgebers, derived from German meaning time (*Zeit*) giver (*Geber*). The primary being Light–Dark cycles due to the rotation of earth around its axis though food, social cues, temperature, etc., can be considered as nonphotic zeitgebers. The circadian rhythms persist even in the absence of zeitgebers under constant conditions such as constant darkness or constant light. That means rhythms are endogenous. The persistence of rhythms in the absence of cues leads to free running situation, that is, rhythms continue to run, but with a slight deviation from 24 h. The rhythms are genetically determined [66] (Fig. 20.2). There are several other factors that act to the external stimuli and the result is the generation of the rhythms collectively known as the circadian timing system (CTS) [26]. These self-sustained, endogenous, and entrainable rhythms of sleep and wakefulness, foraging and feeding, body temperature, enzyme activity, hormonal release, energy metabolism, and several other molecular and behavioral parameters helped the organisms to effectively cope with ever-changing environment, thus improving their survival [56, 137, 150]. The importance of these oscillations on health and diseases has been rightly recognized with 2017 Nobel Prize in physiology and medicine jointly awarded to Jeffrey Hall, Michael Rosbash, and Michael Young *"for their discoveries of molecular mechanisms controlling the circadian rhythm"* [29].

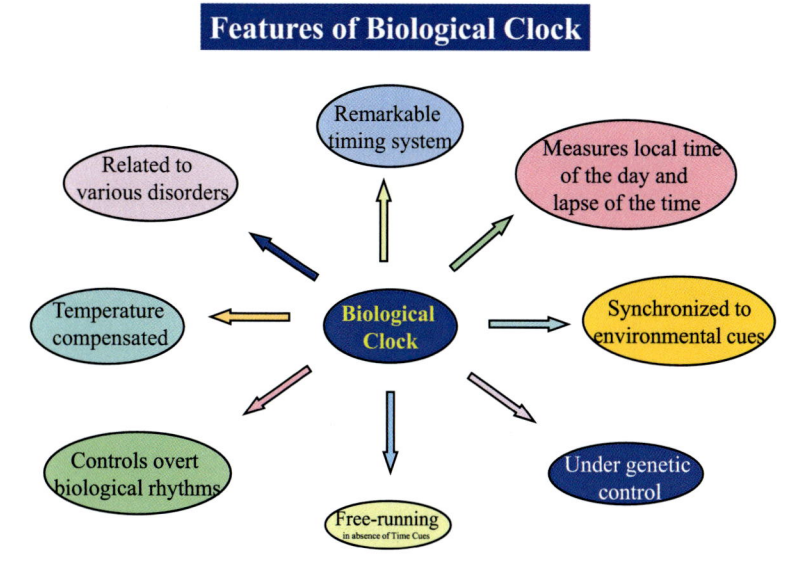

Fig. 20.2 Diagrammatic representation of various features of clock

Circadian Time-Keeping System (CTS)

The presence of temporally regulated rhythms in physiology and behavior hinted at the possible existence of a circadian clock which was confirmed by lesion experiments and was discovered to be located in the hypothalamic suprachiasmatic nucleus (SCN) in mammals [150]. SCN, the principal pacemaker, consisting of nearly 20,000 tightly packed neurons, acts as "master clock" by synchronizing the peripheral oscillators located in all other cells and tissues [1]. This hierarchical architecture of the CTS in mammals function robustly based on specialized input and output pathways.

Input Pathways to the SCN

Three major pathways convey information to the SCN resulting in entrainment of the master clock. Blue light activates photosensitive retinal ganglion cells in retina and will be communicated to SCN via *retinohypothalamic tract* (RHT) with the release of excitatory neurotransmitter glutamate and the neuropeptide pituitary adenylatecyclase-activating protein (PACAP). The release of these neurotransmitters leads to stimulation of signaling pathways involving Ca^{2+} and cAMP and leads to induction of clock genes [1, 22, 55].

RHT also indirectly communicates photic information to SCN via intergeniculate leaflets (IGL) *geniculohypothalamic tract* (GHT) pathway through gamma aminobutyric acid (GABA) and neuropeptide Y (NPY) [64]. However, the IGL also processes non-photic information such as arousal status received via pathway originating from the dorsal raphe nucleus (DRN), suggesting assimilation of photic and non-photic signals to entrain the SCN [30].

Another important input to the SCN comes directly from both the median raphe nucleus (MRN) and dorsal raphe nucleus (DRN) [96]. Here the serotonergic tract participates in entrainment of the circadian clock by non-photic cues such as physical activity and exercise [30].

Output Pathways from the SCN

Within hypothalamus, SCN axons project to the preoptic area, lateral septum, bed nucleus of the striaterminalis, the subparaventricular zone, and also to the arcuate nucleus and the dorsomedial hypothalamus. In thalamus, efferents from the SCN innervate the IGL and paraventricular nucleus (PVN). Glutamate, GABA, peptide neurotransmitters, AVP, VIP, prokineticin 2 (PK2), cardiolipin-like cytokine, and transforming growth factor α (TGFα) have been shown as SCN output signals [30, 54].

Melatonin, a multitasking molecule, also a messenger of darkness, is secreted from pineal gland and is considered to be an internal zeitgeber which communicates the time information from SCN to all other peripheral clocks through circulation. As the photic information reaches SCN via the RHT, the subsequent activation of signaling pathways lead to induction of clock genes which ultimately results in the regulation of biosynthesis and release of melatonin from pineal. The pathway governing melatonin synthesis is triggered during scotophase in the absence of light. In pineal, tryptophan is converted into serotonin (5-HT) by 5-hydroxytryptophan. Serotonin subsequently undergoes N-acetylation and methylation by arylalkylamine N-acetyltransferase (AANAT) and hydroxyindole-O-methyl-transferase (HIOMT) respectively, ultimately resulting in melatonin synthesis (reviewed in Jagota [55] and Reiter et al. [123]).

The neural circuitry from SCN to pineal gland involves a multisynaptic pathway that includes PVN, intermediolateral cell column (ILCC), superior cervical ganglia (SCG), and finally terminate on pinealocytes and release norepinephrine (NE) which acts on both α1- and β-adrenergic receptors potentiating cAMP production to activate protein kinase A (PKA) and stimulating adenylatecyclase (AC) respectively. PKA phosphorylates cAMP response element-binding protein (CREB) which in turn activates N-acetyl transferase (*Nat*) gene that leads to melatonin synthesis and secretion. However, the cAMP also suppresses *Nat* expression by inducible cAMP early repressor (ICER) which competes with P-CREB [55, 138].

Melatonin exerts its effects and influences cellular physiology majorly by membrane bound G-protein-coupled receptors such as MT1 and MT2. These receptors regulate cellular processes by inhibition of adenylatecyclase, followed by a decrease in cAMP levels and modulation of PKA activity [123, 160].

Molecular Mechanisms

The molecular mechanism governing the mammalian CTS involves two main processes such as the posttranslational modifications (PTMs) of proteins (e.g., phosphorylation) and the autoregulatory transcriptional–translational feedback loop (TTFL) that consist of tightly interconnected positive and negative limbs [137].The positive limb composed of transcriptional activators BMAL1 and CLOCK (or NPAS2 in brain), hetero-dimerize and bind to the E-box elements present in the promoter of several clock-controlled genes (CCGs) including the clock genes *Period* (*Per1* and *Per2*) and *Cryptochrome* (*Cry1* and *Cry2*). PER-CRY heterodimers enter the nucleus to repress BMAL-CLOCK activity by deacetylating histones 3 and 4 by recruiting PSF/Sin3-HDAC complex [34].The auxiliary feedback loops consist of nuclear receptors REV-ERBs and RORs which are transcriptionally controlled by the BMAL1-CLOCK. RORs activate the transcription of *Bmal1* while REV-ERBs inhibit the transcription of *Bmal1*, there by regulating their own activator [22, 137].

Moreover, the gene coding for adenine dinucleotide (NAD+) synthesis in the mammalian salvage pathway, nicotinamide phosphoribosyltransferase (*Nampt*), is a CCG.NAD+, a metabolic oscillator, modulates the transcriptional activity of clock through a histone deacetylase, SIRT1 [99]. This indicates that the cellular metabolism via NAD+ can feedback to the clock, suggesting an interplay between the elements of clock output and the clock itself [1].

Transcriptional regulation of circadian clock is also controlled by D-box elements [140]. The PAR-Zip transcription factors such as D-box-binding proteins (DBP) which are under the E-box-mediated transcriptional control bind to these elements, and hence, they indirectly regulate CCGs [72].

In addition to transcriptional regulation, PTMs regulate the subcellular localization and stability of PER–CRY complexes [45] allowing progressive and delayed (circa 24 h) maturation of PER/CRY as transcriptional repressors. CK1ε and mitogen-activated protein kinase play an important role in the activation or repression of BMAL1, while CK2α and GSK-3β help in cellular localization and proteasomal degradation respectively [75]. In case of PER, site-specific phosphorylation at a "priming site" (FASP site) delays phosphorylation at "sites" (PERs or βTrCP site) that would signal for nuclear entry and degradation by proteasomal pathways [141]. In addition, salt inducible kinase 3 (SIK3 kinase) modulates PER2 phosphorylation rhythms and abundance [47]. Similarly, microRNAs (miRNAs) and several RNA-binding protein complexes regulate RNA stabilization and degradation, circadian polyadenylation and splicing [112]. Overall, the molecular mechanisms underlying the regulation of circadian rhythms in a cell involves numerous complex processes.

Aging and Theories of Aging

Aging is an inevitable unidirectional process that eventually leads to the progressive decline of metabolism, physiology, and behavior and ultimately culminates in death. Aging is explained through a couple of theories. Programmed theory deems the timeline of different stages of life linked to change of metabolism, physiology, and behavior. First theory falls into three subcategories: (i) Programmed longevity – programmed switching of particular genes leading to senescence resulting in overt manifestations. (ii) Endocrine theory – evolutionarily conserved hormonal signaling such as insulin/IGF-1 signaling pathway regulates the process of aging. We have vividly discussed the endocrine regulation of aging elsewhere [59]. (iii) Immunological theory – preprogrammed deterioration of immune system.

Second, damage or error theory postulates that the accumulated damages or errors at several levels over the period of time would cause aging and is linked to metabolic disorders, epigenetic alterations, genomic instability, telomere attrition, loss of proteostasis, altered intercellular communications, cellular senescence, deregulated nutrient sensing, stem cell exhaustion and mitochondrial dysfunction, and DNA damage [82]. Of all the macromolecules that are being damaged as the age progresses, DNA is very important because it cannot be replaced like other macromolecules [38] and also slowing down of DNA repair mechanisms progresses aging process.

Age-Associated Circadian Dysfunction

Age has a marked effect on CTS, which impacts the temporal organization of circadian physiology and behavior (Fig. 20.3, Table 20.1). In humans, fragmented sleep and progressive advance in sleep phase has been recorded in elderly [35, 53]. Similarly, the amplitude of feeding rhythms, secretion of hormones, and body

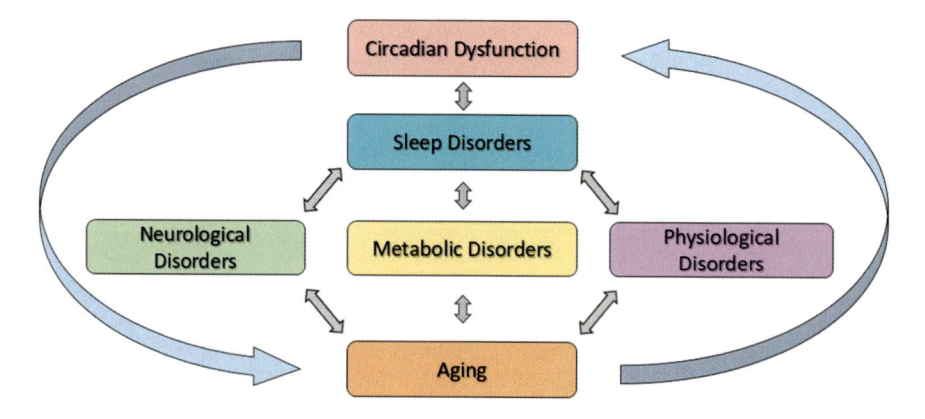

Fig. 20.3 Schematic diagram showing interaction between aging and clock dysfunction

Table 20.1 Effect of clock-associated genes on aging

Animal model	Clock-associated gene	Age-associated phenotypes	Author
Mice	*Clock* mutant	Age-related cataract development	[31]
		Age-related arthropathy	[161]
	Bmal1 mutant	Premature aging	[111]
		Premature aging of hippocampal neurogenic niche	[3]
		Induced severe age-dependent astrogliosis	[98]
		Progression of noninflammatory arthropathy	[15]
		Accelerated prothrombotic phenotype	[133]
		Accelerated age-dependent arthropathy	[158]
	Reduced Sirt1	Reduced amplitude of circadian rhythms in aging	[11]
	Sirt1 mutant	Premature aging	[147]
	Per1/2 mutant	Premature aging	[74]
	Per2 overexpression	Increased expression of aging markers	[147]
Irradiated mice	*Clock* mutant	Accelerated aging	[5]
Drosophila	*Per* mutant	Accelerated aging	[70]

temperature are known to decline with age [149]. In aged animals, decline in locomotor activity rhythms and disrupted sleep–wake cycles suggests age-associated circadian alterations [9]. Reports from mice have demonstrated that the aged animals are more vulnerable to negative effects of photoperiodic phase shifts as the adaptability of circadian system is compromised with aging [7]. In rodents, age-related changes in circadian rhythms have been reported for body temperature, activity–wakefulness, locomotor activity patterns, drinking behavior, and serotonin rhythms [57, 87, 121, 148, 152]. In addition, core clock in aged mice showed diminished response to the external stimuli suggesting CTS deterioration [83].

Influence of Aging on Central and Peripheral Clocks

As the SCN communicates directly and indirectly to various peripheral clocks, circadian clock and aging may be interconnected by pathology at the level of the SCN and SCN output signals [91]. Though there appears no reduction in the total number of cells in aged SCN [50], the age-related loss of amplitude in SCN electrical activity [100] suggests alterations in cellular properties, neuronal circuitry, and clock genes [10]. At single cell level, the neurons of aged SCN showed diminished amplitude of potassium currents and resting membrane potential as a result of possible alterations in large conductance calcium-activated potassium channels (BK channels) [36]. Age-associated alterations in cellular communication in SCN has been evident with reports showing age-dependent loss of neuronal connectivity, marked by decline in synaptic spines and shortened dendrites, altered electrical activity, and altered signaling molecules [107]. Moreover, alterations in the expression of neuropeptide vasoactive-intestinal polypeptide (VIP) and arginine-vasopressin (AVP) reported upon aging would hamper the SCN output as they are essential for intracellular coupling within the SCN [93]. Disrupted GABAergic signaling in aged SCN indicates clock deterioration [107]. Weakened melatoninergic feedback to the SCN is suggested by reports showing diminished MT1 receptor expression in aged human SCN [145]. Age-linked circadian disruption is significantly contributed by desynchrony between SCN and oscillators in peripheral tissues. Phase-shifting studies in elderly involving exposure to different LD regimens showed decline in the ability to re-entrain in several parameters such as rhythms of activity, rest/sleep and body temperature [51]. Similarly, phase advance study using PER2::LUC mouse demonstrated that oscillators in esophagus, thymus gland and lungs in older mice took longer time period to get entrained to specific light–dark schedule, in comparison with younger mice [128].

Aging is also known to be resulting in declined total melatonin secretion. Studies in humans, primates, and hamsters indicate that the normal nighttime peak in elderly is reduced and phase advanced compared to younger adults [51]. In addition, diminished pineal melatonin synthesis and SCN expression of melatonin receptors have been reported in individuals with Alzheimer's or Parkinson's disease [142]. Severe alterations in daily rhythms and levels of serotonin metabolism in SCN of aged rats and a rat model of PD have been reported from our lab [57, 89, 120]. Similarly, cortisol, a hormone under clock control, which also synchronizes peripheral clocks was observed to show age-related reduction in amplitude and advance in phase [51]. In a recent study analyzing hepatic transcriptome, it was observed that 2626 genes (44.8%) were exclusively oscillatory in young mice whereas in old mice only 1626 genes (28.4%) were rhythmic [126]. Further, age-dependent decline in cyclic global protein acetylation was observed in peripheral clock liver [126]. Recently, from our laboratory, we have reported the age-associated day–night variations of proteins in SCN, substantianigra (SN), and pineal gland of rats [61]. In SCN, the number of proteins showing day–night variations were found to have decreased from 32 (in young adults) to 9 (in old age). Similarly, SN also showed a decrease from 59 to 9. However, in pineal, the number of proteins showing oscillations increased from 51 to 62 [61]. Our earlier studies investigating daily rhythms of lipid peroxidation and antioxidant enzyme activities in rats showed age-dependent variations in liver [87]. Further, reports from our laboratory have demonstrated differential alterations in daily rhythms and levels of NO and *Socs1* expression in various peripheral clocks of aged rats [143, 144] suggesting desynchrony [61].

Influence of Aging on Clock Genes and Proteins

The canonical genes and proteins constituting the TTFL of the core clock machinery show drastic variations upon aging (Table 20.2). We have reported severe alterations in rhythms and levels of various clock genes in the SCN of mid- and old-aged rats [90]. Similarly, studies in aged mice SCN showed changes in *Rev-erb a*, *Dec1*, and *Dbp* expression [12]. Earlier studies in mice showed altered expression of the

Table 20.2 Age-induced circadian rhythm disorders

Sleep-associated disorders	
Advanced sleep phase	[24]
Irregular sleep–wake disorder	[162]
Circadian sleep–wake rhythm disorders (CSRD)	[65]
REM sleep behavior disorder (RBD)	[117]
Insomnia	[39]
Co-occurrence of obstructive sleep apnea and insomnia	[2]
Restless leg syndrome/Willis–Ekbom disease (RLS/WED)	[151]
Free running disorder	[156]
Sleep fragmentation	[77]
Reduced total sleep time	[16]
Reduced slow wave sleep time	[86]
Sleep disordered breathing	[62]
Advanced sleep–wake phase disorder (ASWPD)	[73]
Delayed sleep–wake phase disorder (DSWPD)	[73]
Neurological disorders	
Cognitive decline	[13]
Dementia	[91]
Alzheimer's disease	[78]
Parkinson's disease	[44]
Mood and behavior-related disorders	
Adult attention deficit hyperactivity disorder	[8]
Bipolar disorder	[125]
Depression	[97]
Major depressive disorder (MDD)	[92]
Metabolic disorders	
Cancer	[105]
Rheumatoid arthritis	[27]
Type II diabetes mellitus	[119]
Cardiovascular diseases	[13]

CLOCK and BMAL1 proteins in the SCN, hippocampus, amygdala, and several other brain regions by mid-age [153]. Additionally, altered expression of *Bmal1* and *Per2* transcripts in various non-SCN brain regions in aged hamsters has been reported [32]. Studies on Per1:luc rats showed slight age-related changes in *Per1* expression in SCN, whereas robust changes in peripheral oscillators [155]. Though, reports on rhythmic PER2, PER3, and BMAL1 expression in cortex of elderly humans suggest persistence of clock function in old age [76], altered PER1, 2, 3 rhythms in leukocytes indicated desynchrony [49]. A detailed account on age-linked alterations in the core clock gene expression in the SCN is discussed elsewhere [10].

The mechanisms underlying the modulation of clock gene expression in aging might be involving the cross-talk between circadian and metabolic regulatory systems [116]. The NAD+-dependent protein deacetylase SIRT1 is known to play a significant role in the age-related deterioration of the circadian clock [20]. The role of SIRT1 in the modulation of circadian clock is well known [99] and studies in the SCN of aged mice have demonstrated decline in SIRT1 levels concomitantly with levels of BMAL1 and PER2 [20]. Further, *Sirt1* knockout mice resulted in senescent-like phenotype and *Sirt1* overexpression resulted in antiaging phenotype with respect to alterations in clock [20]. In addition, role of SIRTs in age-associated epigenetic changes in the clock has also gained considerable importance [104].

Therapeutic Interventions

Understanding the molecular components and their feedback mechanisms that are involved in aging and circadian dysfunction helps the researchers to identify the target molecules and to develop the therapeutic drugs to delay the progress of aging and age-associated disorders. Here we discuss few of the potential therapeutic strategies that are showing promising results in combating the aging process and restoring the circadian clock.

Effect of Antioxidants on Aging and CTS

(i) **Melatonin**

The primary function of melatonin is to relay the circadian signals to the peripheral clocks and to synchronize them with the central clock [138]. It is also well established that melatonin relays seasonal temporal information [113]. Melatonin is shown to regulate several clock genes and considered as chronobiotic [59]. It is a multitasking molecule with several properties like anti-inflammatory and antioxidant, and can stimulate antioxidant enzymes like glutathione reductase, glutathione peroxidase, and catalase [87, 118]. Melatonin irregularities have been attributed to several circadian dysfunctions that are involved in cancers [122]. The secretary rhythm of melatonin has been linked with the immune changes and thyroid hormone in aging mice [114]. And also, the nocturnal secretion of melatonin is related to the cell-mediated immunity [84]. The peak expression of IL-1b, IL-2, IL-6, and TNF-α is observed shortly after the maximum melatonin serum levels [25].

In humans, melatonin metabolite 6-hydroxymelatonin sulfate showed similar pattern in both young and healthy centenary subjects; this is evident to claim that melatonin can be a proper aging marker [37]. Melatonin reduces the microglial activity in brain which is considered as anti-inflammaging property [46]. It also has shown to alleviate age-induced memory impairments and neuronal degeneration [130]. Cognitive deficits and neurodegeneration were shown to be alleviated with melatonin administration [103]. Melatonin has also shown the beneficiary effects on the sleep deprivation-induced memory deficits [4] and also improved sleep efficacy [135]. Further, melatonin supplementation has slowed down the progress of cognitive deficits and also ameliorated the sundowning syndrome in Alzheimer's disease patients [18]. In correspondence to Alzheimer's disease, melatonin has reduced β-amyloid, anomalous nitration of proteins, β-fibrillogenesis, tau phosphorylation, and also increased survival rate of AD transgenic mice [79, 80, 88]. We have recently demonstrated the restoratory effects of melatonin on age-induced alterations in NO daily rhythms and *Socs1* expression in peripheral oscillators [143, 144]. We further demonstrated that melatonin administration could differentially restore the circadian phase, amplitude and the expression levels of clock genes such as *Bmal1, Per1, Per2, Cry1,* and *Cry2* [89, 90] in aged and a rat model of PD. Additionally, we demonstrated age and PD-related changes in the number of oscillatory proteins which shows day–night variations, and the melatonin administration has resulted in differential restoration of these proteins in both aging and PD [61]. In concordance with it, several other researchers have shown the beneficial effects of melatonin in age-associated disorder like Parkinson's disease [19].

The beneficiary effects of melatonin can be attributed to its amphiphilic nature that can easily cross the blood brain barrier [123]. The effect of melatonin on several clock-associated genes has been extensively discussed elsewhere [59].

(ii) **Resveratrol**

Resveratrol (3,5,4′-trihydroxy-trans-stilbene), a polyphenol purified from natural sources is known to modulate CTS [102] and also rescues from various age-related impairments [41]. Studies in nocturnal primate gray lemur further revealed the influence of resveratrol on circadian clock. Resveratrol improved the synchronization of old animals to light–dark cycles and restored the rhythms of locomotor activity and body temperature [115]. Interestingly, few researchers have demonstrated its role in clock-mediated rescue from disorder of lipid metabolism in a rodent model [134]. Corroborative evidence on beneficial effects of resveratrol came from a report where it attenuated the insulin resistance in liver by modulating core clock elements as well as SIRT1 [163]. Considering the importance of clock in healthy aging [28], an earlier study emphasized SIRT1 and its activator resveratrol's role in BMAL1-CLOCK-mediated transcription of clock genes, highlighting its influence on CTS [109].

(iii) **Curcumin**

Curcumin (diferuloylmethane), a potent antioxidant, is a polyphenol derived from rhizomes of *Curcuma longa* and known for its multiple beneficiary effects [129]. It has been reported that curcumin can cross blood brain barrier [139]. Its role as antioxidative, anti-inflammatory, anticancerous, neuro-protective, and clock restoratory agent is widely explored in various animal models [17]. Several studies have also suggested curcumin as a potential SIRT1 activator [157], it could mediate antiaging effects. Reports showing *Bmal1* and SIRT1 activation by curcumin suggest a modulatory role for curcumin in age and associated circadian disorders [43]. We have shown the influence of curcumin on circadian clock by studying serotonin (5-HT) and its metabolite 5-hydroxy indole acetic acid (5-HIAA) rhythms in the SCN and pineal of rats [58]. Interestingly, a recent study demonstrated a synergistic function of melatonin and curcumin in tumor suppression [131]. Overall, these reports suggest curcumin can be a potential drug candidate in reversing age-linked clock dysfunction.

(iv) *Withania somnifera*

Withania somnifera (WS), known as Ashwagandha, is known to promote health, enhance longevity, and create a sense of well-being [85]. With various biologically active constituents, the leaf, root, and fruit extracts of the plant have potential regenerative properties and is used for the treatment of various disorders [146]. Several studies have explored the free radical scavenging activity, regulation of lipid peroxidation, glutathione-S-transferase activity, and anti-inflammatory property of WS [63, 108]. Similarly, clinical and preclinical experiments have revealed the potential of WS against cancer, insomnia, anxiety, stress, cognitive, and age-associated neurodegenerative disorders such as AD and PD [68, 85, 127]. Evidence for antiaging effects of WS has also come from studies reporting downregulation of senescence in human fibroblasts and lifespan extension in *C. elegance* [71]. Moreover, we have recently reported the differential restoratory effect of Ashwagandha leaf extract on age-induced alterations in SCN core clock transcript expression rhythms [60]. Our results showed an age-specific action of WS, as we observed restorations in the phase of *Per1*, *Cry1,* and *Bmal1* in the SCN of mid-aged (12 m) rats and only *Per1* in old-aged (24 m) rats [60].

Effect of Calorie Restriction (CR) on Aging and CTS

CR is demonstrated to be a potential strategy in lifespan extension and improved health in various organisms [14]. A recent report elucidating the role of CR in rescuing age-dependent circadian alterations by involving SIRT1 activation in peripheral clock liver [126] corroborated the previous knowledge on antiaging effects of CR [81]. CR, a strong metabolic cue and known to function as a zeitgeber for peripheral clocks has been shown to modulate peripheral gene expression [6, 111]. Studies investigating the SCN VIP expression, pineal melatonin, blood glucose, and locomotor activity rhythms suggested the influence of CR on circadian clock. In addition to it, CR could also synchronize the peripheral clocks and influence clock-controlled output systems, such as the food anticipatory activity (FAA) and body temperature. Further, studies exploring the influence of CR on photic responses suggested a role for CR in entrainment of circadian clock [94]. Transcriptome analysis in various peripheral clocks under CR demonstrated *Per2* as the most upregulated gene in majority of the clocks [136]. Emerging studies have also linked CR with elevated expression of several core clock genes including *Per1*, *Per2*, *Cry2*, and *Bmal1*. Similarly, some of the key CCGs which code for transcription factors such as *Dbp, Dec1, Dec2, Hlf, Tef, and E4bp4* were differentially affected in under CR [110]. These observations indicate that CR could directly synchronize central clock as well as peripheral clocks and rescue from the age-associated circadian ailments [126]. Interestingly, a recent study using *Bmal1* knockout mice has highlighted the role of BMAL1 in CR-mediated longevity effects [110].

Effect of Small Molecules as Modulators in Aging and CTS

In the recent years, there is an upsurge in the studies related to usage of small molecules as drugs. Through chemical screening approaches more than 2,00,000 small molecules have been identified as circadian regulators that may act as modifiers of period length, phase delay, phase advance, phase attenuation, and amplitude (Table 20.3) [22]. Small molecules such as CKI inhibitors and synthetic ligands for the nuclear receptors CRY1, REV-ERBs, and RORs have been proposed as therapeutic alternatives for several CTS dysfunction [22]. Studies on small molecules would open a new avenue in modulating age-related circadian dysfunctions toward healthy and slowly progressive aging.

Table 20.3 Small molecules modulating circadian clock

Small molecule	Circadian effect	References
KN-62	Targets CaMKII and attenuates phase shifts	[42]
Lithium	Stabilizes REV-ERBα and lengthens circadian period	[159]
PF-670462	Inhibits CK1δ resulting in period lengthening	[95]
L-methyl selenocysteine	Enhances transcriptional activation of *Bmal1*	[52]
SR9009 and SR9011	REV-ERB agonists Altersgene expression and circadian behavior in obese mice	[132]
Compound 5 and compound 6	Induction of cAMP leading to phase delays	[21]
KL001	Stabilizes CRY resulting in period lengthening	[101]
SSR 149415 and OPC-21268	AVP receptors antagonists Accelerate re-entrainment after shift work and jet lag	[154]
Resveratrol	SIRT1 activator Modulates circadian clock	[20]
2-ethoxypropanoic acid	Targets CRY Activates E-box transcription	[23]
Neoruscogenin	ROR agonist Induces *Bmal1* expression	[48]
SR8278	Targets REV-ERB Reduces anxiety and induces maniac-like behavior	[67]
Nobiletin	Targets ROR receptors Increases the amplitude of target gene expression	[40]

Conclusion

Aging is associated with disruption of the chronobiological cycle. CTS undergoes reduced sensitivity to external cues with aging in various physiological, biochemical, and molecular parameters. Numerous clinical studies have established a direct correlation between abnormal circadian clock functions and the severity of neurodegenerative and sleep disorders. Therapeutic interventions using various pharmacological agents such as antioxidants, CR, and small molecules may help to restore CTS dysfunction in elderly.

Acknowledgments AJ is thankful to Professor Pramod Rath for giving this opportunity and sincere patience during preparation of manuscript. The work is supported by DBT (102/IFD/SAN/5407/2011-2012), ICMR (Ref. No. 55/7/2012-/BMS), and UPE II Grants to AJ. KK is thankful to DST-INSPIRE for SRF and NBT is thankful to UGC for SRF.

References

1. Albrecht U. Timing to perfection: the biology of central and peripheral circadian clocks. Neuron. 2012;74:246–60.
2. Alessi C, Martin JL, Fiorentino L, Fung CH, Dzierzewski JM, Rodriguez Tapia JC, Song Y, Josephson K, Jouldjian S, Mitchell MN. Cognitive behavioral therapy for insomnia in older veterans using nonclinician sleep coaches: randomized controlled trial. J Am Geriatr Soc. 2016;64:1830–8.
3. Ali AA, Schwarz-Herzke B, Stahr A, Prozorovski T, Aktas O, von Gall C. Premature aging of the hippocampal neurogenic niche in adult Bmal1-deficient mice. Aging (Albany NY). 2015;7:435–49.
4. Alzoubi KH, Mayyas FA, Khabour OF, BaniSalama FM, Alhashimi FH, Mhaidat NM. Chronic melatonin treatment prevents memory impairment induced by chronic sleep deprivation. Mol Neurobiol. 2016;53:3439–47.
5. Antoch MP, Gorbacheva VY, Vykhovanets O, Toshkov IA, Kondratov RV, Kondratova AA, Lee C, Nikitin AY. Disruption of the circadian clock due to the clock mutation has discrete effects on aging and carcinogenesis. Cell Cycle. 2008;7:1197–204.
6. Asher G, Sassone-corsi P. Review time for food: the intimate interplay between nutrition, metabolism, and the circadian clock. Cell. 2015;161:84–92.
7. Azzi A, Dallmann R, Casserly A, Rehrauer H, Patrignani A, Maier B, Kramer A, Brown SA. Circadian behavior is light-reprogrammed by plastic DNA methylation. Nat Nurosci. 2014;17:377–82.
8. Baird AL, Coogan AN, Siddiqui A, Donev RM, Thome J. Adult attention-deficit hyperactivity disorder is associated with alterations in circadian rhythms at the behavioural, endocrine and molecular levels. Mol Psychiatry. 2012;17:988–95.
9. Banks G, Heise I, Starbuck B, Osborne T, Wisby L, Potter P, Jackson IJ, Foster RG, Peirson SN, Nolan PM. Genetic background influences age-related decline in visual and nonvisual retinal responses, circadian rhythms, and sleep. Neurobiol Aging. 2015;36:380–93.
10. Banks G, Nolan PM, Peirson SN. Reciprocal interactions between circadian clocks and aging. Mamm Genome. 2016;27:332–40.
11. Belden WJ, Dunlap JC. Aging well with a little wine and a good clock. Cell. 2013;153:1421–2.
12. Bonaconsa M, Malpeli G, Montaruli A, Carandente F, Grassi-Zucconi G, Bentivoglio M. Differential modulation of clock gene expression in the suprachiasmatic nucleus, liver and heart of aged mice. Exp Gerontol. 2014;55:70–9.

13. Bonomini F, Rodella LF, Rezzani R. Metabolic syndrome, aging and involvement of oxidative stress. Aging Dis. 2015;6:109–20.
14. Brandhorst S, Choi IY, Wei M, Cheng CW, Sedrakyan S, Navarrete G, Dubeau L, Yap LP, Park R, Vinciguerra M, Di Biase S, Mirzaei H, Mirisola MG, Childress P, Ji L, Groshen S, Penna F, Odetti P, Perin L, Conti PS, Ikeno Y, Kennedy BK, Cohen P, Morgan TE, Dorff TB, Longo VD. A periodic diet that mimics fasting promotes multi-system regeneration, enhanced cognitive performance, and healthspan. Cell Metab. 2015;22:86–99.
15. Bunger MK, Walisser JA, Sullivan R, Manley PA, Moran SM, Kalscheur VL, Colman RJ, Bradfield CA. Progressive arthropathy in mice with a targeted disruption of the Mop3/Bmal-1 locus. Genesis. 2005;41:122–32.
16. Cagnin A, Fragiacomo F, Camporese G, Turco M, Bussè C, Ermani M, Montagnese S. Sleep-wake profile in dementia with Lewy bodies, Alzheimer's disease, and normal aging. J Alzheimers Dis. 2017;55:1529–36.
17. Calabrese V, Cornelius C, Mancuso C, Pennisi G, Calafato S, Bellia F, Bates TE, Giuffrida Stella AM, Schapira T, Dinkova Kostova AT, Rizzarelli E. Cellular stress response: a novel target for chemoprevention and nutritional neuroprotection in aging, neurodegenerative disorders and longevity. Neurochem Res. 2008;33:2444–71.
18. Cardinali DP, Brusco LI, Liberczuk C, Furio AM. The use of melatonin in Alzheimer's disease. Neuro Endocrinol Lett. 2002;1:20–3.
19. Carocci A, Sinicropi MS, Catalano A, Lauria G, Genchi G. Melatonin in Parkinson's disease. In: Abdul QR, editor. A synopsis of Parkinson's disease. Intech; 2014. pp 71–99.
20. Chang HC, Guarente L. SIRT1 mediates central circadian control in the SCN by a mechanism that decays with aging. Cell. 2013;153:1448–60.
21. Chen Z, Yoo SH, Park YS, Kim KH, Wei S, Buhr E, Ye ZY, Pan HL, Takahashi JS. Identification of diverse modulators of central and peripheral circadian clocks by high-throughput chemical screening. Proc Natl Acad Sci U S A. 2012;109:101–6.
22. Chen Z, Yoo S, Takahashi JS. Development and therapeutic potential of small-molecule modulators of circadian systems. Annu Rev Pharmacol Toxicol. 2018;58:231–52.
23. Chun SK, Jang J, Chung S, Yun H, Kim NJ, Jung JW, Son GH, Suh YG, Kim K. Identification and validation of Cryptochrome inhibitors that modulate the molecular circadian clock. ACS Chem Biol. 2014;9:703–10.
24. Cooke JR, Ancoli-Israel S. Normal and abnormal sleep in the elderly. Handb Clin Neurol. 2011;98:653–65.
25. Couto-Moraes R, Palermo-Neto J, Markus RP. The immune–pineal axis: stress as a modulator of pineal gland function. Ann N Y Acad Sci. 2009;1153:193–202.
26. Curtis J, Burkley E, Burkley M. The rhythm is gonna get you: the influence of circadian rhythm synchrony on self-control outcomes. Soc Personal Psychol Compass. 2014;8:609–25.
27. De Cata A, D'Agruma L, Tarquini R, Mazzoccoli G. Rheumatoid arthritis and the biological clock. Expert Rev Clin Immunol. 2014;10:687–95.
28. Declerck K, Berghe WV. Back to the future: epigenetic clock plasticity towards healthy aging. Mech Ageing Dev. 2018;174:18–29. https://doi.org/10.1016/j.mad.2018.01.002.
29. Dibner C, Schibler U. Body clocks: time for the nobel prize. Acta Physiol (Oxford). 2018;222(2):e13024. https://doi.org/10.1111/apha.13024.
30. Dibner C, Schibler U, Albrecht U. The mammalian circadian timing system: organization and coordination of central and peripheral clocks. Annu Rev Physiol. 2010;72:517–49.
31. Dubrovsky YV, Samsa WE, Kondratov RV. Deficiency of circadian protein CLOCK reduces lifespan and increases age-related cataract development in mice. Aging (Albany NY). 2010;2:936–44.
32. Duncan MJ, Prochot JR, Cook DH, Tyler Smith J, Franklin KM. Influence of aging on Bmal1 and Per2 expression in extra-SCN oscillators in hamster brain. Brain Res. 2013;1491:44–53.
33. Dunlap JC. Molecular bases for circadian clocks. Cell. 1999;96:271–90.
34. Duong HA, Robles MS, Knutti D, Weitz CJ. A molecular mechanism for circadian clock negative feedback. Science. 2011;332:1436–9.
35. Espiritu J. Aging-related sleep changes. Clin Geriatr Med. 2008;24:1–14.

36. Farajnia S, Meijer JH, Michel S. Age-related changes in large-conductance calcium-activated potassium channels in mammalian circadian clock neurons. Neurobiol Aging. 2015;36:2176–83.
37. Ferrari E, Cravello L, Falvo F, Barili L, Solerte SB, Fioravanti M, Magri F. Neuroendocrine features in extreme longevity. Exp Gerontol. 2008;43:88–94.
38. Freitas AA, de Magalhães JP. A review and appraisal of the DNA damage theory of ageing. Mutat Res. 2011;728:12–22.
39. Gleason K, McCall WV. Current concepts in the diagnosis and treatment of sleep disorders in the elderly. Curr Psychiatry Rep. 2015;17:45.
40. Gloston G, Yoo S, Chen Z. Clock-enhancing small molecules and potential applications in chronic diseases and aging. Front Neurol. 2017;8:100.
41. Gocmez S, Gacar N, Utkan T, Gacar G, Scarpace PJ, Tumer N. Protective effects of resveratrol on aging-induced cognitive impairment in rats. Neurobiol Learn Mem. 2016;131:131–6.
42. Golombek DA, Ralph MR. KN-62, an inhibitor of Ca2+/calmodulin kinase II, attenuates circadian responses to light. Neuroreport. 1994;5:1638–40.
43. Grabowska W, Sikora E, Bielak-Zmijewska A. Sirtuins, a promising target in slowing down the ageing process. Biogerontology. 2017;18:447–76.
44. Gruszka A, Hampshire A, Barker RA, Owen AM. Normal aging and Parkinson's disease are associated with the functional decline of distinct frontal-striatal circuits. Cortex. 2017;93:178–92.
45. Gustafson CL, Partch CL. Emerging models for the molecular basis of mammalian circadian timing. Biochemistry. 2015;54:134–49.
46. Hardeland R. Deacceleration of brain aging by melatonin. In: Stephen CB, Campbell A, editors. Oxidative stress in applied basic research and clinical practice. New York: Springer; 2016. p. 345–76.
47. Hayasaka N, Hirano A, Miyoshi Y, Tokuda IT, Yoshitane H, Matsuda J, Fukada Y. Salt-inducible kinase 3 regulates the mammalian circadian clock by destabilizing PER2 protein. elife. 2017;6:e24779. https://doi.org/10.7554/eLife.24779.
48. Helleboid S, Haug C, Lamottke K, Zhou Y, Wei J, Daix S, Cambula L, Rigou G, Hum DW, Walczak R. The identification of naturally occurring neoruscogenin as a bioavailable, potent, and high-affinity agonist of the nuclear receptorRORα (NR1F1). J Biomol Screen. 2014;19:399–406.
49. Hida A, Kusanagi H, Satoh K, Kato T, Matsumoto Y, Echizenya M, Shimizu T, Higuchi S, Mishima K. Expression profiles of PERIOD1, 2, and 3 in peripheral blood mononuclear cells from older subjects. Life Sci. 2009;84:33–7.
50. Hofman MA, Swaab DF. Living by the clock: the circadian pacemaker in older people. Ageing Res Rev. 2006;5:33–51.
51. Hood S, Amir S. The aging clock: circadian rhythms and later life. J Clin Invest. 2017;127:437–46.
52. Hu Y, Spengler ML, Kuropatwinski KK, Comas-Soberats M, Jackson M, Chernov MV, Gleiberman AS, Fedtsova N, Rustum YM, Gudkov AV, Antoch MP. Selenium is a modulator of circadian clock that protects mice from the toxicity of a chemotherapeutic drug via upregulation of the core clock protein, BMAL1. Oncotarget. 2011;2:1279–90.
53. Jagota A. Aging and sleep disorders. Indian J Gerontol. 2005;19:415–24.
54. Jagota A. Suprachiasmatic nucleus: the center for circadian timing system in mammals. Proc Indian Natl Sci Acad. 2006;B71:275–88.
55. Jagota A. Age- induced alterations in biological clock: therapeutic effects of melatonin. In: Thakur M, Rattan S, editors. Brain aging and therapeutic interventions. Dordrecht: Springer; 2012. p. 111–29.
56. Jagota A, de la Iglesia HO, Schwartz WJ. Morning and evening circadian oscillations in the suprachiasmatic nucleus in vitro. Nat Neurosci. 2000;3:372–6.
57. Jagota A, Kalyani D. Effect of melatonin on age induced changes in daily serotonin rhythms in suprachiasmatic nucleus of male Wistar rat. Biogerontology. 2010;11:299–308.

58. Jagota A, Reddy MY. The effect of curcumin on ethanol induced changes in suprachiasmatic nucleus (SCN) and pineal. Cell Mol Neurobiol. 2007;27:997–1006.
59. Jagota A, Thummadi NB. Hormones in clock regulation during ageing. In: Rattan SIS, Sharma R, editors. Hormones and ageing and longevity. Healthy ageing and longevity, vol. 6. Cham: Springer; 2017. p. 243–65.
60. Jagota A, Kowshik K. Chapter 21: therapeutic effects of Ashwagandha in brain aging and clock disfunction (Invited chapter). In: Kaul S, Wadhwa R, editors. Science of Ashwagandha: preventive and therapeutic potentials. Cham: Springer; 2017. p. 437–56.
61. Jagota A, Mattam U. Daily chronomics of proteomic profile in aging and rotenone-induced Parkinson's disease model in male Wistar rat and its modulation by melatonin. Biogerontology. 2017;18:615–30.
62. Kapur VK, Auckley DH, Chowdhuri S, Kuhlmann DC, Mehra R, Ramar K, Harrod CG. Clinical practice guideline for diagnostic testing for adult obstructive sleep apnea: an American Academy of Sleep Medicine Clinical Practice Guideline. J Clin Sleep Med. 2017;13:479–504.
63. Khan MA, Subramaneyaan M, Arora VK, Banerjee BD, Ahmed RS. Effect of *Withania somnifera* (Ashwagandha) root extract on amelioration of oxidative stress and autoantibodies production in collagen-induced arthritic rats. J Complement Integr Med. 2015;12:117–25.
64. Kim HJ, Harrington ME. Neuropeptide Y-deficient mice show altered circadian response to simulated natural photoperiod. Brain Res. 2008;1246:96–100.
65. Kim MJ, Lee JH, Duffy JF. Circadian rhythm sleep disorders. J Clin Outcomes Manag. 2013;20:513–28.
66. Klein DC, Moore RY, Reppert SM, editors. Suprachiasmatic nucleus: the mind's clock. New York: Oxford University Press; 1991.
67. Kojetin DJ, Burris TP. REV-ERB and ROR nuclear receptors as drug targets. Nat Rev Drug Discov. 2014;13:197–216.
68. Konar A, Shah N, Singh R, Saxena N, Kaul SC, Wadhwa R, Thakur MK. Protective role of Ashwagandha leaf extract and its component withanone on scopolamine-induced changes in the brain and brain-derived cells. PLoS One. 2011;6:e27265.
69. Korenčič A, Bordyugov G, Lehmann R, Rozman D, Herzel H. Timing of circadian genes in mammalian tissues. Sci Rep. 2014;4:5782.
70. Krishnan N, Kretzschmar D, Rakshit K, Chow E, Giebultowicz JM. The circadian clock gene period extends healthspan in aging Drosophila melanogaster. Aging (Albany NY). 2009;1:937–48.
71. Kumar R, Gupta K, Saharia K, Pradhan D, Subramaniam JR. *Withania somnifera* root extract extends lifespan of *Caenorhabditis elegans*. Ann Neurosci. 2013;20:13–6.
72. Lavery DJ, Lopez-molina L, Margueron R, Fleury-Olela F, Conquet F, Schibler U, Bonfils C. Circadian expression of the steroid 15 alpha-hydroxylase (Cyp2a4) and coumarin 7-hydroxylase (Cyp2a5) genes in mouse liver is regulated by the PAR leucine zipper transcription factor DBP. Mol Cell Biol. 1999;19:6488–99.
73. Lavoie CJ, Zeidler MR, Martin JL. Sleep and aging. Sleep Sci Pract. 2018;2:3.
74. Lee CC. Tumor suppression by the mammalian period genes. Cancer Causes Control. 2006;17:525–30.
75. Lee Y, Kim K. Posttranslational and epigenetic regulation of the CLOCK/BMAL1 complex in the mammalian. Anim Cells Syst. 2012;16:1–10.
76. Lim ASP, Myers AJ, Yu L, Buchman AS, Duffy JF, De Jager PL, Bennett DA. Sex difference in daily rhythms of clock gene expression in the aged human cerebral cortex. J Biol Rhythm. 2013;28:117–29.
77. Lim ASP, Fleischman DA, Dawe RJ, Yu L, Arfanakis K, Buchman AS, Bennett DA. Regional neocortical gray matter structure and sleep fragmentation in older adults. Sleep. 2016;39:227–35.
78. Lindemer ER, Greve DN, Fischl BR, Augustinack JC, Salat DH. Regional staging of white matter signal abnormalities in aging and Alzheimer's disease. Neuroimage Clin. 2017;14:156–65.

79. Ling ZQ, Tian Q, Wang L, Fu ZQ, Wang XC, Wang Q, Wang JZ. Constant illumination induces Alzheimer–like damages with endoplasmic reticulum involvement and the protection of melatonin. J Alzheimers Dis. 2009;16:287–300.
80. Lin L, Huang QX, Yang SS, Chu J, Wang JZ, Tian Q. Melatonin in Alzheimer's disease. Int J Mol Sci. 2013;14:14575–93.
81. Longo VD, Panda S. Fasting, circadian rhythms, and time-restricted feeding in healthy lifespan. Cell Metab. 2016;23:1048–59.
82. López-Otín C, Blasco MA, Partridge L, Serrano M, Kroemer G. The hallmarks of aging. Cell. 2013;153:1194–217.
83. Lupi D, Semo M, Foster RG. Impact of age and retinal degeneration on the light input to circadian brain structures. Neurobiol Aging. 2012;33:383–92.
84. Maestroni GJ, Sulli A, Pizzorni C, Villaggio B, Cutolo M. Melatonin in rheumatoid arthritis: synovial macrophages show melatonin receptors. Ann N Y Acad Sci. 2002;966:271–5.
85. Manchanda S, Mishra R, Singh R, Kaur T, Kaur G. Aqueous leaf extract of *Withania somnifera* as a potential neuroprotective agent in sleep-deprived rats: a mechanistic study. Mol Neurobiol. 2017;54:3050–61.
86. Mander BA, Winer JR, Walker MP. Sleep and human aging. Neuron. 2017;94:19–36.
87. Manikonda PK, Jagota A. Melatonin administration differentially affects age-induced alterations in daily rhythms of lipid peroxidation and antioxidant enzymes in male rat liver. Biogerontology. 2012;13:511–24.
88. Matsubara E, Bryant–Thomas T, Quinto JP, Henry TL, Poeggeler B, Herbert D, Cruz–Sanchez F, Chyan YJ, Smith MA, Perry G, Shoji M, Abe K, Leone A, Grundke–Ikbal I, Wilson GL, Ghiso J, Williams C, Refolo LM, Pappolla MA. Melatonin increases survival and inhibits oxidative and amyloid pathology in a transgenic model of Alzheimer's disease. J Neurochem. 2003;85:1101–8.
89. Mattam U, Jagota A. Daily rhythms of serotonin metabolism and the expression of clock genes in suprachiasmatic nucleus of rotenone-induced Parkinson's disease male Wistar rat model and effect of melatonin administration. Biogerontology. 2015;16:109–23.
90. Mattam U, Jagota A. Differential role of melatonin in restoration of age-induced alterations in daily rhythms of expression of various clock genes in suprachiasmatic nucleus of male Wistar rats. Biogerontology. 2014;15:257–68.
91. Mattis J, Sehgal A. Circadian rhythms, sleep, and disorders of aging. Trends Endocrinol Metab. 2016;27:192–203.
92. Maurya PK, Noto C, Rizzo LB, Rios AC, Nunes SO, Barbosa DS, Sethi S, Zeni M, Mansur RB, Maes M, Brietzke E. The role of oxidative and nitrosative stress in accelerated aging and major depressive disorder. Prog Neuro-Psychopharmacol Biol Psychiatry. 2016;65:134–44.
93. Maywood ES, Reddy AB, Wong GK, O'Neill JS, O'Brien JA, McMahon DG, Harmar AJ, Okamura H, Hastings MH. Synchronization and maintenance of timekeeping in suprachiasmatic circadian clock cells by neuropeptidergic signaling. Curr Biol. 2006;16:599–605.
94. Mendoza J, Graff C, Dardente H, Pevet P, Challet E. Feeding cues alter clock gene oscillations and photic responses in the suprachiasmatic nuclei of mice exposed to a light/dark cycle. J Neurosci. 2005;25:1514–22.
95. Meng QJ, Maywood ES, Bechtold DA, Lu WQ, Li J, Gibbs JE, Dupré SM, Chesham JE, Rajamohan F, Knafels J, Sneed B, Zawadzke LE, Ohren JF, Walton KM, Wager TT, Hastings MH, Loudon AS. Entrainment of disrupted circadian behavior through inhibition of casein kinase 1 (CK1) enzymes. Proc Natl Acad Sci U S A. 2010;107:15240–5.
96. Moga MM, Moore RY. Organization of neural inputs to the suprachiasmatic nucleus in the rat. J Comp Neurol. 1997;389:508–34.
97. Murri MB, Pariante C, Mondelli V, Masotti M, Atti AR, Mellacqua Z, Antonioli M, Ghio L, Menchetti M, Zanetidou S, Innamorati M, Amore M. HPA axis and aging in depression: systematic review and meta-analysis. Psychoneuroendocrinology. 2014;41:46–62.
98. Musiek ES, Lim MM, Yang G, Bauer AQ, Qi L, Lee Y, Roh JH, Ortiz-Gonzalez X, Dearborn JT, Culver JP, Herzog ED, Hogenesch JB, Wozniak DF, Dikranian K, Giasson BI, Weaver

DR, Holtzman DM, Fitzgerald GA. Circadian clock proteins regulate neuronal redox homeostasis and neurodegeneration. J Clin Invest. 2013;123:5389–400.
99. Nakahata Y, Sahar S, Astarita G, Kaluzova M, Sassone-Corsi P. Circadian control of the NAD+ salvage pathway by CLOCK-SIRT1. Science. 2009;324:654–7.
100. Nakamura TJ, Takasu NN, Nakamura W. The suprachiasmatic nucleus: age-related decline in biological rhythms. J Physiol Sci. 2016;66:367–74.
101. Nangle S, Xing W, Zheng N. Crystal structure of mammalian cryptochrome in complex with a small molecule competitor of its ubiquitin ligase. Cell Res. 2013;23:1417–9.
102. Oike H, Kobori M. Resveratrol regulates circadian clock genes in rat-1 fibroblast cells. Biosci Biotechnol Biochem. 2008;72:3038–40.
103. Olcese JM, Cao C, Mori T, Mamcarz MB, Maxwell A, Runfeldt MJ, Wang L, Zhang C, Lin X, Zhang G, Arendash GW. Protection against cognitive deficits and markers of neurodegeneration by long–term oral administration of melatonin in a transgenic model of Alzheimer disease. J Pineal Res. 2009;47:82–96.
104. Orozco-Solis R, Sassone-Corsi P. Circadian clock: linking epigenetics to aging. Curr Opin Genet Dev. 2014;26:66–72.
105. Palesh O, Aldridge-Gerry A, Zeitzer JM, Koopman C, Neri E, Giese-Davis J, Jo B, Kraemer H, Nouriani B, Spiegel D. Actigraphy-measured sleep disruption as a predictor of survival among women with advanced breast cancer. Sleep. 2014;37:837–42.
106. Palmer J. An introduction to biological rhythms. Saint Louis: Elsevier; 2012.
107. Palomba M, Nygård M, Florenzano F, Bertini G, Kristensson K, Bentivoglio M. Decline of the presynaptic network, including GABAergic terminals, in the aging suprachiasmatic nucleus of the mouse. J Biol Rhythm. 2008;23:220–31.
108. Panchawat S. In vitro free radical scavenging activity of leaves extracts of *Withania somnifera*. Recent Res Sci Technol. 2011;3:40–3.
109. Park I, Lee Y, Kim H, Kim K. Original article effect of resveratrol, a SIRT1 activator, on the interactions of the CLOCK/BMAL1 complex. Endocrinol Metab. 2014;29:379–87.
110. Patel SA, Chaudhari A, Gupta R, Velingkaar N, Kondratov RV. Circadian clocks govern calorie restriction – mediated life span extension through BMAL1- and IGF-1-dependent mechanisms. FASEB J. 2017;30:1634–42.
111. Patel SA, Velingkaar N, Makwana K, Chaudhari A, Kondratov R. Calorie restriction regulates circadian clock gene expression through BMAL1 dependent and independent mechanisms. Sci Rep. 2016;6:25970.
112. Pegoraro M, Tauber E. The role of microRNAs (miRNA) in circadian rhythmicity. J Genet. 2008;87:505–11.
113. Pevet P, Challet E. Melatonin: both master clock output and internal time–giver in the circadian clocks network. J Physiol Paris. 2011;105:170–82.
114. Pierpaoli W, Yi C. The involvement of pineal gland and melatonin in immunity and aging I. Thymus–mediated, immunoreconstituting and antiviral activity of thyrotropin–releasing hormone. J Neuroimmunol. 1990;27:99–109.
115. Pifferi F, Dal-pan A, Languille S, Aujard F. Effects of resveratrol on daily rhythms of locomotor activity and body temperature in young and aged grey mouse lemurs. Oxidative Med Cell Longev. 2013;2013:187301.
116. Popa-Wagner A, Buga AM, Dumitrascu DI, Uzoni A, Thome J, Coogan AN. How does healthy aging impact on the circadian clock? J Neural Transm. 2017;124:89–97.
117. Rabadi MH, Mayanna SK, Vincent AS. Predictors of mortality in veterans with traumatic spinal cord injury. Spinal Cord. 2013;51:784–6.
118. Radogna F, Diederich M, Ghibelli L. Melatonin: a pleiotropic molecule regulating inflammation. Biochem Pharmacol. 2010;80:1844–52.
119. Rakshit K, Thomas AP, Matveyenko AV. Does disruption of circadian rhythms contribute to beta-cell failure in type 2 diabetes? Curr Diab Rep. 2014;14:474.
120. Reddy MY, Jagota A. Melatonin has differential effects on age-induced stoichiometric changes in daily chronomics of serotonin metabolism in SCN of male Wistar rats. Biogerontology. 2015;16:285–302.

121. Reddy VDK, Jagota A. Effect of restricted feeding on nocturnality and daily leptin rhythms in OVLT in aged male Wistar rats. Biogerontology. 2014;15:245–56.
122. Reiter RJ, Tan DX, Korkmaz A, Erren TC, Piekarski C, Tamura H, Manchester LC. Light at night, chronodisruption, melatonin suppression, and cancer risk: a review. Crit Rev Oncog. 2007;13:303–28.
123. Reiter RJ, Tan DX, Galano A. Melatonin: exceeding expectations. Physiology. 2014;29(5):325–33.
124. Reppert SM, Wever DR. Coordination of circadian timing system. Nature. 2002;418:935–41.
125. Robillard R, Naismith SL, Hickie IB. Recent advances in sleep-wake cycle and biological rhythms in bipolar disorder. Curr Psychiatry Rep. 2013;15:402.
126. Sato S, Solanas G, Peixoto FO, Bee L, Symeonidi A, Schmidt MS, Brenner C, Masri S, Benitah SA, Sassone-Corsi P. Circadian reprogramming in the liver identifies metabolic pathways of aging. Cell. 2017;170:664–70.
127. Sehgal N, Gupta A, Valli RK, Joshi SD, Mills JT, Hamel E, Khanna P, Jain SC, Thakur SS, Ravindranath V. *Withania somnifera* reverses Alzheimer's disease pathology by enhancing low-density lipoprotein receptor-related protein in liver. Proc Natl Acad Sci. 2012;109:3510–5.
128. Sellix MT, Evans JA, Leise TL, Castanon-Cervantes O, Hill DD, DeLisser P, Block GD, Menaker M, Davidson AJ. Aging differentially affects the re-entrainment response of central and peripheral circadian oscillators. J Neurosci. 2012;32:16193–202.
129. Shen LR, Parnell LD, Ordovas JM, Lai CQ. Curcumin and aging. Biofactors. 2013;39:133–40.
130. Shin EJ, Chung YH, Le HLT, Jeong JH, Dang DK, Nam Y, Wie MB, Nah SY, Nabeshima YI, Nabeshima T, Kim HC. Melatonin attenuates memory impairment induced by Klotho gene deficiency via interactive signaling between MT2 receptor, ERK, and Nrf2–related antioxidant potential. Int J Neuropsychopharmacol. 2015;18(6):pii: pyu105. https://doi.org/10.1093/ijnp/pyu105.
131. Shrestha S, Zhu J, Wang Q, Du X, Liu F, Jiang J, Song J, Xing J, Sun D, Hou Q, Peng Y. Melatonin potentiates the antitumor effect of curcumin by inhibiting IKK β/NF- κ B/COX-2 signaling pathway. Int J Oncol. 2017;51:1249–60.
132. Solt LA, Wang Y, Banerjee S, Hughes T, Kojetin DJ, Lundasen T, Shin Y, Liu J, Cameron MD, Noel R, Yoo SH, Takahashi JS, Butler AA, Kamenecka TM, Burris TP. Regulation of circadian behaviour and metabolism by synthetic REV-ERB agonists. Nature. 2012;485:62–8.
133. Somanath PR, Podrez EA, Chen J, Ma Y, Marchant K, Antoch M, Byzova TV. Deficiency in core circadian protein Bmal1 is associated with a prothrombotic and vascular phenotype. J Cell Physiol. 2011;226:132–40.
134. Sun L, Wang Y, Song Y, Cheng XR, Xia S, Rahman MR, Shi Y, Le G. Resveratrol restores the circadian rhythmic disorder of lipid metabolism induced by high-fat diet in mice. Biochem Biophys Res Commun. 2015;458:86–91.
135. Sun X, Ran D, Zhao X, Huang Y, Long S, Liang F, Guo W, Nucifora FC Jr, Gu H, Lu X, Chen L, Zeng J, Ross CA, Pei Z. Melatonin attenuates hLRRK2–induced sleep disturbances and synaptic dysfunction in a *Drosophila* model of Parkinson's disease. Mol Med Rep. 2016;13:3936–44.
136. Swindell WR. Comparative analysis of microarray data identifies common responses to caloric restriction among mouse tissues. Mech Ageing Dev. 2008;129:138–53.
137. Takahashi JS. Transcriptional architecture of the mammalian circadian clock. Nat Rev Genet. 2017;18:164–79.
138. Tan DX, Xu B, Zhou X, Reiter RJ. Associated health consequences and rejuvenation of the pineal gland. Molecules. 2018;23(2):pii: E301. https://doi.org/10.3390/molecules23020301.
139. Tsai YM, Chien CF, Lin LC, Tsai TH. Curcumin and its nano-formulation: the kinetics of tissue distribution and blood-brain barrier penetration. Int J Pharm. 2011;416:331–8.
140. Ueda HR, Hayashi S, Chen W, Sano M, Machida M, Shigeyoshi Y, Iino M, Hashimoto S. System-level identification of transcriptional circuits underlying mammalian circadian clocks. Nat Genet. 2005;37:187–92.

141. Vanselow K, Vanselow JT, Westermark PO, Reischl S, Maier B, Korte T, Herrmann A, Herzel H, Schlosser A, Kramer A. Differential effects of PER2 phosphorylation: molecular basis for the human familial advanced sleep phase syndrome (FASPS). Genes Dev. 2006;20:2660–72.
142. Videnovic A, Zee PC. Consequences of circadian disruption on neurologic health. Sleep Med Clin. 2015;10:469–80.
143. Vinod C, Jagota A. Daily NO rhythms in peripheral clocks in aging male Wistar rats: protective effects of exogenous melatonin. Biogerontology. 2016;17:859–71.
144. Vinod C, Jagota A. Daily Socs1 rhythms alter with aging differentially in peripheral clocks in male Wistar rats: therapeutic effects of melatonin. Biogerontology. 2017;18:333–45.
145. von Gall C, Weaver DR. Loss of responsiveness to melatonin in the aging mouse suprachiasmatic nucleus. Neurobiol Aging. 2008;29:464–70.
146. Wadhwa R, Konar A, Kaul SC. Nootropic potential of Ashwagandha leaves: beyond traditional root extracts. Neurochem Int. 2016;95:109–18.
147. Wang RH, Zhao T, Cui K, Hu G, Chen Q, Chen W, Wang XW, Soto-Gutierrez A, Zhao K, Deng CX. Negative reciprocal regulation between Sirt1 and Per2 modulates the circadian clock and aging. Sci Rep. 2016;6:28633.
148. Weinert D. Age-dependent changes of the circadian system. Chronobiol Int. 2000;17:261–83.
149. Weinert D. Circadian temperature variation and ageing. Ageing Res Rev. 2010;9:51–60.
150. Welsh DK, Takahashi JS, Kay SA. Suprachiasmatic nucleus: cell autonomy and network properties. Annu Rev Physiol. 2010;72:551–77.
151. Winkelman JW, Armstrong MJ, Allen RP, Chaudhuri KR, Ondo W, Trenkwalder C, Zee PC, Gronseth GS, Gloss D, Zesiewicz T. Practice guideline summary: treatment of restless legs syndrome in adults: report of the guideline development, dissemination, and implementation Subcommittee of the American Academy of Neurology. Neurology. 2016;87:2585–93.
152. Witting W, Mirmiran M, Bos NP, Swaab DF. The effect of old age on the free-running period of circadian rhythms in rat. Chronobiol Int. 1994;11:103–12.
153. Wyse CA, Coogan AN. Impact of aging on diurnal expression patterns of CLOCK and BMAL1 in the mouse brain. Brain Res. 2010;1337:21–31.
154. Yamaguchi Y, Suzuki T, Mizoro Y, Kori H, Okada K, Chen Y, Fustin JM, Yamazaki F, Mizuguchi N, Zhang J, Dong X, Tsujimoto G, Okuno Y, Doi M, Okamura H. Mice genetically deficient in vasopressin V1a and V1b receptors are resistant to jet lag. Science. 2013;342:85–90.
155. Yamazaki S, Straume M, Tei H, Sakaki Y, Menaker M, Block GD. Effects of aging on central and peripheral mammalian clocks. Proc Natl Acad Sci U S A. 2002;99:10801–6.
156. Yan SS, Wang W. The effect of lens aging and cataract surgery on circadian rhythm. Int J Ophthalmol. 2016;9:1066–74.
157. Yang Y, Duan W, Lin Y, Yi W, Liang Z, Yan J, Wang N, Deng C, Zhang S, Li Y, Chen W, Yu S, Yi D, Jin Z. SIRT1 activation by curcumin pretreatment attenuates mitochondrial oxidative damage induced by myocardial ischemia reperfusion injury. Free Radic Biol Med. 2013;65:667–79.
158. Yang G, Chen L, Grant GR, Paschos G, Song WL, Musiek ES, Lee V, McLoughlin SC, Grosser T, Cotsarelis G, FitzGerald GA. Timing of expression of the core clock gene Bmal1 influences its effects on aging and survival. Sci Transl Med. 2016;8:324ra16.
159. Yin L, Wang J, Klein PS, Lazar MA. Nuclear receptor Rev-erbalpha is a critical lithium-sensitive component of the circadian clock. Science. 2006;311:1002–5.
160. Yonei Y, Hattori A, Tsutsui K, Okawa M, Ishizuka B. Effects of melatonin: basics studies and clinical applications. Anti-Aging Med. 2010;7:85–91.
161. Yu EA, Weaver DR. Disrupting the circadian clock: gene-specific effects on aging, cancer, and other phenotypes. Aging (Albany NY). 2011;3:479–93.
162. Zhou QP, Jung L, Richards KC. The management of sleep and circadian disturbance in patients with dementia. Curr Neurol Neurosci Rep. 2012;12:193–204.
163. Zhou B, Zhang Y, Zhang F, Xia Y, Liu J, Huang R, Wang Y, Hu Y, Wu J, Dai C, Wang H, Tu Y, Peng X, Wang Y, Zhai Q. CLOCK/BMAL1 regulates circadian change of mouse hepatic insulin sensitivity by SIRT1. Hepatology. 2014;59:2196–206.

The Biology of Aging and Cancer: A Complex Association

21

Mohit Rajput, Lalita Dwivedi, Akash Sabarwal, and Rana P. Singh

Introduction

All living organisms reproduce to produce offspring which become young and then over the course of time encounter the universal phenomenon of aging and finally die. Aging is the universal process that occurs in all living organisms. Every species has the desire to live young; however, aging and death are unavoidable. Since ancient times, humans have tried to understand the mechanisms of aging. Aging is accompanied by several age-related disorders that are the major cause of demise of all living beings. It has been found that there is a close association between aging and diseases. Last century has seen remarkable increase in life expectancy in humans due to medical advancements; therefore, the number of old aged people has been increasing worldwide. However, old age also brings various age-associated disorders, including cancer. Cancer is a multifactorial disease characterized by uncontrolled cell proliferation. At first glance, aging and cancer seem to be unrelated, but growing research in this area have indicated that there is close relation between aging and cancer. The association of aging and cancer is widely observed, together with other common age-related disorders. In this chapter, we will try to address questions about aging and its biology. Why do living organisms become aged during a course of time? Could this phenomenon be avoided? What are the genetics and physiology of the aging? Is there any particular set of genes which

---

M. Rajput · L. Dwivedi · A. Sabarwal
Cancer Biology Laboratory, School of Life Sciences, Jawaharlal Nehru University, New Delhi, Delhi, India

R. P. Singh (✉)
Cancer Biology Laboratory, School of Life Sciences, Jawaharlal Nehru University, New Delhi, Delhi, India

School of Life Sciences, Central University of Gujarat, Gandhinagar, Gujarat, India
e-mail: rana_singh@mail.jnu.ac.in

govern the process of aging? Why is life span the characteristic of the species? What is the common biology of aging and cancer?

Aging and Its Hallmarks

Dysregulation of several molecular events during the process of aging causes loss of homeostasis and thus predisposes the human system to age-related disorders such as osteoporosis, sarcopenia, cardiovascular and neurodegenerative disease, diabetes, lipodystrophy, and cancer. Aging also causes social and economic distress to the sufferer. Understanding the complexities of aging will provide assistance in developing strategies to prevent or delay aging and aging-related disorders [1].

With the progression of time from newborn to old, organisms come across several molecular and cellular changes. These changes collectively described as the "hallmarks of aging" usually contribute to aging. López-Otín et al. [2] enlisted nine hallmarks of aging; these are genomic instability, epigenetic alterations, telomere attrition, loss of protein homeostasis, dysregulation of nutrient sensing, mitochondrial dysfunction, cellular senescence, stem cell exhaustion, and altered intercellular communication. Ideally, certain common criterion has to be fulfilled by any factors in aging to be called as hallmark. Each hallmark should manifest during normal aging and experimental modulation in hallmarks should either accelerate or retard the aging [2].

Causes of Aging

Despite extensive research in the field of molecular biology and genetics, enigma that regulates lifespan and aging is unsolved. Many theories have been presented to explain the process of aging, which falls in the category of either programmed or error theory, but no single theory is convincing enough to understand the process completely [3]. Each theory is unique, however, not mutually exclusive, and thus describes the process of aging at different levels alone or in combination with another theory. Hence, these theories of aging have been classified into four groups, evolutionary, molecular, cellular, and systematic. Aging is programmed through both intrinsic and extrinsic factors and their simultaneous interactions at several levels. It is postulated that in the middle of last century, due to improved hygiene and medical advancements, increased longevity has been observed in humans. Evolutionary theory emphasizes on the role of natural selection on aging, which has been further explained by mutation accumulation and antagonistic pleiotropy theory. Mutation accumulation theory argued that the decline in natural selection over successive generations lead to accumulation of deleterious mutations; on the other hand, antagonistic pleiotropy theory emphasizes that pleiotropic effects of genes, where beneficial genes acts in early life, become deleterious at later ages contributing to aging.

Molecular theory suggests that differential expression of certain set of genes in early and late life might promote aging. Lifespan is the characteristics of the selection of genes that promotes longevity. Role of insulin-like signaling pathways in worms, mice, and flies in regulating lifespan indicated that altered gene expression might affect the process of aging [4]. In the cellular theory of aging, cellular senescence explains the accumulation of senescent cells with time either due to loss of telomeres or cell stress. The production of reactive oxygen species (ROS) during oxidative metabolism causes damage to the macromolecules in the cells that might contribute to aging.

System-based theories state that decline in functionality of critical organs that are involved in maintenance and control of other organ systems disrupts the integrity and function thus becomes less efficient [4]. Hence, exploring the root cause of aging is complicated since multiple factors are involved during the process of aging. However, these theories have tried to explain certain aspects of aging, but consensus has not been met.

Aging and Senescence

Probably, some might think that the process of aging and senescence is the same; however, these processes are separate but overlapped. Senescence is a process of cell growth arrest that is irreversible in nature attributed to the fact that any external physiological stimulation cannot sensitize the arrest cells to re-enter into cell cycle. In 1961, Hayflick and Moorhead were the first who observed the phenomenon of senescence. They have observed the irreversible growth arrest of fibroblast cells in in vitro culture. Role of senescence is implicated in protection of cancer as well as in aging [5, 6]. The morphology of the senescent cells is distinctive, usually flat, and enlarged, and these cells are found to have increased expression of several senescence-associated biomarkers such as β-galactosidase and cell cycle proteins p53, p21, p16, p27, and p15. Aging and senescence are linked processes since both limit the lifespan of the cell, but to correlate these two phenomena is challenging due to the complexity of these processes. It is postulated that accumulation of senescent cells in aged tissues over the period of time caused the loss of homeostasis of the system. There are various biological stimuli which lead to senescence increases with aging. Recent studies in mouse model of progeroid syndrome have established connection between aging and senescence which further emphasizes that these processes are related and discrete but not the same. They have targeted the p16-positive senescent cells *in vivo* that alleviate age-related features in mouse models. There are several molecular alterations such as activation of p53 and telomere shortening that link the process of aging and senescence [7]. It is postulated that senescence could promote aging through many possible ways. Senescence depletes the tissue stem or progenitor cells, thus compromising the tissue repair and regeneration processes. Secretory molecules from these cells also affect the vital process of the cells such as cell growth, differentiation, and migration, which further affect the tissue architecture [8].

Potential Animal Models for Investigating Connections Between Aging and Cancer

Several *in vitro* studies have tried to decipher the molecular mechanisms associated with aging and cancer, but more *in vivo* studies are needed for deeper understanding of molecular mechanisms interplaying in aging and cancer. Several organisms have been established in past few decades to study the process of aging. mTOR pathway regulates the process of aging and cancer in several model organisms such as yeast, nematodes, and flies. Another breakthrough in yeast, nematodes, and flies have shown how the overexpression of the Sir2 influences the lifespan of these organisms, indicating a direct connection between aging and cancer. The life extension in mice models was found upon calorie restriction indicated connection with calorie intake and its effects on aging. The most widely studied animal models for aging include nematodes and fruit fly. There are many established vertebrate animal models which have been used to study aging and related diseases, and these include mouse, rats, naked mole-rats, primates, fish, dogs, cats, and birds [1, 9].

Cancer and Its Incidences with Aging

Cancer is a deadly disease characterized by uncontrolled growth of cells. Several properties of cancer cells include loss of growth control, resistance to apoptosis, immortality, angiogenesis, and ability to invade and metastasize. Cancer is generally driven by different changes at genetic and epigenetic levels. Genetic changes include mutations leading to loss of function mutations in tumor suppressor genes and activation of oncogenes. Epigenetic changes include global hypomethylation leading to genome instability and activation of various oncogenes and transposable elements, promoter-specific hypermethylation resulting into suppression of certain tumor suppressor genes, microRNA-based silencing of different oncogenes, and various kinds of histone modifications [10].

Age-related increase in cancer incidence is well documented. It has been reported that more than 60% of new cancer incidences and more than 70% of cancer-related mortality occur in people over the age of 65 years [11]. But this trend is not universal and also depends on the type of cancer. For example, osteosarcoma and acute lymphoblastic leukemia (ALL) peak during early years of life [12, 13]. Also, testicular and cervical cancer declines after middle age (WONDER database).

Study by Veronesi et al. showed that recurrence of breast cancer after mastectomy was higher in younger women compared to old ones [14]. An experiment by Ershler et al. showed that growth of injected Lewis lung carcinoma and B16 melanoma in younger mice was rapid compared to older mice [15], while younger mice coped better than the older ones during chemical or radiation-induced sarcomas [16]. Hence, studies have provided evidence that there is high incidence of cancer with age. The complexity of both the process makes it difficult to study these phenomena simultaneously and their related molecular mechanisms. However, several studies have provided the information on common biology of the cancer and aging.

Therefore, in the next section, we have highlighted all the common aspects of cancer and aging.

Common Biology of Cancer and Aging

Multiple changes at cellular, molecular, and physiological level during aging affect the biology of cancer. There are various phenomena which are common between cancer and aging such as telomere shortening, genomic instability, senescence, global hypomethylation, promoter-specific hypermethylation, metabolism, and autophagy. Hence, the link between common biology of cancer and aging is incontrovertible, but the underlying molecular mechanisms are yet to be fully understood.

Telomere Shortening

Telomeres are the repetitive sequences at the ends of chromosomes. DNA polymerases are unable to completely replicate these telomeres as these polymerases are unidirectional and need a labile primer. Thus, after each cell division, telomeres get shortened [5]. This shortening of telomeres after several cell divisions results in genome instability and activates DNA damage response and leads to irreversible growth arrest known as replicative senescence.

Telomerase, an enzyme that could synthesize DNA repeats *de novo*, and recombination between telomeres known as alternative lengthening of telomeres (ALT) are two ways by which telomeres are elongated. Adult cells have very less expression of telomerase, so the telomeres undergo shortening with age [17]. Telomere shortening has been observed during aging as well as during development of most of the cancers.

Several studies have shown that the shortening of telomeres with age and length of telomeres can predict longevity and age-associated defects in humans. The senescent cells as a result of replicative senescence have also been found to be accumulated during aging. Genetic variants of p16, which is a marker for senescence, are also found to be involved in age-influenced diseases [18]. Mice lacking telomerase have shown decreased lifespan even after first generation and further decrease occurred in subsequent generations. Telomere shortening has been found at old age in *Mus musculus* and *Mus spretus*. Mice with overexpression of telomerase which escape tumor development have shown increased lifespan. Mice deficient in telomerase have shown to age prematurely due to impaired function of stem cells leading to decreased regeneration capacity of tissues which in turn leads to multiple organ defects and is considered as the most important cause of telomere-associated aging [19]. Furthermore, factors that contribute to aging such as obesity or stress have shown to decrease telomere length and activity of telomerase. Various premature aging syndromes are linked to mutations in telomerase, telomere shortening, and chromosome instability. The telomere-associated aging is found to be regulated by p21, downstream target of p53 and PMS2, a mismatch repair protein. In telomere-deficient mice, deletion of p21 increases lifespan [6].

Most of the tumors have also found to have short telomeres as a consequence of excessive proliferation, but later on by activation of telomerase or ALT, they maintain their telomeres and elongate them [19]. The telomere shortening has also been associated with cancer susceptibility. Increased risk of chromosomal aberrations caused by breakage-fusion-bridge cycles after telomere shortening predisposes the cells to cancer. The telomere shortening results in the loss of telomere function which in turn results in the genomic instability leading to tumor development. Higher incidence of cancer has been found in mice with short telomeres which can be attributed to genomic instability. Furthermore, the replicative senescence has been suggested to promote cancer development in old tissues due to secretion of various tumor-promoting factors [18]. Telomerase inactivation which causes telomeres to shorten has been reported to cause cancer and progeroid syndrome in mice which generally have higher telomerase expression and comparatively long telomeres [20].

Incidence of cancer was found to increase in mice which were genetically manipulated to express telomerase postnatally through Tert transgenesis, and rate of cancer protection was increased in mice with telomerase deficiency. These mice were found to be resistant to tumor induction by carcinogens and genetic defects. Two exceptions exist for telomerase-deficient mice models of cancer protection, one is when this deficiency is combined with absence of p53 and another is when this occurs with overexpression of TRF2, a shelterin component which recruits nuclease to telomeres. These two conditions result in genomic instability which in turn promotes cancer development [6, 19]. Furthermore, senescence induced by telomere shortening has been found to inhibit the cancer development in mice deficient in telomerase and with expression of Myc gene in its B cells [6]. Thus, telomere shortening acts as a potent barrier to tumorigenesis, but cancer cells overcome this barrier at some point of cancer development leading to malignant transformation.

Hence, the maintenance of telomeres has been associated with both aging and cancer.

Genomic Instability

Genomic instability is at the heart of both aging and cancer. Genomic instability is caused in various ways, among them the most common is oxidative damage due to ROS generation. Major source of ROS generated in a cell is through mitochondria. The damage caused by ROS to molecules such as DNA, lipids, and protein during aging has also been found to occur in cancer. For example, malondialdehyde (MDA) and 4-hydroxynonenal/4-hydroxy-2-nonenal (HNE), the two byproducts of lipid peroxidation, have been found to be involved both in tumor development and aging [21]. There are few reports linking the inhibition of oxidative damage and increase in lifespan. Mice deficient in p66shc, a redox protein having role in

ROS generation, and mice that overexpress catalase, enzyme responsible for chelating H_2O_2, showed delayed aging. Mice overexpressing human thioredoxin also showed extended lifespan. Similarly, inhibition of oxidative damage is one of the potent strategies in cancer chemoprevention. Use of N-acetylcysteine results in protection against some lymphomas and carcinogen-induced lung cancers. It has also been reported to intervene in cancer development in mice. Moreover, resveratrol has shown anti-aging and anti-cancer effect which could be partly attributed to its antioxidant property [19].

Other causes of genome instability that link cancer and aging are telomere attrition and global hypomethylation. As discussed earlier under telomere shortening section, telomere shortening occurs during aging, and it also predisposes to cancer; this is due to the result of genomic instability and chromosome aberrations after telomere attrition [18]. Global hypomethylation is another phenomenon which results in genomic instability and links cancer and aging. It results in the activation of certain latent retrotransposons leading to genome instability and has been found to occur in cancer as well as in aging [22, 23].

Mutation in various genes which are involved in the maintenance of genomic stability results in premature aging as well as neoplasia. Such genes also provide evidences for the role of genomic instability in linking cancer and aging. Examples of such genes include ATM kinase, BRCA1, p53, WRN, BLM,RTS, ERCC4, BUB1, SIRT1, and ku86 [17, 18]. Mutations in p53, ATM, and BRCA1 result in genetically inherited cancer syndromes such as Li-Fraumeni syndrome, ataxia-telangiectasia, familial breast, and ovarian cancer, respectively. All these genes are involved in DNA damage surveillance and are also linked to aging. Mice expressing super active form of p53 while maintaining its normal regulation results in cancer resistance and normal aging. But the mice with super active isoform of p53 with loss of its regulation are cancer resistant but show premature aging. Mice with overexpression of p53 and p19ARF showed delayed aging due to elimination of DNA damage [17, 19]. Also, the life-extending mutations in *C. elegans* ultimately result in activation of p53. Likewise, mice with mutation in ATM showed rapid aging [19]. Mutation in BRCA1 leading to impairment of its repairing capacity of DNA damage resulted in premature aging in mice, and such mice were also prone to cancer development [24].

WRN, BLM, and RTS encode RecQ helicases which are involved in repairing of double-strand DNA breaks and thus are very important for maintaining genome integrity [24]. WRN gene also interacts with telomere-binding proteins and helps in proper maintenance of telomeres. WRN gene has also been found to be epigenetically silenced in cancers [17]. Mutations in these genes result in premature aging disorders with cancer predisposition, namely, Werner syndrome, Bloom's syndrome, and Rothmund-Thomson syndrome, respectively. Werner syndrome includes multiple features of aging as premature thinning and graying of hairs, wrinkling of skin, type II diabetes, cardiovascular diseases, osteoporosis, cataracts, hypogonadism, and cancer mainly sarcomas. Bloom's syndrome involves immune deficiency, dwarfism, decreased fertility, and development of multiple types of cancers early in life. Rothmund-Thomson syndrome patients show growth deficiency, growth retardation, poikilodermatous changes in the skin, cataracts, graying and thinning of hairs, photosensitivity, and increased predisposition to cancer [25].

ERCC4 encodes an enzyme involved in nucleotide excision repair (NER). Mutation in this gene leads to xeroderma pigmentosum syndrome which predisposes the patients to various solid tumors. Severe mutations in the same gene also result in premature aging as shown by a study on a single patient [17]. BUB1 is a mitotic checkpoint protein which has role in chromosome segregation and maintenance of genome integrity. Overexpression of BUB1 in mice has been found to prolong healthy lifespan and protects against cancer [18].

SIRT1 is a member of sirtuin family of enzymes (NAD-dependent deacetylase) and it is a mammalian orthologue of sir2 gene in *Drosophila*. Various members of sirtuin family have role in gene silencing, recombination, and maintenance of genome stability. Mice with deleted Sirt1 die shortly after birth due to genomic instability [17]. Sir2 was also reported to have noticeable longevity activity and overexpression of sir2 increased lifespan in yeast [26]. Extended lifespan in yeast and worms after caloric restriction was found to be mediated by sir2, and its overexpression was sufficient to cause lifespan extension [17]. Role of SIRT1 in cancer development as well as in cancer prevention has also been studied. Depending upon the context, it can act as tumor suppressor or oncogene. Overexpression of SIRT1 has been studied in some tumors as leukemia and various anticancer therapies target SIRT1. Activators of SIRT1 are undergoing clinical trials for cancer prevention as it has been observed to have protective role against colon and breast cancer in mice [18]. SIRT3 and SIRT6 have also been reported to have role in aging. Mice deficient in SIRT6 resulted in accelerated aging and SIRT3 had role in healthy aging [26].

Ku86 is an important component of nonhomologous end-joining repair mechanism. Mice with deletion of ku86 resulted in decreased lifespan and early onset of cancer, but the incidence of cancer was decreased. This decrease may be due to unavailability of sufficient time for cancer development because of shorter lifespan [27].

Senescence

Senescence is characterized by irreversible growth arrest following various stress signals. Senescence can be induced by progressive shortening of telomeres, DNA damage, epigenetic damage, mitogenic or oncogenic stimuli, and activation of tumor suppressors and inactivation of oncogenes. Senescence due to telomere attrition is called as replicative senescence, and senescence caused by different stresses except telomere shortening is known as stress-induced premature senescence (SIPS) [5, 28].

DNA double-strand break caused by radiations, topoisomerase inhibitors, and other means results in senescence induction. Damage caused by most of the chemotherapies also results in senescence [5]. Oxidative damage caused by ROS generation has also shown to cause senescence. Propagation of human cells in 20% oxygen level which is higher than physiological level or in presence of H_2O_2 resulted in senescence-like condition [28].

The change in epigenome such as chromatin de-condensation by HDAC inhibitors and by excessive activity of p300, a histone acetylase has been reported to cause senescence by derepression of p16 which is a marker of cellular senescence [5]. Various oncogenes have been shown to elicit senescence. Oncogenic form of H-ras which causes activation of MAPK pathways induces senescence in primary cells. Raf, MEK, and some other components of MAPK pathway as well as other oncogenic pathways also trigger senescence [28]. Other mitogenic stimuli which result in senescence include overexpression of ERBB2, a growth factor receptor, truncation of growth factor signaling by loss of PTEN, and stimulation by interferon-γ. p53, p21, p16, and pRB are the tumor suppressor genes that are also involved in regulation of cellular senescence, and overexpression of these genes is sufficient for induction of cellular senescence [5]. Inhibition of several oncogenes has also been observed to cause senescence. Inhibition of CDK4, c-myc, and some embryonic factors having oncogenic properties such as TBX2 and TWIST1/2 also results in senescence [28].

DNA damage caused by various stimuli such as dysfunctional telomeres, oxidative stress, oncogenes, etc. resulted in activation of DNA damage sensors such as ATM/ATR and Chk1/Chk2. These kinases then phosphorylate p53 which further activates p21 which ultimately causes growth arrest; p21 also causes inhibition of phosphorylation of pRb by CDKs which further inhibit cell cycle progression [29]. p21 is found to be upregulated during replicative senescence. Deletion of p21 gene increases lifespan and health in mice deficient in telomeres. Components of mismatch repair pathway, especially PMS2, are found to have role in senescence and aging after telomere loss and act upstream of p21 [6]. p21 overexpression can induce senescence and its deletion delays senescence.

INK4a/ARF locus also plays an important role in regulation of senescence. This locus encodes two different proteins, p16 and ARF; p16 inhibits CDKs and thus activates pRb and ARF sequesters Mdm2 leading to stabilization of p53. p16 expression has been reported to increase during replicative as well as stress-induced senescence, but it is still unclear whether telomere shortening regulates this increase and how different stress signals leading to senescence regulate p16 expression. The INK4a locus is regulated at various levels. Polycomb group of repressors such as Bmi, CBX7/8, and TBX 2/3 represses this locus and thus inhibits senescence, and during senescence these regulators show decreased expression. This inhibition is counteracted by various activators such as MAPKAP which inhibits polycomb complexes, SWI/SNF which causes chromatin remodeling and JMJD3, and a lysine demethylase which activates the INK4 locus. Activation of p38 MAPK in response to ROS, oncogenes, and dysfunctional telomere also regulates this locus [28]. At transcriptional level, p16 is activated by transcription factors Ets1/2 or E47. Id1 forms heterodimers with this transcription factors and thus inhibits activation of p16. Its expression was found to decrease during senescence [29]. P16 and ARF are also regulated posttranslationally. UV treatment leads to the activation of p16 by inhibition of its degradation by SKP2 and ARF is regulated by ULF ubiquitin ligase.

MicroRNAs (miRNAs) are also found to regulate senescence. Some miRNAs act as inhibitor and some as activator of senescence. ROS and enzymes generating and inhibiting ROS have also been found to regulate senescence. Akt also regulate senescence by modulating ROS [29].

Senescence in Cancer and Aging

Senescence also provides a link between cancer and aging. Accumulation of senescent cells has been reported during aging as well as in tumors. Accumulation of senescent cells was found in various tumors such as in neurofibromas, benign lesions of the prostate, and the skin. Accumulation of senescent cells has also been observed in several mouse models of benign tumors and hyperplasia [6]. Senescence has tumor-suppressive function, and at this point, cancer and aging diverge; however, in spite of its role in inhibition of cancer, senescence is also involved in cancer progression. Hence, the biology of cancer and aging merges at this point as aging-derived senescence results in cancer development. Various senescence-associated secretory profile/phenotype (SASP) components are involved in tumor progression.

Inactivation of p53, p16, ARF, p21, SUV39H1, and Rb, the proteins which regulate senescence, results in malignant growth [28, 29]. Various chemotherapy-based strategies also depend on induction of senescence in cancer cells. Different therapies induce senescence by different means as by targeting telomerase, reactivating p53, inactivation of oncogene as c-myc, and induction of DNA damage. All these changes lead to senescence and thus cancer regression [29].

Mouse and human epithelial cells were reported to show increased proliferation after administration of senescent fibroblast, while non-senescent fibroblasts do not have this effect. This stimulation is mediated by different SASP components such as VEGF which help in tumor angiogenesis, MMP3 which has role in invasion, and GRO and amphiregulin which promote cancer cell proliferation. SASP components such as IL6 and IL8 are involved in induction of epithelial to mesenchymal transition (EMT) in cultured cells. EMT helps cancer cells to invade and metastasize to different sites leading to malignant growth. Some other SASP factors have role in stem cell renewal and chemoresistance of cancer cells [5]. Senescent cells also contribute to cancer initiation by generation of ROS and are considered to be a major cause for cancer development [30].

SASP acts as an important factor linking cancer and aging. As senescent cells accumulate during aging, SASP components create a fertile microenvironment for cancer development. Thus, increase in cancer incidences with age may be partly due to SASP-mediated effects which accumulate during aging. Infiltration of leukocytes by SASP components results in inflammation leading to DNA damage which can result in development of cancer as well as ageing phenotypes [5]. All these facts show that senescence can act as double-edged sword; during early stage of life, it helps in tumor suppression, but with aging it causes permissible environment for cancer development.

Cellular senescence apart from having a role in cancer is also considered as an important factor in aging. Accumulation of senescent cells has been reported to occur during aging and certain age-related pathologies as dementia, atherosclerosis, osteoarthritis, benign hyperplastic prostate, and respiratory disease [29, 31]. A number of senescent cells were found to increase with age in various tissues such as the skin of baboons, the liver of mouse, the retina of primates, and human skin and liver [18, 31]. Further, senescent cells were found to be responsible for disruption of normal structure of breast epithelial cells and pulmonary arteries, and these disruptions were hypothesized to cause or contribute to age-related changes mediated by SASP components. Indirect evidences indicate that senescent cells are also implicated in genesis of various age-associated diseases. Direct evidence that senescent cells can drive the age-related pathologies came from a study which involves targeted clearance of senescent cells expressing p16 in a progeroid mouse model. This targeted clearance resulted in delay in age-associated pathologies [5]. Moreover, regulators of senescence have also been linked to aging. CDK2a locus which is expressed at very low levels in young tissues and is a regulator of senescence gets activated during aging resulting in increased p16 levels with age in humans and rodents, and it serves as marker for aging. Single nucleotide polymorphism and variants of p16 have been found to be linked to aging and age-related diseases [17]. Caloric restriction which slows the rate of aging is reported to decrease the age-related increase in p16 levels. Further supporting this link, age-dependent increase in p16 has been shown to decrease regeneration potential of various adult stem cells, and mice lacking p16 showed increased regeneration capacity, while mice deficient in polycomb repressors have been shown to decline in self-renewal capacity of stem cells [17, 29].

Inhibition of p21, another regulator of senescence, results in decreased regeneration capacity of hematopoietic stem cells which is attributed to depletion of stem cells as a consequence of increased cell cycle progression. Deletion of p21 resulted in reversion of progeroid phenotype of mice with mutation in RNA component of telomerase. p53 which is an important regulator of senescence also provides a link between senescence and aging. Mice deficient in p53 showed increase in self-renewal capacity of stem cells and mice with increased p53 activity aged prematurely due to exhaustion of stem cells, while simultaneous increase in p53 and ARF led to increased lifespan [29]. Furthermore, another biomarker for senescence, β-galactosidase, also showed specific pattern in the skin from young and old donor showing age-related increase [32].

Despite these observations, it is still unclear whether cellular senescence is one of the causal mechanisms for aging. There are some indirect evidences that support this notion. For example, the cells derived from the patients having Werner syndrome, a premature aging syndrome, undergo less cell divisions before achieiving senescence as compared to similar aged normal individuals. The studies showed prolonged lifespan of cells cultured in vitro from younger donors compared to old donors and also provide indirect link between senescence and aging, but the direct evidences in this field are still lacking [29].

However, it is considered that senescence might cause aging in different ways. Firstly, excess accumulation of senescent cells results in loss of homeostasis and organ function. Secondly, senescent cells, due to their growth arrest property, deplete the stem/progenitor cell resources gradually leading to decreased renewal capacity. Finally, various SASP components as degradative proteases, cytokines, and growth factors cause tissue disruption [24]. Fibroblasts isolated from skin samples from old animals showed decreased potential to replicate when compared to younger ones under culture conditions, and replicative potential was found to have direct link to lifespan as found in one of the study comparing various species [32].

Few senescent cells also include a heterochromatin foci which silence some pro-proliferative genes, and some senescent cells have also increased levels of proteins with tumor suppressor function as DEC1 (deleted in esophageal cancer) and DcR2 (decoy receptor 2) [17]. Thus, senescence is another important phenomenon which is shared by both cancer and aging.

Epigenetic Modifications

Cancer and aging also share some common biology at the level of epigenetic changes. Global hypomethylation and promoter-specific hypermethylation are the hallmarks of both processes. Various genes have been shown to be hypermethylated during aging and cancer. For example, RASSF1 gene is hypermethylated in various cancers and its hypermethylation increases with age [18]. ESR1 gene which shows age-related hypermethylation has been found to be methylated in colon and prostate cancer [33]. Target genes of polycomb group of proteins have also found to be methylated in aging as well as cancer. It has been found that the hypermethylation which results during aging shows a significant overlap with methylation during carcinogenesis and the CpG sites having age-linked hypermethylation also seem to get coherently methylated in cancer [34]. Age-derived hypermethylation as well as methylation during cancer preferentially takes place at bivalent chromatin sites which are marked in stem cells. Rakyan et al. studied 28 genes which were found to be hypermethylated in various cancers including BRCA1, p14, p15, p16, E-cadherin, cyclin D1, RARβ, RASSF1, APC, and MGMT. They observed that the promoters of these genes have age-associated hypermethylation [35]. Apart from the commonly hypermethylated genes in cancer and aging, there are several genes which show age-dependent methylation and mutations in both aging and cancer [33]. Few genes which are found to be mutated during aging-related syndromes have also shown hypermethylation during cancer development as WRN gene and lamin A/C gene. Both genes are mutated in Werner syndrome and hypermethylated in several cancers [23].

Several studies which were performed in human fibroblast and stem cells have indicated the presence of global hypomethylation with age and generally occurs in repetitive sequences and the regions outside of CpG islands. This hypomethylation is mirrored in cancer cells too and helps in cancer development [10, 33]. Furthermore, mice lacking DNMT3A show reduced lifespan and increased tumor growth [33].

Another example of this category includes a class of histone deacetylases known as sirtuins. Deletion of SIR2 in yeast decreases the lifespan and overexpression extends life. SIRT1 is mammalian homologue of yeast sir2, and it has been found to decrease during aging and to increase in various cancers as lymphoma, soft tissue sarcoma and lung carcinoma of mice, leukemia, lung cancer, and prostate cancer of humans. Colon cancer is the exception in which sirt1 is found to be downregulated.

The acetylation levels of H4-K16 and H3K9 which are targets of sirt1 also show alteration in various cancers. Furthermore, trimethylation at H4-K20 also imposes an example of divergence between aging and cancer as it increases during aging and declines in cancer cells [23]. Levels of p16 show divergence between aging and cancer and it is regulated by various epigenetic mechanisms. SNF5 (chromatin remodular), H2AZ (histone variant), JMJD3 (histone demethylase), MLL1 (histone methylase), p300 (acetyltransferase), and chromatin modulation by c-myc activate p16 and result in decreased carcinogenesis, and many of these factors also activate p16 during aging [5, 10]. EZH2, a histone methylase, deactivates p16 and is downregulated during aging but upregulated in various cancers. Furthermore, a chromatin remodeling complex, NuRD, shows different patterns during aging and cancer. It is targeted to wrong site in acute promyelocytic leukemia (APL) and in aging it is reported to be downregulated [10]. Thus, epigenetic modifications play important role in the regulation of aging and cancer.

Metabolism

Growth hormone/Insulin growth factor (GH/IGF) pathway is one of the important pathways linking aging and cancer at physiological level. Mice having mutations in GH/IGF1 signaling live longer and show decrease in incidence of cancer and exhibit extended latency period for cancer development in comparison with the wild-type mice. Growth hormone receptor null mice were reported to have lesser number of tumors and less severe tumor lesions. Frequency of tumor development and death due to neoplasms was also decreased in these mice compared to wild type, and such mice are also reported to have longer life. Another study in Ames dwarf mice having GH deficiency due to impaired development of pituitary gland also showed delayed tumor development with slightly decreased incidence of tumors. Humans with GHR deficiency had low levels of IGF1 and IGF2 and were found to be resistant to cancer, but no effect on lifespan was observed in such humans [18]. In *C. elegans* and *Drosophila melanogaster* mutation in insulin pathway components, daf-2, age-1 and daf-16 resulted in increased lifespan.

Furthermore, hyperinsulinemia has been considered to have important role in cancer development as well as aging, and antidiabetic biguanides have been reported to prolong lifespan of female rats and mice, and these drugs have also been found to have anticarcinogenic role in various cancer models [36]. A gene named klotho is considered as aging suppressor gene. Mice with defective klotho show premature aging and mice with its overexpression show longer life. Thus, this gene has role in delaying aging. This effect is supposed to be mediated by inhibition of insulin/IGF1 pathway by klotho. Klotho prevents phosphorylation of FOXO 3A and promotes its nuclear translocation leading to upregulated expression of MnSOD by FOXO. This result in decreased ROS and thus decreased IGF1 pathway as ROS are required for modulation of this pathway which in turn results in delayed aging. Klotho has also been reported to downregulate Wnt signaling, ROS, and p53/p21 pathway in addition to IGF1 pathway. Klotho also acts as a tumor suppressor gene, and this function is mediated by inhibition of insulin/IGF1, p53/p21, and Wnt signaling by klotho [19, 37] (Fig. 21.1).

The mammalian homologues for daf2 and daf16 are IGF1 receptor and FOXO, respectively. The longevity effect of daf2 mutant is mediated mainly by the downstream component daf16. Daf16 regulates a number of enzymes involved in stress management and metabolism. Daf16 in worms and drosophila was sufficient to increase the longevity. Mammalian FOXO protein members are also involved in metabolism and stress resistance, and they also act as tumor suppressor and play role in longevity. FOXO1, FOXO3, and FOXO4 have anti-cancerous activities [17, 19]. Thus, IGF/FOXO signaling regulates lifespan and cancer development. Decreased insulin signaling and increased FOXO activity delay aging and prevent cancer development. Two studies considering genetic variations in IGF1R and FOXO3A in human populations have also proved the same. It suggests that this signaling might play important role in longevity and cancer protection in humans too [20]. Another pathway linking metabolism, cancer, and aging is mTOR pathway. During nutrient deprivation, it gets deactivated, while in presence of plenty of nutrients, it gets activated. Regulators of this pathway have altered expressions in

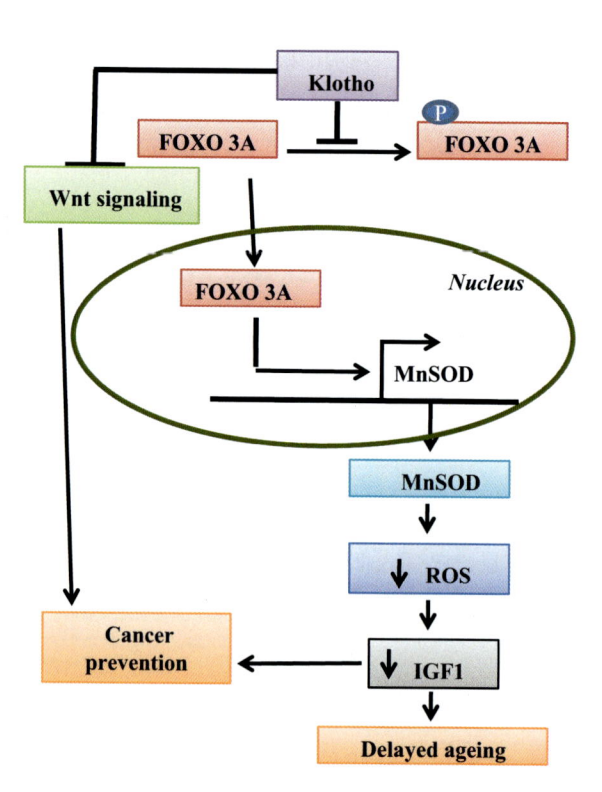

Fig. 21.1 Klotho and Wnt signaling in aging and cancer. Klotho gene is a connecting link between aging and cancer. It regulates FOXO and inhibits Wnt signaling. Through these targets, it prevents aging as well as cancer

various cancers, and rapamycin, an inhibitor of mTOR, is being used against various tumors. This pathway also has role in aging. Various studies in different organisms have shown that inhibition of this signaling results in increased lifespan, and lifespan increase by caloric restriction has been reported to regulate this pathway in yeast. In mammals also this pathway has important role in metabolism. Succinate dehydrogenase, a Krebs cycle enzyme, also links metabolism with aging and cancer. Mutation in it results in the development of certain tumors in humans, and mutations in the same complex also resulted in progeroid phenotype in worms and yeast [17]. SIRT1 also forms a link between cancer and aging by regulating metabolism. During nutrient-deprived condition, it associates with PGC1α, a transcription factor, and increases the efficiency of metabolism and decreases the ROS production. It also activates FOXO protein [19]. Most of these pathways discussed above are interconnected to each other and work together by regulating each other.

Caloric restriction (CR) has been reported to cause lifespan extension in most species with more significant effects in rodents. Various studies including rodents and rhesus monkey reported that it also causes reduction in cancer incidences. Some of these effects of CR may be attributed to decreased fat, as obesity has been found to cause decreased lifespan and increased age-associated diseases such as cancer in humans. The molecular mechanism behind CR-mediated effects is not well understood, but it is thought that GH/IGF1, mTOR, AMPK, and SIRT1 might play role in CR-mediated effects. GH/IGF1 is considered to be the most probable mechanism. During CR, its activity decreases in rodents, and it has been reported to have role in aging as well as cancer; thus both the effects of CR can be explained by this pathway. Furthermore, there is an evidence for IGF1/Insulin pathway as a probable mechanism for CR-mediated effect. In one study of bladder cancer in p53-null mice, delayed incidence of tumor was inhibited after restoration of IGF1 in mice with CR. mTOR pathways have role in longevity and might be associated with CR. AMPK is energy sensor which has been linked to CR as well as longevity [18]. SIRT1 also has role in increasing lifespan and is considered to be an important sensor of CR and regulator of CR-mediated increase in lifespan [19]. Thus, different types of genes which regulate various mechanisms such as metabolism, epigenome and DNA damage repair are important for aging and cancer as shown in Fig. 21.2.

Autophagy and Apoptosis

Autophagy also forms a link between cancer and aging. Autophagy clears damaged and old organelles and proteins for maintenance of homeostasis. During aging, the autophagic potential decreases and inhibition of autophagy results in aging-like phenotypes. Inhibition of autophagy also inhibits the life-prolonging effects of various factors such as caloric restriction, inhibition of insulin/IGF pathway, and sirtuin activation [38]. Autophagy acts as tumor suppressor by clearing damaged proteins and organelles. Mice with deletion of one copy of Beclin1 gene were observed to develop tumors and this locus has been found to be deleted in various human tumors.

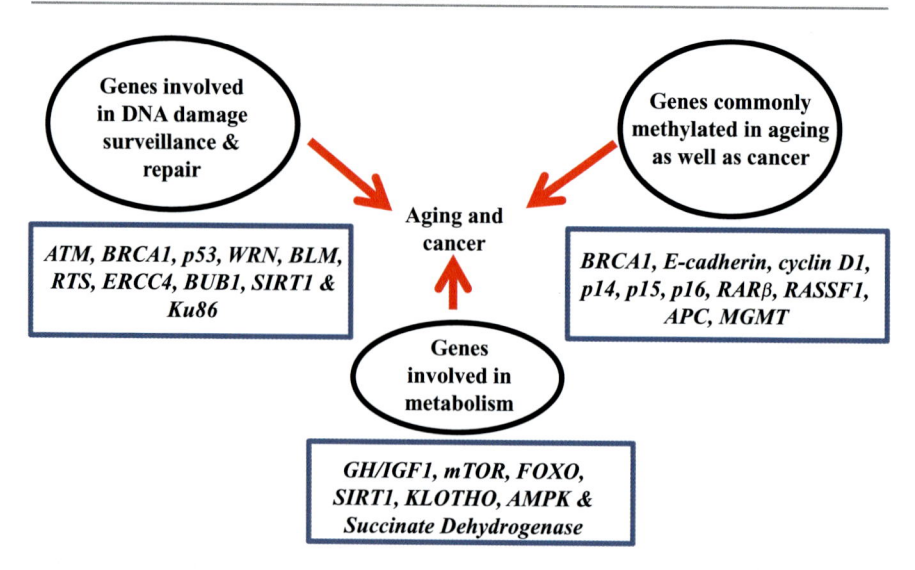

Fig. 21.2 Genes involved in common biology of cancer and aging. Various genes having role in DNA damage response and metabolism are known to represent the genes which are commonly mutated during tumor development as well as aging. In addition to these genes, there is also a set of genes which indicates the common biology between aging and cancer. Most of the genes which are found to be hypermethylated during aging are also reported to show hypermethylation in cancer. The abovementioned categories of genes portrayed in a schematic picture represent common biology between aging and cancer

Tumor suppressor genes such as p53 and PTEN also induce autophagy. Thus, inhibition of autophagy results in aging as well as in cancer development [17]. Apart from its tumor suppressor effect, autophagy has also been found to have oncogenic effect by promoting the tumor growth.

Apoptosis also functions to remove damaged cells to maintain homeostasis. During young age, it is very important to maintain homeostasis and also for tissue differentiation, but with increased age, exhaustion of stem cells occurs due to apoptosis. This exhaustion results in loss of cellularity and tissue function, thus contributing to aging [24]. Thus, apoptosis over a long period of time increases aging. On another side, inhibition of apoptosis results in cancer development and various anticancer therapies act by inducing apoptosis [39]. Hence, apoptosis is one of the divergent pathways between aging and cancer.

Interplay Between Cancer and Aging

As biology of cancer and aging converges at various points as shown in Fig. 21.3, the therapies which target cancer can also act as geroprotector and vice versa since the pathways which are the targets of anticancer therapies as mTOR, MAPK, PI3/Akt, and GH/IGF1 are also targeted for lifespan extension. Rapalogues which include rapamycin and its analogues, Akt/PI3 inhibitors and MAPK inhibitors,

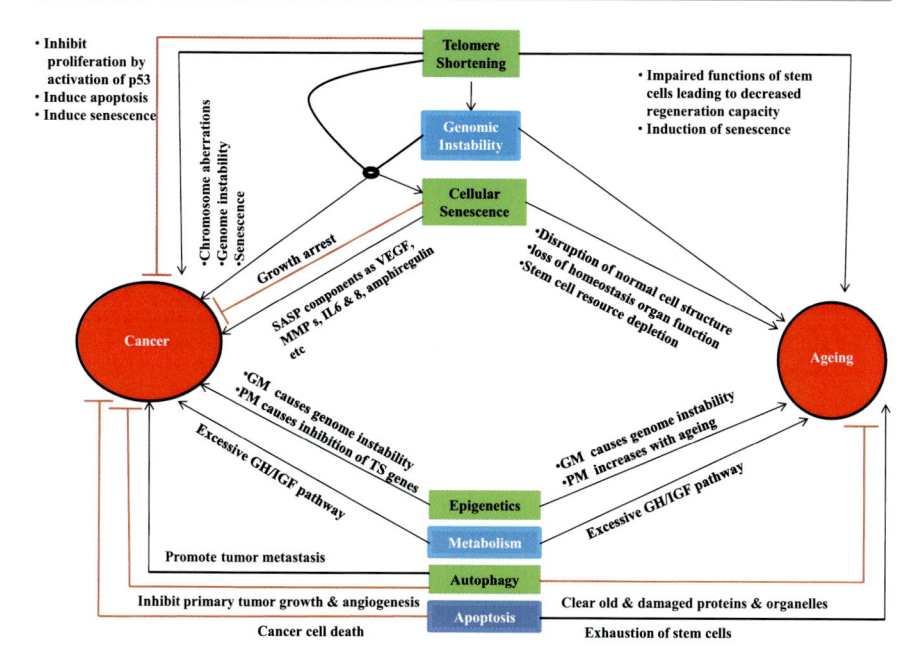

Fig. 21.3 Cellular events linking aging and cancer. Aging and cancer converge at various points, but there are several cellular events at which both the processes seem to converge, but the same events also mark their divergence depending on the cellular context and time, such as cellular senescence. Boxes in the green represent the cellular events which link aging and cancer in convergent as well as divergent manner. Boxes in the blue show the events which indicate toward convergence between the two processes, and box in purple represents the cellular event which marks the divergence between the two processes. GM, global hypomethylation; PM, promoter-specific methylation; TS, tumor suppressor; SASP, senescence-associated secretory profile/phenotype; VEGF, vascular endothelial growth factor; MMPs, matrix metalloproteinases; IL, interleukin; GH/IGF, growth hormone/insulin growth factor

represent the anticancer drugs which can also act as anti-aging agents as shown in Box 21.1 [40]. Class I and class II geroprotectors were found to decrease the tumor incidences in laboratory animals with few exceptions which showed no change in cancer incidence [25]. Caloric restriction which delays aging also helps in cancer prevention. Metformin which slows aging also provides protection against various cancers [40]. Moreover, various caretaker and gatekeeper genes as XPA, ku80, cat, PARP, p53, Bcl-2, RB, TERT, etc. regulate both the processes [25].

Furthermore, carcinogen has been found to cause premature aging in various experimental animals, and premature ageing has also been observed to favor carcinogenesis. Many of the premature aging syndromes as Werner, Bloom, and Rothmund-Thomson are also associated with increased cancer incidences [25].

Various genetically modified mice models have shown that the genetic modifications leading to decreased lifespan were also associated with increased incidence of tumor with few exceptions [25]. Hence, in the next section, we have discussed the diverging and converging aspects associated with cancer and aging.

> **Drugs having anti-ageing as well as anti-cancerous properties**
>
> - Rapalogues
> - Akt/PI3K & MAPK inhibitors
> - Metformin
> - Class I and Class II geroprotectors

Box 21.1: Drugs Having Anti-cancerous as Well as Anti-ageing Properties
Various pathways, genes, and cellular events indicate the link between the common biology of aging and cancer, and this concept is further supported by the drugs which have shown both anti-cancerous and anti-aging properties. This box represents some of these drugs such as rapalogues which are rapamycin and rapamycin-based drugs with anticancer properties and which are also found to increase the lifespan. Other such anticancer drugs having anti-aging effects too include Akt/PI3K and MAPK inhibitors. There are several drugs which have anti-aging property and also inhibit cancer. These include metformin, class I and class II geroprotectors.

Factors Influencing Cancer Incidence with Age

Several studies have shown that the cancer incidence may increase or decrease during aging as shown in Table 21.1. Most of the cancers occur in the old age of 65 or above [41] and rate of cancer mortality is also higher in older patients [32]. However, some tumors decrease with age, and cancers in older people have been found to be less aggressive compared to tumors in younger patients. Rate of growth and metastasis of various cancers decrease in old aged animals under experimental conditions, but this is not universal [32]. There are some factors which can explain the less growth and aggressiveness of tumor during aging as explained in Table 21.1.

Following factors can explain this increase or decrease in cancer incidence and mortality with advanced age.

Accumulation of Damaged Macromolecules

With the increase in age, accumulation of various damages such as oxidative damage and mutations occurs in genome. This is also supported by impaired mechanisms of DNA repair which can predispose cells to cancer development [32]. Protein misfolding also increases with age in some organ which can make it prone for various diseases including cancer. With aging, the increase in lipid peroxidation due to increased ROS has also been reported. Such situation may lead to cellular abnormalities which may even support the growth of cancer.

Table 21.1 Factors associated with the increase/decrease in cancer incidences with aging. Most of the cancers are reported in aged people, but there are also reports that cancer incidences decrease in elder people. This increase or decrease in cancer with age depends on the type of cancer and cellular context. This table indicates certain factors that could attribute to the increase or decrease in cancer during aging

Factors responsible for increase in cancer incidences during aging	Factors responsible for decrease in cancer incidences during aging
Repetitive exposure to carcinogen	Increase in senescence leading to growth arrest initially
Accumulation of damaged DNA with age	Certain immune cells play role in tumor growth and angiogenesis, and hence increased immune senescence with age also plays a role in decreased incidence of cancer with age
Increase in senescence which promotes tumor growth through various SASP components	Derepression of INK4 locus which codes for tumor suppressor proteins p16 and ARF
Immune senescence with age contributes to increase the chances of cancer development	Decrease in angiogenic potential with age
Telomere shortening	
Increase in global hypomethylation and gene-specific hypermethylation	

Carcinogen Exposure

Usually the development of cancer requires several as well as effective hits of carcinogen exposure targeting proto-oncogenes and/or tumor suppressor genes. Further, the fidelity of DNA repair machinery declines with age generating more chances for the accumulation of mutations in older people. Further, longer stay in a particular environment due to a particular occupation may also increase the risk from the repetitive exposure of environmental carcinogens. Therefore, in such situations, aging will increase the risk of cancer development from exposure to physical, chemical, or biological carcinogens [25].

Senescence

As discussed earlier, senescence increases with age, and various SASP components as MMP3, VEGF, IL6, IL8, and Gros stimulate the tumor growth. SASP-mediated oxidative damage and inflammation also stimulate cancer development [5]. Additionally, senescence may also act as tumor suppressor mechanism by arresting cell growth at earlier stages of carcinogenesis. Thus, increased rate of senescence during aging can also protect cells from cancer development.

Immune Senescence

Immune system is involved in cancer prevention. Immune system recognizes and removes the tumor cells based on the antigens displayed by tumor cells, and this phenomenon is called as immune surveillance. Several mouse models have shown that immune surveillance has important role in prevention from cancer. Immunodeficient mice showed increased incidence of spontaneous as well as carcinogen-induced tumor development. There are evidences for role of immune system in prevention of cancer development in humans also. Patients with immunodeficiency and patients taking therapy to suppress immune system in order to prevent transplant rejection have been found to show increased risk of tumor development. Lymphocyte infiltrations at the site of tumor development, autoimmune syndrome associated with neoplasm, and certain tumors with unstable microsatellite provide evidences for role of immune surveillance in cancer prevention. Various tumors as melanoma, lymphoma, mesothelioma, and lung carcinoma have shown spontaneous regression of tumors which could be attributed to immune surveillance as lymphocytes were found to be infiltrated in these cases [42].

Immune surveillance decreases with age due to immune senescence which affects activity of all the components of immune system and mostly occurs during aging. Various age-associated diseases including cancer have been linked to decrease immunity during aging [28, 32]. T-cells help in immune surveillance against cancer by means of natural killer cells which form lymphokine-activated killer cells upon IL2 stimulation and by cytotoxic T-lymphocytes, B-lymphocytes, macrophages, and various cytokines such as IL2 and IFNγ and TNFα which are also found to have role in antitumor immune surveillance. Proliferation and antigen stimulation capacity of T-cells decrease with aging and production of IL2 also decreases; both of these lead to decrease in the expression of p53. Furthermore, aging results in T-cell population deficient in its co-stimulator CD28 which may contribute to decrease antitumor activity of T-cells with aging. During aging, capacity of B-cells to respond to antigens also decreases, while levels of immunoglobulins increase posing a risk for development of lymphomas and myelomas. Functions of monocytes and macrophages and secretion as well as function of various cytokines also get dysregulated in old aged people [41]. Increase in the suppressor T-cells which help tumor to escape immune surveillance and promote their growth has also been reported during aging [41]. Moreover, inflammation and associated cytokines as IL6 and TNF increase during aging and cause risk of cancer development as observed for colon carcinoma and myeloma [18, 32]. Aging also results in decreased activation of T-cells by dendritic cells, dysregulation of TLRs, increase in levels of myeloid-derived suppressor cells and an immune suppressor, IDO, and shift of TH1 to TH2 cells [28]. All these changes in the immune system result in decreased immune surveillance and increased risk of cancer development during aging.

Immune system also has role in cancer development. Certain immune cells as monocytes or lymphocytes provide certain factors which stimulate tumor growth. For example, macrophages promote tumor growth by production of angiogenic and growth-stimulating factors. TNFα has also been shown to behave like growth stimulators for some tumors in spite of their tumor protective role. T-cells also produce angiogenic factors. In some circumstances, this growth stimulation outweighs the inhibition by same cells [32, 41]. In such conditions, immune senescence provides a benefit to older persons by decreasing their susceptibility to tumor formation. Melanoma growth provides a good example of this case. In elder mice, its growth is slower than the younger ones, and this is attributed to decreased production of angiogenic factors by T-cells due to senescence leading to reduced functionality of T-cells in elder mice [41].

Hormones

Estrogens and androgens help in the development of certain cancers such as breast, endometrial, and prostate cancer. Level of these hormones decreases during aging resulting in decreased risk of hormone-dependent tumors in older age. Furthermore, GH/IGF1 signaling also decreases during aging due to decreased levels of GH and IGF1. As this signaling plays important role in cancer development, thus decrease in this pathway might be linked to decreased incidence of tumor during aging [18]. However, aberration in the level of production of these hormones or mutations may increase the risk of associated cancers.

Activation of INK4 Locus

INK4 locus codes for two tumor suppressor proteins, p16 and ARF. During aging, this locus gets de-repressed resulting in increased expression of theses tumor suppressor genes. This increase could also be a factor for protection against cancer during aging. Downregulation of p16 and ARF has been observed in many cancers.

Angiogenesis

Angiogenesis is very important for tumor growth and development. During aging, angiogenic potential of cells declines due to decreased response to angiogenic signals, and this decline inhibits tumor progression in older people. There are few evidences which suggest that the slower rate of some tumors during aging may be due to decreased angiogenesis [18].

Thus, aging and cancer share various pathways in common, but many of these pathways have dual role in linking aging and cancer. Telomere shortening, senescence, and epigenetic modifications represent these pathways which at some point show convergence between aging and cancer, but at another point, they indicate the divergence between aging and cancer. This convergence and divergence is context dependent. Thus, a clear conclusion cannot be drawn about the common biology of aging and cancer. However, from all the points discussed earlier, it can be concluded that a balance between the convergent and divergent pathways (Fig. 21.3) of aging and cancer is required for a healthy and tumor-free lifespan.

Molecular Events Involved in Aging and Cancer

As discussed earlier, aging causes many complexities in the biological system and thus responsible for development of many age-related pathologies, including cancer. As the organism grows older, the population of senescent cells increases in the organ system. This process of cellular senescence prevents the proliferation of the cells that might have experienced certain stress including oncogenic stress. Senescence is a very complex process that could be beneficial or detrimental to living organism depending on the various physiological conditions. The connection between aging and cancer is widely observed in many epidemiological studies. It is observed that there is exponentially rise in incidence of cancer after sexual maturity of the organisms. However, the relationship between biology of aging and cancer is still not clear [5, 18].

The complexity of aging is well established that is accompanied with several molecular and physiological changes. However, there are findings which have indicated that hundreds of genes modulate the process of aging, thus longevity in model organisms. Also, there are approximately 100 genes found in mice which are involved in the process of ageing. Most importantly, it has been found that a dietary and genetic manipulation that affects the ageing in rodents also affects the cancer incidence and progression [18]. Hence, it could be concluded that mechanism of both the processes might overlap; therefore, understanding common molecular events involved might unravel the connection of aging and cancer.

Signaling Pathways Involved in Aging and Cancer

The organized signaling pathways are crucial for the growth of organism and maintaining the homeostasis of the cells. Disruption in signaling network predisposes the organism to many diseases. Several signaling pathways are involved that regulate the process of aging, longevity, and cancer. The signaling network which regulates aging includes hormonal signaling, nutrient sensing signaling, and mitochondrial-mediated ROS signaling.

Over a period of time, these signaling network intensities change that causes many age-related pathologies including cancer. It is found that the molecular changes in signaling pathways that are involved in repair slows down in older ages; this causes decrease in the normal repair process in organs. For example, activation of Notch signaling pathway in skeletal muscle decreases with age, which further diminishes the activation of satellite cells, thus causing defects in repair of muscle injury. Role of Wnt signaling as a pro-aging signaling and cancer progression is well known [43, 44]. The Ras and PI3K signaling affects diverse processes inside the cells including autophagy, cellular senescence, and aging. Both of these pathways are critical for the regulation of cancer as well as aging. It is found that when senescence-induced cells are treated with MEK, PI3K, and mTOR inhibitors, induction of cellular senescence and aging was blocked. Additionally, a drug that targets these signal transduction pathways inhibits the cellular senescence and aging [45].

Microarray-based studies have shown the association between cancer and aging by investigating the expression of aging-related signature genes and by analyzing their expression profiling in various human tumors. They have found several aging-related signature genes; however, only few genes were found to be commonly expressed in multiple tissues. These genes are involved in pathways in cancer and MAPK signaling, hence implicated in cancer. Human aging-related genes are implicated in both cancer and aging. For example, FOS which regulates cell proliferation and transformation is associated with both aging and cancer. FOXO1 and p53 are other two genes found to be associated with aging and cancer. There are certain critical molecular mechanisms such as DNA damage response, metabolic process, p53 signaling pathway, apoptosis, cell cycle regulation, and immune/inflammatory response which underlie both aging and cancer [46, 47]. NF-κB is a critical transcription factor which regulates many physiological processes such as innate immune system, apoptosis, inflammation, and cell proliferation. In one study in gastric mucosa of rat, the activity of NF-κB was found to be increased with age. It is estimated that inflammation contributes approximately 15% of solid cancer with NF-κB playing a key role [48].

Role of Oxidative Stress in Cancer and Aging

Excessive production of ROS in most of the cancer is well known, where ROS modulates several signaling pathways. It is suggested that moderate increase in oxidative stress can stimulate the cancer cell proliferation and survival, while high oxidative stress might cause cell death and apoptosis. Enhanced oxidative stress is found during aging development (Fig. 21.4). Senescent cells are found to have more oxidative damage to DNA and protein and are associated with high intracellular ROS level. The free radical theory of aging which was given by Denham Harman explained the role of free radicals in causing aging. It is stated that aging results from accumulation of deleterious effects caused by free radicals for a long time; hence the organism's lifespan will be defined by their ability to cope up with these radicals. There is an evidence of increased mitochondrial DNA damage and intracellular ROS production in aged tissue. Studies in model organism also have established this theory, where it has been observed that oxidative stress-resistant mutant strain of *C. elegans* has extended lifespan. The increase in ROS production in aged mitochondrion is an adaptive response that provides increased stress resistance [49, 50]. Telomeres are critical for maintaining the DNA structure. The length of telomeres shortened with each division is thus considered as major cause of replicative senescence. It has been observed that telomere shortening is directly related to the cellular oxidative stress, which affects telomere maintenance at various levels. The presence of oxidized base such as 8-oxoguanine in the telomere regions greatly affects the telomerase-mediated elongation process of telomeres [49].

Mitochondria are the major source of ROS in the mammalian cells; thus oxidative damage to mitochondrial DNA is very common due to close proximity. Role of mitochondria in aging is well known as explained by the mitochondrial theory of

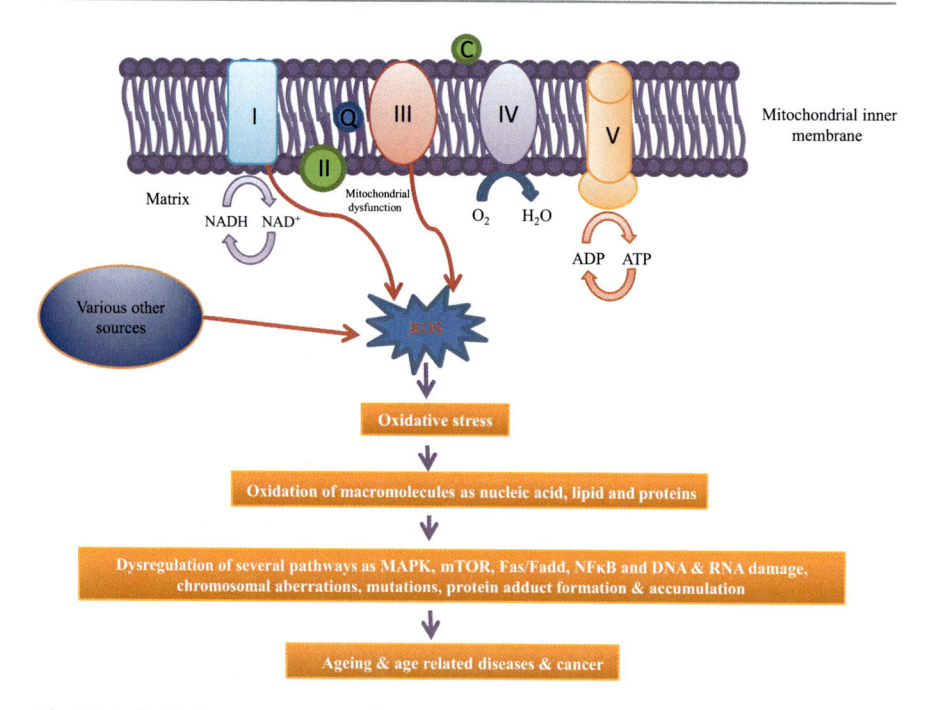

Fig. 21.4 Oxidative stress causes aging as well as cancer. Reactive oxygen species (ROS) generated due to mitochondrial dysfunction and various other endogenous as well as exogenous sources cause oxidative damage leading to damage of various macromolecules such as lipid, protein, and nucleic acids which resulted in dysregulation of various pathways, chromosomal aberrations, and mutations that could lead to cancer and age-related phenotypes

aging. It suggested that oxidative damage to mitochondrial macromolecules such as mtDNA, proteins, and lipids is responsible for aging. The role of mitochondria in cancer progression is prominent as many mutations in mtDNA have been known in neoplastic transformation [49]. Thus, it can be suggested that mitochondrial dysfunction is common important factor for both aging and cancer progression.

Deterioration of Immune System with Aging

Immune system alterations at higher ages are very common which increases the risk factor of individuals to develop many diseases including cancer. With aging, decline in both parts of immune system that is adaptive as well as innate is known as immunosenescence. However, its precise role in development and progression of cancer in elderly people is still controversial. Age-associated changes in function of dendritic cells (DC) which act as a bridge between innate and adaptive immunity have been observed. The modulatory effects of DC on T-cell response are found to be compromised by certain factors such as VEGF, IL10, and TGF-β secreted by tumor cells. Cytotoxic effect of monocytes was found to be decreased against tumor cells found in elderly people, along with change in neutrophils function which might contribute to tumor progression. Natural killer (NK) cells play very important role as they are major killer of cancer cells, but with aging, altered phenotype was observed with less cytotoxic activity. There is an emergence of novel network of mechanism with aging which is capable of suppressing the immune response and thus might contribute in emergence of many diseases including cancer [51].

Role of p53 in Aging and Cancer

p53 regulates various physiological processes inside the cell especially during stress; it triggers cell cycle arrest, apoptosis, or cellular senescence. In case of DNA damage, p53 can cause cell cycle arrest, thus providing a chance to repair machinery to repair the damage; however, if the damage is irreparable, apoptosis or senescence would be triggered. Tumor suppressor gene, p53, also known as longevity assurance gene, promotes longevity through reducing the somatic mutations and survival of mutant cells, thus avoiding the occurrence of cancer. However, increased function of p53 promotes aging. How does p53 increase organismal longevity? The activity of this protein in aging is mostly studied in lower organisms such as nematodes and fruit flies. There is evidence which suggest that p53 regulate the redox status of the cells and thus reduce the oxidative stress in the cells. Role of oxidative stress in aging and cancer is well established; hence p53 promotes longevity by reducing oxidative stress [52]. Studies in various mouse models that contain mutations in p53 genes have established the role of p53 in aging. Animal having conditional and tissue-specific knockouts along with factors that are critical for aging provides information for the role of p53 in aging. Excessive expression of p53 is toxic to the cells and can cause death along with accelerated aging. It has been found that p53 affects the replicative senescence in fibroblasts in cell culture. This gene is also found to affect early aging phenotype Brca1-deficient mice. Although the role of p53 in cancer is prominent, however, its role in aging is very complicated [53]. Hence, there is a need to investigate this area more deeply, so that overlapping factors that contribute to aging and cancer could be explored. These findings pave the way for developing therapy that could target both cancer and aging.

Insulin/IGF1 Signaling Is Intricate to Aging and Cancer

Age-related changes in endocrine hormones provide other hallmarks for aging. However, the link between hormones and aging-associated pathology is controversial, especially with growth hormone (GH). Growth hormone (GH) is a peptide hormone which regulates various critical physiological processes in the body. This hormone is secreted by anterior pituitary and affects the metabolism, aging, and immune system. The growth hormone receptors expressed on the various tissues where GH binds and exerts its effect through activating tyrosine kinase Janus kinase 2 (JAK2). Thus, GH signaling regulates multitude of biological processes such as

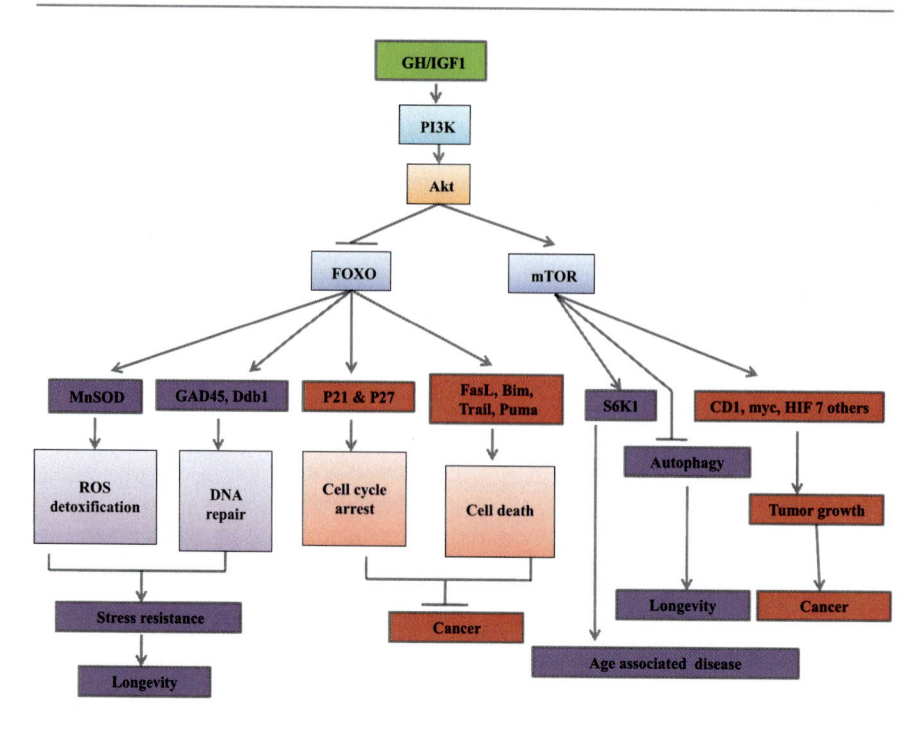

Fig. 21.5 GH/IGF1 pathway results in cancer as well as aging. GH/IGF1 pathway through its downstream targets, mTOR and FOXO, is linked with cancer and aging. Deregulation of this pathway is involved in the development of cancer as well as aging

prevention of apoptosis, cellular proliferation, cytoskeletal reorganization, differentiation and migration, and regulation of metabolic pathways as shown in Fig. 21.5 [54].

The level of GH is found to be very high in early period of life owing to rapid growth period. With time, there is decline in GH in plasma of adult and elderly people. This decline is well documented across the mammalian species primarily due to less active hypothalamus system which further affects secretions from the pituitary. This age-related decline in activity of somatotropic axis hormones (GnRH, GH, IGF-1) is known as "somatopause" [55].

The relationship between aging and hormones came into light from studies in lower organisms where it was found that genetic alteration in hormonal signaling could affect the longevity in these organisms. The studies in *C. elegans* provide the evidence that certain genes which regulate aging are homologous to mammalian genes that regulate the insulin transmission and insulin-like growth factor (IGF) signaling. Further studies in worm, insects, and yeast have suggested that insulin/IGF-like signaling (IIS) pathway controls aging in these species. In one study, knockout mice heterozygous for the igf1r (+/−) have increased lifespan as compared to wild type. Similar results were observed in case of heterozygous females. It was observed that igf1r (+/−) mice were resistant to oxidative damage that is detrimental and a major cause of aging [56, 57]. Increased lifespan in female mice has been observed when the insulin receptors are removed selectively from the adipose tissue; however, mice lacking GH and GH receptors in both sexes resulted in robust increase in life span. The long-lived mice having GH-related mutants have several advantages as compared to wild type. It is found that mutant mice have shown reduced incidence of cancer and delayed onset of cancer and aging with marked increase in lifespan [57].

The role of somatotropin signaling in regulating aging is controversial and complicated. However, there is clear evidence that Laron syndrome, which shows GH resistance, provides the protection from cancer and diabetes.

In several studies, it is observed that genetic variants of somatotropin signaling are found to be reduced in exceptionally long-lived people. The variants of gene FOXO which is regulated by IIS have also shown association with longevity in unrelated human population [57]. Aging is one of the most important risk factors for insulin resistance that could ultimately lead to cancer.

Nutrient Signaling

Several signaling pathways have been evolved which can sense the extracellular and intracellular level of nutrient, collectively termed as nutrient signaling pathways. This is requisite for maintaining the homeostasis of the system and, if disrupted, can cause serious consequences. Sensing the outside and inside nutrient levels in unicellular organisms is not complex. However, in multicellular organisms, most of the cells do not expose the environmental nutrients directly, but they have developed the system to sense fluctuations. It is well known that there are several nutrients sensing signaling mechanisms exist for different nutrient molecules such as lipid, amino acid, and glucose [58]. The prominent effects of nutrients on the life and health span of the living organism from yeast to mammals have been well implicated. Eukaryotic model *Saccharomyces cerevisiae* has been used most commonly to study the process of longevity. Dietary modulation of nutrients in culture media has shown to impact the lifespan of the yeast. Studies in *S. cerevisiae* observed the effect of glucose on lifespan depending upon abundance where higher glucose promotes the pro-aging signaling. There are many nutrients upon which living organism relies for growth and proper maintenance; however, the amino acids and proteins are well-studied component that modulates the longevity. In rodents, dietary intake of protein and certain amino acids such as methionine and tryptophan increased the longevity of the species. The effect of essential amino acids in diet influences the lifespan as observed in fruit flies. Different amino acids might affect different signaling in living organisms; thus varied responses have been observed [59, 60]. The loss of nutrient sensing has been accompanied by mTORC1 activation and tumor cell growth. mTORC1 is integral to local as well as systemic nutrient signals (Fig. 21.6) [61].

Hence, the nutrient regulates both processes of aging as well as cancer. There are some common molecules found to be involved in aging and cancer that suggest

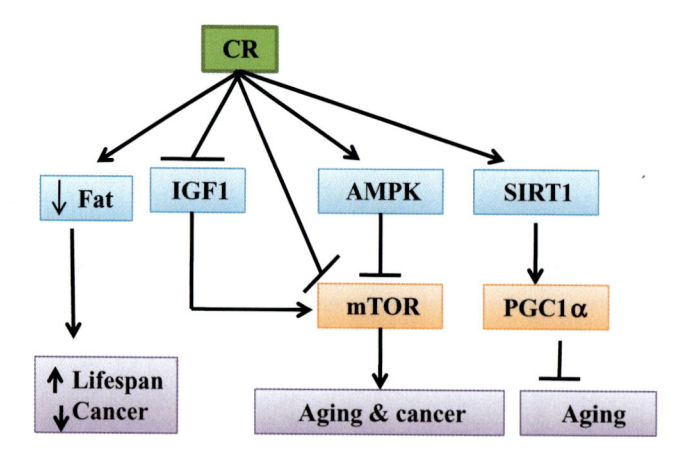

Fig. 21.6 Caloric restriction in aging and cancer. Caloric restriction (CR) extends lifespan and also prevents cancer development. Some of the effects of CR may be attributed to decrease in fat accumulation. CR targets GH/IGF1 pathway to exert its effects, and the other mechanisms through which CR works include AMPK, SIRT1, and mTOR

these two processes are mutually inclusive. Further insights into common biology will help in understanding the connection between these two phenomena.

Role of Stem Cells in Aging and Cancer

The basic properties every stem cell would follow irrespective of their origin are the self-renewal to form daughter cell and give rise to different types of differentiated cells. Minor populations of these cells with such properties are cardinal for proper functionality of the tissue throughout the life time of the organism.

The role of stem cells is aging and cancer has been reviewed extensively. Various studies have supported the notion that aging is a gradual and complex process invariably accompanied by loss of tissue homeostasis maintenance and repair tissue injury. Many pathophysiological conditions that mostly occurred in elderly populations such as anemia, osteoporosis, and sarcopenia are resultant from imbalance between cell renewal and cell loss. It is postulated that the stem cells play a central role in aging. The stem cells decline with time accelerates causing the loss of homeostasis and regenerative potential which attributes to aging. On the other hand, the role of stem cells in cancer is incontrovertible. Cancer acquisition is mostly followed by occurrence of multiple mutations in the genome of the cell. Stem cell renewal is a lifetime process, and thus, these cells are the ideal reservoirs for the accumulation of mutations over the period of time. It is postulated that the cell renewal is integral to the homeostatic maintenance of the tissue system. Since, the mutations in the stem cells checked out in the system through various processes and mutated stem cells removed out through apoptosis or permanent growth arrest. This would lead to the stem cell attrition in the tissue which paves the way for aging due to lack of maintenance of the tissues. On the other hand, if these mutated cells have not been sorted out from the system, then this accumulation of mutations could be fatal and can lead to cancer [62]. Hence, it is germane to discuss aging and cancer closely and deciphering their common molecular events would be beneficial for understanding aging and related pathologies such as cancer.

The stem cells are found at the spot in the tissue having specialized microenvironment, which controls the stem cell behavior, termed as "stem cell niche." It protects the stem cells from depletion and over proliferation. These niches are conserved across the species, and in mammals, several techniques such as genetic and imaging techniques have been used to decipher several stem cell niches. The well-characterized niches in mammals are hematopoietic, small intestine stem cells and hair follicle. There are several growth factors, and cytokines have been secreted by niche which regulates stem cell division and self-renewable capacity. However, it is observed that with age the niches get weakened, and thus normal function of niches gets affected. This would overall affect the functionality of the stem cells and improper maintenance leads to aging and tissue regeneration. The influence in stem cell niches due to various factors could also lead to development of cancer [63]. Therefore, exploring stem cell niches will help in understanding the behavior of stem cells and related altered signaling that might link the process of aging and cancer.

Recent studies have suggested the role of stem cells and progenitors cells in age-dependent increase in cancer. It has been found that mutant clonal hematopoiesis evolves over time and predicts enhanced risk of leukemias and other age-related pathologies. These clones further acquire the additional mutations, and thus clones with multiple mutations were observed in HSC compartment of the acute myeloid leukemia (AML). Several studies have shown that younger individuals have detectable level of cancer-associated genetic alterations; however, dominance of these clones with alterations is found to increase exponentially with age. The aging-dependent increases in mutations in stem and progenitor cells are prompt by increase in mutation initiation rates, and selection of clonal dominance of cells acquired non-neutral mutations [64].

There are a small number of cells which rarely divide found at the top of the hierarchy of the hematopoietic system, and these cells are referred to as quiescent hematopoietic stem cells. It is evident that the process of quiescence protects the HSC cells from accumulating the mutations; however, how these cells linked to mutations of stem cell and clonal dominance is not yet clear. Several molecules are known to be involved in maintenance of HSC quiescence, and these molecules include transcription factor, Nrf2, ATM, and CEBP/a. These molecules have been found to influence the health and lifespan of the organisms. ATM is known to protect the cells and tissues from DNA damage and prevent premature aging in the organisms [64].

Telomere shortening is nearly found in all the human tissues including stem and progenitor cells. This process could contribute to genomic instability in adult cells over time owing to replicative aging. It is found that the level of activity of telomerase, an enzyme required for maintenance of telomeres, is either low or absent in most of the stem cells except embryonic stem cells and cancer stem cells [64, 65]. In recent years, metabolism has been regarded as an emerging hallmark for cancer.

Also, studies have shown the influential role of metabolism in aging and functionality of stem cells. Metabolism and related signaling pathways have shown to affect the rate of aging and longevity along with cancer in many organisms such as *C. elegans* and mice [64].

Overall, studies from past few years have untangled the many aspects of cancer and aging that have shown causal relation between these processes. Several emerging evidences have suggested that aging induces the modifications in several signaling pathways in stem and progenitors cells that could accelerate initiation and clonal dominance of mutations. Therefore, understanding the common cellular and molecular biology of stem and progenitors cells in cancer and aging will undoubtedly help in exploring novel targets that could be used as a therapy and thus improves the early detection and treatment of aging-associated pathologies such as cancer.

Conclusions

In past few decades, with several medical advancements, the life expectancy has increased leading to survival of more aged people. The unhealthy aging population will increase the burden on the medical care and management, as well as decrease the quality of life (Fig. 21.7). It is evident from several studies that incidence of cancer with age is one of the most common pathologies. However, common shared

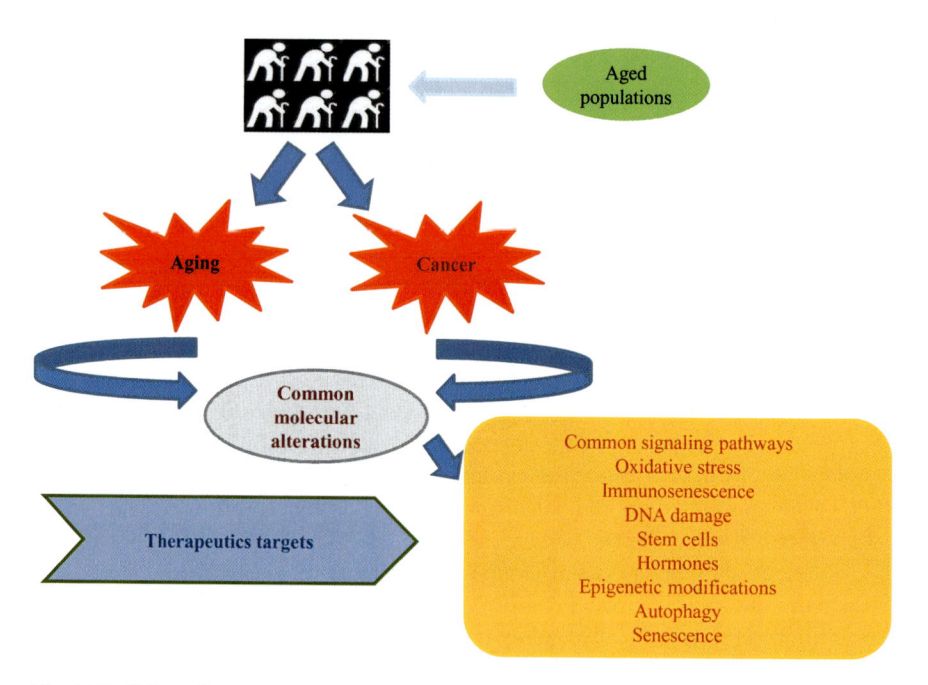

Fig. 21.7 Schematic representation showing the common molecular alterations involved in aging and cancer in aged populations

link between aging and cancer is yet to be explored in-depth. Although several studies have tried to explore the aspects of aging and related pathology such as cancer, there is a need to explore this field for better understanding. The exploration of common molecular mechanisms associated with cancer and aging would help in developing several targets that could simultaneously affect aging and cancer. The specific molecular interventions could delay aging and avoid the occurrence of cancer. Hence, molecular targets developed from combined study of cancer and aging will pave the way for novel therapeutics target and thus improve the human health. Moreover, studies should be directed to explore the links between aging and cancer in order to pave the way for a healthy aging.

References

1. Lees H, Walters H, Cox LS. Animal and human models to understand ageing. Maturitas. 2016;93:18–27.
2. López-Otín C, Blasco MA, Partridge L, Serrano M, Kroemer G. The hallmarks of aging. Cell. 2013;153(6):1194–217.
3. Jin K. Modern biological theories of aging. Aging Dis. 2010;1:72–4.
4. Weinert BT, Timiras PS. Invited review: theories of aging. J Appl Physiol. 2003;95:1706.
5. Campisi J. Aging, cellular senescence, and cancer. Annu Rev Physiol. 2013;75:685–705.
6. Collado M, Blasco MA, Serrano M. Cellular senescence in cancer and aging. Cell. 2007;130:223–33.
7. Rufini A, Tucci P, Celardo I, Melino G. Senescence and aging: the critical roles of p53. Oncogene. 2013;32:5129–43.
8. Rodier F, Campisi J. Four faces of cellular senescence. J Cell Biol. 2011;192:547–56.
9. Mitchell SJ, Scheibye-Knudsen M, Longo DL, de Cabo R. Animal models of aging research: implications for human aging and age-related diseases. Annu Rev Anim Biosci. 2015;3:283–303.
10. Cruickshanks HA, Adams PD. Chromatin: a molecular interface between cancer and aging. Curr Opin Genet Dev. 2011;21:100–6.
11. Yancik R, Ries LA. Cancer in older persons: an international issue in an aging world. Semin Oncol. 2004;31:128–36.
12. Frank SA. Dynamics of cancer: incidence, inheritance, and evolution. Princeton NJ: Steven A Frank; 2007.
13. Anisimov VN. Carcinogenesis and aging 20 years after: escaping horizon. Mech Ageing Dev. 2009;130:105–21.
14. Veronesi U, Marubini E, Mariani L, Galimberti V, Luini A, Veronesi P, et al. Radiotherapy after breast-conserving surgery in small breast carcinoma: long-term results of a randomized trial. Ann Oncol Off J Eur Soc Med Oncol. 2001;12:997–1003.
15. Ershler WB, Stewart JA, Hacker MP, Moore AL, Tindle BH. B16 murine melanoma and aging: slower growth and longer survival in old mice. J Natl Cancer Inst. 1984;72:161–4.
16. Balducci L, Ershler WB. Cancer and ageing: a nexus at several levels. Nat Rev Cancer. 2005;5:655–62.
17. Finkel T, Serrano M, Blasco MA. The common biology of cancer and ageing. Nature. 2007;448:767–74.
18. de Magalhaes JP. How ageing processes influence cancer. Nat Rev Cancer. 2013;13:357–65.
19. Serrano M, Blasco MA. Cancer and ageing: convergent and divergent mechanisms. Nat Rev Mol Cell Biol. 2007;8:715–22.
20. Campisi J, Yaswen P. Aging and cancer cell biology, 2009. Aging Cell. 2009;8:221–5.

21. Kudryavtseva AV, Krasnov GS, Dmitriev AA, Alekseev BY, Kardymon OL, Sadritdinova AF, et al. Mitochondrial dysfunction and oxidative stress in aging and cancer. Oncotarget. 2016;7:44879–905.
22. Das PM, Singal R. DNA methylation and cancer. J Clin Oncol Off J Am Soc Clin Oncol. 2004;22:4632–42.
23. Fraga MF, Agrelo R, Esteller M. Cross-talk between aging and cancer: the epigenetic language. Ann N Y Acad Sci. 2007;1100:60–74.
24. Campisi J. Cancer and ageing: rival demons? Nat Rev Cancer. 2003;3:339–49.
25. Anisimov VN. The relationship between aging and carcinogenesis: a critical appraisal. Crit Rev Oncol Hematol. 2003;45:277–304.
26. Lopez-Otin C, Blasco MA, Partridge L, Serrano M, Kroemer G. The hallmarks of aging. Cell. 2013;153:1194–217.
27. Vogel H, Lim DS, Karsenty G, Finegold M, Hasty P. Deletion of Ku86 causes early onset of senescence in mice. Proc Natl Acad Sci U S A. 1999;96:10770–5.
28. Augert A, Bernard D. Immunosenescence and senescence immunosurveillance: one of the possible links explaining the cancer incidence in ageing population. In: Wang Z, Inuzuka H, editors. Senescence and senescence-related disorders. Rijeka: InTech; 2013.. p Ch. 04.
29. Kong Y, Cui H, Ramkumar C, Zhang H. Regulation of senescence in cancer and aging. J Aging Res. 2011;2011:963172.
30. Falandry C, Bonnefoy M, Freyer G, Gilson E. Biology of cancer and aging: a complex association with cellular senescence. J Clin Oncol Off J Am Soc Clin Oncol. 2014;32:2604–10.
31. Anisimov VN. Biology of aging and cancer. Cancer control: J Moffitt Cancer Cent. 2007;14:23–31.
32. Ershler WB, Longo DL. Aging and cancer: issues of basic and clinical science. J Natl Cancer Inst. 1997;89:1489–97.
33. Johnson AA, Akman K, Calimport SR, Wuttke D, Stolzing A, de Magalhaes JP. The role of DNA methylation in aging, rejuvenation, and age-related disease. Rejuvenation Res. 2012;15:483–94.
34. Lin Q, Wagner W. Epigenetic aging signatures are coherently modified in cancer. PLoS Genet. 2015;11:e1005334.
35. Rakyan VK, Down TA, Maslau S, Andrew T, Yang T-P, Beyan H, et al. Human aging-associated DNA hypermethylation occurs preferentially at bivalent chromatin domains. Genome Res. 2010;20:434–9.
36. Anisimov VN. Biology of aging and Cancer. Cancer Control. 2017;14(1):23–31.
37. Xie B, Chen J, Liu B, Zhan J. Klotho acts as a tumor suppressor in cancers. Pathol Oncol Res: POR. 2013;19:611–7.
38. Rubinsztein DC, Marino G, Kroemer G. Autophagy and aging. Cell. 2011;146:682–95.
39. Gerl R, Vaux DL. Apoptosis in the development and treatment of cancer. Carcinogenesis. 2005;26:263–70.
40. Blagosklonny MV. Selective anti-cancer agents as anti-aging drugs. Cancer Biol Ther. 2013;14:1092–7.
41. Malaguarnera L, Ferlito L, Di Mauro S, Imbesi RM, Scalia G, Malaguarnera M. Immunosenescence and cancer: a review. Arch Gerontol Geriatr. 2001;32:77–93.
42. Swann JB, Smyth MJ. Immune surveillance of tumors. J Clin Invest. 2007;117:1137–46.
43. Carlson ME, Silva HS, Conboy IM. Aging of signal transduction pathways, and pathology. Exp Cell Res. 2008;314:1951–61.
44. Kaur A, Webster MR, Weeraratna AT. In the Wnt-er of life: Wnt signalling in melanoma and ageing. Br J Cancer. 2016;115:1273–9.
45. Steelman LS, Chappell WH, Abrams SL, Kempf RC, Long J, Laidler P, et al. Roles of the Raf/MEK/ERK and PI3K/PTEN/Akt/mTOR pathways in controlling growth and sensitivity to therapy-implications for cancer and aging. Aging (Albany NY). 2011;3:192–222.
46. Wang X. Microarray analysis of ageing-related signatures and their expression in tumors based on a computational biology approach. Genomics Proteomics Bioinformatics. 2012;10:136–41.

47. Wang X. Discovery of molecular associations among aging, stem cells, and cancer based on gene expression profiling. Chin J Cancer. 2013;32:155–61.
48. Ahmad A, Banerjee S, Wang Z, Kong D, Majumdar AP, Sarkar FH. Aging and inflammation: etiological culprits of cancer. Current Aging Sci. 2009;2:174–86.
49. Cui H, Kong Y, Zhang H. Oxidative stress, mitochondrial dysfunction, and aging. J Signal Transduction. 2012;2012:646354.
50. Afanas'ev I. Reactive oxygen species signaling in cancer: comparison with aging. Aging Dis. 2011;2:219–30.
51. Fulop T, Larbi A, Kotb R, de Angelis F, Pawelec G. Aging, immunity, and cancer. Discov Med. 2011;11:537–50.
52. Rodier F, Campisi J, Bhaumik D. Two faces of p53: aging and tumor suppression. Nucleic Acids Res. 2007;35:7475–84.
53. Hasty P, Christy BA. p53 as an intervention target for cancer and aging. Pathobiol Aging Age Relat Dis. 2013;3:1–11.
54. Lanning NJ, Carter-Su C. Recent advances in growth hormone signaling. Rev Endocr Metab Disord. 2006;7:225–35.
55. Bartke A. Growth hormone and aging: a challenging controversy. Clin Interv Aging. 2008;3:659–65.
56. Holzenberger M, Dupont J, Ducos B, Leneuve P, Geloen A, Even PC, et al. IGF-1 receptor regulates lifespan and resistance to oxidative stress in mice. Nature. 2003;421:182–7.
57. Anisimov VN, Bartke A. The key role of growth hormone-insulin-IGF-1 signaling in aging and cancer. Crit Rev Oncol Hematol. 2013;87:201–23.
58. Efeyan A, Comb WC, Sabatini DM. Nutrient-sensing mechanisms and pathways. Nature. 2015;517:302–10.
59. Santos J, Leitao-Correia F, Sousa MJ, Leao C. Dietary restriction and nutrient balance in aging. Oxidative Med Cell Longev. 2016;2016:4010357.
60. Gallinetti J, Harputlugil E, Mitchell JR. Amino acid sensing in dietary-restriction-mediated longevity: roles of signal-transducing kinases GCN2 and TOR. Biochem J. 2013;449:1–10.
61. Menon S, Manning BD. Cell signalling: nutrient sensing lost in cancer. Nature. 2013;498:444–5.
62. Rossi DJ, Jamieson CH, Weissman IL. Stems cells and the pathways to aging and cancer. Cell. 2008;132:681–96.
63. Singh SR. Stem cell niche in tissue homeostasis, aging and cancer. Curr Med Chem. 2012;19:5965–74.
64. Adams PD, Jasper H, Rudolph KL. Aging-induced stem cell mutations as drivers for disease and cancer. Cell Stem Cell. 2015;16:601–12.
65. Hiyama E, Hiyama K. Telomere and telomerase in stem cells. Br J Cancer. 2007;96:1020–4.

Printed in the United States
By Bookmasters